The ARRL
Extra Class
License Manual
For Ham Radio

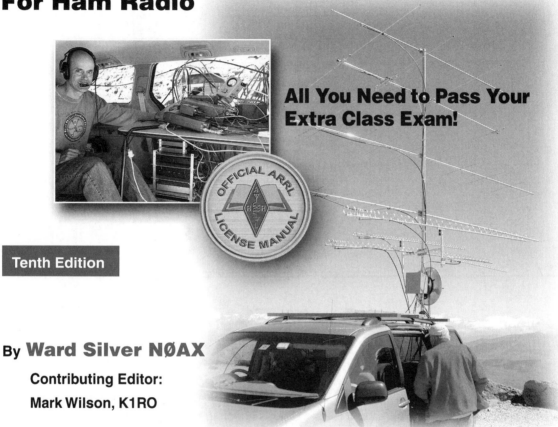

**All You Need to Pass Your
Extra Class Exam!**

Tenth Edition

By **Ward Silver NØAX**

Contributing Editor:
Mark Wilson, K1RO

Production Staff: **Maty Weinberg, KB1EIB,** Editorial Assistant

Michelle Bloom, WB1ENT, Production Supervisor, Layout

Jodi Morin, KA1JPA, Assistant Production Supervisor, Layout

Carol Michaud, KB1QAW, Production Assistant

David Pingree, N1NAS, Senior Technical Illustrator

Sue Fagan, KB1OKW, Graphic Design Supervisor, Cover Design

ARRL *The national association for*
AMATEUR RADIO®
225 Main Street, Newington, CT 06111-1494

Cover Photos
It's not unusual to find someone operating from
the top of Mt. Washington, New Hampshire. On
this trip up the mountain in September 2011,
Frank Pingree, K1RAR found Ed Parsons, K1TR,
operating the September VHF QSO Party.
Photos by David Pingree, N1NAS

Copyright © 2012 by

The American Radio Relay League, Inc

Copyright secured under the Pan-American Convention

All rights reserved. No part of this work may be reproduced in any form except by written permission of the publisher.
All rights of translation are reserved.

Printed in USA

Quedan reservados todos los derechos

ISBN: 978-0-87259-517-0

Tenth Edition
Third Printing

This book may be used for Extra license exams given beginning July 1, 2012. *QST* and the ARRL website (**www.arrl.org**) will have news about any rules changes affecting the Extra class license or any of the material in this book.

We strive to produce books without errors. Sometimes mistakes do occur, however. When we become aware of problems in our books (other than obvious typographical errors), we post corrections on the ARRL website. If you think you have found an error, please check **www.arrl.org/extra-class-license-manual** for corrections. If you don't find a correction there, please let us know, either using the Feedback Form at the back of this book or by sending e-mail to **pubsfdbk@arrl.org**.

The ARRL Extra Class License Manual ON THE WEB

www.arrl.org/extra-class-license-manual

Visit *The ARRL Extra Class License Manual* home on the Web for additional resources.

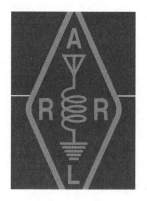

Contents

Foreword

Welcome to the tenth edition of *The ARRL Extra Class License Manual*. You are holding the key to your final step up the Amateur Radio license ladder! With full access to the entire Amateur Radio frequency spectrum, you will be permitted to operate using every privilege granted to amateurs by the Federal Communications Commission. This includes the frequencies reserved in the US for holders of the Extra class license.

With your increased privileges comes the challenge of increased responsibility to fulfill the Basis and Purpose of the Amateur Service as stated in Part 97.1 of the FCC's rules and regulations:

- Engaging in public service
- Advancing the radio art
- Enhancing your technical and operating skills
- Providing trained operators and technicians
- Enhancing international goodwill

That last point is significant in that even in this age of instant worldwide connectivity — no other group has access to direct personal communication without any intervening networks or equipment! As an Extra class licensee, you'll be able to take full advantage of those communication opportunities.

The ARRL Extra Class License Manual may seem huge, but it covers every one of the more than 700 questions included in the question pool. Each topic is addressed in sufficient detail that you can learn the "why" and "how" behind each answer. This will help you retain the information after you pass the exam and you will get more benefit and enjoyment from your upgraded license.

The book includes numerous examples to help you become comfortable with the necessary calculations. Graphics are included to help you visualize the concepts and explanations. If you would like a more concise study guide, *ARRL's Extra Q&A* is a companion to this book, presenting each question and a short explanation of the correct answer.

As with the license manuals for the Technician and General licenses, this manual organizes the material into a natural progression of topics. Each topic is followed by a list of questions from the exam on that subject. This makes the material easier to learn, remember, and use. Along with the printed manuals, additional resources are provided on the ARRL website at **www.arrl.org/extra-class-license-manual**. These web pages list supplemental references, such as a math tutorial, and links to resources you can use to go beyond the exam questions.

Of course, you aren't just studying to pass a license exam and we aren't satisfied just to help you pass. We want you to enjoy Amateur Radio to its fullest and that's why the ARRL provides opportunities for continued education, experimentation, and growth through technical and operating training resources, in both print and electronic form.

Be sure to take advantage of the technical and operating aids in the many books and supplies that make up our "Radio Amateur's Library." Make the Technology section of the ARRL's website (**www.arrl.org/technology**) a frequent stop on your web travels. If you're not yet an ARRL member, there are hardly better reasons to join! Check *QST* each month for new material or browse the ARRL Publications Catalog on-line at **www.arrl.org/arrl-store**. You can also request a printed catalog or place an order by phone, 888-277-5289; by fax, 860-594-0303; by e-mail, **pubsales@arrl.org**

This tenth edition of *The ARRL Extra Class License Manual* is not just the work of one author. It builds on the excellent material developed by authors and editors of previous editions, many ARRL staff members, and readers of the earlier editions. You can help make this manual better, too. After you've used the book to prepare for your exam, e-mail your suggestions (including any corrections you think need to be made) to us at **pubsfdbk@arrl.org** or use the Feedback Form at the back of this book and mail the form. Your comments are welcome!

Upgrading your license is only the beginning of your adventure – you'll have access to the complete palette of the amateur experience with every one of ham radio's tool available. You can use these for personal enjoyment as well as apply them for the benefit of other amateurs and the public.

Thanks for making the decision to upgrade and reach the highest level of achievement in Amateur Radio. You won't regret it — good luck!

David Sumner, K1ZZ
Chief Executive Officer
Newington, Connecticut
March 2012

When to Expect New Books

A Question Pool Committee (QPC) consisting of representatives from the various Volunteer Examiner Coordinators (VECs) prepares the license question pools. The QPC establishes a schedule for revising and implementing new question pools. The current question pool revision schedule is as follows:

Question Pool	Current Study Guides	Valid Through
Technician (Element 2)	*The ARRL Ham Radio License Manual*, 2nd edition *ARRL's Tech Q&A*, 5th Edition	June 30, 2014
General (Element 3)	*The ARRL General Class License Manual*, 7th edition *ARRL's General Q&A*, 4th Edition	June 30, 2015
Amateur Extra (Element 4)	*The ARRL Extra Class License Manual*, 10th Edition *ARRL's Extra Q&A*, 3rd Edition	June 30, 2016

As new question pools are released, ARRL will produce new study materials before the effective date of the new pools. Until then, the current question pools will remain in use, and current ARRL study materials, including this book, will help you prepare for your exam.

As the new question pool schedules are confirmed, the information will be published in *QST* and on the ARRL website at **www.arrl.org**.

 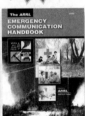

The Adventure Continues!

Congratulations! You've taken the next step in your journey through Amateur Radio. As an Extra class licensee, you'll experience a whole new dimension of operating enjoyment — new frequencies, new modes and new activities. To help you make the most of Amateur Radio, your national association — the ARRL — offers a wide range of services. Here are some that may be interesting and useful to you.

ARRL The national association for Amateur Radio: What's in it for You?

♦ **Help for New Hams:** Are you a beginning ham looking for help in getting started in your new hobby? The hams at ARRL HQ in Newington, Connecticut, will be glad to assist you. Call 800-32-NEWHAM. ARRL maintains a computer database of ham clubs and ham radio "helpers" from across the country who've told us they're interested in helping beginning hams. There are probably several clubs in your area! Contact us for more information.

♦ **Licensing Classes:** If you're going to become a ham, you'll need to find a local license exam opportunity sooner or later. ARRL Registered Instructors teach licensing classes all around the country, and ARRL-sponsored Volunteer Examiners are right there to administer your exams. To find the locations and dates of Amateur Radio licensing classes and test sessions in your area, visit **www.arrl.org** or call the New Ham Desk at 800-32-NEWHAM.

♦ **Clubs:** As a beginning ham, one of the best moves you can make is to join a local ham club. Whether you join an all-around group or a special-interest club (repeaters, DXing, and so on), you'll make new friends, have a lot of fun, and you can tap into a ready reserve of ham radio knowledge and experience. To find the ham clubs in your area, visit **www.arrl.org** or call HQ's New Ham Desk at 800-32-NEWHAM.

♦ **Technical Information Service:** Do you have a question of a technical nature? (What ham doesn't?) Contact the Technical Information Service (TIS) at HQ or on the web at **www.arrl.org/ technology**. Our resident technical experts will help you over the phone or by email, send you specific information on your question (antennas, interference and so on) or refer you to your local ARRL Technical Coordinator or Technical Specialist. It's expert information — and it doesn't cost Members an extra cent!

♦ **Regulatory Information:** Need help with a thorny antenna zoning problem? Having trouble understanding an FCC regulation? Vacationing in a faraway place and want to know how to get permission to operate your ham radio there? HQ's Regulatory Information Specialist (**www.arrl.org/regulatory-advocacy**) has the answers you need!

♦ **Operating Awards:** Like to collect "wallpaper"? The ARRL sponsors a wide variety of certificates and Amateur Radio achievement awards. For information on awards you can qualify for, visit **www.arrl.org** or contact the Membership and Volunteer Programs Department at HQ.

♦ **QSL Service and Logbook of the World (LoTW):** With your expanded HF privileges, you're likely to try your hand at working DX stations in other countries. You can confirm these exciting contacts by exchanging paper QSL cards through ARRL's QSL service. Many hams also collect confirmations through ARRL's secure on-line Logbook of the World. Visit **www. arrl.org** to learn more about these services.

♦ **Equipment Insurance:** When it comes to protecting their Amateur Radio equipment investments, ARRL Members travel First Class. ARRL's "all-risk" equipment insurance plan protects you from loss or damage to your station. (It can protect your ham radio computer, too.) It's comprehensive and cost effective, and it's available only to ARRL Members. Why worry about losing your valuable radio equipment when you can protect it for only a few dollars a year?

♦ **Amateur Radio Emergency Service:** If you're interested in providing public service and emergency communications for your community, you can join more than 25,000 other hams who have registered their communications capabilities with local Emergency Coordinators. Your EC will call on you and other ARES members for vital assistance if disaster should strike your community. Visit **www.arrl.org** or contact the Membership and Volunteer Programs Department at HQ for information.

♦ **Audio-Visual Programs:** Need a program for your next ham club meeting, informal get-together or public display? ARRL offers many programs to choose from. Visit **www.arrl.org** for a complete list.

♦ **Blind, Disabled Ham Help:** For a list of available resources and information on the Courage HANDI-HAM System, contact the ARRL Program for the Disabled at HQ.

With your membership you also receive the monthly journal, *QST*. Each colorful issue is packed with valuable information you can use. You'll find technical information, weekend projects, operating tips, news, ads for the latest equipment and much more. *QST* Product Reviews are the most respected source of information to help you get the most for your Amateur Radio equipment dollar. (For many hams, *QST* alone is worth far more than the cost of ARRL membership.)

The ARRL also publishes newsletters and dozens of books covering all aspects of Amateur Radio. Our Headquarters station, WIAW, transmits bulletins of interest to radio amateurs and Morse code practice sessions.

When it comes to representing Amateur Radio's best interests, ARRL's team in Washington, DC, is constantly working with the FCC, Congress and industry to protect and foster your privileges as a ham operator.

Regardless of your Amateur Radio interests, ARRL Membership is relevant and important. We will be happy to welcome you as a Member. Use the membership form to **join today**. And don't hesitate to contact us if you have any questions!

What is Amateur Radio?

Perhaps you've just picked up this book in the library or from a bookstore shelf and are wondering what this Amateur Radio business is all about. Maybe you have a friend or relative who is a "ham" and you're interested in becoming one, as well. In that case, a short explanation is in order.

Amateur Radio or "ham radio" is one of the longest-lived wireless activities. Amateur experimenters were operating right along with Marconi in the early part of the 20th century. They have helped advance the state-of-the-art in radio, television and dozens of other communications services since then, right up to the present day. There are more than 700,000 Amateur Radio operators or "hams" in the United States alone and several million more around the world!

Amateur Radio in the United States is a formal *communications service*, administered by the Federal Communications Commission or FCC. Created officially in its present form in 1934, the Amateur Service is intended to foster electronics and radio experimentation, provide emergency backup communications, encourage private citizens to train and practice operating, and even spread the goodwill of person-to-person contact over the airwaves.

John, W1RT operates this multi-band "rover" during VHF+ contests, driving to hilltops around New England such as from Mohawk Mountain where he is shown giving the antennas a little "hands on" adjustment.

Taking ham radio on a vacation trip can be a lot of fun! During a recent trip to Puerto Rico, Sean KX9X made contacts through amateur satellite AO-27 from beach while using a handheld radio!

Who Is a Ham and What Do Hams Do?

Anyone can be a ham — there are no age limits or physical requirements that prevent anyone from passing their license exam and getting on the air. Kids as young as 6 years old have passed the basic exam, and there are many hams out there over the age of 100. You probably fall somewhere in the middle of that range.

Once you get on the air and start meeting other hams, you'll find a wide range of capabilities and interests. Of course, there are many technically skilled hams who work as engineers, scientists or technicians. But just as many don't have a deep technical background. You're just as likely to encounter writers, public safety personnel, students, farmers, truck drivers — anyone with an interest in personal communications over the radio.

The activities of Amateur Radio are incredibly varied. Amateurs who hold the Technician class license — the usual first license for hams in the US — communicate primarily with local and regional amateurs using relay stations called repeaters. Known as "Techs," they sharpen their skills of operating while portable and mobile, often joining emergency communications teams. They may instead focus on the burgeoning wireless data networks assembled and used by hams around the world. Techs can make use of the growing number of Amateur Radio satellites, built and launched by hams along with the commercial "birds." Technicians transmit their own television signals, push the limits of signal propagation through the atmosphere and experiment with microwaves. Hams hold most of the world records for long-distance communication on microwave frequencies, in fact!

Hams who advance or *upgrade* to General and then to Extra class are granted additional privileges with each step to use the frequencies usually associated with shortwave operation. This is the traditional Amateur Radio you probably encountered in movies or books. On these frequencies, signals can travel worldwide and so amateurs can make direct contact with foreign hams. No Internet, phone systems, or data networks are required. It's just you, your radio, and the ionosphere — the upper layers of the Earth's atmosphere!

Many hams use voice, Morse code, computer data modes and even image transmissions to communicate. All of these signals are mixed together where hams operate, making the experience of tuning a radio receiver through the crowded bands an interesting experience.

One thing common to all hams is that all of their operation is noncommercial, especially the volunteers who provide emergency communications. Hams pursue their hobby purely for personal enjoyment and to advance their skills, taking satisfaction from providing valuable services to their fellow citizens. This is especially valuable after natural disasters such as hurricanes and earthquakes when commercial systems are knocked out for a while. Amateur operators rush in to provide backup communication for hours, days, weeks or even months until the regular systems are restored. All this from a little study and a simple exam!

Sisters Autumn and Hannah operate K5LBJ during the bi-annual School Club Roundup competition that features the radio clubs at schools across the country.

Arie, PA3A was one of four Dutch hams who participated in a Mercy Ships project in Sierra Leone during 2011. During their free time, they operated as 9L5MS on the HF bands.

Want to Find Out More?

If you'd like to find out more about Amateur Radio in general, there is lots of information available on the Internet. A good place to start is on the American Radio Relay League's (ARRL) hamradio introduction page at

Participating in a "radiosport" competition is a great way to build up your radio skills. The W2GD team specializes in 160 meter operation as shown here during the 2010 ARRL 160 Meter Contest.

www.helloradio.org. Books like *Ham Radio for Dummies* and *Getting Started With Ham Radio* will help you "fill in the blanks" as you learn more.

Along with books and Internet pages, there is no better way to learn about ham radio than to meet your local amateur operators. It is quite likely that no matter where you live in the United States, there is a ham radio club in your area — perhaps several! The ARRL provides a club lookup web page at **www. arrl.org** where you can find a club just by entering your Zip code or state. Carrying on the tradition of mutual assistance, many clubs make helping newcomers to ham radio a part of their charter.

If this sounds like hams are confident that you'll find their activities interesting, you're right! Amateur Radio is much more than just talking on a radio, as you'll find out. It's an opportunity to dive into the fascinating world of radio communications, electronics, and computers as deeply as you wish to go. Welcome!

Lots of hams enjoy using digital modes to communicate. Melanie, KD0LRC is shown using PSK31 during the KO0A Field Day operation in St Charles, MO.

About the ARRL

The seed for Amateur Radio was planted in the 1890s, when Guglielmo Marconi began his experiments in wireless telegraphy. Soon he was joined by dozens, then hundreds, of others who were enthusiastic about sending and receiving messages through the air—some with a commercial interest, but others solely out of a love for this new communications medium. The United States government began licensing Amateur Radio operators in 1912.

By 1914, there were thousands of Amateur Radio operators—hams—in the United States. Hiram Percy Maxim, a leading Hartford, Connecticut inventor and industrialist, saw the need for an organization to band together this fledgling group of radio experimenters. In May 1914 he founded the American Radio Relay League (ARRL) to meet that need.

Today ARRL, with approximately 155,000 members, is the largest organization of radio amateurs in the United States. The ARRL is a not-for-profit organization that:
• promotes interest in Amateur Radio communications and experimentation
• represents US radio amateurs in legislative matters, and
• maintains fraternalism and a high standard of conduct among Amateur Radio operators.

At ARRL headquarters in the Hartford suburb of Newington, the staff helps serve the needs of members. ARRL is also International Secretariat for the International Amateur Radio Union, which is made up of similar societies in 150 countries around the world.

ARRL publishes the monthly journal *QST*, as well as newsletters and many publications covering all aspects of Amateur Radio. Its headquarters station, W1AW, transmits bulletins of interest to radio amateurs and Morse code practice sessions. The ARRL also coordinates an extensive field organization, which includes volunteers who provide technical information and other support services for radio amateurs as well as communications for public-service activities. In addition, ARRL represents US amateurs with the Federal Communications Commission and other government agencies in the US and abroad.

Membership in ARRL means much more than receiving *QST* each month. In addition to the services already described, ARRL offers membership services on a personal level, such as the Technical Information Service—where members can get answers by phone, email or the ARRL website, to all their technical and operating questions.

Full ARRL membership (available only to licensed radio amateurs) gives you a voice in how the affairs of the organization are governed. ARRL policy is set by a Board of Directors (one from each of 15 Divisions). Each year, one-third of the ARRL Board of Directors stands for election by the full members they represent. The day-to-day operation of ARRL HQ is managed by an Executive Vice President and his staff.

No matter what aspect of Amateur Radio attracts you, ARRL membership is relevant and important. There would be no Amateur Radio as we know it today were it not for the ARRL. We would be happy to welcome you as a member! (An Amateur Radio license is not required for Associate Membership.) For more information about ARRL and answers to any questions you may have about Amateur Radio, write or call:

ARRL — The national association for Amateur Radio
225 Main Street
Newington CT 06111-1494
Voice: 860-594-0200
Fax: 860-594-0259
E-mail: **hq@arrl.org**
Internet: **www.arrl.org/**

Prospective new amateurs call (toll-free):
800-32-NEW HAM (800-326-3942)
You can also contact us via e-mail at **newham@arrl.org**
or check out *ARRLWeb* at **www.arrl.org/**

Chapter 1
Introduction

In this chapter, \you'll learn about:
- **New frequencies and activities enjoyed by Extra licensees**
- **Reasons to upgrade from General or Advanced**
- **Requirements and study materials for the Extra exam**
- **How to prepare for your exam**
- **How to find an exam session**
- **Where to find more resources**

Welcome to *The ARRL Extra Class License Manual*! You're about to begin the final chapter of your Amateur Radio licenses studies. By earning both a Technician and a General class license, you've learned a tremendous amount about the technology of radio and the operating practices that make it useful and effective. By upgrading to Extra, you'll complete that journey, with full access to everything Amateur Radio has to offer, joining more than 125,000 other "Extras."

1.1 The Extra Class License and Amateur Radio

Just about every ham thinks about obtaining the Extra class "ticket" at one point or another in their ham career — and why not? Extra class licensees have complete access to all frequencies available to the Amateur Service. There's also the good feeling of knowing that you've demonstrated broad and useful knowledge of technology, operating practices, and the

Table 1-1
Exam Elements Needed to Qualify for an Extra Class License

Current License	Exam Requirements	Study Materials
None or Novice	Technician (Element 2)	*The ARRL Ham Radio License Manual* or *ARRL's Tech Q&A*
	General (Element 3)	*The ARRL General Class License Manual* or *ARRL's General Q&A*
	Amateur Extra (Element 4)	*The ARRL Extra Class License Manual* or *ARRL's Extra Q&A*
Technician (issued on or after March 21, 1987)*	General (Element 3)	*The ARRL General Class License Manual* or *ARRL's General Q&A*
	Amateur Extra (Element 4)	*The ARRL Extra Class License Manual* or *ARRL's Extra Q&A*
General or Advanced	Amateur Extra (Element 4)	*The ARRL Extra Class License Manual* or *ARRL's Extra Q&A*

*Individuals who qualified for the Technician license before March 21, 1987, will be able to receive credit for Element 3 (General class) by providing documentary proof to a Volunteer Examiner Coordinator.

If you're thumbing through this book wondering what Amateur Radio is all about, welcome to a unique and valuable hobby! More than 700,000 people in the United States and almost six million around the world have a license to operate on the radio wavelengths allocated to amateurs. You'll find them on the traditional shortwave bands sending signals around the world, just as they have for almost a century. Using portable and handheld equipment, they communicate with local and regional friends, too.

Amateurs or "hams" are experimenters and innovators. Many design and build their own equipment and antenna systems. Hams have created novel and useful hybrids of computer and Internet technology along with the traditional radio. It's possible today to use a tiny handheld radio to access local systems connected to the Internet that relay the signals to a similarly-equipped ham halfway around the world. Along with the traditional Morse code and voice signals, hams are continually inventing ways to send digital data over radio. In fact, you may have used a technology adapted from an amateur invention!

It's not necessary to be an electrical engineer to be a ham, either! There are hams from all walks of life. During times of emergency, hams step in to assist by adapting their communications systems to the situation at hand. The ham spirit of volunteerism and "can do" helps relief and public safety agencies as well as private citizens around the world. If you can learn to operate a radio and follow some simple "rules of the road," you can participate. All it takes is an interest in radio and a willingness to learn and to help others.

Amateur Radio is regulated by the Federal Communications Commission (FCC) and amateurs must be licensed to use the public airwaves that are allocated to their use. To get the license, prospective amateurs must pass a simple question-and-answer exam. In the United States, there are three license classes: Technician (entry-level), General and Amateur Extra (top-level). Each successive license grants additional privileges. This book is the study guide for the Extra class license exam.

Amateur Radio has been around for quite a while — since before World War I. It's still a vital service today, at the forefront of emergency communications, technical and operating innovation, and spreading goodwill around the world, one contact at a time. To learn more about Amateur Radio, log on to **www.hello-radio.org** for a guided tour of a truly fascinating hobby!

Many hams find it satisfying to give something back to their community. One way is to become involved in emergency communications by your local Amateur Radio Emergency Service (ARES®) group or becoming a NOAA SKYWARN volunteer.

FCC rules and regulations. Having an Extra class license doesn't mean you know everything there is to know — quite the contrary — but you will be better prepared to learn and grow within the service.

If you have been hesitant to study for the exam because you feel the theory is too difficult or your math and electronics background is rusty, take heart. If you study patiently and make use of the available resources for Extra class students, you *will* succeed in passing the exam. It may take more than one attempt, but this is hardly unusual. You can continue to study and try again later. You might also try just taking a different version of the exam at the very same test session. Sooner or later, you'll be able to add "/AE" to your call sign on the air! The key is to make the personal commitment to passing the exam and be willing to study.

Table 1-1 shows what you need to qualify for Extra class, depending on your current license. Remember how you felt when you started studying for your Technician license and again when you hit the upgrade trail to General? Now you know that it's just a matter of persistent study and practice to obtain that coveted Extra class "ticket"! Congratulations for taking the first step — let's get started!

1.2 Extra Class Overview

Most of this book's readers will have already earned their General license, whether they have been a ham for quite awhile or are new to the hobby. Some may have passed their Advanced class exam years ago (Advanced licenses are no longer issued but they can be renewed). No matter which type you might be, you're to be commended for making the effort to pass the Extra class exam!

UPGRADING TO EXTRA

As you begin your studies remember that you've already overcome the hurdles of passing not just one, but at least two license exams! The Extra class exam questions are certainly more difficult, but you already know all about the testing procedure and the basics of ham radio. You can approach the process of upgrading with confidence!

The Extra class licensee has access to additional spectrum on four of the most popular amateur HF bands. Located at the lower edges of the CW and phone segments of the 80, 40, 20 and 15 meter bands, the "Extra sub-bands" are prime frequencies for DXing and contest operating. At other times, the additional elbow room found there makes operating more en-

---

Morse Code — Very Much Alive!

Part of Amateur Radio since its beginnings, Morse code has been part of the rich amateur tradition for 100 years and many hams still use it extensively. At one time, Morse code proficiency at 20 words per minute (WPM) was required for an Extra class license! That requirement was lowered to a one-speed-fits-all 5 WPM and then dropped entirely in February 2007. Morse is likely to remain part of the amateur experience well into the 21st century, however.

Morse (also known as CW, for *continuous wave* transmission) remains popular for solid reasons! It's easy to build CW transmitters and receivers. There is no more power-efficient mode of communications that is copied by the human ear. The extensive set of prosigns and abbreviations allow amateurs to communicate a great deal of information even if they don't share a common language.

Extra class licensees have access to some prime Morse code frequencies at the bottom of the 80, 40, 20 and 15 meter bands. For this reason, obtaining an Extra class ticket should also be a powerful incentive to learn the code or improve your code speed to take advantage of what those frequencies have to offer.

To help you learn Morse, the ARRL has a complete set of resources listed on its web page at **www.arrl.org/learning-morse-code**. Computer software and on-the-air code practice sessions are available for personal training and practice. Organizations such as FISTS (**www.fists.org**) — an operator's style of sending is referred to as his or her "fist" — help hams learn Morse code and will even help you find a "code buddy" to share the learning with you.

CW operation is particularly effective when operating as a portable station with temporary antennas. Author NØAX took his mini-paddle along when activating Emerald Island, North Carolina, for the Islands On the Air (IOTA) awards program.

Books to Help You Learn

As you study the material on the licensing exam, you will have lots of other questions about the how and why of Amateur Radio. The following references, available from your local bookstore or the ARRL (**www.arrl.org/catalog**) will help "fill in the blanks" and give you a broader picture of the hobby:

✔ *Ham Radio for Dummies* by Ward Silver, NØAX. Written for new hams and hams interested in new activities, this book supplements the information in study guides with an informal, friendly approach to the hobby.

✔ *ARRL Operating Manual.* With in-depth chapters on the most popular ham radio activities, this is your guide to nets, award programs, DXing and more. It even includes a healthy set of reference tables and maps.

✔ *Understanding Basic Electronics* by Walter Banzhaf, WB1ANE. Students who want more technical background about electronics should take a look at this book. It covers the fundamentals of electricity and electronics that are the foundation of all radio.

✔ *Basic Radio* by Joel Hallas, W1ZR. This book goes beyond electronic circuits to explain how radios are designed and perform. It covers the key building blocks of receivers, transmitters, antennas and propagation.

✔ *ARRL Handbook.* This is the grandfather of all Amateur Radio references and belongs on the shelf of hams. Almost any topic you can think of in Amateur Radio technology is represented here.

✔ *ARRL Antenna Book.* After the radio itself, all radio depends on antennas. This book provides information on every common type of amateur antenna, feed lines and related topics, and practical construction tips and techniques.

joyable. The improved technical understanding of upgrading to Extra will make you a more knowledgeable and skilled operator, too.

Once you've earned Extra class status, you'll be able to administer license exam sessions for any license class! Extra class VEs are needed to give both General and Extra class exams — an excellent opportunity to repay some of the assistance you've received. Technician and General class hams may turn to you for guidance, too. If you have a desire to teach other hams, as an Extra there's no limit to what you can do.

CALL SIGNS

A call sign is a very personal identification and many hams keep theirs for a lifetime. Will you *have* to change calls when you receive your Extra class license? No — just as you could keep your call sign when you upgraded to General, so it is when you upgrade to Extra. However, you might want your on-the-air identity to reflect your new license status! Extra class licensees can choose Vanity call signs from the coveted "one-by-two" or 1 × 2 series of call signs, such as K1EA or N6TR, or a 2 × 1 such as NN1N or KX9X. Extras are also allocated the set of 2 × 2 calls that begin with the letter combinations AA through AK, such as AB1FM. If you would like to know more about what calls are available, it's fun to browse through N4MC's "Vanity HQ" website at **www.vanityhq.com**. Pick your favorite and put it in a prominent location as an incentive to keep studying!

Bernie, W3UR, combined a family vacation to the Barbados with some holiday-style operating as 8P9UR.

Entering as a multioperator team, Michael, K2KR, and Kristin, KCØINX, shared the excitement of "radiosport" by activating the West Texas section in the annual ARRL November Sweepstakes.

FOCUS ON HF AND ADVANCED MODES

Where the Technician and General class licenses introduced you to whole new areas of radio and electronics, the Extra class license is more focused. You'll be expected to build on what you learned for the first two licenses. If you still have your study guides for those licenses handy, they may provide useful background material. The sidebar "Books To Help You Learn" lists several additional reference books that provide more than enough information for the Extra class student. Here are some examples of topics you'll be studying:

- The different types of station control
- AC impedance, resonance and filters
- Contest and DX operating procedures
- Digital modes such as PSK31
- Test instruments such as the oscilloscope
- Semiconductor devices and basic electronic circuits
- Special topics on antennas, feed lines and propagation

Not every ham uses every mode and frequency, of course. By learning about this wider range of ideas, it helps hams to make better choices for regular operating. You will become aware of just how wide and deep ham radio really is. Better yet, the introduction of these new ideas may just get you interested in giving them a try!

QS0708-Hutch03

90
120 · 60
2-Ele. Yagi
150 · Dipole · −10 · 30
−20
−30
180 · 0

Max. Gain = 10.87 dBi   Freq. = 18.1179 MHz

By using inexpensive antenna modeling software, comparing and optimizing antenna designs is within the reach of every ham. This graphic depicts the elevation pattern of a 2-element Yagi.

1.3 The Volunteer Testing Process

The procedure for upgrading to Extra class is identical to the one you followed for other license exams. You must attend an exam session administered by Volunteer Examiners (VEs) accredited by a Volunteer Examiner Coordinator (VEC) and pass a 50-question written exam consisting of questions drawn from the Element 4 Question Pool. (You'll also have to have passed the elements for the Technician and General class licenses.) This is an increase of 15 questions from the General and Technician 35-question exams (see **Table 1-2**).

When you're ready, you'll need to find a test session. If you're in a licensing class, the instructor will help you find and register for a session. Otherwise, you can find a test session by using the ARRL's web page for finding exams, **www.arrl.org/find-an-amateur-radio-license-exam-session**. If you can register for the test session in advance, do so. Other sessions, such as those at hamfests or conventions, are usually available to anyone who shows up, also known as *walk-ins*. You may have to wait for an available space though, so go early!

Bring the *original* of your current FCC-issued license and a photocopy to send with the application. You'll need a photo ID, such as a driver's license, passport or employer's identity card, or two forms of identification if no photo ID is available. Know your Social Security Number (SSN) or FCC-issued Federal Registration Number (FRN). You can bring pencils or pens, blank scratch paper and a calculator with the memory erased but any kind of electronic devices or phone that can access the Internet are prohibited. For a complete list of items to bring or that are prohibited on exam day please see the ARRL's web page **www.arrl.org/what-to-bring-to-an-exam-session**. (If you have a disability and need these devices to take the exam, contact the session sponsor ahead of time as described in the sidebar later in this chapter.)

Once you're signed in, you'll need to fill out a copy of the National Conference of Volunteer Examiner Coordinator's NCVEC Quick Form 605 (**Figure 1-1**). This is an application for a new or upgraded license. It is only used:

- At test sessions
- For a VEC to process a license renewal
- For a VEC to process a license change
- For a VEC to process a new club license or a club license change

Do not use an NCVEC Quick Form 605 for any kind of application directly to the FCC — it will be rejected. Use a regular FCC 605-Main Form. After filling out the form, pay the test fee and get ready. Check the ARRL VEC website **www.arrl.org/arrl-vec-exam-fees** or the website of the VEC certifying the exam for the current amount.

You will be given a question booklet and an answer sheet. Be sure to read the instructions, fill in all the necessary information and sign your name wherever

Volunteer Examiners will orchestrate every aspect of "Exam Day" and will let you know how you did right away. After you pass the exam, they'll issue you a Certificate of Successful Completion of Examination (CSCE), which allows you to start using your new privileges right away. Just be sure to sign "temporary AE" after your call sign until your upgrade appears in the official FCC online database or your new license shows up in the mail.

Table 1-2
Amateur License Class Examinations

License Class	Elements Required	Number of Questions
Technician	2 (Written)	35 (passing is 26 correct)
General	3 (Written)	35 (passing is 26 correct)
Extra	4 (Written)	50 (passing is 37 correct)

NCVEC QUICK-FORM 605 APPLICATION FOR
AMATEUR OPERATOR/PRIMARY STATION LICENSE

SECTION 1 - TO BE COMPLETED BY APPLICANT

PRINT LAST NAME	SUFFIX (Jr., Sr.)	FIRST NAME	INITIAL	STATION CALL SIGN (IF ANY)
MORIN		Joanne	B	KA1JPA

MAILING ADDRESS (Number and Street or P.O. Box)
225 Main St.

SOCIAL SECURITY NUMBER (SSN) or (FRN) FCC FEDERAL REGISTRATION NUMBER
987-654-321

CITY	STATE CODE	ZIP CODE (5 or 9 Numbers)	E-MAIL ADDRESS (OPTIONAL)
Newington	CT	06067	

DAYTIME TELEPHONE NUMBER (Include Area Code) OPTIONAL

FAX NUMBER (Include Area Code) OPTIONAL

ENTITY NAME (IF CLUB, MILITARY RECREATION, RACES)

Type of Applicant: [X] Individual [ ] Amateur Club [ ] Military Recreation [ ] RACES (Modify Only)

CLUB, MILITARY RECREATION, OR RACES CALL SIGN

I HEREBY APPLY FOR (Make an X in the appropriate box(es))

SIGNATURE OF RESPONSIBLE CLUB OFFICIAL (not trustee)

[ ] EXAMINATION for a **new** license grant

[X] EXAMINATION for **upgrade** of my license class

[ ] CHANGE my **name** on my license to my new name

Former Name: _____
(Last name) (Suffix) (First name) (MI)

[ ] CHANGE my mailing address to **above** address

[ ] CHANGE my station **call sign** systematically

Applicant's Initials: _____

[ ] RENEWAL of my license grant.

Do you have another license application on file with the FCC which has not been acted upon?	PURPOSE OF OTHER APPLICATION	PENDING FILE NUMBER (FOR VEC USE ONLY)

I certify that:
- I waive any claim to the use of any particular frequency regardless of prior use by license or otherwise;
- All statements and attachments are true, complete and correct to the best of my knowledge and belief and are made in good faith;
- I am not a representative of a foreign government;
- I am not subject to a denial of Federal benefits pursuant to Section 5301 of the Anti-Drug Abuse Act of 1988, 21 U.S.C. § 862;
- The construction of my station will NOT be an action which is likely to have a significant environmental effect (See 47 CFR Sections 1.1301-1.1319 and Section 97.13(a));
- I have read and WILL COMPLY with Section 97.13(c) of the Commission's Rules regarding RADIOFREQUENCY (RF) RADIATION SAFETY and the amateur service section of OST/OET Bulletin Number 65.

Signature of applicant (Do not print, type, or stamp. Must match applicant's name above.) (Clubs: 2 different individuals must sign)

X *Joanne B Morin*   Date Signed: 2/23/07

SECTION 2 - TO BE COMPLETED BY ALL ADMINISTERING VEs

Applicant is qualified for operator license class:

[ ] NO NEW LICENSE OR UPGRADE WAS EARNED

[ ] TECHNICIAN Element 2

[ ] GENERAL Elements 2 and 3

[X] AMATEUR EXTRA Elements 2, 3 and 4

DATE OF EXAMINATION SESSION
EXAMINATION SESSION LOCATION
VEC ORGANIZATION
VEC RECEIPT DATE

I CERTIFY THAT I HAVE COMPLIED WITH THE ADMINISTERING VE REQUIRMENTS IN PART 97 OF THE COMMISSION'S RULES AND WITH THE INSTRUCTIONS PROVIDED BY THE COORDINATING VEC AND THE FCC.

1st VEs NAME (Print First, MI, Last, Suffix)	VEs STATION CALL SIGN	VEs SIGNATURE (Must match name)	DATE SIGNED
Steven R. Ewald	WV1X	*Steven R. Ewald*	2/23-07
Rose-Anne Lawrence	KB1DMW	*Rose-Anne Lawrence*	2-23-07
Penny E Harts	N1NAC	*Penny E Harts*	2.23.07

DO NOT SEND THIS FORM TO FCC – THIS IS NOT AN FCC FORM.
IF THIS FORM IS SENT TO FCC, FCC WILL RETURN IT TO YOU WITHOUT ACTION.

NCVEC FORM 605 - February 2007
FOR VE/VEC USE ONLY - Page 1

ARRL0138

Figure 1-1 — This sample NCVEC Quick Form 605 shows how your form will look after you have completed your upgrade to Extra.

it's required. Check to be sure your booklet has all the questions and be sure to mark the answer in the correct space for each question.

While this may be review if you've already passed the General class exam, it's useful (and calming) to review. You don't have to answer the questions in order — skip the hard ones and go back to them. If you read the answers carefully, you'll probably find that you can eliminate one or more "distracters." Of the remaining answers, only one will be the best. If you can't decide on the correct answer, go ahead and guess. There is no penalty for an incorrect guess. When you're done, go back and check your answers and double-check your arithmetic — there's no rush!

Once you've answered all 50 questions, the VEs will grade and verify your test results. Assuming you've passed (your last amateur exam ever!) you'll fill out a *Certificate of Successful Completion of Examination* (CSCE) and the VE team will complete your NCVEC FCC Form 605. The exam organizers will submit your results to the FCC while you keep the CSCE as evidence that you've passed your Extra test.

You'll be more than ready to start using those new privileges as soon as you get home! When you give your call sign, append "/AE" (on CW or digital modes) or "temporary AE" (on phone). As soon as your license class is upgraded in the FCC's database of licensees, typically a week or two later, you can stop adding the suffix. The CSCE is good for 365 days or until receiving your paper license.

If you don't pass, don't be discouraged! You might be able to take another version of the test right then and there if the session organizers can accommodate you. Even if you decide to try again later, you now know just how the test session feels — you'll be more relaxed and ready next time. The Extra class sub-bands are full of hams who took their Extra test more than once before passing. You'll be in good company!

FCC AND ARRL VEC LICENSING RESOURCES

After you pass your exam, the examiners will file all of the necessary paperwork so that your license will be granted by the Federal Communications Commission (FCC). In a week or two, you will be able see your new license status — and your new call sign if you requested one — in the FCC's database via the ARRL website. Later, you'll receive a paper license by mail. The ARRL VEC can also process license renewals and modifications for you as described at **www.arrl.org/call-sign-renewals-or-changes**.

When you initially passed your Technician exam, you may have applied for your FCC Federal Registration Number (FRN) or may have been issued one automatically by FCC from the information gathered from your form. This allows you to access the information for any FCC licenses you may have and to request modifications to them. These functions are available via the FCC's Universal Licensing System (ULS) website (**wireless.fcc.gov/uls**) and complete instructions for using the site are available at **www.arrl.org/universal-licensing-system**. When accessing the ULS, your FRN will allow you to watch the database for your license upgrade!

1.4 How to Use This Book

Designed to help the student really learn the material, the topics in this study guide build on one another. The first sections cover operating techniques and FCC regulations, while later chapters progress from simple electronics through radio signals. The book concludes with antennas, propagation and safety topics. Each section references the exact questions that could be on the exam, so you'll have a chance to test your understanding as soon as you complete each topic. Online references and sources supplement the information in the study guide, so you can get extra help on a topic or just read for interest. You'll have a lot of resources on your side during your studies!

This study guide will provide the necessary background and explanation for the answers to the exam questions. By learning this material, you will go beyond just learning the answers. You'll understand the fundamentals behind them and this makes it easier to learn, remember

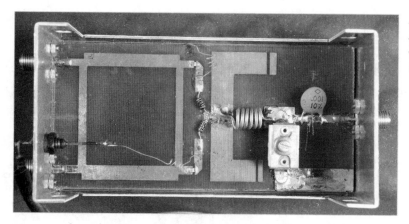

This dual-balanced mixer for the 1296 MHz band was designed and built by Paul, W1GHZ.

and use what you know. This book also contains many useful facts and figures that you can use in your station and on the air.

If you are taking a licensing class, help your instructors by letting them know about areas in which you need help. They want you to learn as thoroughly and quickly as possible, so don't hold back with your questions. Similarly, if you find the material particularly clear or helpful, tell them that, too, so it can be used in the next class!

Just before the Question Pool section of this book you'll find a large Glossary of radio terminology. The Question Pool contains the complete set of exam questions and answers and it is followed by the Index and an advertising section with displays from some of Amateur Radio's best-known vendors of equipment and supplies.

WHAT WE ASSUME ABOUT YOU

You don't have to be a technical guru or an expert operator to upgrade to Extra class! The topics you will encounter build from the basic science of radio and electricity that you mastered for Technician and General. The math is a little more involved than for General class topics and tutorials are available at the *ARRL Extra Class License Manual* website, **www.arrl.org/extra-class-license-manual**. You should have a calculator capable of doing logarithms and trigonometric functions. You'll also be allowed to use it during the license exam, of course.

ADVANCED STUDENTS

If you have some background in radio, perhaps as a technician or trained operator, you may be able to short-circuit some of the sections. To find out, locate the shaded boxes in the text that list the exam questions for each topic. Turn to the Question Pool and if you can answer the questions correctly, move to the next topic in the text. It's common for technically-minded students to need help with the rules and regulations, while students with an operating background tend to need more help with the technical material. Regardless of your previous knowledge and experience, be sure that you can answer the questions because they will certainly be on the test!

SELF-STUDY OR CLASSROOM STUDENTS

The ARRL Extra Class License Manual can be used either by an individual student studying on his or her own, or as part of a licensing class taught by an instructor. If you're part of a class, the instructor will guide you through the book, section by section. The solo student can move at any pace and in any convenient order. You'll find that having a friend to study with makes learning the material more fun as you help each other over the rough spots.

Don't hesitate to ask for help! Your instructor can provide information on anything you find difficult. Classroom students may find asking their fellow students to be helpful. If you're studying on your own, there are resources for you, too! If you can't find the answer in the book or at the website, e-mail your question to the ARRL's New Ham Desk, **newham@arrl.org**. You may not be a new ham, but your question will be routed to the appropriate person. The ARRL's experts will answer directly or connect you with another ham who can answer your questions.

Connie, K5CM, leads a team of stations equipped with professional-quality gear to transmit precisely-controlled signals during the annual ARRL Frequency Measuring Tests. Hams across North America attempt to be the most accurate in measuring the signal frequencies.

USING THE QUESTION POOL

The Element 4 Question Pool is divided into 10 subelements, E1 through E0. Each subelement is further divided into sections that focus on specific topics. For example, E1 on the Commission's Rules has six sections, E1A through E1F. Each question is then numbered E1Axx, with 'xx' representing the number of the question in the section. The questions associated with each topic are listed in the book immediately following the text about the topic. Sometimes questions from several question pool sections may be discussed in a combined section of text. Material that addresses a question is indicated in the text by the question number in square brackets, such as [E1A01].

As you complete each topic be sure to review each of the exam questions highlighted in the shaded text boxes. This will tell you which areas need a little more study time. When you understand the answer to each of the questions, move on. Resist the temptation to just memorize the answers. Doing so leaves you without the real understanding that will make your new Extra class privileges enjoyable and useful. *The ARRL Extra Class License Manual* covers every one of the exam questions, so you can be sure you're ready at exam time.

When using the Question Pool, cover or fold over the answers at the edge of the page to be sure you really do understand the question. Each question is accompanied by a cross-reference back to the page on which that topic is discussed. If you don't completely understand the question or answer, please go back and review that material. The ARRL's condensed guide *Extra Q&A* also provides short explanations for each one of the exam questions.

ONLINE REVIEW AND PRACTICE EXAMS

Use this book with ARRL Exam Review for Ham Radio to review chapter-by-chapter. Take randomly-generated practice exams using questions from the actual examination question pool. You won't have any surprises on exam day! Go to w**ww.arrl.org/examreview**.

EXTRA CLASS LICENSE MANUAL WEB SITE

The ARRL also maintains a special web page for Extra class students at **www.arrl.org/extra-class-license-manual**. It provides or links to helpful supplements and clarifications to the material in the book. The useful and interesting online references listed there put you one click away from additional information on many topics.

FOR INSTRUCTORS

If you're an instructor, note that this edition of the study guide has the same organization as the 9th edition. Topics are presented in a sequence intended to be easier for the student to learn. For example, the section on Radio Signals and Equipment comes after the Components and Circuits section. Because the Extra class exam topics are more sophisticated than for General class, it's more important to develop the context and background for each topic.

The ARRL has also created supporting material for instructors such as graphics files, handouts and a detailed topics list. Check **www.arrl.org/resources-for-license-instruction** for support materials.

CONVENTIONS AND RESOURCES

Throughout your studies keep a sharp eye out for words in *italics*. These words are important so be sure you understand them. Many are included in the extensive Glossary in the back of the book. Another thing to look for are the addresses or URLs for web resources in **bold**, such as **www.arrl.org/extra-class-license-manual**. By browsing these web pages while you're studying, you will accelerate and broaden your understanding.

Question numbers in square brackets, such as [E1A01] indicates material that addresses a specific question. This will help you review or examine specific topics.

Throughout the book, there are many short sidebars that present topics related to the subject you're studying. These sidebars may just tell an interesting story or they might tackle a subject that needs separate space in the book. The information in sidebars helps you understand how the information you're studying relates to ham radio in general.

TIME TO GET STARTED

By following these instructions and carefully studying the material in this book, soon you'll be joining the rest of the Extra class licensees! Each of us at the ARRL Headquarters and every ARRL member looks forward to the day when you join the fun. 73 (best regards) and good luck!

Table 1-3
Extra Class (Element 4) Syllabus

Valid July 1, 2012 through June 30, 2016

SUBELEMENT E1 — COMMISSION'S RULES
[6 Exam Questions – 6 Groups]

E1A — Operating Standards: frequency privileges; emission standards; automatic message forwarding; frequency sharing; stations aboard ships or aircraft

E1B — Station restrictions and special operations: restrictions on station location; general operating restrictions, spurious emissions, control operator reimbursement; antenna structure restrictions; RACES operations

E1C — Station control: definitions and restrictions pertaining to local, automatic and remote control operation; control operator responsibilities for remote and automatically controlled stations

E1D — Amateur Satellite service: definitions and purpose; license requirements for space stations; available frequencies and bands; telecommand and telemetry operations; restrictions, and special provisions; notification requirements

E1E — Volunteer examiner program: definitions, qualifications, preparation and administration of exams; accreditation; question pools; documentation requirements

E1F — Miscellaneous rules: external RF power amplifiers; national quiet zone; business communications; compensated communications; spread spectrum; auxiliary stations; reciprocal operating privileges; IARP and CEPT licenses; third party communications with foreign countries; special temporary authority

SUBELEMENT E2 – OPERATING PROCEDURES
[5 Exam Questions – 5 Groups]

E2A — Amateur radio in space: amateur satellites; orbital mechanics; frequencies and modes; satellite hardware; satellite operations

E2B — Television practices: fast scan television standards and techniques; slow scan television standards and techniques

E2C — Operating methods: contest and DX operating; spread-spectrum transmissions; selecting an operating frequency

E2D — Operating methods: VHF and UHF digital modes; APRS

E2E — Operating methods: operating HF digital modes; error correction

SUBELEMENT E3 — RADIO WAVE PROPAGATION
[3 Exam Questions – 3 Groups]

E3A — Propagation and technique, Earth-Moon-Earth communications; meteor scatter

E3B — Propagation and technique, trans-equatorial; long path; gray-line; multi-path propagation

E3C — Propagation and technique, Aurora propagation; selective fading; radio-path horizon; take-off angle over flat or sloping terrain; effects of ground on propagation; less common propagation modes

SUBELEMENT E4 — AMATEUR PRACTICES
[5 Exam Questions – 5 Groups]

E4A — Test equipment: analog and digital instruments; spectrum and network analyzers, antenna analyzers; oscilloscopes; testing transistors; RF measurements

E4B — Measurement technique and limitations: instrument accuracy and performance limitations; probes; techniques to minimize errors; measurement of "Q"; instrument calibration

E4C — Receiver performance characteristics, phase noise, capture effect, noise floor, image rejection, MDS, signal-to-noise-ratio; selectivity

E4D — Receiver performance characteristics, blocking dynamic range, intermodulation and cross-modulation interference; 3rd order intercept; desensitization; preselection

E4E — Noise suppression: system noise; electrical appliance noise; line noise; locating noise sources; DSP noise reduction; noise blankers

SUBELEMENT E5 — ELECTRICAL PRINCIPLES
[4 Exam Questions — 4 Groups]

E5A — Resonance and Q: characteristics of resonant circuits: series and parallel resonance; Q; half-power bandwidth; phase relationships in reactive circuits

E5B — Time constants and phase relationships: RLC time constants: definition; time constants in RL and RC circuits; phase angle between voltage and current; phase angles of series and parallel circuits

E5C — Impedance plots and coordinate systems: plotting impedances in polar coordinates; rectangular coordinates

E5D — AC and RF energy in real circuits: skin effect; electrostatic and electromagnetic fields; reactive power; power factor; coordinate systems

SUBELEMENT E6 — CIRCUIT COMPONENTS
[6 Exam Questions – 6 Groups]

E6A — Semiconductor materials and devices: semiconductor materials germanium, silicon, P-type, N-type; transistor types: NPN, PNP, junction, field-effect transistors: enhancement mode; depletion mode; MOS; CMOS; N-channel; P-channel

E6B — Semiconductor diodes

E6C — Integrated circuits: TTL digital integrated circuits; CMOS digital integrated circuits; gates

E6D — Optical devices and toroids: cathode-ray tube devices; charge-coupled devices (CCDs); liquid crystal displays (LCDs); toroids: permeability, core material, selecting, winding

E6E — Piezoelectric crystals and MMICs: quartz crystals; crystal oscillators and filters; monolithic amplifiers

E6F — Optical components and power systems: photoconductive principles and effects, photovoltaic systems, optical couplers, optical sensors, and optoisolators

SUBELEMENT E5 — ELECTRICAL PRINCIPLES
[4 Exam Questions — 4 Groups]

E5A — Resonance and Q: characteristics of resonant circuits: series and parallel resonance; Q; half-power bandwidth; phase relationships in reactive circuits

E5B — Time constants and phase relationships: RLC time constants: definition; time constants in RL and RC circuits; phase angle between voltage and current; phase angles of series and parallel circuits

E5C — Impedance plots and coordinate systems: plotting impedances in polar coordinates; rectangular coordinates

E5D — AC and RF energy in real circuits: skin effect; electrostatic and electromagnetic fields; reactive power; power factor; coordinate systems

SUBELEMENT E6 — CIRCUIT COMPONENTS
[6 Exam Questions – 6 Groups]

E6A — Semiconductor materials and devices: semiconductor materials germanium, silicon, P-type, N-type; transistor types: NPN, PNP, junction, field-effect transistors: enhancement mode; depletion mode; MOS; CMOS; N-channel; P-channel

E6B — Semiconductor diodes

E6C — Integrated circuits: TTL digital integrated circuits; CMOS digital integrated circuits; gates

E6D — Optical devices and toroids: cathode-ray tube devices; charge-coupled devices (CCDs); liquid crystal displays (LCDs); toroids: permeability, core material, selecting, winding

E6E — Piezoelectric crystals and MMICs: quartz crystals; crystal oscillators and filters; monolithic amplifiers

E6F — Optical components and power systems: photoconductive principles and effects, photovoltaic systems, optical couplers, optical sensors, and optoisolators

SUBELEMENT E7 — PRACTICAL CIRCUITS
[8 Exam Questions – 8 Groups]

E7A — Digital circuits: digital circuit principles and logic circuits: classes of logic elements; positive and negative logic; frequency dividers; truth tables

E7B — Amplifiers: Class of operation; vacuum tube and solid-state circuits; distortion and intermodulation; spurious and parasitic suppression; microwave amplifiers

E7C — Filters and matching networks: filters and impedance matching networks: types of networks; types of filters; filter applications; filter characteristics; impedance matching; DSP filtering

E7D — Power supplies and voltage regulators

E7E — Modulation and demodulation: reactance, phase and balanced modulators; detectors; mixer stages; DSP modulation and demodulation; software defined radio systems

E7F — Frequency markers and counters: frequency divider circuits; frequency marker generators; frequency counters

E7G — Active filters and op-amps: active audio filters; characteristics; basic circuit design; operational amplifiers

E7H — Oscillators and signal sources: types of oscillators; synthesizers and phase-locked loops; direct digital synthesizers

SUBELEMENT E8 — SIGNALS AND EMISSIONS
[4 Exam Questions – 4 Groups]

E8A — AC waveforms: sine, square, sawtooth and irregular waveforms; AC measurements; average and PEP of RF signals; pulse and digital signal waveforms

E8B — Modulation and demodulation: modulation methods; modulation index and deviation ratio; pulse modulation; frequency and time division multiplexing

E8C — Digital signals: digital communications modes; CW; information rate vs. bandwidth; spread-spectrum communications; modulation methods

E8D — Waves, measurements, and RF grounding: peak-to-peak values, polarization; RF grounding

SUBELEMENT E9 — ANTENNAS AND TRANSMISSION LINES
[8 Exam Questions — 8 Groups]

E9A — Isotropic and gain antennas: definition; used as a standard for comparison; radiation pattern; basic antenna parameters: radiation resistance and reactance, gain, beamwidth, efficiency

E9B — Antenna patterns: E and H plane patterns; gain as a function of pattern; antenna design; Yagi antennas

E9C — Wire and phased vertical antennas: beverage antennas; terminated and resonant rhombic antennas; elevation above real ground; ground effects as related to polarization; take-off angles

E9D — Directional antennas: gain; satellite antennas; antenna beamwidth; losses; SWR bandwidth; antenna efficiency; shortened and mobile antennas; grounding

E9E — Matching: matching antennas to feed lines; power dividers

E9F — Transmission lines: characteristics of open and shorted feed lines: ⅛ wavelength; ¼ wavelength; ½ wavelength; feed lines: coax versus open-wire; velocity factor; electrical length; transformation characteristics of line terminated in impedance not equal to characteristic impedance

E9G — The Smith chart

E9H — Effective radiated power; system gains and losses; radio direction finding antennas

SUBELEMENT E0 — SAFETY
[1 exam question — 1 group]

E0A — Safety: amateur radio safety practices; RF radiation hazards; hazardous materials

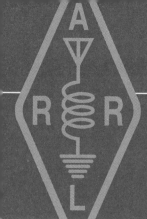

Chapter 2

Operating Practices

In this chapter, you'll learn about:
- **Frequencies available to Extra class licensees**
- **DX and contest operating**
- **VHF and HF digital operating**
- **Satellite orbital mechanics and signals**
- **Satellite transponders, frequencies and modes**

Congratulations, you're going to be an Extra and no amateur frequency will be denied to your transmissions! You may have picked up quite a bit of operating experience as a Technician and General. If so, you can probably just scan this section for the items on which you need to brush up. If you have spent most of your operating time on VHF+ frequencies, or if you're wondering if the new HF frequencies are really that special, read on. This chapter also covers amateur satellite operations.

2.1 General Operating

EXTRA CLASS HF FREQUENCIES

The small additional portions of the HF spectrum available only to Extra class licensees don't look like much on the allocation charts in **Figure 2-1**, but they are prime real estate for the HF operator. The 80, 40, 20 and 15 meter frequencies are major DX and contesting bands. The 160, 30, 17 and 12 meter bands are too narrow to justify a license class-based subband, while 10 meters is so big that there's no need for a subband.

Why are these small slices of spectrum worth upgrading for? Traditional operating practice is for DX contacts to take place closer to the bottom edge of the bands. The higher ends of the bands are where most of the domestic QSOs and net operations take place. You'll find that nearly all serious DXers (and contest operators) hold Extra class licenses because that's where the contacts are most likely to be made! Even without DX contacts going on, the Extra class segments are less crowded.

Don't forget, you're still bound by the FCC-defined subbands for CW, data and phone! You're also required to keep your emissions completely inside the subband. For example, you may wonder why no one is using 3.601 MHz to call "CQ DX." Well, it's because with the carrier on 3.601 MHz, the sideband energy in a lower-sideband (LSB) signal occupies from 3.601 down to 3.598 MHz, 2 kHz out of the subband! This is discussed in detail in the chapter on Rules and Regulations.

Frequency Selection

As an Extra class operator, you will be expected to accommodate all sorts of band conditions. Be flexible. Know how to complete a scheduled contact, even if the planned frequency is busy. Always have a "plan B" of an alternate frequency or time. Use the resources avail-

US Amateur Radio Bands

The national association for
ARRL AMATEUR RADIO®
Published by:
www.arrl.org
225 Main Street, Newington, CT USA 06111-1494

Effective Date
March 5, 2012

US AMATEUR POWER LIMITS

FCC 97.313 An amateur station must use the minimum transmitter power necessary to carry out the desired communications. (b) No station may transmit with a transmitter power exceeding 1.5 kW PEP.

KEY

Note:
CW operation is permitted throughout all amateur bands.

MCW is authorized above 50.1 MHz, except for 219-220 MHz.

Test transmissions are authorized above 51 MHz, except for 219-220 MHz.

= RTTY and data
= phone and image
= CW *only*
= SSB phone
= USB phone,CW, RTTY, and data.
= Fixed digital message forwarding systems *only*

E = Amateur Extra
A = Advanced
G = General
T = Technician
N = Novice

See *ARRLWeb* at *www.arrl.org* for detailed band plans.

ARRL
We're At Your Service

ARRL Headquarters:
860-594-0200 (Fax 860-594-0259)
email: hq@arrl.org

Publication Orders:
www.arrl.org/catalog
Toll-Free 1-888-277-5289 (860-594-0355)
email: orders@arrl.org

Membership/Circulation Desk:
www.arrl.org/catalog
Toll-Free 1-888-277-5289 (860-594-0338)
email: membership@arrl.org

Getting Started in Amateur Radio:
Toll-Free 1-800-326-3942 (860-594-0355)
email: newham@arrl.org

Exams: 860-594-0300 email: vec@arrl.org   rev. 2/29/2012
Copyright © ARRL 2011

160 Meters (1.8 MHz)
Avoid interference to radiolocation operations from 1.900 to 2.000 MHz

E,A,G
1.800   1.900   2.000 MHz

80 Meters (3.5 MHz)
E / A / G / N,T (*200 W*)
3.500 3.600 3.700 3.800 4.000 MHz
3.525 3.600

60 Meters (5.3 MHz)
2.8 kHz
E,A,G (*100 W*)
5330.5 5346.5 5357.0 5371.5 5403.5 kHz
5330.5 5346.5 5357.0 5371.5 5403.5 kHz

General, Advanced, and Amateur Extra licensees may use the following five channels on a secondary basis with a maximum effective radiated power of 100 W PEP relative to a half wave dipole. Only upper sideband suppressed carrier voice transmissions, CW, RTTY and data such as PACTOR III. The frequencies are 5330.5, 5346.5, 5357.0, 5371.5 and 5403.5 kHz. The occupied bandwidth is limited to 2.8 kHz centered on 5332, 5348, 5358.5, 5373, and 5405 kHz respectively.

40 Meters (7 MHz)
E / A / G / N,T (*200 W*)
7.000 7.125 7.175 7.300 MHz
7.025 7.125

Phone and image modes are permitted between 7.075 and 7.100 MHz for FCC licensed stations in ITU Regions 1 and 3 and by FCC licensed stations in ITU Region 2 West of 130 degrees West longitude or South of 20 degrees North latitude. See Sections 97.305(c) and 97.307(f)(11). Novice and Technician licensees outside ITU Region 2 may use CW only between 7.025 and 7.075 MHz and between 7.100 and 7.125 MHz. 7.200 to 7.300 MHz is not available outside ITU Region 2. See Section 97.301(e). These exemptions do not apply to stations in the continental US.

30 Meters (10.1 MHz)
Avoid interference to fixed services outside the US.
200 Watts PEP
E,A,G
10.100   10.150 MHz

20 Meters (14 MHz)
E / A / G / N,T (*200 W*)
14.000 14.025 14.150 14.175 14.225 14.350 MHz
14.150 14.225

17 Meters (18 MHz)
E,A,G
18.068 18.110 18.168 MHz

15 Meters (21 MHz)
E / A / G / N,T (*200 W*)
21.000 21.025 21.200 21.225 21.275 21.450 MHz
21.200

12 Meters (24 MHz)
E,A,G
24.890 24.930 24.990 MHz

10 Meters (28 MHz)
E,A,G / N,T (*200 W*)
28.000 28.300 28.500 29.700 MHz

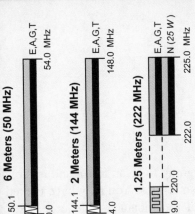

6 Meters (50 MHz)
E,A,G,T
50.0   50.1   54.0 MHz

2 Meters (144 MHz)
E,A,G,T
144.0 144.1 148.0 MHz

1.25 Meters (222 MHz)
E,A,G,T / N (*25 W*)
219.0 220.0 222.0 225.0 MHz

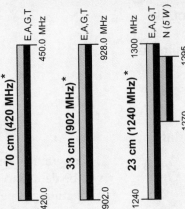

*Geographical and power restrictions may apply to all bands above 420 MHz. See *The ARRL Operating Manual* for information about your area.

70 cm (420 MHz)*
E,A,G,T
420.0   450.0 MHz

33 cm (902 MHz)*
E,A,G,T
902.0   928.0 MHz

23 cm (1240 MHz)*
E,A,G,T / N (*5 W*)
1240 1270 1295 1300 MHz

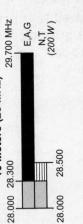

All licensees except Novices are authorized all modes on the following frequencies:

2300-2310 MHz	10.0-10.5 GHz	122.25-123.0 GHz
2390-2450 MHz	24.0-24.25 GHz	134-141 GHz
3300-3500 MHz	47.0-47.2 GHz	241-250 GHz
5650-5925 MHz	76.0-81.0 GHz	All above 275 GHz

Figure 2-1 — US Amateur Bands

able to all hams for planning your activities. For example, there are plenty of "contest calendars" on the web, such as those at **www.arrl.org/contests** or **hornucopia.com/contestcal**. Take a look to see if there is a major contest before making a schedule. Special event stations (**www.arrl.org/special-event-stations**) and DXpeditions may appear on your favorite frequency for a few days, as well. If you're a net control station or net manager, be sure that your net has an alternate frequency to accommodate a busy band or poor conditions — practice using Plan B so that you'll be ready when you really do need to use it!

While an amplifier may cut through the QRM on a busy band, it's especially important to learn how to use your antennas and receiver to hear well. Study this manual's sections on receiver linearity to see how to help your receiver handle the strong signals you'll encounter. Know where the nulls in your antenna pattern are so that you can use them to reduce interference both to and from your signals. You'll be an Extra — you can do it!

DXING

Why chase DX? Why chase those elusive distant contacts through noisy bands and unruly pileups? DXing exemplifies a big part of what ham radio is about: the continuous improvement of equipment, antennas, propagation understanding and operating skills. The quest for "a little more distance" or that special propagation opportunity drives a lot of technical and operating advances. There are many awards to demonstrate one's personal skills, providing an incentive to get on the air and operate!

Your definition of DX will probably depend upon the bands on which you are operating. For HF operators, DX usually means any stations outside of your own country. On the VHF and UHF bands, however, DX may mean stations more than 50 or 100 miles away — beyond your radio horizon. DX extends into space, too — amateurs have been bouncing signals off the Moon to work nearly halfway around the world on microwave frequencies!

Once you've made some DX contacts, perhaps you'll want to chase an award or two. To confirm your contacts, you can use an on-line service such as the ARRL's online Logbook of The World (LoTW — **www.arrl.org/logbook-of-the-world**) or exchange a traditional paper QSL card. While mailing the card directly is quick and using the QSL bureau system is inexpensive, many DX stations use the services of a *QSL manager* who confirms contacts and sends out responding QSL cards for a DX station. [E2C05] This is a good compromise for active DXers.

DX Windows and Watering Holes

If you've done some operating already, you may have made some DX contacts — perhaps via sporadic-E ("E-skip") on 6 meters or regular F-layer skip on HF. Did you notice where those contacts were on the bands? As you may have already discovered as a Technician or General, DXing and contests on VHF and UHF bands take place in the "weak signal" allocations at the low end of the band. [E2C06] HF DXing is similar, with DX activity most common toward the low end of the bands, concentrated around calling frequencies and *DX windows*. **Table 2-1** shows some of the "watering holes" used by DXers.

A DX window is a narrow range of frequencies in which QSOs take place

Table 2-1

DX Windows and "Watering Holes"

Band	Frequency (MHz)
160 meters*	1.830-1.835
80 meters	3.505
75 meters†	3.795-3.800
40 meters	7.005 (CW)
20 meters	14.005 and 14.020 (CW)
	14.190-200 (Phone)
15 meters	21.005 and 21.020 (CW)
	21.195 and 21.295 (Phone)
10 meters	28.495 (Phone)
6 meters‡	50.100-130
2 meters‡	144.200

*This window is being phased out as allocations continue to align around the world
†The recent change in US amateur allocations and continued realignment of allocations make this window less important
‡Operation begins around the calling frequency, then moves to adjacent channels

between countries or continents that may not share frequency allocations or that have very narrow amateur bands. For example, 160 meters once had many different frequency allocations around the world because of conflicts with radionavigation systems. The 1830 to 1835 kHz window was a narrow slice of the band common to many different countries. As these conflicts disappear and allocations become worldwide, so will the need for a DX window.

If you want to work DX, windows and calling frequencies are good places to start, but don't just park there! Tune around, listening for signals with that "DX sound" — audio colored by flutter and multiple paths. Pileups are the best sign of a DX station. They occasionally form in the General class subbands, but more commonly in the Extra class frequencies. While you may find the DX station in the middle of the pileup, it's also common for them to work "split" (see the sidebar, "Where Are They Listening?"). In such cases, the DX station is probably transmitting 5 to 20 kHz below the callers on phone and 1 to 5 kHz lower on CW.

Pileup Productivity

You've tuned across a bunch of stations giving their call signs repeatedly. Is it a DX pileup? If there is a pause after a station gives his call and a signal report, then many stations call, you've almost certainly found a DX pileup! Listen for someone giving the DX station's call during a contact, such as "T77C from W1JR, you're five-nine" on phone or "T77C DE W1JR 5NN" on CW. While the temptation is great to jump in there and call, don't! You have to be able to *hear* the DX station *before* you start calling. Otherwise, you are just causing QRM.

Where Are They Listening?

Chasing DX is a popular activity on the HF bands. For this reason a frequency can become extremely crowded with amateurs calling the DX station, called a *pileup*. When more than a few stations are calling, propagation is difficult, or the DX station is weak, it can become quite difficult for everyone to stay "in sync" so that contacts can be made efficiently. In other situations, the callers and the DX station may not have a common frequency on which to use simplex communications — such as on 40 meter phone, for example.

Many DX operators use a technique called *operating split* to control such a pileup. This technique makes it possible for them to hear the calling stations and exchange information. It also makes it possible for the calling stations to hear the DX station in the clear, away from the calling stations. The DX station transmits on one frequency, and listens elsewhere, over another range of frequencies. If you hear the DX station announcing "up 5" or "listening 200 to 210" or something similar, it means they are listening on a frequency away from their transmit frequency. Calling them on their transmit frequency won't do any good — they're not listening there!

On phone, most split operations use a frequency separation of 5 to 20 kHz and on CW 1 to 5 kHz. Only in the cases of the rarest DX stations will a broader range be used. It is almost always the case that the DX station listens above the transmit frequency.

Start by learning how to set up your transceiver for split-frequency operation. Many transceivers have an "auto split" function or plenty of memories and VFO registers so that you can be ready to go in seconds. Practice so that you can do this quickly, as soon as you find the DX station. Learn the configuration well so that you will be sure to transmit on the frequency on which the DX station is listening, rather than on the "wrong VFO," right on top of the DX station. During the excitement of the chase, it is easy to make a mistake! Oops!

Does this sound complicated? With a little practice, it will become second nature. What's the key? Listening! If you pay attention to the DX station, you'll never have to ask "Where are they listening?" You'll be able to tune and find the right frequency for yourself. And while other stations are transmitting blindly or waiting for someone else to announce the listening frequency, you'll be entering the contact details in your log and preparing your QSL!

If all of the stations are on one frequency or close to one frequency, the DX station is probably working simplex. If the stations are spread out over a few kilohertz, the DX station is probably working *split*. Look for the DX station down a few kilohertz or more if the DX station is operating on a frequency unavailable to you, perhaps outside the US band entirely. This practice separates the signals of the calling stations from the DX station, reducing interference and improving efficiency. [E2C10]

Many DX stations like to be on "even" frequencies such as 14220 kHz as opposed to 14220.5 kHz. If you can find the DX station (and remember that propagation may not allow you to hear them), listen for operating instructions. The DX station may be calling stations from a particular region or with a specific number in their call sign. Then follow the instructions.

Listen to the stations that get through — how are they operating? Are they from your area? Is the DX station staying on one frequency or tuning around for callers? Follow that pattern. Remember that a little bit of listening will pay big dividends — if you are transmitting you *can't* be listening!

In general, give your full call sign once or twice (using standard phonetics on phone) and then pause to listen for the DX station. [E2C11] Remember, if you do not properly identify your station, then you are in violation of the FCC rules! Repeat if necessary, but if you don't hear the DX, stop calling and get back in sync with the pileup. *Only* respond to the DX station when you are called or a partial call overlaps with yours. If the DX asks for "six alpha?" and your call doesn't contain either six or A, stand by! Listen for the station being called so you can adjust your timing or frequency, or both. Resist the urge to call out of turn.

DXing Propagation

The DXer soon learns the truth of the old adage, "You can't work 'em if you can't hear 'em!" That makes understanding propagation crucial to DXing success. During the years that this edition of the *Extra Class License Manual* is in print (2012-2016), the sunspot cycle will build toward a peak in 2013 that will last for a year or two, then gradually begin to fall off. This will be a period of rapid changes in HF conditions. It's important to understand the basic variations of long-distance propagation that occur hour-to-hour, day-to-day and season-to-season.

Every day there are big changes as night turns to day and vice versa. Bands open and close quickly, sometimes in minutes. By knowing the "band basics" you can plan your operating periods and react to the conditions you find on the air. For example, if you're making HF contacts with stations in Europe after sunset, soon the signals all across the band will start to get weaker. When the band is about to close, signals will begin to exhibit the rapid fading that gives them a distinctive fluttery sound. What can you do to keep making those contacts? By learning how HF

Resources for DXing

This section of the *Extra Class License Manual* only touches on the basics of DX operating. To get the most out of your station and your operating time, there are many resources to turn to:

✔ *The ARRL Operating Manual* —this reference book has a detailed section on DXing discussing everything from DXing basics to maps and QSLing directions.

✔ *The Complete DXer*, by Bob Locher W9KNI — now in its third edition, this easy-to-read book provides valuable guidance with real-life stories and discussions.

✔ *The ARRL DX Bulletin* (**www.arrl.org/w1aw-bulletins-archive-dx**), the *OP DX Bulletin* (**www.papays.com/opdx.html**), *Daily DX* (**www.dailydx.com**), and *QRZ DX* (**www.dxpub.com**) are e-mailed to subscribers around the world.

✔ Several DX-themed e-mail reflectors are virtual DX communities. The most popular are **mailman.qth.net/mailman/listinfo/dx** and **groups.yahoo.com/group/dx-list**.

propagation works, you know that the MUF between your station and Europe is dropping. In response, change to a lower-frequency band. [E2C12] This is called "following the bands" and it works in reverse as the MUF moves higher through the morning.

The best strategy for learning and becoming more skilled is to tune, listen and operate. There is no substitute for personal experience! For information about solar conditions, visit websites such as **www.spaceweather.com** and **www.hfradio.org**. You quickly learn what to watch for as ionospheric and solar conditions change. By listening to worldwide on-the-air beacons, such as those that are part of the International Beacon Project sponsored by Northern California DX Foundation and International Amateur Radio Union (**www.ncdxf.org**), you can correlate your expectations with actual behavior.

Subscribe to ARRL propagation bulletins via your member information web page and read the columns on propagation in magazines and on websites. Supplement your tuning with DX spotting information from around the world to give you an idea about propagation in other areas.

Use propagation prediction software, but remember it is statistical in nature and actual conditions may vary dramatically from the predictions. The software is only as good as its models of the Earth's geomagnetic field, so it may not predict unusual openings.

> *Before you go on, study test questions E2C05, E2C06, E2C10, E2C11 and E2C12. Review this section if you have difficulty.*

CONTESTING

Why contest? What is a radio contest, anyway? Contests, also known as *radiosport*, are on-the-air operating events, usually held on weekends, in which operators try to make as many contacts as possible within defined time limits and according to a detailed set of rules. Contests provide a competitive outlet, enable quick additions to your state or DXCC totals, and offer a level of excitement that's hard to imagine until you've tried it!

While each contest has its own particular purpose and operating rules, the main purpose for all Amateur Radio contests is to enhance communication and operating skills. When you optimize your station for best operating efficiency "in the heat of battle" and learn to pull out those weak stations to make the last available contacts, you are honing useful skills and building a station that can make a big difference in public service or emergency operating. The best way to maintain both is to use them on a regular basis. Contests provide a fun way to keep a keen edge on your equipment capabilities and your operating skills.

Every contest contact includes its own particular set of information (called an *exchange*) that participants send to each other. Most exchanges include call signs, signal report, contact sequence number (called a *serial number*), or location. *QST*'s Contest Corral and the ARRL website (**www.arrl.org/contests**) list contests for each month, showing the sponsor's website and the contest's exchange. There are many excellent online contest event calendars such as those maintained by WA7BNM (**hornucopia.com/contestcal**) and SM3CER (**www.sk3bg. se/contest**). Check the rules on the sponsor's website to see what what you need to do to participate correctly. Even if you're not a contester, these are good resources to let you know when the bands will be busy!

Contest activity is permitted by the FCC regulations on any frequency available to the station licensee in accordance with the mode subband divisions. HF contest activity, like DXing with which it shares many traits, is usually concentrated toward the lowest frequencies of a phone or CW band, expanding upward according to activity levels. Digital contest activity centers around the digital calling frequencies for that mode. HF contest activity by general agreement does not take place on the 60, 30, 17 and 12 meter bands, giving non-contest operators some room during busy weekend events. [E2C03]

VHF and UHF contests, sometimes called VHF+ contests, are conducted in the "weak signal" areas at the low end of the VHF, UHF and microwave bands. Most activity is close to the calling frequencies for CW and SSB activity. Contacts via repeaters are usually not allowed, since the goal is to exercise the skill of the operators in making contacts without using an intermediate station. Although there may be a considerable amount of FM simplex activity during the larger VHF+ contests, it is discouraged on the FM simplex calling frequencies — 50.3, 146.52, 223.5 and 446.0 MHz. [E2C04]

Operating in a Contest

There are two basic ways to operate in a contest: *search-and-pounce* and *running*. Search and pounce describes exactly the technique of tuning up and down a band for stations to contact (searching), then working them (pouncing). As you tune, you'll hear some stations running, which is calling "CQ Contest" or simply "CQ Test" and then working stations that call in. To get up to speed with the style of operating, just listen for a while. You'll quickly learn the cadence of the calls and responses, what information is exchanged, and in what order.

Contest operating is also a good activity to help you learn more about radio-frequency propagation and plan an operating strategy. For example, if you want to participate in a worldwide DX contest, you might think immediately about operating on the higher frequency bands, such as 20, 15 and even 10 meters. During a sunspot (solar) minimum, however, a more ideal operating strategy would be to emphasize operation on the lower-frequency bands, such as 160, 80 and 40 meters during the evening and nighttime hours, and then changing to 20 meters during daylight hours.

You can maximize your contest scores and your operating enjoyment by paying attention to the band conditions as you operate. This will help you decide when to change bands or modify your strategy. It may be especially important to notice openings and change bands during a period of low sunspot activity because the higher-frequency bands may close entirely after dark, with no DX stations to be heard. Some contests also have *single-band* categories in which the operator stays on one band for the whole contest. There's no better way to learn about propagation on that band!

Submitting a Contest Log

You are encouraged to make contacts whether you intend to submit a log to the contest sponsors or not — no entry is required. [E2C01] But you've put a lot of effort into making contacts and working multipliers — don't sell yourself short by not submitting your log of contest QSOs! Even if you've just entered the contest casually, making a handful of QSOs, go ahead and submit a log. The sponsor will appreciate your efforts and it allows them to more accurately gauge contest activity and score the results.

Each contest has a deadline for submission of entries, found in the rules on the sponsor's website. If you send in a written log, make certain that your entry is postmarked by the deadline date.

By far the most common method of submitting a log is via e-mail as a log file in *Cabrillo format*. The Cabrillo format is a standard for organizing the information in a submitted contest log so that the sponsor can check and score the QSOs. [E2C07] Cabrillo-formatted files are composed of printable ASCII characters with the log information in fixed-position columns. Cabrillo also adds a number of standardized "header" lines that contain information about the log, such as the operator's name, call, and location, contest category, power, and so forth. **Figure 2-2** shows an example of a Cabrillo log file's header and QSO section.

Using Cabrillo format allows sponsors to automate the process of log collection, sorting and checking even though there may be many different programs generating the submitted log files. The Cabrillo log file should be named Yourcall.log (for example, n0ax.log). You can read about the Cabrillo format at **www.kkn.net/~trey/cabrillo**. All popular computer software for contest logging will generate files in the Cabrillo format for submission to the contest sponsor.

Once you have created your log, e-mail it to the sponsors before the published deadline. The rules will provide the address to use. Attach your log to an e-mail with the subject line containing just the call you used during the contest, nothing else. The e-mail may be processed by a human or by a software "robot" program that automatically scans the received log files. Before sending the e-mail, double-check the attached file to be sure it is the file you intended to send!

Using Spotting Networks

Virtually all contests require entrants in the single operator ("single op") category to do everything themselves — finding stations to contact, operating the radio, logging the contacts and so forth. When two or more operators work together, that falls into a multioperator or "multiop" category.

Many contests have added a *Single-Operator Assisted (SOA)* or *Single-Operator Unlimited (SOU)* category. This category usually has the same rules as the Single-Operator category with one major exception: The operator may use information from DX spotting networks. This usually consists of "spots" listing the call sign and frequency of stations operating in the contest.

The SOA category has opened up a whole new world for many contesters, allowing access to information about band conditions and stations and multipliers available during the contest like never before. This category allows operators who enjoy monitoring their spotting network for new countries to participate in contests without missing anything. Other operators like working cooperatively with their friends while operating contests. Still others feel that the use of spotting networks can help them score more points for their club by increasing their multiplier totals. Being part of the network can keep some operators motivated to push

```
START-OF-LOG: 2.3
CREATED-BY: Manual
CALLSIGN: N0AX
CONTEST: QRP ARCI Fall QSO Party
CATEGORY: All-band, >5 watts
LOCATION: WA
CLAIMED-SCORE: 5772
NAME: Ward Silver
ADDRESS: PO Box 927
ADDRESS: Vashon, WA 98070
OPERATORS: N0AX
SOAPBOX: Lots of strong signals from QRP stations!
QSO:    Freq   Mode   Date          Time   Call   RST   QTH   QRP    SentCall   RST   QTH   QRP RCV
QSO:    14     CW     2007-Oct-20   1952   N0AX   599   WA    3692   N0UR       559   MN       6846
QSO:    14     CW     2007-Oct-20   1953   N0AX   599   WA    3692   KO7X       559   WY         5W
QSO:    14     CW     2007-Oct-20   1957   N0AX   599   WA    3692   W5KDJ      599   TX      11325
QSO:    14     CW     2007-Oct-20   1958   N0AX   599   WA    3692   W9BOK      579   WI         5W
QSO:    14     CW     2007-Oct-20   1959   N0AX   599   WA    3692   K5KJP      549   FL       4759
END-OF-LOG:
```

Figure 2-2 —This is a sample of a Cabrillo-formatted contest log that can be e-mailed to the contest sponsors. The header section contains information about the entry and operator. Each QSO is listed on a separate line with all of the information in fixed columns for easy processing. After the last QSO the line "END-OF-LOG" concludes the log.

on to bigger scores. Being "connected" in many ways makes SOA closer to a multiop category than single op.

Feel free to make use of the information from spotting networks, as long as it is permitted by the rules of the contest for your entry category. The only way in which spotting networks may *not* be used is to *self-spot*, that is announcing your own call sign and frequency on the spotting networks. [E2C02] This is because self-spotting is another form of CQing. Contest rules require that soliciting contacts be done only on the bands of the contest and without intermediate stations or the Internet. Self-spotting can result in disqualification of your entry.

Before you go on, study test questions E2C01 through E2C04 and E2C07. Review this section if you have difficulty.

2.2 Digital Mode Operating

The number of digital modes is growing by leaps and bounds as amateurs apply their considerable ingenuity to the design of protocols and codes and signals. For a number of technical and operating reasons, some of these modes are most commonly used on the HF bands and others are more commonly used on the VHF and UHF bands. This section of the *License Manual* deals with the operating aspects of the digital modes. You'll learn more of the technical details in the Radio Modes and Equipment chapter.

PACKET RADIO

Digital communication via *packet radio* or just "packet" is used most often on the VHF and UHF bands. Packet radio on the 2 meter band uses AFSK (audio frequency shift keying) modulation to send data at 1200 baud (see the Radio Modes and Equipment chapter for a discussion of baud) over an FM link. The data is exchanged in packets called *frames* by using the methods defined by the AX.25 protocol standard. On the VHF and UHF bands, packet operation is allowed wherever FM voice operations are allowed. (In addition to these FCC rules, however, you also should follow the accepted band plans when selecting a packet operating frequency.) Packet radio uses the American National Standard Code for Information Interchange (ASCII) to represent each character, no matter what bands the stations operate on. Complete information on packet radio and the AX.25 protocol are available via the Tucson Amateur Packet Radio (TAPR) website, **www.tapr.org**.

Packet is also used on HF with a few differences from packet on VHF and UHF. HF packet typically uses FSK (frequency shift keying) modulation of an SSB signal with a data rate of 300 baud. On the HF bands, packet radio is limited by FCC rules to those band segments reserved for CW and data modes. Because of the characteristics of HF propagation, packet on HF performs quite poorly compared to VHF.

Figure 2-3 shows the main components of a typical packet radio station. Most amateurs use a terminal node controller (TNC) for VHF packet although a multimode communications processor (MCP) that works on several digital modes will also do the job. Another alternative is software using the digital signal processing capabilities of a personal computer sound card to eliminate the TNC or MCP by connecting your radio and computer sound card directly, using a simple hardware interface.

A stand-alone TNC contains microprocessor-controlled circuitry that translates the data from the host computer to and from the audio tones that modulate the FM link. The computer runs a *terminal emulator* program that allows it to send and receive the ASCII characters used by the packet radio system for controlling the TNC and communicating with other stations.

The most common use for packet radio is making connections to *store-and-forward* systems. This refers to a computer system that receives messages, stores them, and makes them available to other users or forwards them to other computer systems. Strictly speaking, connections with such a system can be made using any suitable digital mode or protocol. The most common method of creating these systems on VHF and UHF is to use packet radio. Three such systems are PacketCluster, PACSAT satellites and Winlink.

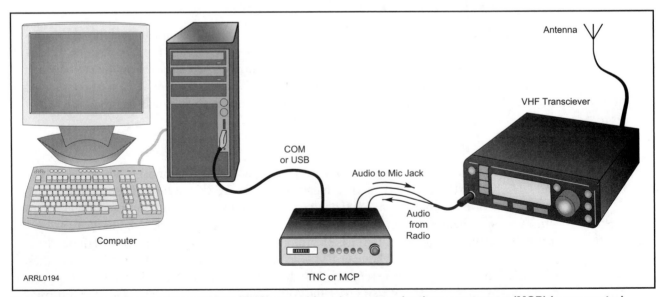

Figure 2-3 — A terminal node controller (TNC) or multimode communications processor (MCP) is connected between a computer and a VHF radio for VHF packet radio communications.

PacketCluster

Many amateurs interested in chasing DX and operating in DX contests use a special type of packet bulletin board software called *PacketCluster*. The name refers to the cluster of users who remain connected to the PacketCluster host. These systems allow many stations to connect to the cluster and communicate with each other and all stations on the cluster. The primary function of a PacketCluster is to relay "DX spots" to all current stations. These spots include a station call sign, operating frequency, the time the station was spotted and other information. (Spots are also mentioned in the sections on DX and contest operating earlier in this chapter.) One station sends a spot to the cluster host and the information is rebroadcast to all of the listening stations. Many PacketCluster systems share information with other such systems and support direct user connections via the Internet. DX spotting networks span the globe, providing worldwide notice of DX and contest activity. A list of spotting networks you can access by TELNET is maintained by NG3K at **www.ng3k.com/Misc/cluster.html**, or you can log on to the DX Summit network using a web browser at **www.dxsummit.fi**.

PACSATs

Packet radio systems are not limited to terrestrial locations. There are a number of packet radio store-and-forward systems in space called PACSATs that provide an interesting mix of satellites and packet-radio technology. These satellites are in low Earth orbits (LEO) that nearly circle the north and south poles. These orbits are called "sun synchronous" because they tend to be in range of any particular spot on the Earth at about the same time every day.

These small satellites function as packet bulletin board store-and-forward systems. A terrestrial station can send a message through a PACSAT by uploading it to the satellite for another station to download when the satellite is in view. [E2D04, E2D05]

The most widely used store-and-forward satellite is the packet system on-board the International Space Station. Other short-lived "CubeSats" often provide temporary digital capabilities on an experimental basis. (See the "Satellite Status" page at **www.amsat.org** for more information on these satellites.)

> *Before you go on, study test questions E2D04 and E2D05. Review this section if you have difficulty.*

AUTOMATIC PACKET REPORTING SYSTEM (APRS)

The Automatic Packet Reporting System (APRS) is another interesting packet radio application that makes use of packet radio beacon functions. (The system is also referred to by its original name, the Automatic Position Reporting System.) Bob Bruninga, WB4APR, developed the APRS system as a result of trying to use packet radio for real-time public-service communications. Position and other data from a station is broadcast to other stations in the region and Internet servers. The network of APRS stations is created by individual stations using packet radio to relay APRS data between other stations in the network.

An APRS station typically uses a 2 meter FM radio operating on the national APRS frequency of 144.39 MHz. [E2D06] The station also includes a packet radio terminal node controller (TNC) and computer system running APRS software. A Global Positioning System (GPS) receiver can then be connected to the station to supply position information.

APRS stations transmit a beacon packet containing the station's location. Information such as weather conditions and short text messages can also be transmitted. Other stations act as *digipeaters* to relay the packets to other stations and to *gateways* that relay the station's information to Internet APRS servers. APRS software running on a personal computer can be used to display the station call sign on a map with a user-selected icon and the transmitted data.

APRS uses the AX.25 Amateur Packet Radio Protocol. [E2D07] This means you can use your standard packet radio hardware to operate in the APRS network. To simplify the communication between stations, an APRS beacon is transmitted as an unnumbered information (UI) frame. [E2D08]

APRS packets are not directed to a specific station and receiving stations do not acknowledge correct receipt of the packets. This simplifies APRS operation because the stations do not all have to remain "connected" for the network to function and transfer data.

A typical APRS station consists of a TNC and VHF transceiver as in Figure 2-3 with the addition of a GPS receiver connected to the TNC. (The TNC must have an NMEA-compatible interface as discussed below.) When mobile, the station does not have to display data from any other APRS station. The TNC must be configured to allow digipeating so that it can relay APRS packets. This allows gateway stations to receive and forward them to a central server accessible at **www.findu.com**. By using a web browser, anyone can see the last reported position (and other information) of any station. APRS networks can be used to support public service or emergency communications by providing event managers and organizers continuously-updated location and other information from an APRS-equipped station. [E2D10]

Position data is sent to the APRS network as latitude and longitude. [E2D11] You can obtain this information:

- From an accurate map in degrees, minutes and seconds, entering it manually into a host computer running APRS software, or
- As an NMEA-0183-formatted text string from a Global Positioning System (GPS) satellite receiver or other suitable navigation system.

NMEA-0183 refers to a data formatting standard from the National Marine Electronics Association. A number of commercially available TNCs support direct connections to a GPS receiver via an RS-232 serial data interface.

Before you go on, study test questions E2D06, E2D07, E2D08, E2D10 and E2D11. Review this section if you have difficulty.

2.3 Amateur Satellites

The spherical shape of the Earth and other factors limit terrestrial communication at VHF and UHF. Long-haul communication at VHF and UHF may require the use of higher effective radiated power (ERP) or may not be possible at all. The communication range of amateur stations is increased greatly by using repeaters, transponders, or store-and-forward equipment onboard satellites orbiting the Earth. There are several operational amateur satellites providing communications plus the amateur equipment on board the International Space Station (ISS) and a number of experimental CubeSats. Most of the satellites can be accessed or used to relay signals with very modest equipment. (More information on using amateur satellites is available at **www.amsat.org**.)

UNDERSTANDING SATELLITE ORBITS

Two factors affect a body in orbit around the Earth: forward motion (inertia) and gravitational attraction. Forward motion, the inertia of a body, tends to keep a body moving in a straight line in the direction it is moving at that instant. If the body is above the surface of the Earth, that straight line heads away into space. Gravity, on the other hand, tends to pull the body toward the Earth. When inertia and gravity are balanced, the object's path is a stable *orbit* around the Earth. One orbit is defined as one complete revolution about the Earth. [E2A03]

Johannes Kepler was the first to mathematically describe the mechanics of the orbits of the planets. His three laws of planetary motion also describe the lunar orbit and the orbits of artificial Earth satellites. A brief summary of Kepler's Laws will help you understand the motion of artificial Earth satellites.

Kepler's First Law tells us that all satellite orbits are shaped like an ellipse with the center of the Earth at one of the ellipse's focal points (**Figure 2-4** describes the geometry of an ellipse). The *eccentricity* of an ellipse (or an orbit) is equal to the distance from the center to one of the focal points divided by the semimajor axis. Notice that when the focal point is at the center, the eccentricity is 0, corresponding to a circle. The larger the eccentricity the "thinner" the ellipse. The eccentricity of an elliptical orbit ranges between 0 and 1.

Kepler's Second Law is illustrated in **Figure 2-5**. The time required for a satellite to move in its orbit from point A to point B is the same as the time required to move from A' to B'. What this means is that a satellite moves faster in its elliptical orbit when it is closer to the Earth, and slower when it is farther away. The area of section AOB is the same as the area of section A'OB'.

Kepler's Third Law tells us that the greater the average distance from the Earth, the longer it takes for a satellite to complete each orbit. The time required for a satellite to make a complete orbit around the Earth is called the *orbital period*. Low-flying amateur satellites typically have periods of approximately 90 minutes. The International Space Station flies in a fairly low, nearly circular orbit, and has a period of about 90 minutes. Satellite AO-40 had a high, elliptical orbit and a period of more than 19 hours. If the orbit is high enough and circular, the satellite's orbit will be *geosynchronous* and take exactly the same time that it takes the Earth to rotate once about its axis. If the geosynchronous orbit is over the Earth's equator, the satellite appears to stay in the same position. This special type of orbit is called *geostationary*. [E2A13]

Kepler's laws can be expressed mathematically, and if you know the values of a set of measurements of the satellite orbit (called *Keplerian elements*), you can calculate the posi-

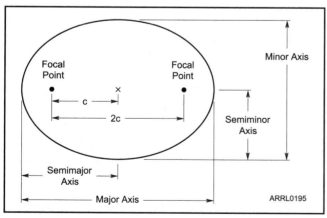

Figure 2-4 — This drawing illustrates the geometry of an ellipse. The ratio of distance c to the length of the semimajor axis is called the eccentricity of the ellipse. Eccentricity can vary from 0 (a circle) to 1 (a straight line).

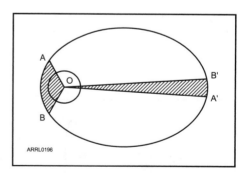

Figure 2-5 — A graphical representation of Kepler's second and third laws. The two shaded sections have equal areas. The time for the satellite to move along its orbit from A to B or from A´ to B´ is the same.

tion of the satellite at any time. [E2A12] This tells you where to point your antennas to send your signal directly toward the satellite. A computer program is used to perform these orbital calculations. The results even can be displayed over a map on the computer monitor, with position updates calculated and shown every few seconds. You should get the current set of Keplerian elements about once a month for satellites you want to track since satellite orbits change slowly over time. Keplerian elements for the current amateur satellites are available online from the AMSAT website.

The earliest amateur satellites simply orbited the Earth and transmitted information about the satellite battery voltage, temperature and other physical conditions. These had relatively low, nearly circular orbits, and are sometimes referred to as the Phase 1 satellites. Next came active two-way communications satellites, also with relatively low, nearly circular orbits. Several satellites in current use fit this category, sometimes called Phase 2 satellites. Their orbits are approximately 500 to 700 km above the Earth. The orbital period of Phase 1 and Phase 2 satellites is around 90 minutes. A third generation has very elliptical orbits that take them more than 35,000 km from Earth at apogee and about 800 km at perigee. Known as Phase 3 satellites, this group has orbital periods of 8 to 20 hours and provides greater communications range because they are in view of a larger area at one time.

Orbital Mechanics

Inclination is the angle of a satellite orbit with respect to Earth. Inclination is measured between the plane of the orbit and the plane of the equator (**Figure 2-6**). If a satellite is always over the equator as it travels through its orbit, the orbit has an inclination of 0 degrees. If the orbit path takes the satellite over the poles, the inclination is 90 degrees. (If it goes over one pole, it will go over the other.) The inclination angle is always measured from the equator counterclockwise to the satellite path. **Figure 2-7** gives some examples of orbits with different inclinations.

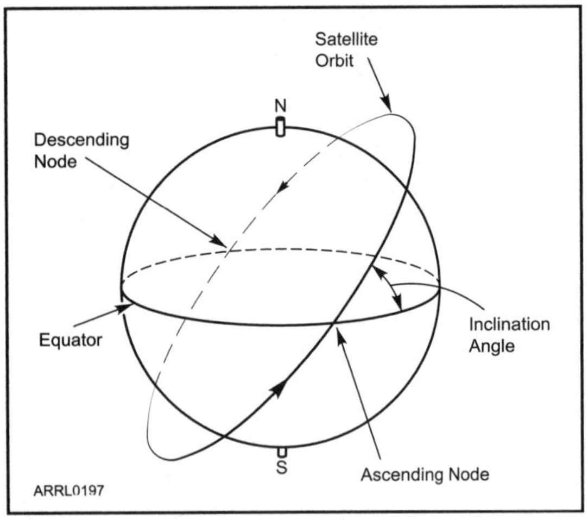

A *node* is the point where a satellite's orbit crosses the equator. The *ascending node* is where it crosses the equator when the satellite is traveling from south to north. Inclination is specified at the ascending node. *Equator crossing* (EQX) is usually specified in time (UTC) of crossing and in degrees west longitude. The *descending node* is the point where the orbit crosses the equator traveling from north to south. When the satellite is within range of your location, it is common to describe the pass as either an *ascending pass* (traveling south to north over your area) or a *descending pass* (traveling north to south). [E2A01, E2A02]

Figure 2-6 — An illustration of basic satellite orbit terminology.

The point of greatest height in a satellite orbit is called the *apogee* as shown in **Figure 2-8**. *Perigee* is the point of least height. Half the distance between the apogee and perigee is equal to the *semimajor axis* of the satellite orbit.

Faraday Rotation and Spin Modulation

The polarization of a radio signal passing through the ionosphere does not remain constant. A "horizontally

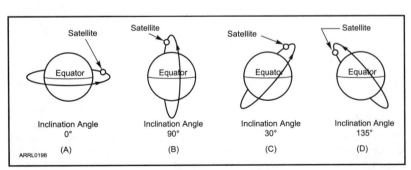

Figure 2-7 — Satellite orbit inclination angles are measured at an ascending node. The angle is measured from the equator to the orbital plane, on the eastern side of the ascending node.

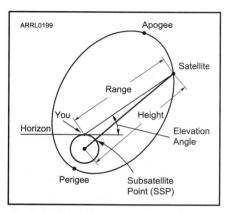

Figure 2-8 — An illustration of the very elliptical orbit of a Phase 3 satellite.

polarized" signal leaving a satellite will not be horizontally polarized when it reaches Earth. The signal will seem to be changing polarization at a receiving station. This effect is called *Faraday rotation* and it is caused by the effect of the ionosphere on the signal passing through it. The best way to deal with Faraday rotation is to use circularly polarized antennas for transmitting and receiving.

Satellites are often stabilized by being spun about an axis like a gyroscope. This stabilizes the satellite and keeps it oriented in the same direction as it travels around the Earth. When the spacecraft's spin axis is not pointed directly at your ground station, you are likely to experience amplitude changes and possibly polarization changes resulting from the spacecraft rotation. These changes, which affect the uplinks and downlinks, occur at an integer multiple of the satellite's spin frequency. You will notice the effect as a fairly rapid, pulsed signal fading. [E2A10] This condition is called *spin modulation*. It is important to note that the signal is not electronically modulated in the sense to which amateurs are accustomed; rather, the apparent modulation is an effect of physical rotation. Using linear antennas (horizontal or vertical polarization) will deepen the spin-modulation fades to a point where they may become annoying. Circularly polarized antennas will minimize the effect, just as they do for Faraday rotation. [E2A11]

> Before you go on, study test questions E2A01 through E2A03 and E2A10 through E2A13. Review this section if you have difficulty.

SATELLITE OPERATION

Of the many components of a satellite — batteries, solar panels, controllers — the most important to the Amateur Radio operator trying to use the satellite are the radio components. Satellites can have three different types of radio equipment onboard — repeaters, transponders and store-and-forward systems.

Repeaters

A satellite-borne repeater operates in the same way as a terrestrial repeater. It receives FM voice signals on a single frequency or channel and retransmits what it receives on another channel. Satellite repeaters typically operate with their input and output frequencies on different bands (called a *cross-band repeater*) to allow them to dispense with the heavy and bulky cavity duplexers required for same-band operation. Otherwise, accessing a satellite repeater is just the same as a terrestrial repeater. With some satellites, you can even use low power, handheld transceivers with small beam antennas to make contacts. Repeaters are installed on the International Space Station and on satellites such as AO-27. (See the AMSAT website for operational status and new satellites.)

Transponders

By convention, *transponder* is the name given to any linear translator that is installed in a satellite. In contrast to a repeater, a transponder's receive passband includes enough spectrum for many channels. An amateur satellite transponder does not use channels in the way that voice repeaters do. Received signals from an entire segment of a band are amplified, shifted to a new frequency range and retransmitted by the transponder. See **Figure 2-9**.

The major hardware difference between a repeater and a transponder is signal detection. In a repeater, the signal is reduced to *baseband* (audio) before it is retransmitted. In a transponder, signals in the passband are shifted to an intermediate frequency (IF) for amplification and retransmission.

Operationally, the contrast is much greater. An FM voice repeater is a one-signal, one-mode-input and one-signal, one-mode-output device. A transponder can receive several signals at once and convert them to a new range. Further, a transponder can be thought of as a multimode repeater. Whatever mode is received is retransmitted. [E2A07] The same transponder can simultaneously handle SSB and CW signals.

The use of a transponder rather than a channelized repeater allows more stations to use the satellite at one time. In fact, the number of different stations using a transponder at any one time is limited only by mutual interference, and the fact that the output power of the satellite is divided among the users. (On the LEO satellites, the output power is a couple of watts. For Phase 3 satellites, transponder outputs range from about 50 W on the 2.4 and 10 GHz bands to about 1 W on the 24 GHz band.) Because all users must share the power output, continuous-carrier modes such as FM and RTTY generally are not used through amateur satellites and all users should limit their transmitting ERP to allow as many stations as possible to use the transponder. [E2A08]

Satellite Operating Frequencies

Satellites used for two-way communication generally use one amateur band to receive signals from Earth (the *uplink*) and another to transmit back to Earth (the *downlink*). Amateur satellites use a variety of uplink and downlink bands, ranging from HF through microwave frequencies. When describing satellite transponders, the input (uplink) band is given, followed by the corresponding output (downlink) band. For example, a 2 meter/70 cm transponder would have an input passband centered near 145 MHz and an output passband centered near 435 MHz.

Transponders usually are identified by *mode* — but not mode of transmission such as SSB or CW or RTTY. Mode has an entirely different meaning in this case. The operating mode of a satellite identifies the uplink and downlink frequency bands that the satellite is using. [E2A04]

Satellite operating modes are specified by letter designators that correspond to the frequency range of the uplink and downlink bands. For example, the letter V designates a satellite uplink or downlink in the 2 meter band and U designates uplinks and downlinks in the 70 cm band. So a satellite operating with an uplink on 70 cm and a downlink on 2 meters is

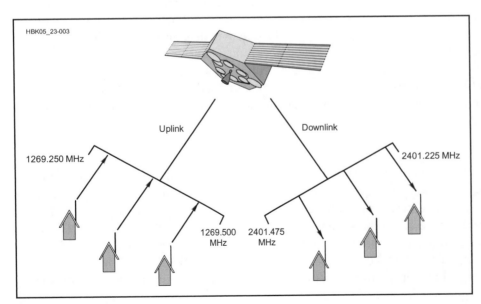

HBK05_23-003

Uplink   Downlink

1269.250 MHz

1269.500 MHz

2401.475 MHz

2401.225 MHz

Figure 2-9 — A linear transponder acts much like a repeater, except that it relays an entire group of signals, not just one signal at a time. In this example, the satellite is receiving three signals on its 23 cm uplink passband and retransmitting them on its 13 cm downlink.

Table 2-2

Satellite Operating Modes

Frequency Band	Letter Designation
HF	
15 and 10 meters (21-30 MHz)	H
VHF	
2 meters (144-146 MHz)	V
UHF	
70 cm (435-438 MHz)	U
23 cm (1.26-1.27 GHz)	L
13 cm (2.4-2.45 GHz)	S
5 cm (5.8 GHz)	C
3 cm (10.45 GHz)	X
1.5 cm (24 GHz)	K

Operating Example

Satellite Receive, First Letter Satellite Transmit, Second Letter	Operating Frequency Range
Mode U/V	435-438 MHz / 144-146 MHz
Mode V/U	144-146 MHz / 435-438 MHz
Mode L/U	1.26-1.27 GHz / 435-438 MHz
Mode V/H	144-146 MHz / 21-30 MHz
Mode H/S	21-30 MHz / 2.40-2.45 GHz
Mode L/S	1.26-1.27 GHz / 2.40-2.45 GHz
Mode L/X	1.26-1.27 GHz / 10.45 GHz
Mode C/X	5.8 GHz / 10.45 GHz

in "Mode U/V" (remember, the uplink band is always the first letter). **Table 2-2** lists various satellite operating modes by their frequency bands and letter designations. [E2A05, E2A06, E2A09]

An amateur satellite may operate on several of these modes, sometimes simultaneously. Some of the more flexible communications satellites can use various band combinations at different times. You will have to check the operating schedule for the particular satellite you plan to use.

Before you go on, study test questions E2A04 through E2A09. Review this section if you have difficulty.

Table 2-3

Questions Covered in This Chapter

2.1 General Operating	*2.2 Digital Mode Operating*	*2.3 Amateur Satellites*
E2C01	E2D04	E2A01
E2C02	E2D05	E2A02
E2C03	E2D06	E2A03
E2C04	E2D07	E2A04
E2C05	E2D08	E2A05
E2C06	E2D10	E2A06
E2C07	E2D11	E2A07
E2C10		E2A08
E2C11		E2A09
E2C12		E2A10
		E2A11
		E2A12
		E2A13

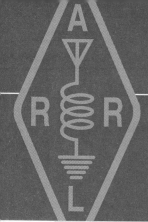

Chapter 3

Rules and Regulations

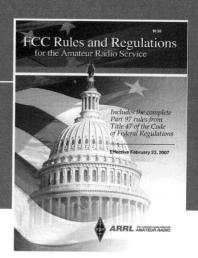

In this chapter, you'll learn about:
- **Operating regulations and frequency privileges**
- **Restrictions on station operation, location and antennas**
- **Local, remote and automatic station control**
- **Amateur-satellite service rules**
- **Rules for exams and volunteer examiners**
- **Auxiliary stations, power amplifier certification, spread spectrum requirements and non-US operating agreements**

The Amateur Radio Extra class license exam (Element 4) includes six questions about the FCC's Part 97 rules for the amateur service. These questions cover a wide range of topics, including operating standards, station restrictions and special operations, station control, the amateur satellite service, the Volunteer Examiner program and miscellaneous rules.

The *Extra Class License Manual* does not contain a complete listing of the Part 97 rules, but focuses on the specific rules on the exam. Every amateur should have an up-to-date copy of the complete Part 97 rules in their station or easily available from a website for reference. The ARRL's booklet *FCC Rules and Regulations for the Amateur Radio Service* contains the complete printed text of Part 97. The text of Part 97 is also available on the ARRL website at **www.arrl.org/part-97-amateur-radio**.

3.1 Operating Standards

Having already passed your Technician and General class license exams, you're pretty familiar with the basic structure of the FCC rules. As an Extra class licensee, however, your growing amount of experience will expose you to a wider variety of operating situations and circumstances. While the exam can't cover every possible niche of the rules, it can touch on a few areas that you might encounter. The most obvious new experiences will be your new and exclusive HF operating privileges! The exam also covers the responsibilities of control operators when engaged in automatic message forwarding, operation aboard ships and planes, and operation under RACES rules.

FREQUENCY AND EMISSION PRIVILEGES

As an Extra class licensee, you will have all of the frequency privileges available in Amateur Radio. The privileges above 50 MHz are listed in Section 97.301(a) of the FCC rules and those below 30 MHz in §97.301(b). (Section is often represented by the § character when referencing a specific portion of the FCC rules and we'll use that convention in this manual.) **Table 3-1** shows the HF section of the rules, along with the applicable notes from §97.303.

As an Extra class licensee you have *exclusive* privileges on segments of the 80, 75, 40, 20 and 15 meter bands. (Many amateurs consider the 80 and 75 meter designations to be the

Table 3-1

Amateur Extra Class HF Bands and Sharing Requirements

(This table excerpts just the 160 through 10 meter privileges for Extra class licensees from §97.301 and accompanying notes and restrictions from §97.303. Operating privileges on the 60 meter band consist of five channels and specific emission types. See the FCC rules for more information.)

§97.301 Authorized frequency bands.

The following transmitting frequency bands are available to an amateur station located within 50 km of the Earth's surface, within the specified ITU Region, and outside any area where the amateur service is regulated by any authority other than the FCC.

(b) For a station having a control operator who has been granted an Amateur Extra Class operator license, who holds a CEPT radio amateur license, or who holds a Class 1 IARP license:

Wavelength band	ITU Region 1	ITU Region 2	ITU Region 3	Sharing requirements See §97.303, Paragraph:
MF	kHz	kHz	kHz	
160 m	1810-1850	1800-2000	1800-2000	(a), (b), (c)
HF	MHz	MHz	MHz	
80 m	3.50-3.60	3.50-3.60	3.50-3.60	(a)
75 m	3.60-3.80	3.60-4.00	3.60-3.90	(a)
40 m	7.0-7.2	7.0-7.3	7.0-7.2	(a), (t)
30 m	10.10-10.15	10.10-10.15	10.10-10.15	(d)
20 m	14.00-14.35	14.00-14.35	14.00-14.35	
17 m	18.068-18.168	18.068-18.168	18.068-18.168	
15 m	21.00-21.45	21.00-21.45	21.00-21.45	
12 m	24.89-24.99	24.89-24.99	24.89-24.99	
10 m	28.0-29.7	28.0-29.7	28.0-29.7	

§97.303 Frequency sharing requirements.

The following is a summary of the frequency sharing requirements that apply to amateur station transmissions on the frequency bands specified in §97.301 of this Part. (For each ITU Region, each frequency band allocated to the amateur service is designated as either a secondary service or a primary service. A station in a secondary service must not cause harmful interference to, and must accept interference from, stations in a primary service. See §§2.105 and 2.106 of the FCC Rules, United States Table of Frequency Allocations for complete requirements.)

(a) Where, in adjacent ITU Regions or sub-Regions, a band of frequencies is allocated to different services of the same category (i.e., primary or secondary allocations), the basic principle is the equality of right to operate. Accordingly, stations of each service in one Region or sub-Region must operate so as not to cause

CW and phone portions of the same band.) **Table 3-2** shows that you will have a total of 250 kHz of exclusive spectrum on 75/80, 40, 20 and 15 meters! The table also shows the emission types (transmitted signals) authorized for use on certain portions of these bands. These emission types are defined in §97.3(c)(1)(2)(3)(5) and (7). While the exam does not contain questions about specific frequency privileges, it is still important to know what they are! (**Figure 2-1**, in Chapter 2, shows a handy chart of all US amateur privileges.)

Special Restrictions

While most amateur bands are divided into Phone and CW/Data segments and have no other special restrictions on operating privileges, two bands are exceptions.

The 30 meter band was allocated to amateurs on a secondary basis at the World Administrative Radio Conference in 1977. Amateurs may only transmit CW and data signals on this band and are limited to 200 W output power — phone and image signals are not allowed.

harmful interference to any service of the same or higher category in the other Regions or sub-Regions. (See ITU Radio Regulations, edition of 2004, No. 4.8.)

(b) No amateur station transmitting in the 1900-2000 kHz segment, the 70 cm band, the 33 cm band, the 23 cm band, the 13 cm band, the 9 cm band, the 5 cm band, the 3 cm band, the 24.05-24.25 GHz segment, the 76-77.5 GHz segment, the 78-81 GHz segment, the 136-141 GHz segment, and the 241-248 GHz segment shall not cause harmful interference to, nor is protected from interference due to the operation of, the Federal radiolocation service.

(c) No amateur station transmitting in the 1900-2000 kHz segment, the 3 cm band, the 76-77.5 GHz segment, the 78-81 GHz segment, the 136- 141 GHz segment, and the 241-248 GHz segment shall cause harmful interference to, nor is protected from interference due to the operation of, stations in the non-Federal radiolocation service.

(d) No amateur station transmitting in the 30 meter band shall cause harmful interference to stations authorized by other nations in the fixed service. The licensee of the amateur station must make all necessary adjustments, including termination of transmissions, if harmful interference is caused.

(t) (1) The 7-7.1 MHz segment is allocated to the amateur and amateur-satellite services on a primary and exclusive basis throughout the world, except that the 7-7.05 MHz segment is:
    (i) Additionally allocated to the fixed service on a primary basis in the countries listed in 47 CFR 2.106, footnote 5.140; and
    (ii) Alternatively allocated to the fixed service on a primary and exclusive basis (i.e., the segment 7-7.05 MHz is not allocated to the amateur service) in the countries listed in 47 CFR 2.106, footnote 5.141.

(2) The 7.1-7.2 MHz segment is allocated to the amateur service on a primary and exclusive basis throughout the world, except that the 7.1-7.2 MHz segment is additionally allocated to the fixed and mobile except aeronautical mobile (R) services on a primary basis in the countries listed in 47 CFR 2.106, footnote 5.141B.

(3) The 7.2-7.3 MHz segment is allocated to the amateur service on an exclusive basis in Region 2 and to the broadcasting service on an exclusive basis in Region 1 and Region 3. The use of the 7.2-7.3 MHz segment in Region 2 by the amateur service shall not impose constraints on the broadcasting service intended for use within Region 1 and Region 3.

Table 3-2
Exclusive HF Privileges for Amateur Extra Class Operators

Band	Operating Privileges	Frequency Privileges
80 m	CW, RTTY, Data	3500-3525 kHz
75 m	CW, Phone, Image	3600-3700 kHz
40 m	CW, RTTY, Data	7000-7025 kHz
20 m	CW, RTTY, Data	14.000-14.025 MHz
	CW, Phone, Image	14.150-14.175 MHz
15 m	CW, RTTY, Data	21.000-21.025 MHz
	CW, Phone, Image	21.200-21.225 MHz

You must know where your transmitted signal is with respect to the displayed frequency on your transceiver! This is essential for both compliance with your license privileges and for operating convenience. What frequency is your radio displaying? This varies with operating mode, as shown in **Table 3-3** and **Figure 3-1**.

On CW, the radio displays the frequency of the transmitted signal. On AM, SSB and FM the radio displays the carrier frequency of the signal. On AM and FM, the carrier frequency is in the middle of your transmitted signal. (The displayed frequency of some transceivers can be configured differently — check the operating manual of your radio.)

The bandwidth of a properly-adjusted amateur USB or LSB signal is from 2.5 to 3 kHz. That means the sidebands of an SSB signal extend up to 3 kHz from the carrier frequency as shown in Table 3-3. You must keep your sidebands entirely within the band so it's important to understand how close you can operate to the edge. On USB, it's prudent to set your carrier frequency no closer than 3 kHz below the band edge and on LSB no closer than 3 kHz above the band edge. [E1A01, E1A02] If your carrier is closer than that to the band edge — say 14.349 MHz USB or 3.601 MHz LSB — your sidebands will be outside the band! [E1A03, E1A04]

Remember that CW signals have sidebands, too, even though the signal is quite narrow compared to SSB. A typical CW signal has a bandwidth of 50-150 Hz, so setting the carrier frequency exactly on a band edge, such as 3500 kHz, will result in your signal extending outside the band even if your displayed carrier frequency is exactly accurate. [E1A12]

Table 3-3
How Mode Affects the Frequency the Transceiver Displays

Mode	Transmitted Carrier Frequency	Transmitted Signal Occupies	The Radio Displays
CW	14,040.00 kHz	14,039.75-14,040.25 kHz Assuming a 500 Hz-wide signal	14,040.0 kHz
LSB	14,200.00 kHz	14,199.70-14,197.00 kHz Assuming 300-3000 Hz audio	14,200.00 kHz
USB	14,200.00 kHz	14,200.30-14,203.00 kHz Assuming 300-3000 Hz audio	14,200.00 kHz
AM	14,200.00 kHz	14,197.00-14,203.00 kHz Assuming 300-3000 Hz audio	14,200.00 kHz
FM	29,600.00 kHz	29,603.00-29,597.00 kHz Assuming 300-3000 Hz audio	29,600.00 kHz

Operation on the 60 meter band is restricted to certain channels and emission types [E1A05, E1A06, E1A07, E1B05, E1B07]:

- Amateurs may only use five 2.8 kHz-wide channels centered on 5332, 5348, 5358.5, 5373, and 5405 kHz.
- USB (upper sideband) voice is the only phone emission allowed. The carrier must be located 1.5 kHz below the center of the channel. This centers a properly-adjusted USB signal on the specified channel frequency.
- RTTY and data emissions are permitted with signals centered on the specified channel frequency (not just anywhere within the channel) as described above.
- CW is permitted with the carrier frequency set to the specified channel center frequency.
- Output power is limited to 100 W ERP (effective radiated power) relative to a dipole.
- Automatic control of RTTY and data emissions is not permitted.

These restrictions are necessary to coexist with other government fixed service stations. The 60 meter band is allocated on a country-by-country basis and is not shown as an amateur allocation by the ITU.

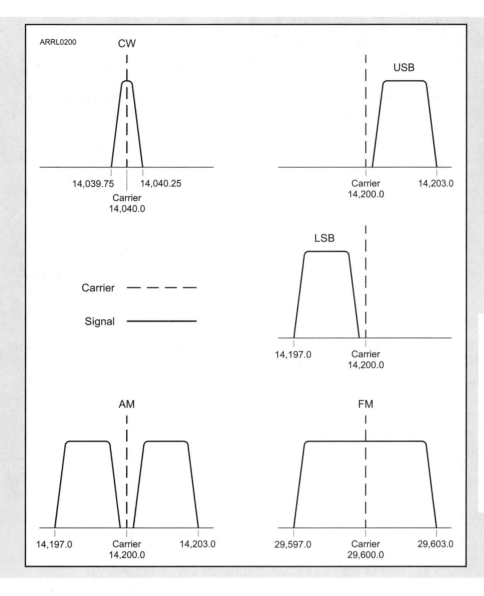

Figure 3-1 — An illustration of the relationship between carrier frequency and the actual signal energy. Most rigs are configured to show the carrier frequency of a signal.

Before you go on, study test questions E1A01 through E1A04, and E1A12. Review this section if you have difficulty.

Before you go on, study test questions E1A05, E1A06, E1A07, E1B05, and E1B07. Review this section if you have difficulty.

AUTOMATIC MESSAGE FORWARDING

The FCC rules establish standards governing the operation of Amateur Radio stations. Within those standards, amateurs relish experimenting with methods of communication, such as types of modulation and digital protocols. Over the years, amateurs have developed or adapted technology to create radio-based data networks capable of automatically forwarding messages, such as packet radio using the AX.25 protocol.

Amateurs have always been held accountable for any message transmitted from their stations. When all third-party communications were relayed by individual amateurs, such rules made sense and were easy to follow. With the advent of automatic message forwarding, requiring every individual station owner to personally screen every message was not feasible.

(a) Any amateur station may participate in a message forwarding system, subject to the privileges of the class of operator license held.

(b) For stations participating in a message forwarding system, the control operator of the station originating a message is primarily accountable for any violation of the rules in this Part contained in the message.

(c) Except as noted in paragraph (d) of this section, for stations participating in a message forwarding system, the control operators of forwarding stations that retransmit inadvertently communications that violate the rules in this Part are not accountable for the violative communications. They are, however, responsible for discontinuing such communications once they become aware of their presence.

(d) For stations participating in a message forwarding system, the control operator of the first forwarding station must:

(1) Authenticate the identity of the station from which it accepts communication on behalf of the system; or

(2) Accept accountability for any violation of the rules in this Part contained in messages it retransmits to the system.

In response, the rules were changed such that the FCC holds the originating station primarily responsible. [E1A08] The first station to forward an illegal message also bears some responsibility but other stations that automatically forward the message are not responsible. If you become aware that your station has inadvertently forwarded a communication that violates FCC rules, you should, of course, immediately discontinue forwarding that message to other stations. [E1A09]

Before you go on, study test questions E1A08 and E1A09. Review this section if you have difficulty.

RACES OPERATION

The Radio Amateur Civil Emergency Service (RACES) is a radio service of amateur stations for civil defense communications during periods of local, regional or national civil emergencies. [§97.3(a)(37)] The complete RACES rules are covered in §97.407.

RACES is a formal FCC service and is not to be confused with the ARRL Amateur Radio Emergency Service (ARES®). The ARES® (**www.arrl.org/ares**) is administered by the ARRL's Membership and Volunteer Programs Department to support public safety and nongovernmental organizations.

Before you can register your Amateur Radio station with RACES you must register with your local civil defense organization. Any licensed amateur station may be operated in RACES, as long as the station is certified by the civil defense organization responsible for the area served. [E1B09] Likewise, any licensed amateur may be the control operator of a RACES station if they are also certified by the local civil defense organization.

RACES station operators do not receive any additional operator privileges because of their RACES registration. So a Technician class operator may only use Technician frequencies when serving as the control operator. Extra class operators must also follow the operator privileges granted by their license. In general, all amateur frequencies are available to stations participating in RACES operation. [E1B10] RACES stations may communicate with any RACES station as well as certain other stations as authorized by the responsible civil defense official.

Before you go on, study test questions E1B09 and E1B10. Review this section if you have difficulty.

STATIONS ABOARD SHIPS OR AIRCRAFT

You may have worked a maritime mobile station signing /MM or an aeronautical mobile with an /AM suffix. That sounds pretty exciting and there is absolutely no reason not to try it yourself except for one small caveat: If you want to operate on board a ship or aircraft you must have the radio installation approved by the master of the ship or the pilot in command of the aircraft. [E1A10] The sidebar "§97.11 — Stations Aboard Ships or Aircraft" lists the specific requirements for such an installation. You don't need any other special permit or permission from the FCC for such an operation, though. Your FCC Amateur Radio license or a reciprocal permit for an alien amateur licensee is all that's needed. [E1A11]

You should also be aware that to operate from international waters (or international air space) from a vessel or plane registered in the US you must have an FCC-issued amateur license or reciprocal operating permit. [E1A13] This is because FCC rules apply to US-registered ships and aircraft. If the boat or plane travels into an area controlled by another country then you must have a license or permission from that country to operate. In international waters or air space, however, any amateur license is sufficient to permit you to operate. You must also obey the frequency restrictions that apply to the ITU Region in which you are operating.

Before you go on, study test questions E1A10, E1A11, and E1A13. Review this section if you have difficulty.

§97.11 — Stations Aboard Ships or Aircraft

(a) The installation and operation of an amateur station on a ship or aircraft must be approved by the master of the ship or pilot in command of the aircraft.

(b) The station must be separate from and independent of all other radio apparatus installed on the ship or aircraft, except a common antenna may be shared with a voluntary ship radio installation. The station's transmissions must not cause interference to any other apparatus installed on the ship or aircraft.

(c) The station must not constitute a hazard to the safety of life or property. For a station aboard an aircraft, the apparatus shall not be operated while the aircraft is operating under Instrument Flight Rules, as defined by the FAA, unless the station has been found to comply with all applicable FAA rules.

3.2 Station Restrictions

In general, you can operate your Amateur Radio station anytime you want to. You can operate from nearly any location in the US. Circumstances, however, may lead to some restrictions on your ability to operate.

OPERATING RESTRICTIONS

Under certain conditions spelled out in §97.121, the FCC may modify the terms of your amateur station license. These have to do with interference between a properly operating amateur station and users of other licensed services. Here's what §97.121(a) says about interference to broadcast signals:

"If the operation of an amateur station causes general interference to the reception of transmissions from stations operating in the domestic broadcast service when receivers of good engineering design, including adequate selectivity characteristics, are used to receive such transmissions, and this fact is made known to the amateur station licensee, the amateur station shall not be operated during the hours from 8 p.m. to 10:30 p.m., local time, and on Sunday for the additional period from 10:30 a.m. until 1 p.m., local time, upon the frequency or frequencies used when the interference is created."

Where the interference from the amateur station is causing a sufficient amount of interference, the FCC can impose limited quiet periods on the amateur station on the frequencies that cause interference. [E1B08] This is not a blanket injunction against interference to broadcast radio and TV signals. Note that the receiver must be "of good engineering design, including adequate selectivity characteristics." The majority of broadcast receivers, manufactured under stringent price constraints, omit key interference-rejection features, particularly filtering. The result is that quiet periods are rarely imposed by the FCC.

The flip side of the coin is that the amateur station must be operating properly without violating any rules, especially those regarding *spurious emissions*. §97.3(a)(42) defines a spurious emission as "an emission, on frequencies outside the necessary bandwidth of a transmission, the level of which may be reduced without affecting the information being transmitted." [E1B01] For stations installed in 2003 or later, spurious emissions must be at least 43 dB below the mean power of the fundamental signal. [E1B11]

It's important to realize that nearly *all* transmissions have some associated spurious emissions. For example, even the purest CW transmitter still generates small amounts of noise and harmonic energy, even though it may be well below the regulatory limit for spurious signals.

The spurious signals may be strong enough to interfere with an over-the-air signal from a distant transmitter. The amateur may be unable to reduce the spurious emissions to zero; a station operating completely legally may still transmit very low level spurious signals that are within the regulatory limits.

> Before you go on, study test questions E1B01, E1B08 and E1B11. Review this section if you have difficulty.

STATION LOCATION AND ANTENNA STRUCTURES

Unless restricted by private agreements such as deed restrictions and covenants, or by local zoning regulations, you are free to install and build whatever antennas and support structures you want — almost. There are certain conditions that may restrict the physical location of your amateur station and the height of the structures associated with it — your towers and antennas.

Restrictions on Location

If the land on which your station is located has environmental importance, or is significant in American history, architecture or culture you may be required to take action as described in §97.13(a). [E1B02] For example, if your station will be located within the boundaries of an officially designated wilderness area, wildlife preserve, or an area listed in the National Register of Historical Places, you must submit an Environmental Assessment to the FCC. [E1B04]

If your station will be located within 1 mile of an FCC monitoring facility, you must protect that facility from harmful interference. [E1B03] If you do cause interference to such a facility, the FCC Engineer in Charge may impose operating restrictions on your station. **Table 3-4** contains a list of these FCC facilities that must be protected from interference.

Restrictions on Antenna Structures

The FCC rules also place some limitations and restrictions on the construction of antennas for amateur stations. Without prior FCC approval and notification to the Federal Aviation Administration (FAA), you may not build an antenna structure, including a tower or other support structure, higher than 200 feet. This applies even if your station is located in a valley or canyon. Unless it is approved, the antenna structure may not be more than 200 feet above the ground level at the station location. This includes all of the tower and antenna. (If

§97.13 — Restrictions on Station Location

(a) Before placing an amateur station on land of environmental importance or that is significant in American history, architecture or culture, the licensee may be required to take certain actions prescribed by §§ 1.1305-1.1319 of this chapter.

(b) A station within 1600 m (1 mile) of an FCC monitoring facility must protect that facility from harmful interference. Failure to do so could result in imposition of operating restrictions upon the amateur station by a District Director pursuant to §97.121 of this Part. Geographical coordinates of the facilities that require protection are listed in §0.121(c) of this chapter.

Table 3-4
Protected FCC Locations

The following FCC field offices are listed in §0.121(b)and must be protected from interference:

Allegan, MI	Kingsville, TX
Belfast, ME	Laurel, MD
Canandaigua, NY	Livermore, CA
Douglas, AZ	Powder Springs, GA
Ferndale, WA	Santa Isabel, PR
Grand Island, NE	Vero Beach, FL
Kenai, AK	Waipahu, HI

you happen to live in sequoia country, this restriction does not apply to trees.) The rule is needed to maintain aviation safety. Structures over 200 feet may be approved, but will be required to have warning markings and tower lighting.

If your antenna is located near a public-use airport, then further height limitations may apply. You must obtain approval from the FAA in such cases. [E1B06] The FAA rules limit antenna structure height based on the distance from the nearest active runways. If your antenna structure is no more than 20 feet above any natural or existing man-made structure then you do not need approval. See **wireless.fcc.gov/antenna/** for more information.

Before you go on, study test questions E1B02, E1B03, E1B04 and E1B06. Review this section if you have difficulty.

3.3 Station Control

The more experience you gain in Amateur Radio, the more varied and interesting the methods you'll find of putting a functioning station together. With the Internet connected directly to the radio and all manner of "smart" interfaces that allow all the equipment to work in concert, the possibilities are literally endless. Even so, all of the different configurations can be reduced to one of three types of control: *local*, *remote* and *automatic* as defined in the next few sections. FCC regulations must be followed under all methods of control.

What's important is that you, the *control operator*, understand the rules for each type of station. The FCC definition for a control operator [§97.3(a)(13)] is "an amateur operator designated by the licensee of a station to be responsible for the transmissions from that station to assure compliance with the FCC Rules." The control operator doesn't have to be the station owner and doesn't even have to be physically present at the transmitter, but someone must be responsible for all amateur transmissions, whether the equipment is directly supervised or not.

LOCAL CONTROL

The FCC defines *local control* [§97.3(a)(30)] as "the use of a control operator who directly manipulates the operating adjustments in the station to achieve compliance with the FCC Rules." The key phrase is "directly manipulates." That means the control operator is physically present at the station. It doesn't matter whether the operator adjusts the equipment directly by hand or uses a computer to make changes or even uses a voice-activated

(a) Owners of certain antenna structures more than 60.96 meters (200 feet) above ground level at the site or located near or at a public use airport must notify the Federal Aviation Administration and register with the Commission as required by Part 17 of this chapter.

(b) Except as otherwise provided herein, a station antenna structure may be erected at heights and dimensions sufficient to accommodate amateur service communications. [State and local regulation of a station antenna structure must not preclude amateur service communications. Rather, it must reasonably accommodate such communications and must constitute the minimum practicable regulation to accomplish the state or local authority's legitimate purpose. See PRB-1, 101 FCC 2d 952 (1985) for details.]

speech system. If you are present at the station and control its operation, that's local control. [E1C07]

Local control is the classic form of radio operation. If you are in your shack, turning the VFO knob and pressing the PTT switch, that's local control. You can use a long stick to turn a knob and that's local control, too.

REMOTE CONTROL

Operating a station by remote control means that the *control point* is no longer at the radio — it's where the control operator is. The control point is where the station's control function is performed and a control operator must be present to control the equipment. Under local control, the control point is at the transmitter and the control operator physically manipulates the controls of the transmitter.

If you're not in direct contact with the radio, but are managing to operate it by means of some intermediary system, that's *remote control*. The exact FCC definition [§97.3(a)(38)] is "the use of a control operator who indirectly manipulates the operating adjustments in the station through a control link to achieve compliance with the FCC Rules." The intermediary system that allows you to operate the radio without being in direct contact with it — that's the *control link*. [E1C01, E1C06] The control link may be through a wire, such as a telephone-line link or a separated front panel in another room. Remote control through a radio link is called *telecommand*.

Because it's possible that the control link could fail, you are expected to have some control backup systems that will keep the transmitter from being left on the air. If the control link malfunctions, §97.213 requires that backup control equipment should limit continuous transmissions to no more than three minutes. [E1C08]

It's important to be aware of the rules for remote control because more and more radio equipment is designed to support remote control. A popular example of stations under remote control are the digital Winlink RMS PACTOR stations (**www.winlink.org**) that wait for a station to call them before responding. The RMS station is considered to be remotely controlled by the calling operator.

Another good example of remote control is found in several new radios that can be operated over a sufficiently fast network connection as if the operator were at the front panel. If a radio is connected digitally to a computer program that serves as the control point, what's the difference between it being under the desk and it being in another state? This type of operation will become much more popular as it becomes a standard feature for radios.

Before you go on, study test questions E1C01, E1C06, E1C07 and E1C08. Review this section if you have difficulty.

Using Automatic Forwarding Systems

More and more hams are building, operating and using automated message forwarding systems, such as Winlink 2000 (**www.winlink.org**), or networked systems, such as D-STAR (**www.icomamerica. com/amateur/dstar**). These systems are relatively new to Amateur Radio and require both a new on-the-air etiquette and a careful attention to the rules to ensure that the communication procedures and message content comply with FCC rules.

These systems use protocols to automate the process of calling, connecting, exchanging data, and disconnecting. The first such amateur system was packet radio, using the AX.25 protocol over VHF and UHF links. (Packet can also be used on HF, but at much slower data rates.) These protocols do a good job of getting the data from one station to another. Unfortunately, the protocols aren't built to recognize signals from other modes that may be operating on or near the same frequency.

This is a crucial difference between digital and analog communications modes in which the information is copied "by ear." It is easy for a human to recognize signals from other modes, even signals that may not be on the same frequency, just nearby. Digital systems have not yet developed that capability. To an HF digital protocol, CW or SSB voice signals are just an interfering tone or noise. A station using digital protocols is likely to react to them as interference or ignore them as it tries to connect with another digital station. This often disrupts the analog communications and is the reason why human supervision of digital stations is so important, particularly on HF where many modes share the same limited band space.

One of the most common sources of behavior-based conflict between modes is the use of semi-automated digital systems on HF. Amateurs have constructed "mailbox" stations that wait silently on a published frequency until called by another digital station using the same protocol. The mailbox station "will not transmit unless transmitted to" and so does not cause interference to signals from other modes. Unless, that is, another digital station calls in, causing the mailbox station to start the connection and transfer process. Thus it is incumbent on the operator of the calling station to listen carefully for other signals on the frequency.

If the operator does not listen, this allows the protocol controller to make the decisions about when to transmit. This can and does enable harmful interference to occur! Another cause of interference is the "hidden transmitter" problem caused by propagation in which the calling station can't hear the other stations on frequency but the mailbox station is heard by everyone. In this situation, even a human operator will not hear the other stations. Nevertheless, the best solution for everyone is for operators of stations attempting to connect to other digital stations to listen by ear first whenever operating on frequencies where other signals are likely to be present. It is not enough to watch for a BUSY light on a modem or controller — that light may only signify the presence of another recognizable digital signal.

Digital messages can also run afoul of content regulations when they are generated by non-amateurs. For example, it's fine to exchange e-mail messages about personal topics but not about work or financial matters. Remember that a non-ham sender is probably unaware of the restrictions on content with which amateurs must comply. Third-party regulations and agreements also apply — know the rules! Higher-speed digital systems may even support direct Internet access. Advertisements are commonplace on many web pages but are not allowed in amateur communications. For this reason alone, web browsing via Amateur Radio is not a good idea. It's important to follow the rules for our service even when they limit what we can do compared to online activities from home over non-amateur networks. It's important that Amateur Radio remain amateur in fact as well as spirit!

AUTOMATIC CONTROL

Does a human control operator have to be present at the control point to supervise every amateur transmission? No — repeaters are a very good example of stations operating under *automatic control* with no control operator present. Automatic control is defined in §97.3(a)(6) as "the use of devices and procedures for control of a station when it is transmitting so that compliance with the FCC Rules is achieved without the control operator being present at a control point." [E1C02, E1C03] The FCC limits the frequencies on which automatically-controlled stations may operate to make sure the amateur bands are primarily used by human operators. **Table 3-5** shows the frequencies on which ground-based stations may operate under automatic control. [E1C09]

When a station is operating under automatic control, it is still required that the station comply with the FCC rules, of course. That means the control operator (usually the station owner) must set up the station so that it can't be made to violate the rules. The control operator is responsible, in this case, for the station complying. In the case of repeater operation where the repeater is operated by stations other than the owner, both the repeater users and the control operator share responsibility for the repeater's proper operation. In addition to repeaters, retransmitting the signals of other amateur stations is only permitted for auxiliary and space stations. [E1C10]

Along with the familiar voice repeaters, there are two other types of amateur stations that may be automatically controlled: *beacon* stations and *auxiliary* stations (discussed in the Miscellaneous Rules section later in this chapter). Beacon stations transmit continuously (limited to 100 W of output power) so that other amateurs can tell when propagation exists between their location and that of the beacon station.

There are special rules about the third-party messages and automatically controlled stations because of the power of message forwarding networks. Third-party traffic is not limited to radiograms — it can be e-mail, digital files, or even keyboard-to-keyboard chat sessions if the content is transferred on behalf of someone who is not a licensed amateur. Automatically controlled stations may only relay third-party communications as RTTY or data emissions and are never allowed to originate the messages. [E1C04, E1C05] These restrictions are in place to be sure Amateur Radio does not become an extension of the commercial data networks.

Table 3-5
Allocations for Ground-Based Repeater Stations

Ground-based repeater stations are authorized in the following band segments:

29.5-29.7 MHz	222.15-225.0 MHz
51.0-54.0 MHz	420.0-431.0 MHz
144.5-145.5 MHz	433.0-435.0 MHz
146-148 MHz	438.0-450.0 MHz

Before you go on, study test questions E1C02 through E1C05, E1C09 and E1C10. Review this section if you have difficulty.

3.4 Amateur-Satellite Service

Just like RACES, the amateur-satellite service is a radio service within the rules that define the Amateur Service. It applies to amateur stations on satellites orbiting the Earth providing Amateur Radio communications. [E1D02] **Table 3-6** defines some of the terms used in the amateur-satellite service rules. According to those rules, amateur-satellite service stations engaging in satellite communications that are on or within 50 km of the Earth's surface are called *Earth stations*. [E1D04]

Stations more than 50 km above the Earth's surface are called *space stations*. The term can be confusing since it is also commonly used to refer to the International Space Station (ISS). In the FCC rules (§97.207) "space stations" refers to *all* amateur stations located

Table 3-6

Amateur-Satellite Service Definitions

The following definitions are included in section §97.3(a) of the FCC rules:

(3) *Amateur-satellite service*. A radiocommunication service using stations on Earth satellites for the same purpose as those of the amateur service.

(16) *Earth station*. An amateur station located on, or within 50 km of the Earth's surface intended for communications with space stations or with other Earth stations by means of one or more other objects in space.

(40) *Space station*. An amateur station located more than 50 km above the Earth's surface.

(43) *Telecommand*. A one-way transmission to initiate, modify, or terminate functions of a device at a distance.

(44) *Telecommand station*. An amateur station that transmits communications to initiate, modify, or terminate functions of a space station.

(45) *Telemetry*. A one-way transmission of measurements at a distance from the measuring instrument.

50 km or more above the Earth's surface. Doubly confusing, the amateur equipment aboard the ISS is itself a space station aboard the International Space Station!

TELECOMMAND AND TELEMETRY

Since most space stations are not operated by amateurs under local control, amateurs must have some way to control the various functions of the satellite. The process of transmitting communications to a satellite to initiate, modify or terminate the various functions of a space station is called *telecommand operation*. [E1D03] Stations that transmit these command communications are *telecommand stations*. Any amateur station that is designated by the space station licensee may serve as a telecommand station. §97.211 describes what telecommand stations are and what they may do.

Obviously, sending telecommand communications to a satellite should not be something any amateur can do. Unauthorized telecommand signals would likely disrupt or even damage the satellite. For this reason, the FCC allows telecommand stations to use special codes that are intended to obscure the meaning of telecommand messages. This is one of the few times an amateur may intentionally obscure the meaning of a message. Otherwise, anyone who copied the transmission could learn the control codes for the satellite.

Space stations provide amateurs with a unique opportunity to learn about conditions in space. For example, a satellite might record the temperature, amount of solar radiation or other measurements and then transmit that information back to Earth. It is also important for the satellite operators to know the status of important parameters such as the state of battery charge, transmitter temperature or other spacecraft conditions. When transmitted back to Earth stations, this information is called *telemetry*, the general term for any one-way transmission of measurements to a receiver located at a distance from the measuring instrument. [E1D01]

Before you go on, study test questions E1D01 through E1D04. Review this section if you have difficulty.

SATELLITE LICENSING AND FREQUENCY PRIVILEGES

Any licensed Amateur Radio operator may be the control operator of a space station — no special license is required. [E1D05] Of course there is a little more involved in operating a space station than there is in regular ground-based communications. Any Amateur Radio station can also be a space station — assuming you can get it into space somehow! Let's assume that your organization has found a way to get your satellite launched and why not? Many organizations have accomplished this feat. You'll need telecommand stations to control the satellite. Again, any amateur station may be a satellite telecommand station subject to the restrictions of the control operator's license class, of course. [E1D10] You just have to authorize them to do it and give them the telecommand information. There is one special telecommand requirement — a space station must have incorporated the ability for its transmitter to be turned off by telecommand. [E1D06] This is not only good operating practice; the FCC requires it in case the satellite is causing interference.

On the HF bands, a space station may only operate on the 17, 15, 12 and 10 meter bands and on portions of the 40 and 20 meter bands. [E1D07] Segments of 2 meters, 70 cm, 23 cm, 13 cm and some microwave bands are also available for space station operation. [E1D08]

And then there are the satellite users who will be clamoring to "squirt your bird"! Once again, no special license is required — any amateur station can operate as an Earth station if the privileges of the license allow the operator to use the frequencies and modes on which the satellite operates. [E1D11] The only reason a satellite would not be usable by operators of every class of US license would be if it had an uplink frequency in either a General or Extra class subband on 40, 20 or 15 meters. There are no satellites currently active using those frequencies.

Before you go on, study test questions E1D05 through E1D11. Review this section if you have difficulty.

3.5 Volunteer Examiner Program

By now, you've passed at least two amateur exams. You should be an expert on the Volunteer Examiner (VE) program! Just kidding, of course, but since you're about to pass your third and final exam (Right? Right!) why do you need to learn more about the program? Well, because as an Extra class licensee, you'll have the opportunity to administer those exams yourself! Extra class VEs are needed to give both General and Extra class exams, so this is a little advertisement for you to become a Volunteer Examiner and help others get their licenses.

The questions on the Extra class exam are designed to familiarize you with the FCC rules in Part 97, Section F that pertain to the "examiner side" of the exam. We'll start with some review of the program's structure.

THE VOLUNTEER EXAMINER COORDINATOR

A Volunteer Examiner Coordinator (VEC) is an organization that has entered into an agreement with the FCC to coordinate amateur license examinations. [E1E03] The organization must meet certain criteria before it can become a VEC. As described in §97.521, the organization must exist for the purpose of furthering the amateur service, but should be more than just a local radio club or group of hams. A VEC is expected to coordinate exams at least throughout an entire call district. The organization must also agree to coordinate exams for all classes of amateur operator license, and to ensure that anyone desiring an amateur license can register and take the exams without regard to race, sex, religion or national origin.

THE VOLUNTEER EXAMINER

A VEC does not administer or grade the actual examinations. The VEC accredits licensed Amateur Radio operators — the Volunteer Examiners (VEs) — to administer exams. Each

VEC is responsible for recruiting and training Volunteer Examiners to administer amateur examinations under their program. The Volunteer Examiners determine where and when examinations for amateur operator licenses will be administered.

ACCREDITATION

When a VEC accredits a Volunteer Examiner, it is certifying that the amateur is qualified to perform all the duties of a VE as required by §97.509 and §97.525. The accreditation process is simply the steps that each VEC takes to ensure their VEs meet all the FCC requirements to serve in the Volunteer Examiner program. [E1E04] Each VEC has its own accreditation process. A VEC has the responsibility to refuse to accredit a person as a VE if the VEC determines that the person's integrity or honesty could compromise amateur license exams.

The ARRL VEC coordinates exams in all regions of the US, and would be pleased to have you apply for accreditation. You do not have to be an ARRL member to serve as an ARRL VE. In fact, one of the requirements of VECs is that they not demand membership in any organization as a prerequisite to serving as a VE!

If you are at least 18 years of age and hold at least a General class license, you meet the basic FCC requirements to be a VE. In addition, you must never have had your amateur license suspended or revoked. **Figure 3-2** shows the application form to become an ARRL VE.

> Before you go on, study test questions E1E03 and E1E04. Review this section if you have difficulty.

EXAM PREPARATION

Coordinating amateur exams involves a bit more responsibility than simply recruiting amateurs to administer the exams. (§97.519 states the requirements for coordinating an exam session.) A VEC coordinates the preparation and administration of exams. Some VECs actually prepare the exams and provide their examiners with the necessary test forms, while others require their VEs to prepare their own exams or purchase exams from a qualified supplier. After the test is completed, the VEC must collect the application documents (NCVEC Form 605) and test results. After reviewing the materials to ensure accuracy, the VEC must forward the documentation to the FCC for applicants that qualify for a new license or a license upgrade.

In addition, all of the VECs must cooperate in the development and maintenance of the questions used on the exams. [§97.523] The set of all the questions available to be asked on an exam is called the *question pool*. [E1E02] A Question Pool Committee (QPC) works regularly to update the questions for each exam element. Exams are made up of questions selected from the question pool.

Volunteer Examiners may prepare written exams for all classes of Amateur Radio operator license. Section 97.507 of the FCC rules gives detailed instructions about who may prepare the various examination elements. You must hold a General, Advanced or Amateur Extra license to prepare an Element 2 written exam for the Technician class license. Only Advanced and Amateur Extra licensees may prepare the Element 3 exam (General) and you must hold an Amateur Extra license to prepare an Element 4 exam (Amateur Extra).

If the VEC or a qualified supplier prepares the exams, they must still use amateurs with the proper license class to prepare the exams. In every case, the exams are prepared by selecting questions from the appropriate question pool.

EXAM SESSION ADMINISTRATION

Every Amateur Radio license exam session must be coordinated by a VEC, and must be administered by a VE team consisting of at least three VEs accredited by the VEC coordi-

ARRL VEC, 225 Main Street, Newington, CT 06111
Phone: 1-860-594-0300 web: arrl.org/arrlvec

VOLUNTEER EXAMINER APPLICATION FORM

PLEASE Type or Print Clearly in Ink

(check one)
☐ General
☐ Advanced
☐ Extra   License Expiration Date: _____

Call Sign: _____

Name: _____
                    (first,   MI,   last)

Mailing address (street or POB): _____

_____

City: _____ State: _____ ZIP: _____ Country: _____

Day phone: (____)_____Night phone: (____)_____ Email address: _____

Has your FCC license ever been suspended or revoked? .... ☐ YES ☐ NO
Have you ever been disaccredited by another VEC? .... ☐ YES ☐ NO
  If yes, which VEC(s) and when? _____

Do you have a call sign change (or Vanity call sign) pending with the FCC? ☐ YES ☐ NO
Do you have any Form 605 application pending with the FCC? .... ☐ YES ☐ NO

Who can we contact to reach you, if you cannot be reached? _____
                                                              (name)   (phone)

For Instant Accreditation, have you participated as a VE in another VEC program
  and is your accreditation in that program current?................................ ☐ YES ☐ NO

  If yes, which VEC coordinated the test session? (enter VEC name here) _____

  You MUST attach a copy of your credentials from that VEC to this form as proof.

CERTIFICATION

*By signing this Application Form, I certify that to the best of my knowledge
that the above information AND the following statements are true:*

1) I am at least 18 years of age.
2) I agree to comply with the FCC Part 97 Amateur Radio Service Rules, especially Subpart F (§97.509).
3) I agree to comply with examination procedures established by the ARRL as Volunteer Examiner Coordinator.
4) I understand that the ARRL as my coordinating VEC, or I as an accredited ARRL VE, may terminate this
   relationship at any time, with or without any reason or cause.
5) I understand that violation of the FCC Rules or willful noncompliance with the VEC will result in the loss of my
   VE accreditation, and could result in loss of my Amateur Radio operator or station licenses, or both.
6) I understand that, even though I may be accredited as a VE, if I am not able or competent to perform certain VE
   functions required for any particular examination, I should not administer that examination (§97.525).

_____   _____   _____
         (signature)   (call sign)   (date)

Look over your form for completeness, make sure it is signed and then send it or fax it to the ARRL VEC.
If *instant accreditation* is sought, you MUST indicate which VEC program you served as an administering VE and attach a copy
of your other VEC credentials to this application. Otherwise your application *must include your completed open-book review.*

ARRL VEC -- VE APPLICATION 04/2007

Figure 3-2 — This is the ARRL VEC Application Form. It is available online in the *ARRL Volunteer Examiner Manual* (www.arrl.org/ve-manual). The *Manual* contains all you need to know to become an ARRL VEC-accredited Volunteer Examiner.

Recovering Exam Expenses

The Amateur Radio license-exam system depends on the services of volunteers. As licensed amateurs and accredited VEs, these volunteers may not charge a fee to administer exams or receive any type of payment for the services they provide. VECs may not charge a fee for coordinating exam sessions, either. Printing exams and forms, mailing paperwork and securing a suitable location to administer exams may all cost money. Neither the VEC nor the VEs should have to bear these costs out of their own pockets, however. FCC rule §97.527 provides a means for those being examined to reimburse the VEs and VEC for certain costs involved with the program. These costs include actual out-of-pocket expenses involved with preparing, processing and administering license exams. [E1E14]

nating the session. [E1E01] The requirements for VEs administering an exam are stated in §97.509.

The general rule for determining who may administer the various exam elements is that you must hold a higher class of license than the elements you are administering. Extra class exams must be administered by VEs holding Extra class licenses. To administer a General exam, you must hold an Advanced or Amateur Extra license. To administer a Technician license exam, you must hold a General, Advanced or Amateur Extra license. VEs are prohibited from administering exams to close relatives as defined by the FCC. [E1E08]

Before actually beginning to administer an examination, the VEs should determine what exam credit, if any, the candidates should be given as described in §97.505. For example, any candidates who already hold an amateur operator license must receive credit for having passed all of the exam elements necessary for that class of license. In addition, any candidate who presents a valid Certificate of Successful Completion of Examination (CSCE) must be given credit for each exam element that the CSCE indicates the examinee has passed. The combination of element credits and exam elements passed at the current exam session will determine if a candidate qualifies for a higher class of license.

During the Exam

All three VEs are responsible for supervising the exam session and must be present during the entire exam session, observing the candidates to ensure that the session is conducted properly. [E1E06] During the exam session, the candidates must follow all instructions given by the Volunteer Examiners. If any candidate fails to comply with a VE's instructions during an exam, the VE Team should immediately terminate that candidate's exam. [E1E07]

When the candidates have completed their exams, the VEs must collect the test papers and grade them immediately. A score of 74% is the minimum to pass the exam. [E1E05] They then notify the candidates whether they passed or failed the exam. If any candidates did not pass all the exam elements needed to complete their license upgrade, then the examiners must return their applications to those candidates and inform them of the grades. [E1E12]

After grading the exams of those candidates who do pass the exam, the entire VE Team must certify their qualifications for new licenses and that they have complied with the VE requirements on their application forms and issue each a CSCE (**Figure 3-3**) for their upgrade. [E1E11]

After the Exam

After a successful exam session, the VE Team must submit the application forms and test papers for all the candidates who passed to the coordinating VEC. [E1E10] They must do this within 10 days of the test session. This is to ensure that the VEC can review the paperwork and forward the information to the FCC in a timely fashion.

Before you go on, study test questions E1E01, E1E02, E1E05, E1E06, E1E07, E1E08, E1E10, E1E11 E1E12 and E1E14. Review this section if you have difficulty.

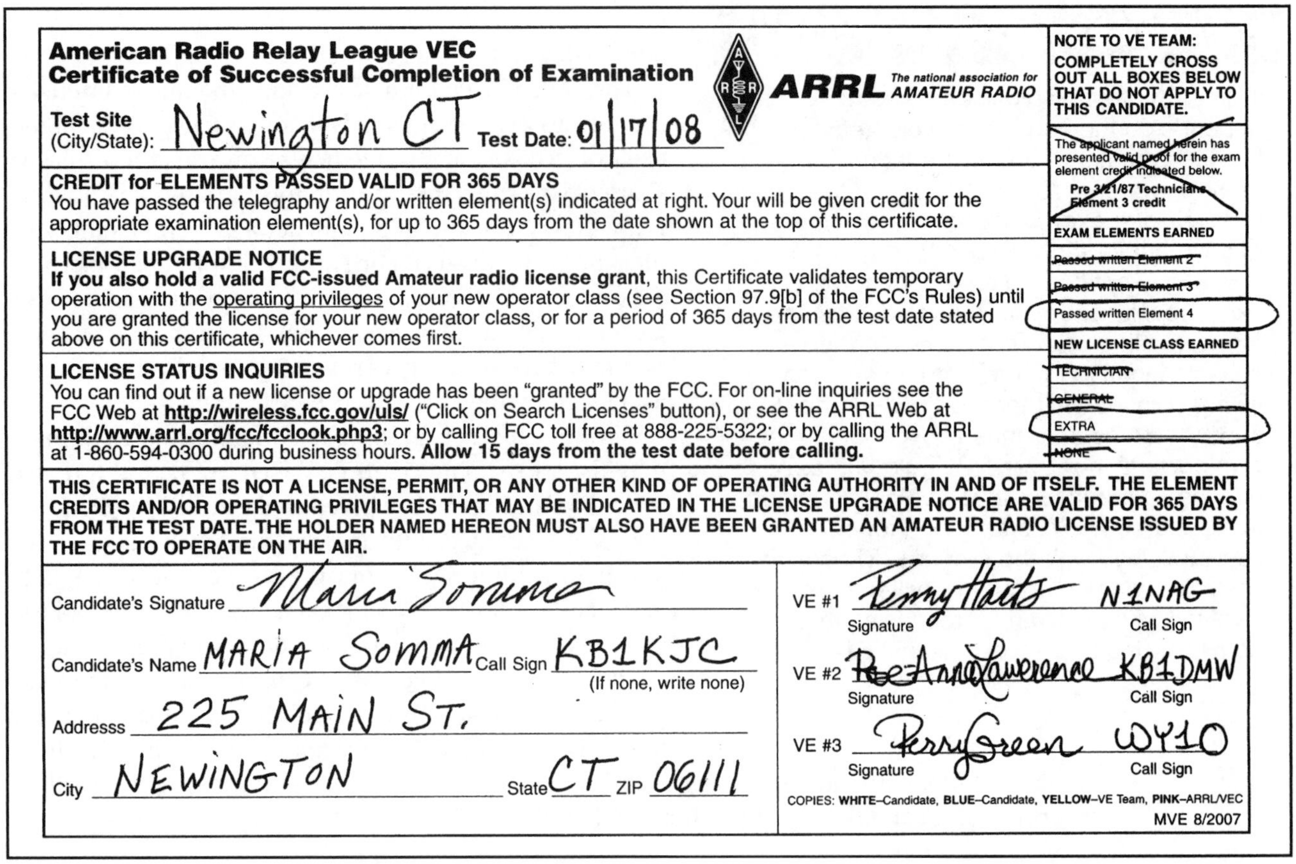

American Radio Relay League VEC
Certificate of Successful Completion of Examination

ARRL *The national association for* AMATEUR RADIO

Test Site (City/State): _Newington CT_ Test Date: _01_ | _17_ | _08_

NOTE TO VE TEAM: COMPLETELY CROSS OUT ALL BOXES BELOW THAT DO NOT APPLY TO THIS CANDIDATE.

The applicant named herein has presented valid proof for the exam element credit indicated below.
Pre 3/21/87 Technician Element 3 credit

EXAM ELEMENTS EARNED
Passed written Element 2
Passed written Element 3
Passed written Element 4

NEW LICENSE CLASS EARNED
TECHNICIAN
GENERAL
EXTRA
NONE

CREDIT for ELEMENTS PASSED VALID FOR 365 DAYS
You have passed the telegraphy and/or written element(s) indicated at right. Your will be given credit for the appropriate examination element(s), for up to 365 days from the date shown at the top of this certificate.

LICENSE UPGRADE NOTICE
If you also hold a valid FCC-issued Amateur radio license grant, this Certificate validates temporary operation with the operating privileges of your new operator class (see Section 97.9[b] of the FCC's Rules) until you are granted the license for your new operator class, or for a period of 365 days from the test date stated above on this certificate, whichever comes first.

LICENSE STATUS INQUIRIES
You can find out if a new license or upgrade has been "granted" by the FCC. For on-line inquiries see the FCC Web at **http://wireless.fcc.gov/uls/** ("Click on Search Licenses" button), or see the ARRL Web at **http://www.arrl.org/fcc/fcclook.php3**; or by calling FCC toll free at 888-225-5322; or by calling the ARRL at 1-860-594-0300 during business hours. **Allow 15 days from the test date before calling.**

THIS CERTIFICATE IS NOT A LICENSE, PERMIT, OR ANY OTHER KIND OF OPERATING AUTHORITY IN AND OF ITSELF. THE ELEMENT CREDITS AND/OR OPERATING PRIVILEGES THAT MAY BE INDICATED IN THE LICENSE UPGRADE NOTICE ARE VALID FOR 365 DAYS FROM THE TEST DATE. THE HOLDER NAMED HEREON MUST ALSO HAVE BEEN GRANTED AN AMATEUR RADIO LICENSE ISSUED BY THE FCC TO OPERATE ON THE AIR.

Candidate's Signature _Maria Somma_

Candidate's Name _MARIA SOMMA_ Call Sign _KB1KJC_ (If none, write none)

Addresss _225 MAIN ST._

City _NEWINGTON_ State _CT_ ZIP _06111_

VE #1 _Kenny Hartz_ Signature _N1NAG_ Call Sign

VE #2 _Ree Anna Lawrence_ Signature _KB1DMW_ Call Sign

VE #3 _Perry Green_ Signature _WY1O_ Call Sign

COPIES: **WHITE**–Candidate, **BLUE**–Candidate, **YELLOW**–VE Team, **PINK**–ARRL/VEC
MVE 8/2007

Figure 3-3 — The Certificate of Successful Completion of Examination (CSCE) documents that the holder has passed one or more amateur license exam elements. If you upgrade your license by passing an exam, the CSCE allows you to start using your new privileges immediately.

RE-ADMINISTRATION OF EXAMINATIONS

The FCC has the authority to administer examination elements and to re-administer any exam element to any licensee. The FCC may administer those exams itself, or may designate a VEC to supervise the re-administration of the exam elements. If you receive a notice from the FCC that you are to appear for the re-administration of an examination, you must follow the directions given in the notice. If you fail to appear for the exam, the FCC will cancel your license and they may issue a new license that is consistent with the exam elements not cancelled. For example, suppose that after passing the Element 4 exam, the FCC notifies you to appear for a re-administration of that exam. If you fail to appear, the FCC will cancel your license and reissue a license based on the elements for which you previously passed an exam. [E1E13]

There are a variety of reasons that the FCC might make such a request. There may have been some testing irregularity at the test session you attended. Although you may not have done anything wrong, the FCC might decide to invalidate *all* of the results from that exam session. In that case you would have to take the exam again to receive proper credit. If the FCC determines that a VE has fraudulently administered or certified an exam, that VE can lose their amateur station license and have their operator privileges suspended. [E1E09] Such problems are extremely rare because of the high integrity of the amateur volunteer licensing program.

Before you go on, study test questions E1E09 and E1E13. Review this section if you have difficulty.

3.6 Miscellaneous Rules

The following sections cover topics of narrow interest. As an Extra class licensee, you'll be expected to know about lesser-visited areas of the FCC rules and have a more complete knowledge of important rules.

AUXILIARY STATIONS

Auxiliary stations [§97.3(a)(7)] are amateur stations, other than in a message forwarding system, that transmit communications point-to-point within a system of cooperating amateur stations. Amateurs are allowed to use auxiliary stations to provide point-to-point communications and control links between a remotely controlled station and its control point. Repeater systems may use point-to-point links to relay audio and control signals from one repeater to all other repeaters in the system. Control operators of auxiliary stations must hold a Technician, General, Advanced, or Extra class license. [E1F12]

You might set up a mobile rig as a cross-band repeater to act as an auxiliary station relaying your signals to and from a nearby low-power handheld transceiver. §97.201(b) limits auxiliary station transmissions to the 2 meter and shorter wavelength bands, excluding the 144.0-144.5 MHz, 145.8-146.0 MHz, 219-220 MHz, 222.0-222.15 MHz, 431-433 MHz, and 435-438 MHz segments.

Before you go on, study test question E1F12. Review this section if you have difficulty.

EXTERNAL POWER AMPLIFIERS

RF power amplifiers capable of operating on frequencies below 144 MHz may require FCC certification. Sections 97.315 and 97.317 describe the conditions under which certification is required, and set out the standards to be met for certification. Many of these rules apply to manufacturers of amplifiers or kits, but several points are important for individual amateurs.

Amateurs may build their own amplifiers or modify amplifiers for use in an Amateur Radio station without concern for the certification rules. An unlicensed person may not build or modify any amplifier capable of operating below 144 MHz without a grant of FCC certification.

To receive a grant of certification, an amplifier must satisfy the spurious emission standards specified in §97.307(d) or (e) when operated at full power output or 1500 W, whichever is less. [E1F11] In addition, the amplifier must meet the spurious emission standards when it is placed in the "standby" or "off" position but is still connected to the transmitter. The amplifier must not be capable of amplifying the input signal by more than 15 dB, and it must exhibit no amplification between 26 and 28 MHz. (This is to prevent the amplifier from being used illegally on the Citizen's Band frequencies.)

A manufacturer must obtain a separate grant of certification for each amplifier model. The FCC maintains a database of all certificated amplifier models, and an amplifier must be on that list before it can be marketed or sold for use in the US amateur service. Dealers may also sell non-certificated amplifiers if they were purchased in used condition and resold to another amateur for use in their station. [E1F03] FCC certification may be denied if the amplifier can be used in a telecommunication service other than the amateur service, or if it can be easily modified to operate between 26 and 28 MHz.

Before you go on, study test questions E1F03 and E1F11. Review this section if you have difficulty.

LINE A AND NATIONAL QUIET ZONES

The 420 to 430-MHz band segment is allocated to the fixed and mobile services in the international allocations table on a primary basis worldwide. Canada has allocated this band segment to its fixed and mobile services, so US amateurs along the Canadian border are not permitted to transmit on these frequencies. An imaginary line, called Line A, runs roughly parallel to and south of the US-Canadian border. [E1F04] US stations north of this line may not transmit on the 420 to 430-MHz band. [E1F05] Section 97.3(a)(29) gives an exact definition of Line A. See **Figure 3-4**.

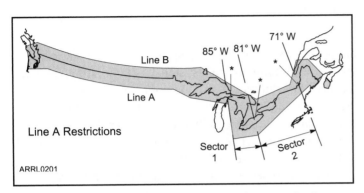

Figure 3-4 — Line A runs parallel to, and just south of, the US/Canadian border.

The FCC has also allocated portions of the 421- 430 MHz band to the Land Mobile Service within a 50-mile radius centered on Buffalo, New York; Detroit, Michigan; and Cleveland, Ohio. Amateurs in these areas must not cause harmful interference to the Land Mobile or government radiolocation users.

Certain other restrictions may apply if your station is within specific geographical regions. For example, there is an area in Maryland, West Virginia and Virginia surrounding the National Radio Astronomy Observatory. This area is known as the *National Radio Quiet Zone*. [E1F06] The NRQZ serves to protect the interests of the National Radio Astronomy Observatory in Green Bank, West Virginia, and also Naval Research Laboratory at Sugar Grove, West Virginia.

If you plan to install an automatically controlled beacon station within the NRQZ, you will have to obtain permission from the National Radio Astronomy Observatory. §97.203 lists the details of this rule point and the address to contact.

Before you go on, study test questions E1F04 through E1F06. Review this section if you have difficulty.

BUSINESS AND PAYMENT

Business communications rules for Amateur Radio are based on the principle that no transmissions are permitted in which you or your employer have a pecuniary (monetary) interest. This is, after all, Amateur Radio, and there are plenty of radio services available for commercial activities. That prohibition is the same for voice and digital messages, including those sent and received via message forwarding networks.

However, your own personal activities don't count as "business" communications. For example, it's perfectly okay for you to use ham radio to talk to your spouse about doing some shopping or to confer about what to pick up at the store. You can even send a message to a business over the air, to order something for example, as long as you don't do it regularly and as part of your normal income-making activities [E1F07]

Note that when you are contacting stations in other countries, communications are limited to remarks of a personal nature or incidental to Amateur Radio. No exception to the non-business rule is made for communications on behalf of non-profit organizations. [§97.117] [E1F13]

Another broad prohibition is receiving compensation for communications via Amateur Radio — either being paid directly "for hire" or in a trade of some sort, such as equipment or services. [E1F08] You should be especially careful when making contributions toward the maintenance of a club station or repeater. It's perfectly okay to share expenses or donate to a club that maintains a station. Normal out-of-pocket expenses can be reimbursed, as well. It is not permitted, however, to provide payment for communications services. For example, the owner of an Amateur Radio repeater cannot accept any form of payment for providing communications services.

Before you go on, study test questions E1F07, E1F08, and E1F13. Review this section if you have difficulty.

SPREAD SPECTRUM OPERATION

Spread spectrum (SS) communication is a signal-transmission technique in which the transmitted carrier is spread out over a wide bandwidth. The FCC refers to this as *bandwidth-expansion modulation*. If you want to learn more about these fascinating modes you can find the details in *The ARRL Handbook*.

The idea behind this communications mode is to spread a little power over a wide bandwidth to minimize interference rather than concentrating a lot of power in a narrow bandwidth. The FCC rules allow US amateurs to communicate using SS with other amateurs in areas regulated by the FCC and with other stations in countries that permit SS communications. Amateurs are permitted to use SS as long as it does not cause harmful interference to other stations using authorized emissions. In addition, the SS transmission must not be used to obscure the meaning on any communication. [E1F09]

The FCC limits the maximum transmitter power for spread spectrum communications to 10 W. [E1F10] Operation using spread spectrum techniques is restricted to frequencies above 222 MHz. [E1F01]

Before you go on, study test questions E1F01, E1F09, and E1F10. Review this section if you have difficulty.

NON-US OPERATING AGREEMENTS

There are three basic types of agreements that allow amateurs licensed in one country to operate from the territory of another country.

- European Conference of Postal and Telecommunications Administrations (CEPT) radio-amateur license — allows US amateurs to travel to and operate from most European countries and their overseas territories without obtaining an additional license or permit. (This does not automatically confer permission to enter restricted areas.) Amateurs from countries that participate in CEPT may also operate in the US. [E1F02]

- International Amateur Radio Permit (IARP) — For operation in certain countries of Central and South America, the IARP allows US amateurs to operate without seeking a special license or permit to enter and operate from that country.

- ITU Reciprocal Permit — a reciprocal agreement between the US and a country that does not participate in either CEPT or IARP agreements.

The complete rules and procedures for obtaining permission to use your license elsewhere in the world are available on the ARRL's International Operating web page at **www.arrl.org/international-operating** and **www.arrl.org/international-regulatory**.

Amateurs who are citizens of a foreign country and are operating in the US under the terms of an operating agreement between the US and the other country will have the same privileges as a US Amateur Extra class licensee if they hold a full-privilege license from their country. This includes Canadians with an Advanced Qualification as well as operators with a Class 1 license issued by the European Conference of Posts and Telecommunications (CEPT) or a Class 1 International Amateur Radio Permit (IARP).

Before you go on, study test questions E1F02. Review this section if you have difficulty.

SPECIAL TEMPORARY AUTHORITY

Occasionally, a new method or technique of communicating comes along that isn't covered by the FCC rules or would be in violation of the rules. If sufficiently good reasons are provided to the FCC, a *Special Temporary Authority* or STA may be granted to provide for experimental amateur communications. [E1F14] For example, STAs were granted to allow amateurs to experiment with spread spectrum communications before that mode was generally permitted on amateur frequencies. An STA was also granted allowing amateurs to experiment with communications on 500 kHz.

STAs are temporary, lasting long enough for experiments to be performed and information accumulated. For particularly ambitious projects the STA might last for six months or longer. During this time, the STA does not grant amateurs the exclusive use of a frequency, nor does it waive all rules — only the ones explicitly covered in the STA. STAs can also be terminated by the FCC at any time if the operation is found to be causing interference, for example. STAs may result in changes to the FCC rules, but they are not permanent waivers of any rule.

Before you go on, study test question E1F14. Review this section if you have difficulty.

Table 3-7

Questions Covered in This Chapter

3.1 Operating Standards
E1A01
E1A02
E1A03
E1A04
E1A05
E1A06
E1A07
E1A08
E1A09
E1A10
E1A11
E1A12
E1A13
E1B05
E1B07
E1B09
E1B10

3.2 Station Restrictions
E1B01
E1B02
E1B03
E1B04
E1B06
E1B08
E1B11

3.3 Station Control
E1C01
E1C02
E1C03
E1C04
E1C05
E1C06
E1C07
E1C08
E1C09
E1C10

3.4 Amateur-Satellite Service
E1D01
E1D02
E1D03
E1D04
E1D05
E1D06
E1D07
E1D08
E1D09
E1D10
E1D11

3.5 Volunteer Examiner Program
E1E01
E1E02
E1E03
E1E04
E1E05
E1E06
E1E07
E1E08
E1E09
E1E10
E1E11
E1E12
E1E13
E1E14

3.6 Miscellaneous Rules
E1F01
E1F02
E1F03
E1F04
E1F05
E1F06
E1F07
E1F08
E1F09
E1F10
E1F11
E1F12
E1F13
E1F14

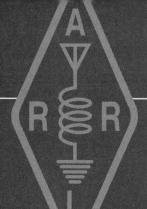

Chapter 4

Electrical Principles

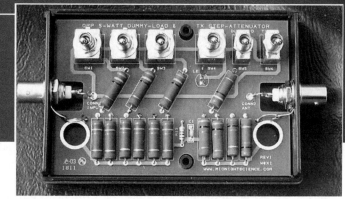

In this chapter, you'll learn about:
- **Rectangular and polar coordinates**
- **Complex numbers**
- **Electric and magnetic fields**
- **Time constants**
- **Complex impedance and calculations**
- **Reactive power and power factor**
- **Resonant circuits, Q and bandwidth**
- **Inductors with magnetic cores**

The Amateur Radio Extra class license exam will include questions about electrical principles from question pool Subelement 5 (four groups of questions) and Subelement 6 (six groups of questions). This chapter begins with some background in the way we use math to talk about electrical signals. Once prepared, we move on to the relationship between voltage and current in real circuits, impedance, resonance, and even magnetic materials used in inductors.

It would be easy to make this book very, very large by trying to teach fundamental subjects from the ground up. To keep this manual focused, you'll be given references on where to find supplemental information on math and electronics. The *Extra Class License Manual* web page (**www.arrl.org/extra-class-license-manual**.) is a good place to start looking, and it should be added to the bookmarks in your Internet browser. Let's get going!

4.1 Radio Mathematics

You learned about frequency, phase, impedance and reactance to pass your Technician and General exams. For the Extra exam, we go a little farther and learn how to work with those values in simple equations. By learning these techniques, you'll be able to work with any combination of resistance (R), inductance (L) and capacitance (C). The very same techniques are used to describe signals in later sections.

RECTANGULAR AND POLAR COORDINATES

We can't actually see the electrons flowing in a circuit, or look at voltage or impedances, so we use math equations or graphs to describe what's happening. Graphs are drawings of what equations describe with symbols — they're both saying the same thing. The way in which mathematical quantities are positioned on the graph is called the *coordinate system*. *Coordinate* is another name for the numeric scale that divides the graph into regular units. The location of every point on the graph is described by a set of *coordinates*.

The two most common coordinate systems used in radio are the *rectangular-coordinate* system shown in **Figure 4-1** (sometimes called *Cartesian coordinates*) and the *polar-coordinate* system shown in **Figure 4-2**. (Additional information on coordinate systems is available

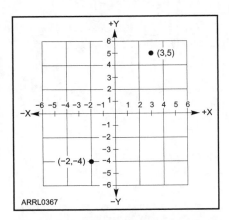

Figure 4-1 — Rectangular-coordinate graphs use a pair of axes at right angles to each other, each calibrated in numeric units. Any point on the resulting grid can be expressed in terms of its horizontal (X) and vertical (Y) values, called coordinates.

on the Math Supplement page listed on the ECLM website.)

The line that runs horizontally through the center of a rectangular coordinate graph is called the *X axis*. The line that runs vertically through the center of the graph is called the *Y axis*. Every point on a rectangular coordinate graph has two coordinates that identify its location, X and Y, also written as (X,Y). [E5C11] Every different pair of coordinate values describes a different point on the graph. The point at which the two axes cross — where the numeric values on both axes are zero — is called the *origin*, written as (0,0).

In Figure 4-1, the point with coordinates of (3,5) is located 3 units to the right of the origin along the X axis and 5 units above the origin along the Y axis. Another point at (−2,−4) is found 2 units to the left of the origin along the X axis and 4 units below the origin along the Y axis. Initially, you may confuse the "X" that refers to position along the X-axis with the X that refers to reactance. Don't worry — with a little practice, you'll be able to keep them straight!

In the polar-coordinate system, points on the graph are also described by a pair of numeric values called *polar coordinates*. In this case we use a length, or *radius*, measured from the origin, and an *angle* from 0° to 360° measured counterclockwise from the 0° line as shown in Figure 4-2. The symbol r is used for the radius and θ for the angle. A number in polar coordinates is written r∠θ. So the two points described in the last paragraph could also be written as (5.83, ∠59.0°) and (4.5, ∠243.4°) and are drawn as polar coordinates in Figure 4-2. Remember that unlike maps, the 0° direction is always to the *right* and not to the top. In mathematics, 0° is not north!

A negative angle essentially means, "turn the other way." With positive angles measured counterclockwise from the 0° axis, the polar coordinates of the point at lower left in Figure 4-2 would be (4.5, ∠−116.6°). When you encounter a negative value for the angle, it means to measure the angle clockwise from 0°. For example, −270° is equivalent to 90°; −90° is equivalent to 270°; 0° and −360° are equivalent; and +180° and −180° are equivalent. An angle can also be specified in *radians* (1 radian = 360 / 2π = 57.3 degrees) but all angles are in degrees in this book and on the exam.

In electronics, it's common to use both the rectangular and polar-coordinate systems when dealing with impedance problems. The examples in the next few sections of this book should help you become familiar with these coordinate systems and the techniques for changing between them.

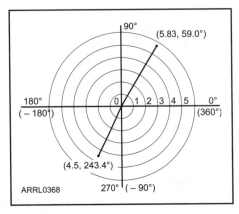

Figure 4-2 — Polar-coordinate graphs use a radius from the origin and an angle from the 0° axis to specify the location of a point. Thus, the location of any point can be specified in terms of a radius and an angle.

COMPLEX COORDINATES

So far in your radio career, you've dealt exclusively with *real numbers* such as π (pi), $\sqrt{2}$, 5 Ω, 2.5 mH, or 53.2 MHz. In solving mathematical equations that describe phases and angles, however, you will encounter numbers that contain the square root of minus one ($\sqrt{-1}$). This is an entirely new type of number! It can't be a real number since no real number multiplied by itself equals −1. As a result, any number that contains $\sqrt{-1}$ is called an *imaginary number*. For convenience, $\sqrt{-1}$ is represented as j in electronics. For example, 2j, 0.1j, 7j/4, and 457.6j are all imaginary numbers. (Mathematicians use i for the same

Working With Polar and Rectangular Coordinates

Complex numbers representing electrical quantities can be expressed in either rectangular form $(a + jb)$ or polar form $(r \angle \theta)$. Adding complex numbers is easiest in rectangular form:

$(a + jb) + (c + jd) = (a + c) + j(b + d)$

Multiplying and dividing complex numbers is easiest in polar form:

$a\angle\theta_1 \times b\angle\theta_2 = (a \times b)\,(\theta_1 + \theta_2)$

and

$$\frac{a\angle\theta_1}{b\angle\theta_2} = \left(\frac{a}{b}\right)\angle(\theta_1 - \theta_2)$$

Converting from one form to another is useful in some kinds of calculations. For example, to calculate the value of two complex impedances in parallel you use the formula

$$Z = \frac{Z_1 Z_2}{Z_1 + Z_2}$$

To calculate the numerator $(Z_1 Z_2)$ you would write the impedances in polar form. To calculate the denominator $(Z_1 + Z_2)$ you would write the impedances in rectangular form. So you need to be able to convert back and forth from one form to the other. There is a good explanation of this process, with examples, on the **www.intmath.com/Complex-numbers/4_Polar-form.php** web page.

Here is the short procedure you can save for reference:
To convert from rectangular $(a + jb)$ to polar form $(r \angle \theta)$:

$r = \sqrt{(a^2 + b^2)}$

$\theta = \tan^{-1}\left(\dfrac{b}{a}\right)$

To convert from polar to rectangular form:

$a = r\cos\theta$

$b = r\sin\theta$

Many calculators have polar-rectangular conversion functions built-in and they are worth learning how to use. Be sure that your calculator is set to the angle units you prefer, radians or degrees.

Example
Convert $3 \angle 30°$ to rectangular form:

$a = 3\cos 30° = 3\,(0.866) = 2.6$

$b = 3\sin 30° = 3\,(0.5) = 1.5$

$3 \angle 30° = 2.6 + j1.5$

Example
Convert $0.8 + j0.6$ to polar form:

$r = \sqrt{(0.8^2 + 0.6^2)} = 1$

$\theta = \tan^{-1}\left(\dfrac{0.6}{0.8}\right) = 36.8°$

$0.8 + j0.6 = 1 \angle 36.8°$

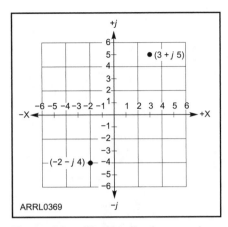

Figure 4-3 — The Y axis of a complex-coordinate graph represents the imaginary portion of complex numbers. This graph shows the same numbers as in Figure 4-1, graphed as complex numbers.

purpose, but I is used to represent current in electronics.) j also has another interesting property that you'll use: $1/j = -j$.

Real and imaginary numbers can be combined by using addition or subtraction. Combining real and imaginary numbers creates a hybrid called a *complex number*, such as $1 + j$ or $6 - j7$. (The convention in complex numbers is for j to be first in the imaginary part of the number.) These numbers come in very handy in radio, describing impedances, relationships between voltage and current, and many other phenomena.

If the complex number is broken up into its real and imaginary parts, those two numbers can also be used as coordinates on a graph using *complex coordinates*. This is a special type of rectangular-coordinate graph that is also referred to as the *complex plane*. By convention, the X axis coordinates represent the real number portion of the complex number and the Y axis represents the imaginary portion. For example, the complex number $6 - j7$ would have the same location as the point $(6,-7)$ on a rectangular-coordinate graph. **Figure 4-3** shows the same points as Figure 4-1, but now they are representing the complex numbers $3 + j5$ and $-2 - j4$, respectively.

Before you go on, study test question E5C11. Review this section if you have difficulty.

4.2 Electrical Principles

This section of the book covers the fundamental physical processes of how electrical circuits work. We'll get into the relationship between voltage and current when inductance and capacitance are involved — that's when things get interesting! Understanding these basic ideas leads you directly to resonance, tuned circuits, Q, and all sorts of great radio know-how. For good measure, we'll finish with a look at some neat behavior of electric and magnetic fields.

Remember that this book doesn't attempt to be an electronic textbook — there are plenty of good references available for that job. If you find yourself missing some crucial background, step back and read through one of the references on the ECLM website.

ELECTRIC AND MAGNETIC FIELDS

Electrical and magnetic energy are invisible — you can't detect them with any of your senses. All you can do is observe their effects such as when a resistor gets hot, a motor spins, or an electromagnet picks up iron or steel. The energy exists as a *field* — a region of space in which energy is stored and through which electrical and magnetic forces act. The metric system's basic unit for electrical and magnetic energy is the *joule* (pronounced with a long u sound, similar to jewel) and abbreviated J. (For serious inquiries as to the nature of fields, see **en.wikipedia.org/wiki/Electric_field** and **en.wikipedia.org/wiki/Magnetic_field**.)

You are already quite familiar with fields in the form of gravity. You are being pulled down toward the Earth as you read this (even you astronaut-hams up there studying for your Extra exam) because you are in the Earth's *gravitational field*. Because your body has mass it interacts with the gravitational field in such a way that the Earth attracts you. (You have your own gravitational field, too, but many orders of magnitude smaller than that of the Earth.) Think of a bathroom scale as a "gravitational voltmeter" that instead of reading "volts," reads "pounds." The heavier something is, the stronger the Earth is attracting it. Weight is the same as force. (Metric scales provide readings in kilograms, a unit of mass. To do so, the scales assume a standard strength for gravity in order to convert weight (a force) to an equivalent mass in kilograms.)

This field makes you do work, such as when you climb stairs. Work has a precise definition when it comes to fields: Work equals force times distance moved in the direction of the field's force. For example, let's say you pick up a mass — a stone that weighs 1 pound — and lift it to a shelf 10 feet above where it previously lay. How much work did you do on the stone? You moved a weight of 1 pound a distance of 10 feet against the attraction of the field, so you have done 10 foot-pounds of work. (It doesn't count if you move the stone sideways instead of vertically.)

What did that work accomplish? You stored gravitational energy in the stone equal to the amount of work that you performed! This stored energy is called *potential energy*, whether gravitational, electrical or magnetic. You could store the same amount of gravitational energy by lifting a 10-pound stone 1 foot or by lifting a stone that weighs 1/10th of a pound 100 feet. If you drop the stone (or push it off the shelf), the potential energy is converted back to *kinetic energy* as the stone moves toward the Earth in the gravitational field.

In electronics we are interested in two types of fields: *electric fields* and *magnetic fields*. Electric fields can be detected as voltage differences between two points. The electric field's analog to gravitational mass is electric charge. Every electric charge has its own electric field, just as every mass has its own gravitational field. The more charged a body is, the "heavier" it is in terms of an electric field. Just as a body with mass feels a force to move in a gravitational field, so does an electric charge in an electric field. Electrical energy is stored by moving electrical charges apart so that there is a voltage between them. If the field does not change with time, it is called an *electrostatic field*.

Magnetic energy is detected by its effects on moving electrical charges or current. The atoms of magnetic materials, such as iron or nickel, have special arrangements of moving electrons (the electric charges) that allow them to interact with magnetic fields. The magnetic analog of gravitational mass is *magnetic moment*, a quantity related to the motion of electric charge. Magnetic energy is stored by moving electric charges, such as current creating a magnetic field. [E5D05] Magnetic fields that don't change with time, such as from a stationary permanent magnet, are called *magnetostatic fields*.

In both cases, potential energy (measured in joules) is stored by moving electric charges in an electric or magnetic field. [E5D04, E5D08] The potential energy is released by allowing the charges to move in response to the field. For example, electric energy is released when a current flows from a charged-up capacitor. Magnetic energy stored by current flowing in an inductor is released when a magnetic material in the field is allowed to move, such as a relay's armature does when the coil is energized.

Before you go on, study test questions E5D04, E5D05 and E5D08. Review this section if you have difficulty.

RC AND RL TIME CONSTANTS

In electronic circuits, electric and magnetic energy are constantly being stored and released by various components. This exchange is observed as the relationship between voltage and current in components that store electrical and magnetic energy.

Electrical Energy Storage

Electrical energy can be stored in a capacitor by applying a dc voltage across its terminals. **Figure 4-4** illustrates a simple circuit for charging and discharging a capacitor. Assuming no energy is stored in the capacitor, there will be an instantaneous inrush of current as charge moves into one capacitor terminal and out of the other when S1 is moved to the A position. The only limit on current is the resistance of the voltage source, the wires connecting the capacitor and the capacitor's internal conducting electrodes (sometimes called "plates"). The capacitor builds up a voltage (indicating electric field strength) as one set of electrodes accumulates an excess of electrons and the other set loses an equal number. Eventually, the voltage at the capacitor terminals is equal to the source voltage and the current ceases to flow.

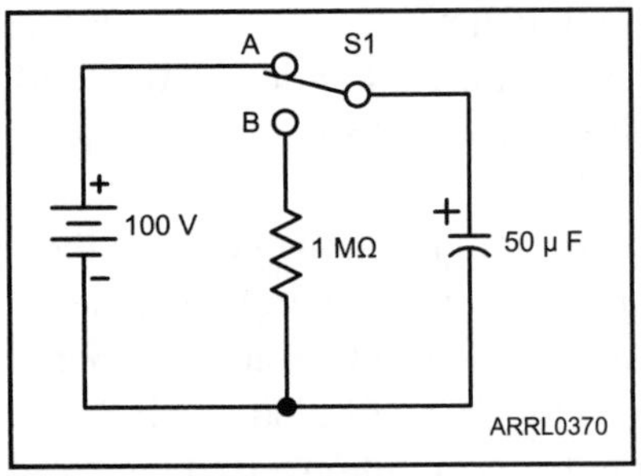

If the voltage source is disconnected, the capacitor will remain charged at that voltage. The charge will stay on the capacitor electrodes as long as there is no path for the electrons to travel from one terminal to the other. The voltage across the capacitor is an indicator of how much electrical energy is stored in the electric field inside the capacitor. Since the charges (electrons) are no longer moving, the energy is stored in the electrostatic field between the electrodes. [E5D03] If a resistor is then connected across the capacitor terminals by moving S1 to the B position, the stored energy will be released as current flows through the resistor, dissipating the stored energy as heat.

Figure 4-4 — A simple circuit for charging a capacitor (S1 in A position) and then discharging it through a resistor (S1 in B position).

Note that the discussion in this section deals with ideal components, such as resistors that have no stray capacitance or inductance associated with the leads or composition of the resistor itself. Ideal capacitors exhibit no losses and there is no resistance in the leads or capacitor electrodes. Ideal inductors are made of wire that has no resistance and there is no stray capacitance between turns. In practice there are no ideal components, so the behavior described here is modified a bit in real-life circuits. Even so, components can come pretty close to the ideal conditions. For example, a capacitor with very low leakage can hold a charge for days or even weeks.

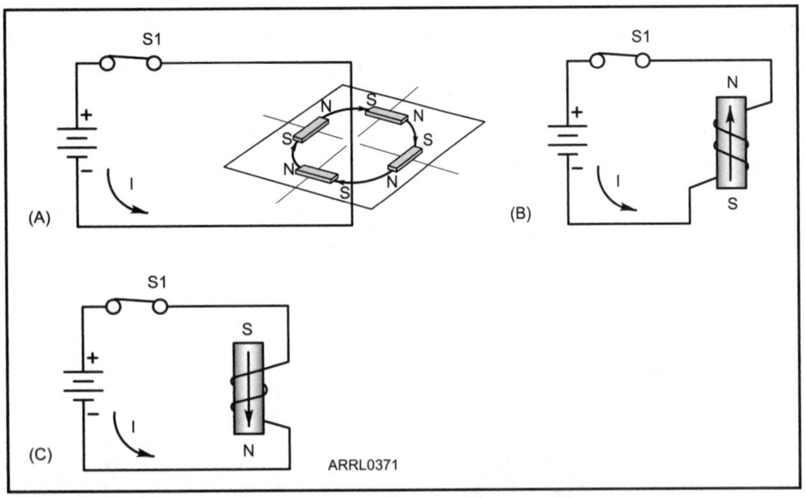

Figure 4-5 — A circuit showing the direction of a magnetic field around a straight wire (A) and in two coils wound in opposite directions (B,C). Note that electronic current (negative to positive) is used, not conventional current (positive to negative).

Magnetic Energy Storage

When electrons flow through a conductor a magnetic field is produced. This magnetic field exists in the space around the conductor, and magnetic energy is stored in this space. The field can be detected by bringing a compass near a wire carrying dc current and watching the needle deflect. **Figure 4-5** illustrates the magnetic field around a wire connected to a battery. If the wire is wound into a coil, so the fields from adjacent turns add together, then a much stronger magnetic field can be produced. The direction of the magnetic field is wrapped around the current at right angles to electronic current flow and can be determined by the *left-hand rule*. [E5D06] (This is shown at **www.allaboutcircuits.com/vol_1/chpt_14/2.html**.)

The use of the left-hand rule is required because we are referring to *electronic current* (the flow of electrons) and not to the more common *conventional current* (the flow of positive charge) that flows in the opposite direction. If electronic current is being discussed, use the left-hand rule and if conventional current, use the *right-hand rule*. The exam question refers

How can you tell what direction a magnetic field is "pointing" or do they even point? Yes, they certainly do have a direction — think of the north and south poles of a magnet! For a permanent magnet, determining direction is pretty easy. All you have to do is observe the way the magnet wants to move when placed near another permanent magnet. Opposite poles (north-south and south-north) will attract each other.

In the case of magnetic fields created by current flowing, it's not so easy. Figure 4-5 shows how a magnetic field is oriented based on the direction of the electrons flowing in the circuit. (The figure shows *electronic current* — the true motion of the electrons from negative to positive. Most electronics uses *conventional current* that flows from positive to negative, exactly opposite electronic current.) In order to find the direction of the field, the *left-hand rule* is used as described at **www.studyphysics. ca/30/Solenoidelectro.pdf**. You just have to know how the inductor is wound and in which direction the current is flowing.

to the flow of electrons and electronic current so use the left-hand rule.

The strength of the magnetic field depends on the amount of current and is stronger when the current is larger. [E5D07] Electrical energy from the voltage source is transferred to the magnetic field in the process of creating the field. So we are storing energy by building up a magnetic field and that means work must be done against some opposing force. That opposing force is created by a voltage that is *induced* (created) in the circuit whenever the magnetic field (or current) is changing.

When you first connect a dc source to a coil of wire, a current begins to flow and a magnetic field begins to build up. The field is changing very rapidly during that time so a large opposing voltage is created, preventing a large current from flowing. As more energy is stored in the field its rate of change drops and so does the opposing voltage. This allows the current to increase to a maximum value limited only by the resistance of the wire in the coil and by any internal resistance of the voltage source. If the current remains constant, then the magnetic field strength is also constant and no more energy is stored.

If the current decreases, then the voltage induced in the wire will act in reverse to oppose the decrease as the stored energy is returned to the circuit. As the magnetic field decreases, the effect of the induced voltage is to keep current flowing.

This induced voltage, also called *induced electromotive force* is sometimes called a *back EMF*, since it acts to oppose any *change* in the amount of current. The faster current tries to change, the higher the back EMF. When the switch in the circuit of Figure 4-5B is first closed, the back EMF will prevent a sudden surge of current through the inductor. Notice that this is complementary to the conditions of a charging capacitor. Likewise, if you open the switch to break the circuit, back EMF will be produced in the opposite direction. In this case the back EMF tries to keep the current flowing, again preventing any sudden change in the magnetic field strength.

Time Constant

If you connect a dc voltage source directly to a capacitor through a resistance, the higher the resistance's value, the longer it will take to charge the capacitor because the resistor limits current flow in the circuit. **Figure 4-6** shows an *RC circuit* (RC means resistor-capacitor) that alternately charges and discharges a capacitor (C) through a resistor (R). With the switch in position A, current through the resistor charges the capacitor to the battery voltage. When the switch is moved to position B, the capacitor returns its stored energy to the circuit as a current through the resistor. The amount of time it takes to charge the capacitor to a certain percentage of the applied voltage or discharge to a certain percentage of its maximum

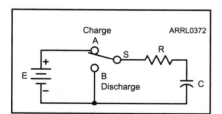

Figure 4-6 — The capacitor in this circuit charges and discharges through the resistor, illustrating the principle of an RC circuit's time constant.

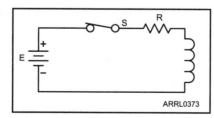

Figure 4-7 — The inductor current flowing through the resistor increases and decreases as voltage is applied, illustrating the principle of an RL circuit's time constant.

voltage is called the circuit's *time constant*.

The same general process applies to an *RL circuit* (RL means resistor-inductor) with an inductor in series with a resistor as in **Figure 4-7**. Instead of voltage increasing toward some maximum value when voltage is applied, it is the current through the inductor that increases instead. The time constant for this type of circuit is the amount of time it takes inductor current to increase or decrease to a specific percentage of its maximum or minimum value.

RC Circuit Time Constant Calculations

The time constant for a simple RC circuit as shown in Figure 4-6 is:

$$\tau = RC \qquad \text{(Equation 4-1)}$$

where:

τ is the Greek letter tau, used to represent the time constant.
R is the total circuit resistance in ohms.
C is the capacitance in farads.

Note that if R is in megohms and C is in microfarads, then τ is in seconds! Remember "megohms times microfarads equals seconds" and it will save you a lot of calculating time.

The capacitor charges and discharges according to an equation known as an *exponential curve*. **Figure 4-8** illustrates the charge and discharge curves, where the time axis is shown in terms of τ and the vertical axis is expressed as a percentage of the applied voltage. These graphs are true for any RC circuit.

If you know the time constant and the applied voltage, you can calculate the voltage on the capacitor at any instant of time. For a charging capacitor:

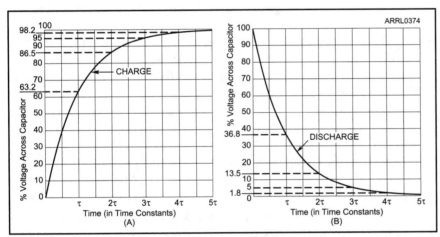

Figure 4-8 — The graph at A shows how the voltage across a capacitor rises with time when charged through a resistor as in Figure 4-4. Graph B shows how the voltage decreases with time as the capacitor is discharged through a resistor. From a practical standpoint, a capacitor is considered to be charged or discharged after a period of 5τ.

$$V(t) = E\,(1 - e^{-t/\tau}) \qquad \text{(Equation 4-2)}$$

where:

V(t) is the voltage across the capacitor at time t.
E is the applied voltage (the battery voltage in Figure 4-4)
t is the time in seconds since the capacitor began charging or discharging.
e is the base for natural logarithms, 2.718.
τ is the time constant for the circuit, in seconds.

If the capacitor is discharging, we have to write a slightly different equation:

$$V(t) = E\,(e^{-t/\tau}) \qquad \text{(Equation 4-3)}$$

These equations can be solved fairly easily with an inexpensive calculator that is able to work with natural logarithms (it will have a key labeled LN or LN X). In that case you could calculate the value for $e^{-t/\tau}$ as the inverse natural log of $-t / \tau$, written as $\ln^{-1}(-t / \tau)$.

(For more information on natural logarithms and exponential functions, refer to the math supplement on the ECLM website.)

Actually, you do not have to know how to solve these equations if you are familiar with the results at a few important points. We'll show you how to use the solutions to the equations at these points as short cuts to most problems associated with time constants.

As shown on the graphs of Figure 4-8, it is common practice to think of charge or discharge time in terms of multiples of the circuit's time constant. If we select times of zero (starting time), one time constant (1τ), two time constants (2τ), and so on, then the exponential term in Equations 4-2 and 4-3 simplifies to e^0, e^{-1}, e^{-2}, e^{-3} and so forth. Then we can solve the equations for those values of time. Let's pick a value for battery voltage of E = 100 V so that the answers will be in the form of a percentage of any applied voltage.

V(0)	=	100 V $(1 - e^0)$	=	100 V $(1 - 1)$	=	0 V,	or	0%
V(1τ)	=	100 V $(1 - e^{-1})$	=	100 V $(1 - 0.368)$	=	63.2 V,	or	63.2%
V(2τ)	=	100 V $(1 - e^{-2})$	=	100 V $(1 - 0.135)$	=	86.5 V,	or	86.5%
V(3τ)	=	100 V $(1 - e^{-3})$	=	100 V $(1 - 0.050)$	=	95.0 V,	or	95%
V(4τ)	=	100 V $(1 - e^{-4})$	=	100 V $(1 - 0.018)$	=	98.2 V,	or	98.2%
V(5τ)	=	100 V $(1 - e^{-5})$	=	100 V $(1 - 0.007)$	=	99.3 V,	or	99.3%

After a time equal to five time constants has passed, the capacitor is charged to 99.3% of the applied voltage. This is fully charged for all practical purposes.

The equation used to calculate the capacitor voltage while it is discharging is slightly different from the one for charging. The exponential term is not subtracted from 1 in Equation 4-3, as it is in Equation 4-2. For values of time equal to multiples of the circuit time constant, the solutions to Equation 4-3 have a close relationship to those for Equation 4-2.

t = 0,	e^0	=	1,	so V(0)	=	100 V,	or	100%
t = 1τ,	e^{-1}	=	0.368,	so V(1τ)	=	36.8 V,	or	36.8%
t = 2τ,	e^{-2}	=	0.135,	so V(2τ)	=	13.5 V,	or	13.5%
t = 3τ,	e^{-3}	=	0.050,	so V(3τ)	=	5 V,	or	5%
t = 4τ,	e^{-4}	=	0.018,	so V(4τ)	=	1.8 V,	or	1.8%
t = 5τ,	e^{-5}	=	0.007,	so V(5τ)	=	0.7 V,	or	0.7%

Here we see that after a time equal to five time constants has passed, the capacitor has discharged to less than 1% of its initial value. This is fully discharged for all practical purposes.

From the calculations for a charging capacitor we can define the time constant of an RC circuit as the time it takes to charge the capacitor to 63.2% of the supply voltage. [E5B01] From the calculations of a discharging capacitor we can also define the time constant as the time it takes to discharge the capacitor to 36.8% of its initial voltage. [E5B02]

Another way to think of these results is that the discharge values are the complements of the charging values. Subtract either set of percentages from 100 and you will get the other set. You may also notice another relationship between the discharging values. If you take 36.8% (0.368) as the value for one time constant, then the discharged value is $0.368^2 = 0.135$ after two time constants, $0.368^3 = 0.05$ after three time constants, $0.368^4 = 0.018$ after four time constants and $0.368^5 = 0.007$ after five time constants. [E5B03] You can change these values to percentages, or just remember that you have to multiply the decimal fraction times the applied voltage. If you subtract these decimal values from 1, you will get the values for the charging equation. In either case, by remembering the percentage 63.2% you can generate all of the other percentages without logarithms or exponentials!

In many cases, you will want to know how long it will take a capacitor to charge or discharge to some particular voltage. Probably the easiest way to handle such problems is to first calculate what percentage of the maximum voltage you are charging or discharging to. Then compare that value to the percentages listed for either charging or discharging the capacitor. Often you will be able to approximate the time as some whole number of time constants.

Suppose you have a 0.01-µF capacitor and a 2-MΩ resistor wired in parallel with a battery. The capacitor is charged to 20 V, and then the battery is removed. How long will it take for the capacitor to discharge through the resistor to reach a voltage of 7.36 V? First, calculate the percentage decrease in voltage:

$$\frac{7.36 \text{ V}}{20 \text{ V}} = 0.368 = 36.8\%$$

You should recognize this as the value for the discharge voltage after 1 time constant. Now calculate the time constant for the circuit using Equation 4-1.

$$\tau = RC = (2 \times 10^6 \, \Omega) \times (0.01 \times 10^{-6} \text{ F}) = 0.02 \text{ second}$$

It will take 0.02 second, or 20 milliseconds to discharge the capacitor to 7.36 V. [E5B05]

The circuit shown in Figure 4-6 is a series circuit. It is common to have a circuit with several resistors and capacitors connected either in series or parallel. If the components are wired in series we can still use Equation 4-1, but we must first combine all of the resistors into one equivalent resistor, and all of the capacitors into one equivalent capacitor. Then calculate the time constant using Equation 4-1, as before.

If the components are connected in parallel, there is an added complication when the circuit is charging, but for a discharging circuit you can still calculate a time constant. Again combine all of the resistors and all of the capacitors into equivalent values, and calculate the time constant using Equation 4-1. (If you have forgotten how to combine resistors and capacitors in series and parallel, review the appropriate sections of *The Ham Radio License Manual* or *The ARRL Handbook*.)

RC Circuit Examples

Let's look at an example of calculating the time constant for a circuit like the one in Figure 4-6, using values of 220 µF and 470 kΩ for C and R. To calculate the time constant, τ, multiply the R and C values, in ohms and farads.

$$\tau = RC = (470 \times 10^3 \, \Omega) \times (220 \times 10^{-6} \text{ F}) = 103.4 \text{ seconds}$$

You can calculate the time constant for any RC circuit in this manner.

If you have two 100-µF capacitors and two 470-kΩ resistors, all in series, first combine the resistor values into a single resistance and the capacitor values into a single capacitance.

$$R_T \text{ (series)} = R_1 + R_2 = 470 \text{ k}\Omega + 470 \text{ k}\Omega = 940 \text{ k}\Omega = 940 \times 10^3 \, \Omega$$

$$C_T \text{(series)} = \frac{1}{(1/C_1 + 1/C_2)} = \frac{1}{2/100} = 50 \text{ µF} = 50 \times 10^{-6} \text{ F}$$

Then the time constant is:

$$\tau = RC = (940 \times 10^3 \, \Omega) \times (50 \times 10^{-6} \text{ F}) = 47 \text{ seconds}$$

Suppose you have two 220-µF capacitors and two 1-MΩ resistors all in parallel. Again, first combine the values into a single resistance and a single capacitance.

$$R_T \text{ (parallel)} = \frac{R_1 \times R_2}{R_1 + R_2} = \frac{1 \text{ M}\Omega \times 1 \text{ M}\Omega}{1 \text{ M}\Omega + 1 \text{ M}\Omega} = \frac{1 \text{ G}\Omega}{2 \text{ M}\Omega} = 0.5 \text{ M}\Omega = 5 \times 10^5 \, \Omega$$

$$C_T \text{ (parallel)} = C_1 + C_2 = 220 \text{ μF} + 220 \text{ μF} = 440 \text{ μF} = 440 \times 10^{-6} \text{ F}$$

Then the time constant is:

$$\tau = RC = (5 \times 10^5 \text{ Ω}) \times (440 \times 10^{-6} \text{ F}) = 220 \text{ seconds [E5B04]}$$

Suppose you have a 450-μF capacitor and a 1-MΩ resistor wired in parallel with a power supply. The capacitor is charged to 800 V, and then the power supply is removed. How long will it take for the capacitor to discharge to 294 V? First, calculate the percentage decrease in voltage:

$$\frac{294 \text{ V}}{800 \text{ V}} = 0.368 = 36.8\%$$

This is the value for the discharge voltage after one time constant. Now calculate the time constant for the circuit using Equation 4-1.

$$\tau = RC = (1 \times 10^6 \text{ Ω}) \times (450 \times 10^{-6} \text{ F}) = 450 \text{ seconds [E5B06]}$$

Or you could have recalled "megohms times microfarads equals seconds" and made the calculation that way.

RL Circuit Time Constant Calculations

When resistance and inductance are connected in series there is a situation similar to what happens in an RC circuit. Figure 4-7 shows a circuit for storing magnetic energy in an inductor. When the switch is closed, a current will try to flow immediately. The instantaneous transition from no current to a value limited only by the voltage source and resistance represents a very large change in current and a back EMF is developed by the inductance. The back EMF is proportional to the rate of change of the current and its polarity is opposite to that of the applied voltage, meaning that it will oppose the change in current. The result is that the initial current is very small but increases quickly, gradually approaching the final current value given by Ohm's Law (I = E / R) as the back EMF decreases toward zero.

Figure 4-9 shows how the current through the inductor of Figure 4-7 increases as time passes. At any given instant, the back EMF will be equal to the difference between the voltage drop across the resistor and the battery voltage. You can see that when the switch is closed initially and there is no current, the back EMF is equal to the full battery voltage. Later on, the current will increase to a steady value and there will be no voltage drop across the inductor. The full battery voltage then appears across the resistor and the back EMF is zero. In practice, the current is essentially equal to the final value after 5 time constants. The curve looks just like the one we found for a charging capacitor.

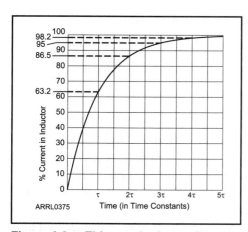

Figure 4-9 — This graph shows the current buildup in an RL circuit. Notice that the curve is identical to the voltage curve for a charging capacitor.

The time constant depends on the circuit components, as for the RC circuit. For an RL circuit, the time constant is given by:

$$\tau = \frac{L}{R} \qquad \text{(Equation 4-4)}$$

The equation for inductor current is another exponential curve, with an equation similar to Equation 4-2.

$$I(t) = \frac{E}{R} \times (1 - e^{-t/\tau})$$

(Equation 4-5)

where:

I(t) is the current in amperes at time t.

E is the applied voltage.

R is the circuit resistance in ohms.

t is the time in seconds after the switch is closed.

τ is the time constant for the circuit in seconds.

If we choose values of time equal to multiples of the circuit time constant, as we did for the RC circuit, then we will find that the current will build up to its maximum value in the same fashion as the voltage does when a capacitor is being charged. This time let's pick a value of 100 A for the maximum current, so that our results will again come out as a percentage of the maximum current for any RL circuit.

t = 0,	e^{-0}	=	1,	so I(0)	=	100 A (1 − 1)	=	0 A,	or 0%
t = 1τ,	e^{-1}	=	0.368,	so I(1τ)	=	100 A (1 − 0.368)	=	63.2 A,	or 63.2%
t = 2τ,	e^{-2}	=	0.135,	so I(2τ)	=	100 A (1 − 0.135)	=	86.5 A,	or 86.5%
t = 3τ,	e^{-3}	=	0.050,	so I(3τ)	=	100 A (1 − 0.050)	=	95.0 A,	or 95%
t = 4τ,	e^{-4}	=	0.018,	so I(4τ)	=	100 A (1 − 0.018)	=	98.2 A,	or 98.2%
t = 5τ,	e^{-5}	=	0.007,	so I(5τ)	=	100 A (1 − 0.007)	=	99.3 A,	or 99.3%

Notice that the current through the inductor will increase to 63.2% of the maximum value during 1 time constant. After 5 time constants, the current is within 1% of the maximum value.

Before you go on, study test questions E5B01 through E5B06, E5D03, E5D06, and E5D07. Review this section if you have difficulty.

PHASE ANGLE

Having learned that voltage and current don't rise and fall together in capacitors and inductors, you're ready to look at the situation when ac voltage is applied, instead of dc. The storage and release of electrical and magnetic energy is what creates the opposition to flow of ac current, the property of capacitors and inductors called *reactance*. (You learned about reactance when studying for your General class exam.)

To understand how ac voltage and current behave in inductors and capacitors, we'll take a look at the amplitudes of ac signals throughout the complete cycle. The relationship between the current and voltage waveforms at a specific instant is expressed as the *phase* of the waveforms. Phase essentially means time or the time interval between two events taking place. The event that occurs first is said to *lead* the second, while the trailing event *lags* the first.

Since all ac cycles of the same frequency have the same *period* (or length in time), we can use the cycle's period as a basic unit of time. This makes phase measurement independent of the waveform frequency, relating only to position relative to the waveform's cycle. If two or more different frequencies are being considered, phase measurements are usually made with respect to the lowest frequency.

It is convenient to relate one complete cycle of the wave to a circle, and to divide the cycle into 360 degrees. **Figure 4-10** shows one com-

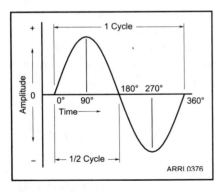

Figure 4-10 — An ac cycle is divided into 360° that are used as a measure of time or phase. Each degree corresponds to 1/360 of the cycle's period.

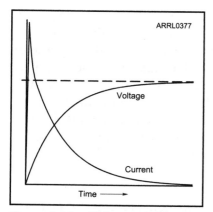

Figure 4-11 — This graph illustrates how the voltage across a capacitor changes as it charges with a dc voltage applied. The charging current is also shown.

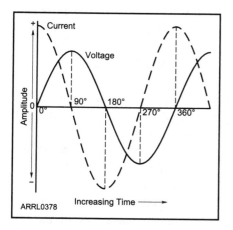

Figure 4-12 — Voltage and current phase relationships when an ac voltage is applied to a capacitor.

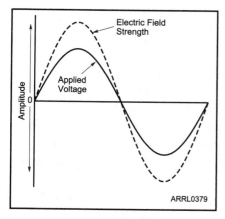

Figure 4-13 — The applied voltage, electric field strength, and stored energy in a capacitor are in phase.

plete cycle of a sine-wave voltage or current, with the wave broken into four quarters of 90° each. Each degree corresponds to 1/360 of the cycle's period. So a phase measurement is usually specified as an angle. Because we know the period of the waveform, we can convert degrees to time.

The *phase angle* between two waveforms is a measurement of the offset in time between similar points on each waveform — maximum-to-maximum, zero-crossing-to-zero-crossing, and so on. One of the waveforms is designated as the reference. A leading phase angle is positive and a lagging phase angle is negative.

AC Voltage-Current Relationship in Capacitors

Figure 4-11 shows the voltage across a capacitor as it charges and the charging current that flows into a capacitor with a dc voltage applied. As soon as a voltage is applied across an ideal capacitor, there is a sudden inrush of current as the capacitor begins to charge. That current tapers off as the capacitor is charged to the full value of applied voltage. By the time the applied voltage is reaching a maximum, the capacitor is also reaching full charge, and so the current into the capacitor goes to zero. A maximum amount of energy has been stored in the electric field of the capacitor at this point.

The situation is different when an ac voltage is applied because the applied voltage is not constant. **Figure 4-12** graphs the relative current and voltage amplitudes when an ac sine wave signal is applied. The scale does not represent specific current or voltage values. Here's what the graph of the two waveforms is telling us during intervals of one-quarter cycle of the voltage waveform:

0° to 90° — Voltage is zero, so no energy is stored in the capacitor. The applied voltage begins increasing and a large inrush of charging current occurs, just as is the case for an applied dc voltage. Current slows as more energy is stored in the capacitor.

90° to 180° — Applied voltage has reached a peak, so no additional charge flows into the capacitor and current flow stops — stored energy is at a maximum. As voltage begins to drop, that is the same as discharging the capacitor, so current reverses and energy is returned to the circuit.

180° to 270° — As the voltage reaches zero it is dropping at its fastest rate, so the discharge current in the reverse direction is at a maximum. Now the applied voltage is increasing again but with the opposite polarity. Energy is being stored in the capacitor again but with the voltage reversed. Charging current is now in the opposite direction, too, but decreases as more energy is stored in the capacitor.

270° to 360° — Once again, applied voltage has reached a peak but with reverse polarity. Charging current ceases as the voltage peaks and begins to drop, repeating the situation between 180° and 270° but with the opposite polarity. When 360° arrives, voltage and current have the same relationship as at 0° and the cycle begins again.

Note that energy is stored in and discharged from the capacitor *twice* during each cycle — once with positive voltage across the capacitor and once with negative voltage. The current waveform describes electrons flowing in and out of the capacitor in response to the applied voltage. Energy storage is at a peak when voltage is maximum as shown in **Figure 4-13**. This occurs at 90° and 270° when current is zero. Note also that current reaches a peak 90° ahead of the voltage waveform. We say that the current through a capacitor *leads* the applied voltage by 90°. [E5B09] You could also say that the voltage applied to a capacitor *lags* the current through it by 90°. To help you remember this relationship, think of the word ICE. This will remind you that the current (I) comes before (leads) the voltage (E) in a capacitor (C). By convention, voltage is the reference waveform for phase angle so in a capacitor the phase angle is –90° (negative).

AC Voltage-Current Relationship in Inductors

The relationship between ac voltage and current in an inductor complements that in a capacitor. **Figure 4-14** shows that instead of stored energy being in phase with applied voltage, it is in phase with the inductor current. This causes the phase relationship between voltage and current to be reversed from that of the capacitor.

In the section on magnetic energy, you learned about back EMF. Back EMF or induced voltage is greatest when the magnetic field is changing the fastest. Furthermore, it is generated with a polarity that opposes the change in current or magnetic-field strength. So when the current is crossing zero on the way to a positive peak, back EMF is at its greatest negative value. When the current is at the positive peak back EMF is zero and so on.

As before, let's examine the situation during each quarter cycle of the applied voltage waveform as shown in **Figure 4-15**. Along with applied voltage and inductor current, a new waveform has been added — *induced voltage* or back EMF. This will help explain the relationship between applied voltage and inductor current.

0° to 90° — Beginning at maximum applied voltage, the opposing induced voltage that resists changes in current flow is also at a maximum so current must increase slowly. As applied voltage falls, the change in current is also reduced and so induced voltage also decreases. As applied voltage reaches zero no additional current flows and induced voltage is zero. Stored energy is a maximum at this point.

90° to 180° — Applied voltage begins to increase in the reverse direction causing a reduction in current and stored energy. Induced voltage increases opposing the change in current. When applied voltage reaches a maximum with reverse polarity, current is now completely stopped and stored energy is zero.

180° to 270° — Reversed from the situation between 0° and 90°, current is now increasing in the opposite direction. Applied voltage is falling and so the rate of change of current is also falling, causing induced voltage to fall as well. As applied voltage reaches zero again, current and stored energy has reached a maximum.

270° to 360° — As between 90° and 180°, applied voltage is increasing in the opposite polarity to current, causing current to drop. The change in current also causes induced voltage to rise in opposition to the change in current. As applied voltage reaches a maximum, current and stored energy once again reach zero.

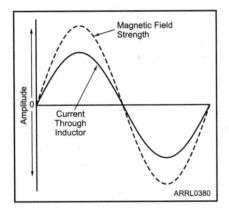

Figure 4-14 — The inductor current, magnetic field strength, and stored energy in an inductor are in phase.

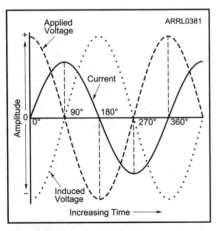

Figure 4-15 — Phase relationships between voltage and current when an alternating voltage is applied to an inductor.

The phase relationship between applied ac voltage and current through an inductor is the opposite from their relationship in a capacitor. Current through an inductor *lags* the applied voltage by 90°. You could also say that the voltage applied to an inductor *leads* the current through it by 90°. [E5B10] A useful mnemonic for remembering these relationships is, "ELI the ICE man." The L and C represent the inductor and capacitor, and the E and I stand for voltage and current. Right away you can see that E (voltage) comes before (leads) I (current) in an inductor and that I comes before (leads) E in a capacitor. Using the same convention as for a capacitor, the phase angle in an inductor is 90° (positive).

Combining Reactance with Resistance

Up to this point we have studied the phase relationships between voltage and current only in inductors and capacitors. Actual circuits include resistance, either as a separate component or as part of the inductor or capacitor. This affects the phase angle between the voltage and current waveforms. The voltage across a resistor is in phase with the current through it, so if a circuit contains both resistance and reactance from either an inductance or capacitance, the resulting phase angle of current through all of the components will be less than 90°. The exact phase angle depends on the relative amounts of resistance and reactance in the circuit.

Revisiting reactance for a moment, reactance is defined as the opposition to ac current flow through an inductance or capacitance. A resistor opposes any type of current flow — ac or dc. You've just seen that inductors develop a back EMF that opposes changes in current flow which is the same thing as resisting changes in stored energy. Capacitors also resist changes in energy by opposing changes in voltage across them that would cause current to flow. This opposition to ac current flow is reactance. To combine reactances in series and parallel, use the same equations as when combining resistances.

Inductive reactance (X_L) increases with increasing frequency because as frequency goes up, so does the rate of change of the applied voltage and of inductor current. A higher rate of change increases the back EMF and thus the opposition to current flow. Similarly, higher inductance also increases inductive reactance. The equation for X_L is:

$$X_L = 2\pi f L$$

where:
   X_L is reactance in ohms.
   f is frequency in hertz.
   L is inductance in henries.

Capacitive reactance increases with decreasing frequency because the longer cycle period means more current will flow, resulting in more energy change during each cycle. More energy change requires the voltage source to overcome wider swings in capacitor voltage and that has the same effect of opposing current flow. Lower capacitance also increases capacitive reactance. The equation for X_C is:

$$X_C = \frac{1}{2\pi f C}$$

where:
   X_C is reactance in ohms.
   f is frequency in hertz.
   C is capacitance in farads.

Before you go on, study test questions E5B09 and E5B10. Review this section if you have difficulty.

COMPLEX IMPEDANCE

When a circuit contains both resistance and reactance the combined effect of the two is called *impedance*, symbolized by the letter Z. Impedance is a more general term than either resistance or reactance. The term is often used with circuits that have only resistance or reactance.

The reactance and resistance comprising an impedance may be connected in series or in parallel, as shown in **Figure 4-16**. In these circuits, the reactance is shown as a box (X), to indicate that it can be either inductive or capacitive. In the series circuit shown at A, the current is the same through both elements but with different voltages appearing across the resistance and reactance. In the parallel circuit shown at B, the same voltage is applied to both elements but different currents may flow in the two branches.

You can see, then, that the phase relationship between current and voltage for the whole circuit can be anything between zero and ±90°. The phase angle depends on the relative amounts of resistance and reactance in the circuit.

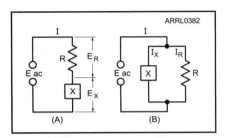

Figure 4-16 — Series and parallel circuits may contain resistance and reactance.

It's important to realize that if there is more than one resistor in the circuit, you must combine them to get one equivalent resistance value. Likewise, if there is more than one reactive element, they must be combined into one equivalent reactance. If there are several inductors and several capacitors, combine all the like elements and calculate the resulting capacitive and inductive reactance.

Capacitive and inductive reactances resist the flow of ac current in different ways and have opposite phase angles so that they cancel each other. This is reflected in the convention of capacitive reactance being treated as a negative value and inductive reactive as a positive value so that adding them together results in a smaller total reactance. If the two opposing reactances have equal values, the resulting cancellation means no reactance is present in the circuit at all!

Combining resistance and reactance is a little more complicated. When the resistance and reactance are in series, the two values can be combined in a relatively straightforward manner. The current is the same in all parts of the circuit ($I = I_R = I_X$), and the voltage is different across each component. We can write an equation for the impedance in the form:

(Equation 4-6)

$$Z = \frac{E}{I} = \frac{E_R + E_X}{I} = \frac{E_R}{I} + \frac{E_X}{I}$$

This is really just Ohm's Law written for impedance instead of resistance, as we are used to seeing it. This equation also shows that we can consider the voltage and current associated with the resistive and reactive elements separately.

Can we take one more step, converting E/I back to R or X and say that the value of Z equals the sum of the R + X values? No! Because the phase angle of the current is different in the resistance (phase angle = 0°) and reactance (phase angle = ±90°). To find the actual impedance of the combined R and X, we need to take phase angle into account. This can be done graphically. Start by drawing the axes for a rectangular coordinate graph as shown in **Figure 4-17**. Resistance values correspond to the X axis and reactance values to the Y axis.

Start by assuming a current of 1 A flowing in the circuit of Figure 4-17A, so the voltage and impedance have the same numeric values of ohms and volts. (If I = 1 A, then numerically Z = E/1.)

Current is the same in both components so use the voltage across the resistor as a reference and draw the voltage across the reactance in the direction of positive reactance on the graph. It is helpful to remember that the reason we label inductive reactance as + and capacitive reactance as – is because of

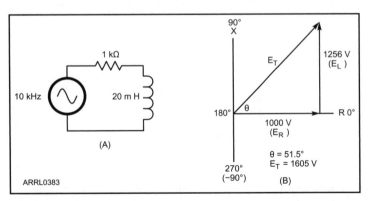

Figure 4-17 — A series RL circuit is shown at A. B shows the right triangle used to calculate the phase angle between the circuit current and voltage. The graph also shows the resulting impedance of the series RL circuit.

this leading and lagging current-voltage relationship.

In the circuit of Figure 4-17, R = 1000 Ω so the voltage across it will be 1000 V. Draw this voltage as a line from the origin along the 0° axis (labeled "R 0°") to the point (1000,0). L = 20 mH and f = 10 kHz so

$$X_L = 6.28 \times (20 \times 10^{-3}\,H) \times (10 \times 10^3\,Hz) = 1256\ \Omega.$$

Draw this voltage as a line pointing straight up in the 90° direction parallel to the reactance axis (labeled "90° X") from the end of the previous line to the point (1000,1256). Because the reactance is inductive, the phase angle between voltage and current is 90° and the voltage line extends upward from the X axis. If the reactance had been capacitive the phase angle would have been –90° and the voltage line would extend down from the X axis.

Complete the figure by drawing a line from the origin to the point (1000,1256). This represents the combination of voltages across 1000 Ω of resistance and 1256 Ω of inductive reactance.

The right triangle you just created represents the solution to the problem. The length of the hypotenuse from the origin to (1000,1256) represents the magnitude of the voltage, $|E_T|$, across the combination of R and L for 1 A of current at 10 kHz. The angle θ is the phase angle between the voltage and the current. Once you know the impedance for the entire circuit, the phase angle for the circuit is the same as the angle of the impedance.

The length of the hypotenuse and the angle can be calculated using trigonometry. (If you are unfamiliar with trigonometry, use the review references listed in the math supplement on the ECLM web page.)

$$|E_T| = \sqrt{1000^2 + 1256^2} = 1605\ V$$

$$\theta = \tan^{-1}\left(\frac{1256}{1000}\right) = 51.5°$$

If frequency or inductance increased, X_L would increase with the result that both E_T and θ would increase. If frequency or inductance decreased, X_L would also decrease, as would E_T and θ.

Completing the impedance calculation, remember that Z = E_T / I. We know I = 1 A and since it is the reference, it can be written as 1∠0° A. We just determined that E_T = 1605 V with a phase angle of 51.5°, so E_T can be written as 1605∠51.5° V. Therefore,

$$Z = \frac{E_T}{I} = \frac{1605\ \angle 51.5°}{1\angle 0°} = 1605\ \angle 51.5°\,\Omega$$

If the reactance in the circuit had been capacitive (negative), the final impedance would have been 1605 ∠–51.5° Ω.

Writing and Graphing Impedance and Phase Angle

The form of impedance written in rectangular form is Z = R ± jX. You can plot complex impedance and the associated phase angle using rectangular coordinates by using the horizontal axis for the value of R and the vertical axis for the value of X. [E5C09, E5C10, E5C13] The j denotes that the reactance value (such as j2 or –j3/2) is a vertical distance along the Y axis.

Impedance can also be written in polar coordinates as $|Z|\angle\theta$, where $|Z|$ is the magnitude of the impedance and θ is its phase angle. Impedances in polar coordinates are plotted with the right side of the horizontal axis indicating 0°, the top half of the vertical axis indicating 90°, and so forth. [E5C14]

Either the rectangular or polar-coordinate system can be used to describe an impedance and the associated phase angle. You might choose to express an impedance value in rectangular coordinates if you want to visualize the resistive and reactive parts. You might choose to express an impedance in polar coordinates if you want to visualize the magnitude and the phase angle of the impedance.

When plotting impedances using complex coordinates, any point that falls on the horizontal axis from 0° to 180° is a *pure resistance* and has no reactive component. [E5C12] Any point that falls on the vertical axis from 90° to –90° (or 270°) is a *pure reactance* and has no resistive component.

Before you go on, study test questions E5C09, E5C10, E5C12, E5C13 and E5C14. Review this section if you have difficulty.

Calculating Impedances and Phase Angles

Let's get some practice working with impedances, admittances, and in determining phase angles in simple circuits by using the following basic rules:

1) Impedances in series add together
2) Admittance is the reciprocal of impedance ($Y = 1/Z$)
3) Admittances in parallel add together
4) Inductive and capacitive reactance in series cancel
5) $1/j = -j$

(See the sidebar, "Working With Polar and Rectangular Coordinates," presented earlier in this chapter for the method of converting complex numbers between rectangular and polar forms.)

As a refresher, remember that *conductivity* is the reciprocal of R ($G = 1/R$) and *susceptance* is the reciprocal of reactance ($B = 1/X$). Both have units of siemens (S). Remember that when taking the reciprocal of an angle, the sign is changed from positive to negative or vice versa.

The following example calculations give the phase angle as a signed value (such as –14 degrees in example 4-18). To apply the examples to the questions on the exam, remember that voltage is the reference for phase angle polarity so that if the phase angle is negative, voltage lags current.

Each of the following examples is found in Subelement E5C of the Extra class question pool:

Example 4-1

Write the impedance $100 - j100 \ \Omega$ in polar form. [E5C15]

Step 1 — $r = \sqrt{100^2 + (100)^2} = 141$

Step 2 — $\theta = \tan^{-1}(-100/100) = -45°$

Step 3 — $Z = 141\angle{-45°} \ \Omega$

Example 4-2

Convert the admittance $7.09\angle45°$ mS (millisiemens) to impedance in polar form. [E5C16]

Step 1 — Use rule 2 to find:

$$|Z| = 1 / 0.00709 = 141 \ \Omega$$

$$\theta = 0° - 45° = -45°$$

Step 2 — $Z = 141\angle-45° \ \Omega$

Example 4-3

Convert the admittance $5\angle-30°$ mS to impedance in rectangular form. [E5C17]

Step 1 — Use rule 2 to find:

$$|Z| = 1 / 0.005 = 200 \ \Omega$$

$$\theta = 0° - (-30°) = 30°$$

Step 2 — $R = |Z| \cos 30° = 173 \ \Omega$

Step 3 — $X = |Z| \sin 30° = 100 \ \Omega$

Step 4 — $Z = 173 + j100 \ \Omega$

Example 4-4

What is the impedance of the circuit in **Figure 4-18** consisting of a 100-Ω resistor in series with an inductor that has 100 Ω of reactance? Give the answer in rectangular form, polar form, and state the phase angle of the circuit. [E5C01]

Step 1 — Use rule 1 to add the resistance and reactance together:

$$Z = 100 + j100 \ \Omega$$

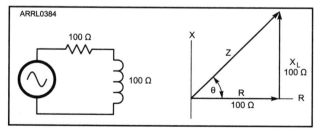

Figure 4-18 — A 100-Ω resistor is connected to an inductor that has a reactance of 100 Ω at some frequency. The impedance triangle shown at B shows the solution, given in the text.

Step 2 — Convert to polar form:

$$r = \sqrt{100^2 + (100)^2} = 141$$

$$\theta = \tan^{-1}(100/100) = 45°$$

$$Z = 141\angle45° \ \Omega$$

Step 3 — The phase angle is equal to the angle of the impedance: $\theta = 45°$

Example 4-5

What is the impedance of the circuit consisting of a 100-Ω resistor in series with a capacitor that has 100 Ω of reactance? Give the answer in rectangular form, polar form, and state the phase angle of the circuit. [E5C06]

Step 1 — Use rule 1 to add the resistance and reactance together:

$$Z = 100 - j100 \ \Omega$$

Step 2 — Convert to polar form:

$$r = \sqrt{100^2 + (100)^2} = 141$$

$$\theta = \tan^{-1}(-100/100) = -45°$$

$$Z = 141\angle{-45°} \ \Omega$$

Step 3 — The phase angle is equal to the angle of the impedance: $\theta = -45°$

Example 4-6

What is the impedance of the circuit consisting of a 400-Ω resistor in series with an inductor that has 300 Ω of reactance? Give the answer in rectangular form, polar form, and state the phase angle of the circuit. [E5C08]

Step 1 — Use rule 1 to add the resistance and reactance together:

$$Z = 400 + j300 \ \Omega$$

Step 2 — Convert to polar form:

$$r = \sqrt{400^2 + 300^2} = 500$$

$$\theta = \tan^{-1}(300/400) = 37°$$

$$Z = 500\angle37° \ \Omega.$$

Step 3 — The phase angle is equal to the angle of the impedance: $\theta = 37°$

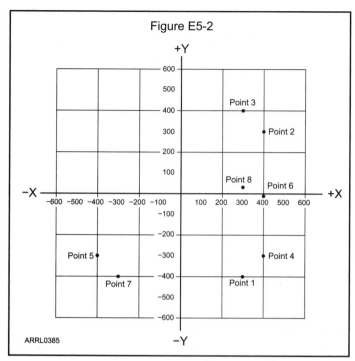

Figure E5-2

Figure 4-19 — This graph, Figure E5-2 from the Extra Class Question Pool, is used for several exam questions.

ARRL0385

Example 4-7

Using the graph in **Figure 4-19**, which point represents the impedance of a circuit consisting of a 300-Ω resistor in series with an 18-µH inductor at 3.505 MHz? [E5C20]

Step 1 — Calculate the inductor's reactance:

$$X_L = 2\pi fL = 400 \ \Omega$$

Step 2 — Use rule 1 to add the resistance and reactance together:

$$Z = 300 + j400 \ \Omega$$

Step 3 — Locate the point on the graph, 300 units along the X (horizontal) axis and +400 units on the Y (vertical) axis. This is Point 3.

Example 4-8

What is the impedance of a circuit consisting of a 40-Ω resistor in series with an 10-µH inductor at 500 MHz? Give the answer in rectangular form. [E5C22]

Step 1 — Calculate the inductor's reactance:

$$X_L = 2\pi fL = 31,400 \ \Omega$$

Step 2 — Use rule 1 to add the resistance and reactance together:

$$Z = 40 + j31,400 \ \Omega$$

Example 4-9

What is the impedance of the circuit consisting of a 300-Ω resistor in series with a capacitor that has 400 Ω of reactance? Give the answer in rectangular form, polar form, and state the phase angle of the circuit. [E5C04]

Step 1 — Use rule 1 to add the resistance and reactance together:

$Z = 300 - j400 \ \Omega$

Step 2 — Convert to polar form:

$$r = \sqrt{300^2 + (-400)^2} = 500$$

$$\theta = \tan^{-1}(-400/300) = -53°$$

$$Z = 500\angle -53° \ \Omega.$$

Step 3 — The phase angle is equal to the angle of the impedance: $\theta = -53°$

Example 4-10

Using the graph in Figure 4-19, which point represents the impedance of a circuit consisting of a 400-Ω resistor in series with a 38-pF capacitor at 14 MHz? [E5C19]

Step 1 — Calculate the capacitor's reactance:

$$X_C = \frac{1}{2\pi fC} = -300 \ \Omega \text{ (capacitive reactance is assigned a negative value)}$$

Step 2 — Use rule 1 to add the resistance and reactance together:

$Z = 400 - j300 \ \Omega$

Step 3 — Locate the point on the graph, 400 units along the X (horizontal) axis and −300 units on the Y (vertical) axis. This is Point 4.

Example 4-11

Using the graph in Figure 4-19, which point represents the impedance of a circuit consisting of a 300-Ω resistor in series with a 19-pF capacitor at 21.200 MHz? [E5C21]

Step 1 — Calculate the capacitor's reactance:

$$X_C = \frac{1}{2\pi fC} = -400 \ \Omega \text{ (capacitive reactance is assigned a negative value)}$$

Step 2 — Use rule 1 to add the resistance and reactance together:

$Z = 300 - j400 \ \Omega$

Step 3 — Locate the point on the graph, 300 units along the X (horizontal) axis and −400 units on the Y (vertical) axis. This is Point 1.

Example 4-12

What is the impedance of the circuit consisting of a 4 Ω resistor in series with an inductor with 4 Ω of reactance and a capacitor with 1 Ω of reactance? (This is called an *RLC circuit*.) Give the answer in rectangular form, polar form, and state the phase angle of the circuit. [E5C18]

Step 1 — Use rules 1 and 4 to add the resistance and reactances together:

$$Z = 4 + j4 - j1 = 4 + j3 \ \Omega$$

Step 2 — Convert to polar form

$$r = \sqrt{4^2 + 3^2} = 5$$

$$\theta = \tan^{-1}(3/4) = 37°$$

$$Z = 5\angle 37° \ \Omega$$

Step 3 — The phase angle is equal to the angle of the impedance: $\theta = 37°$

Example 4-13

What is the impedance of the series circuit consisting of a 100-Ω resistor, 100-Ω inductive reactance, and a 100-Ω capacitive reactance? Give the answer in rectangular form, polar form, and state the phase angle of the circuit. [E5C02]

Step 1 — Use rules 1 and 4 to add the resistance and reactances together:

$$Z = 100 + j100 - j100 = 100 \ \Omega$$

Step 2 — Since the result is a pure resistance, the impedance falls on the horizontal R axis and $\theta = 0°$, so $Z = 100\angle 0°$

Step 3 — The phase angle is $\theta = 0°$

Example 4-14

What is the impedance of the series circuit consisting of a 400-Ω resistor, 600-Ω inductive reactance, and a 300-Ω capacitive reactance? Give the answer in rectangular form, polar form, and state the phase angle of the circuit. [E5C03]

Step 1 — Use rules 1 and 4 to add the resistance and reactances together:

$Z = 400 + j600 - j300 = 400 + j300 \ \Omega$

Step 2 — Convert to polar form:

$r = \sqrt{400^2 + 300^2} = 500$

$\theta = \tan^{-1}(300/400) = 37°$

$Z = 500\angle 37° \ \Omega$

Step 3 — The phase angle is equal to the angle of the impedance: $\theta = 37°$

Example 4-15

What is the impedance of the circuit consisting of a 300-Ω resistor in parallel with an inductor that has 400 Ω of reactance? Give the answer in rectangular form, polar form, and state the phase angle of the circuit. [E5C05]

In Step 1, B_L is negative because $1/jX_L = -j(1/X_L)$. That means Y will have a negative angle in an inductive circuit. Taking the reciprocal of Y causes the angle to be inverted and so the angle of $Z = 1/Y$ in an inductive circuit will be positive.

Step 1 — Use rules 2 and 5 to convert the impedances to admittances:

$G = 1/R = 1/300 = 0.003 \ S$

$B_L = 1/X_L = 1/j400 = -j0.0025 \ S$

Step 2 — Use rule 3 to add the admittances together:

$Y = 0.0033 - j0.0025 \ S$

Step 3 — Write the total admittance in polar form:

$Y = 0.00414\angle -37° \ S$

Step 4 — Use rule 2 to convert the admittance back to impedance:

$Z = 1/Y = (1/.00414) \ \angle(0° - (-37°)) = 240\angle 37° \ \Omega$

Step 5 — The phase angle is equal to the angle of the impedance: $\theta = 37°$

Example 4-16

What is the impedance of the circuit in **Figure 4-20** consisting of a 100-Ω resistor in parallel with a capacitor that has $-j100$ Ω of reactance? Give the answer in rectangular form, polar form, and state the phase angle of the circuit. [E5C07]

Step 1 — Use rules 2 and 5 to convert the impedances to admittances:

$$G = 1/R = 1/100 = 0.01 \text{ S}$$

$$B_C = 1/X_C = 1/-j100 = j0.01 \text{ S}$$

Step 2 — Use rule 3 to add the admittances together:

$$Y = 0.01 + j0.01 \text{ S}$$

Step 3 — Write the total admittance in polar form:

$$Y = 0.0141\angle 45° \text{ S}$$

Step 4 — Use rule 2 to convert the admittance back to impedance:

$$Z = 1/Y = (1/0.0141) \angle(0°-45°) = 71\angle-45° \text{ Ω}$$

Step 5 — The phase angle is equal to the angle of the impedance: $\theta = -45°$

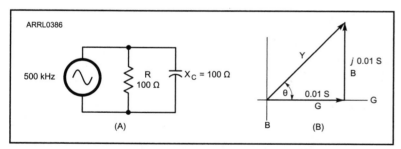

Figure 4-20 — Part A shows a 100-Ω resistor connected in parallel with a capacitor that has 100 Ω of reactance (X_c). Part B shows an "admittance triangle" to help you visualize the solution described in the text.

Example 4-17

Using the graph in Figure 4-19, which point represents the impedance of a circuit consisting of a 300-Ω resistor in series with a 0.64-μH inductor and an 85-pF capacitor at 24.900 MHz? [E5C23]

Step 1 — Calculate the capacitor's reactance:

$$X_C = \frac{1}{2\pi f C} = -75 \ \Omega$$

Step 2 — Calculate the inductor's reactance:

$$X_L = 2\pi f L = 100 \ \Omega$$

Step 3 — Use rules 1 and 4 to add the resistance and reactances together:

$$Z = 300 + -j75 + j100 = 300 + j25 \ \Omega$$

Step 4 — Locate the point on the graph, 300 units along the X (horizontal) axis and +25 units on the Y (vertical) axis. This is Point 8.

Example 4-18

What is the phase angle between voltage and current in a series RLC circuit if X_C is 500 Ω, R is 1 kΩ, and X_L is 250 Ω? [E5B07]

Step 1 — Use rules 1 and 4 to add the resistance and reactances together:

$$Z = 1000 + j250 - j500 = 1000 - j250 \ \Omega$$

Step 2 — Convert Z to polar form:

$$r = \sqrt{(1000)^2 + (-250)^2} = 1030.77$$

$$\theta = \tan^{-1}(-250/1000) = -14°$$

$$Z = 1031 \angle -14° \ \Omega$$

Step 3 — The phase angle is equal to the angle of the impedance: $\theta = -14°$. Since phase angle is from voltage to current, the negative angle indicates that voltage lags the current.

Example 4-19

What is the phase angle between voltage and current in a series RLC circuit if X_C is 100 Ω, R is 100 Ω, and X_L is 75 Ω? [E5B08]

Step 1 — Use rules 1 and 4 to add the resistance and reactances together:

$$Z = 100 + j75 - j100 = 100 - j25 \ \Omega$$

Step 2 — Convert Z to polar form:

$$r = \sqrt{100^2 + (-25)^2} = 103$$

$$\theta = \tan^{-1}(-25/100) = -14°$$

$$Z = 103\angle -14° \ \Omega$$

Step 3 — The phase angle is equal to the angle of the impedance: $\theta = -14°$. Since phase angle is from voltage to current, the negative angle indicates that voltage lags the current.

Compare this to the previous example and note that even though impedance changed by a factor of 10, phase angle was unchanged because the relative amounts of resistance and reactance are the same in both circuits.

Example 4-20

What is the phase angle between voltage and current in a series RLC circuit if X_C is 25 Ω, R is 100 Ω, and X_L is 50 Ω? [E5B11]

Step 1 — Use rules 1 and 4 to add the resistance and reactances together:

$$Z = 100 + j50 - j25 = 100 + j25 \ \Omega$$

Step 2 — Convert Z to polar form:

$$r = \sqrt{100^2 + (25)^2} = 103$$

$$\theta = \tan^{-1}(25/100) = 14°$$

$$Z = 103\angle 14° \ \Omega$$

Step 3 — The phase angle is equal to the angle of the impedance: $\theta = 14°$. Since phase angle is from voltage to current, the positive angle indicates that voltage leads the current.

Example 4-21

What is the phase angle between voltage and current in a series RLC circuit if X_C is 75 Ω, R is 100 Ω, and X_L is 50 Ω? [E5B12]

Step 1 — Use rules 1 and 4 to add the resistance and reactances together:

$$Z = 100 + j50 - j75 = 100 - j25 \text{ Ω}$$

Step 2 — Convert Z to polar form:

$$r = \sqrt{100^2 + (-25)^2} = 103$$

$$\theta = \tan^{-1}(-25/100) = -14°$$

$$Z = 103\angle{-14°} \text{ Ω}$$

Step 3 — The phase angle is equal to the angle of the impedance: $\theta = -14°$. Since phase angle is from voltage to current, the negative angle indicates that voltage lags the current.

Compare this to Example 4-19 and note that the same phase angle is created by different combinations of resistance and reactance if the relative amounts of resistance and reactance are the same in both circuits.

Example 4-22

What is the phase angle between voltage and current in a series RLC circuit if X_C is 250 Ω, R is 1 kΩ, and X_L is 500 Ω? [E5B13]

Step 1 — Use rules 1 and 4 to add the resistance and reactances together:

$$Z = 1000 + j500 - j250 = 1000 + j250 \text{ Ω}$$

Step 2 — Convert Z to polar form:

$$r = \sqrt{(1000)^2 + (250)^2} = 1031$$

$$\theta = \tan^{-1}(250/1000) = 14°$$

$$Z = 1031\angle{14°} \text{ Ω}$$

Step 3 — The phase angle is equal to the angle of the impedance: $\theta = 14°$. Since phase angle is from voltage to current, the positive angle indicates that voltage leads the current.

As in Example 4-20, the phase angle is unchanged for much higher impedances as long as the same relative amounts of resistance and reactance are present.

Before you go on, study test questions E5B07, E5B08, E5B11, E5B12, E5B13 and E5C01 through E5C08, and E5C15 through E5C23. Review this section if you have difficulty.

REACTIVE POWER AND POWER FACTOR

Power is the rate of doing work or using energy per unit of time. Going back to our example at the beginning of this chapter, if you did 10 foot-pounds of work in 5 seconds, then you provided power of 2 foot-pounds per second. If you generate 550 foot-pounds per second of power, you have generated 1 horsepower. So power is a way to express not only how much work you are doing (or how much energy is being stored); it also tells you how fast you are doing it. In the metric system, power is expressed in terms of the watt (W) — 1 watt means energy is being stored or work being done at the rate of 1 joule per second.

You learned earlier that when current increases through an inductor, energy is stored in the inductor's magnetic field. Energy is stored in the electric field of a capacitor when the voltage across it increases. That energy is returned to the circuit when the current through the inductor decreases or when the voltage across the capacitor decreases.

You also learned that the voltages across and currents in these components are 90° out of phase with each other. In one half of the cycle some energy is stored in the inductor or capacitor, and the same amount of energy returned on the next half cycle. A perfect capacitor or inductor does not dissipate or consume any energy, but current does flow in the circuit when a voltage is applied to it. If no energy is consumed in a perfect capacitor or inductor, then no work is done and no power is consumed.

Definition of Reactive Power

To pass the General class license exam, you learned that electrical power is equal to the RMS values of current multiplied by voltage:

$$P = I E \hspace{6cm} \text{(Equation 4-7)}$$

There are certainly voltage and current present for the inductor and capacitor. Why is no power consumed? There is one catch in Equation 4-7 — it is only true when the current and voltage are in phase such as in a resistor where the phase angle is zero. The larger the phase angle, the smaller the amount of work done by the power source supplying the voltage and current. When the phase angle reaches ±90°, no work is being done at all and so the rate (or power) is equal to zero!

In a circuit's inductive or capacitive reactance, energy may be stored in and returned from the magnetic field in the inductor or the electric field in the capacitor but it will not be consumed as power. Only the resistive part of the circuit consumes and dissipates power as heat. [E5D09]

An ammeter and a voltmeter connected in an ac circuit to measure voltage across and current through an inductor or capacitor will both show non-zero RMS values but multiplying them together does not give the true indication of the power being dissipated in the component. The meters do not account for the phase difference between voltage and current.

If you multiply the RMS values of voltage and current from these meters, you will get a quantity that is referred to as *apparent power* — a clue that multiplying RMS values of voltage and current doesn't always give the true picture! Apparent power is expressed in units of *volt-amperes* (VA) rather than watts. The apparent power in an inductor or capacitor is called *reactive power* or *nonproductive, wattless power*. [E5D14] Reactive power is expressed in volt-amperes-reactive (VAR). The apparent power in a resistor is called *real power* because voltage and current are in phase so that the power is dissipated as heat or causes work to be done.

Definition and Calculation of Power Factor

You can account for reactive power in a circuit by using phase angle to calculate the *power factor*. Power factor (abbreviated PF) relates the apparent power in a circuit to the real power. You can find the real power in a circuit containing resistance from Equations 4-8 and 4-9.

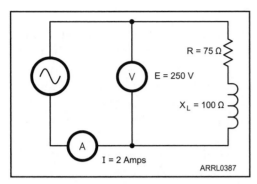

Figure 4-21 — Only the resistance actually dissipates power. The voltmeter and ammeter read the proper RMS value for the circuit, but their product is apparent power, not real average power.

For a series circuit:

$$P = I^2 R \qquad \text{(Equation 4-8)}$$

where I is the RMS current.

For a parallel circuit:

$$P = E^2 / R \qquad \text{(Equation 4-9)}$$

where E is the RMS voltage.

Both of these equations are easily derived by using Ohm's Law to solve for either voltage or current, ($E = I \times R$ and $I = E / R$) and replacing that term with the Ohm's Law equivalent.

One way to calculate power factor is simply to divide the real power by the apparent power:

$$PF = \frac{P_{REAL}}{P_{APPARENT}} \qquad \text{(Equation 4-10)}$$

If PF = 1, then the voltage and current are in phase and all of the apparent power is real power. If PF = 0, then the voltage and current are 90° out of phase and all of the apparent power is reactive power.

Figure 4-21 shows a series circuit containing a 75-Ω resistor and an inductor with an inductive reactance of 100 Ω at the signal frequency. The voltmeter reads 250 V RMS and the ammeter indicates a current of 2 A RMS. This is an apparent power of 250 V × 2 A = 500 VA. Use Equation 4-8 to calculate the power dissipated in the resistor:

$$P_{REAL} = I^2 R = (2 \text{ A})^2 \times 75 \ \Omega = 4 \text{ A}^2 \times 75 \ \Omega = 300 \text{ W}$$

Now by using Equation 4-10, calculate power factor:

$$PF = \frac{300 \text{ W}}{500 \text{ VA}} = 0.6$$

Another way to calculate the real power, if you know power factor, is given by:

$$P_{REAL} = P_{APPARENT} \times PF \qquad \text{(Equation 4-11)}$$

In our example,

$$P_{REAL} = 500 \text{ VA} \times 0.6 = 300 \text{ W [E5D10]}$$

Of course the value found using Equation 4-11 must agree with the value found by either Equation 4-8 or 4-9, depending on whether the circuit is series or parallel.

Phase angle can also be used to calculate power factor and real power. You learned how to calculate the phase angle of either a series or a parallel circuit in the previous section. The power factor can be calculated from the phase angle by taking the cosine of the phase angle:

$$\text{Power factor} = \cos \theta \qquad \text{(Equation 4-12)}$$

where θ is the phase angle between voltage and current in the circuit. PF is positive whether the phase angle is positive or negative

You can see that for a circuit containing only resistance, where the voltage and current are in phase, the power factor is 1, and the real power is equal to the apparent power. For a circuit containing only pure capacitance or pure inductance, the power factor is 0 so there is no real power! For most practical circuits, which contain resistance, inductance and capacitance, and the phase angle is some value greater than or less than 0°, the power factor will be something less than one. In such a circuit, the real power will always be something less than the apparent power. This is an important point to remember.

Let's try some sample problems assuming that you can find the phase angle as described in the previous section. (If you need review of the cosine function, use the math supplement on the *Extra Class License Manual* website, **www.arrl.org/extra-class-license-manual.**)

Example 4-23

What is the power factor for an R-L circuit having a phase angle of 30°? 45°? 60°? Use Equation 4-12 to answer this question: [E5D11, E5D15, E5D16]

PF for phase angle of 30° = cos 30° = 0.866

PF for phase angle of 45° = cos 45° = 0.707

PF for phase angle of 60° = cos 60° = 0.500

Example 4-24

Suppose you have a circuit that draws 4 amperes of current when 100 V ac is applied. The power factor for this circuit is 0.2. What is the real power (how many watts are consumed) for this circuit? [E5D12]

Start by calculating apparent power using Equation 4-7.

$$P_{APPARENT} = 100 \text{ V} \times 4 \text{ A} = 400 \text{ VA}$$

Real power is then found using Equation 4-11:

$$P_{REAL} = 400 \text{ VA} \times 0.2 = 80 \text{ W}$$

Example 4-25

How much power is consumed in a circuit consisting of a 100-Ω resistor in series with a 100-Ω inductive reactance and drawing 1 ampere of current? [E5D13]

Because only the resistance consumes power:

$$P_{REAL} = I^2 R = 1^2 \text{ A}^2 \times 100 \text{ }\Omega = 100 \text{ W}$$

Example 4-26

How many watts are consumed in a circuit having a power factor of 0.6 if the input is 200 V ac at 5 amperes? [E5D17]

First, find apparent power using Equation 4-7:

$$P_{APPARENT} = I \text{ E} = 5 \text{ A} \times 200 \text{ V} = 1000 \text{ VA}$$

Then multiply by the power factor as in Equation 4-11:

$$P_{REAL} = P_{APPARENT} \times PF = 1000 \times 0.6 = 600 \text{ W}$$

Example 4-27

How many watts are consumed in a circuit having a power factor of 0.71 if the apparent power is 500 VA? [E5D18]

Use Equation 4-11 to find P_{REAL}:

$$P_{REAL} = P_{APPARENT} \times PF = 500 \times 0.71 = 355 \text{ W}$$

Before you go on, study test questions E5D09 through E5D18. Review this section if you have difficulty.

RESONANT CIRCUITS

With all of the problems so far, we have used inductor and capacitor values that give different inductive and capacitive reactances. This results in a bigger voltage across one of the series components or a larger current through one of the parallel components. Have you wondered about what happens when both reactances are equal?

In a series circuit with an inductor and a capacitor, voltage leads the current by 90° in the inductor; in the capacitor, voltage lags the current by 90°. Since this is a series circuit the current through all of the components is the same. That means the voltages across the inductor and capacitor are 180° out of phase. Those voltages then cancel, leaving only the voltage across the resistance of the circuit which is in phase with the current.

In a parallel circuit containing inductance and capacitance, voltage is the same across both but it is the currents that are 180° out of phase. The current in the inductor lags the applied voltage by 90° and the current in the capacitor leads by 90°. The cancellation of the current leaves a parallel resistance as the only component in which current can flow and the remaining current is in phase with the voltage.

Whether the components are connected in series or parallel, we say the circuit is *resonant* or is at *resonance* when the inductive reactance value is the same as the capacitive reactance value. [E5A02] Remember that inductive reactance increases as frequency increases and that capacitive reactance decreases as frequency increases. The frequency at which the two are equal is the circuit's *resonant frequency*.

Calculation of Resonant Frequency

Figure 4-22 is a graph of inductive and capacitive reactances with frequency. Neither the exact frequency scale nor the exact reactance scale is important. What is important is that the two lines cross at only one point — the resonant frequency of the circuit using those two components. Every combination of a capacitor and an inductor will be resonant at some frequency.

Since resonance occurs when the reactances are equal, we can derive an equation to calculate the resonant frequency of any capacitor-inductor pair:

$$X_L = 2\pi f L$$

$$X_C = \frac{1}{2\pi f C}$$

Set $X_L = X_C$ at resonance, so:

$$2\pi f L = \frac{1}{2\pi f C} \quad \text{so}$$

$$(2\pi f)(2\pi f) = \frac{1}{LC} \quad \text{and}$$

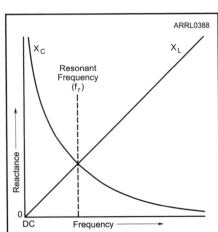

Figure 4-22 — A graph showing the relative change in inductive reactance and capacitive reactance as frequency increases. For any specific inductor-capacitor pair, there is only one frequency at which $X_L = X_C$, the resonant frequency, f_r.

$$4\pi^2 f^2 = \frac{1}{LC}$$

This leads to the formula for resonant frequency:

$$f_r = \frac{1}{2\pi\sqrt{LC}}$$

(Equation 4-13)

Impedance of Resonant Circuits Versus Frequency

Figure 4-23 shows a signal generator connected to a series RLC circuit. The signal generator produces a variable-frequency current through the circuit, which will cause a voltage to appear across each component. As discussed above, the voltage drops across the inductor and capacitor are always 180° out of phase.

When the signal generator produces an output signal at the resonant frequency of the circuit, the voltages across the inductor and capacitor are equal as well as out of phase. This means an equal amount of energy is stored in each component and is transferred between them on alternate half-cycles. If the components have low amounts of resistive loss, the energy continually supplied by the signal generator will cause the voltages across the inductor and capacitor to build to levels several times larger than the voltage applied to the circuit! [E5A01] The current that flows back and forth between the inductor and capacitor to exchange the stored energy is called *circulating current* and it is a maximum at resonance. Circulating currents are sometimes called *tank currents* because the resonant circuit is thought of as a "tank" in which energy is stored.

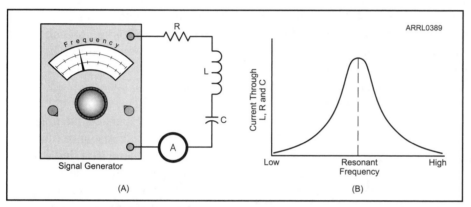

Figure 4-23 — A series-connected LC or RLC circuit presents a minimum value of resistance at the resonant frequency. Therefore, at resonance, the current passing through the circuits reaches a maximum.

In analogy, consider pushing a playground swing. Even though the additional push on each swing is small, if friction is low, the pushes can cause the amplitude of the swing's travel to be much larger than would be caused by any single push. In practical resonant circuits the voltages across the inductor and capacitor are sometimes at least 10 times as large and may be as much as a few hundred times as large as the applied voltage!

With the voltages across the inductor and capacitor canceling each other the only impedance presented to the signal generator is that of the resistance, R. [E5A03] With perfect components and no resistance in the circuit, there would be nothing to restrict the current in the circuit. An ideal series-resonant circuit, then, "looks like" a short circuit to the signal generator. There is always some resistance in a circuit but if the total resistance is small, the current will be large as shown by Ohm's Law. The change in current with frequency is shown in Figure 4-23B. It reaches a maximum at the resonant frequency, f_r. [E5A05]

Let's see how you do with a series-resonant-circuit problem. What frequency should the signal generator in Figure 4-23 be tuned to for resonance if the resistor is 22 Ω, the coil is 50 µH, and the capacitor has a value of 40 pF? Probably the biggest stumbling block of these calculations will be remembering to convert the inductor value to henrys and the capacitor value to farads. After you have done that, use Equation 4-13 to calculate the resonant frequency. [E5A14]

$$50\ \mu H = 50 \times 10^{-6}\ H$$

$$40\ pF = 40 \times 10^{-12}\ F$$

$$f_r = \frac{1}{6.28\sqrt{(50\times10^{-6})(40\times10^{-12})}} = 3.56\times10^6 = 3.56\,\text{MHz}$$

What would happen to f_r if the resistor value was changed to 47 Ω? Nothing! The value of the circuit's resistance does not affect the resonant frequency. This is because the resistor does not store electrical or magnetic energy.

What is the resonant frequency if the resistor is 56 Ω, the inductor is 40 μH, and the capacitor has a value of 200 pF? [E5A15]

$$f_r = \frac{1}{6.28\sqrt{(40\times10^{-6})(200\times10^{-12})}} = 1.78\times10^6 = 1.78\,\text{MHz}$$

In a parallel-resonant circuit there are several current paths, but the same voltage is applied to the components. **Figure 4-24** shows a parallel LC circuit connected to a signal generator. The applied voltage causes current to flow in each of the three circuit branches. At resonance the current through the inductor will be 180° out of phase with the current in the capacitor and again they add up to zero.

As a result, the parallel resonant circuit has a high impedance and can appear to be an open circuit to the signal generator because the current from the signal generator is quite small. At resonance, the magnitude of the impedance of a circuit with a resistor, inductor and capacitor all connected in parallel will be approximately equal to the circuit resistance. [E5A04, E5A07]

Figure 4-24B is a graph of the relative generator current. The current at the input of a parallel RLC circuit is a minimum at resonance. It is a mistake to assume, however, that because the generator current is small the current flowing through the capacitor and inductor is also small. As in a series resonant circuit, the energy being exchanged between the inductor and capacitor as circulating currents can build up to large values. At resonance, the circulating currents will be at maximum, limited only by resistive losses in the components. [E5A06] While the total current from the generator is small at resonance, the voltage measured across the tank reaches a maximum value at resonance.

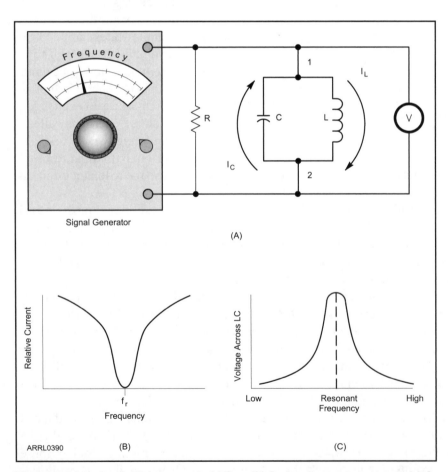

Signal Generator

(A)

Relative Current

f_r

Frequency

(B)

ARRL0390

Voltage Across LC

Low   Resonant   High
       Frequency

(C)

Figure 4-24 — A parallel-connected LC or RLC circuit presents a very high resistance at the resonant frequency. Therefore, at resonance, the voltage across the circuit reaches a maximum.

Figure 4-24C is a graph of the voltage across the inductor and capacitor.

It is also interesting to consider the phase relationship between the voltage across a resonant circuit and the current through that circuit. Because the inductive reactance and the capacitive reactance are equal but opposite their effects cancel each other. The resulting current and voltage in a resonant circuit are in phase. This is true for both a series resonant circuit and a parallel resonant circuit. [E5A08, E5A09]

Calculating the resonant frequency of a parallel circuit is exactly the same as for a series circuit. For example, what is the resonant frequency of a parallel RLC circuit if R is 33 Ω, L is 50 µH and C is 10 pF? [E5A16]

$$f_r = \frac{1}{6.28\sqrt{(50 \times 10^{-6})(10 \times 10^{-12})}} = 7.12 \times 10^6 = 7.12 \, \text{MHz}$$

What is the resonant frequency of a parallel RLC circuit if R is 47 Ω, L is 25 µH and C is 10 pF? [E5A17]

$$f_r = \frac{1}{6.28\sqrt{(25 \times 10^{-6})(10 \times 10^{-12})}} = 10.1 \times 10^6 = 10.1 \, \text{MHz}$$

Using the same technique that we used to derive Equation 4-13, we can easily derive equations to calculate either the inductance or capacitance to resonate with a certain component at a specific frequency:

$$L = \frac{1}{(2\pi f_r)^2 C} = \frac{1}{(2\pi)^2 (f_r)^2 C} \qquad \text{(Equation 4-14)}$$

and

$$C = \frac{1}{(2\pi f_r)^2 L} = \frac{1}{(2\pi)^2 (f_r)^2 L} \qquad \text{(Equation 4-15)}$$

Let's try a couple of practical examples:

What value capacitor is needed to make a circuit that is resonant in the 80 meter band if you have a 20 µH inductor? Choose a frequency in the 80 meter band to work with, such as 3.6 MHz. Then convert to fundamental units: $f_r = 3.6 \times 10^6$ Hz and L = 20 × 10⁻⁶ H. Use Equation 4-15, since that one is written to find capacitance, the quantity we are looking for:

$$C = \frac{1}{(2\pi)^2 (3.6 \times 10^6)^2 (20 \times 10^{-6})}$$

$$C = \frac{1}{(39.48)(1.3 \times 10^{13})(20 \times 10^{-6})}$$

$$C = \frac{1}{1.03 \times 10^{10}} = 9.7 \times 10^{-11} = 97 \times 10^{-12} = 97 \, \text{pF}$$

You can use a 100-pF capacitor. If you try solving this problem for both ends of the 80 meter band, you will find that you need a 103-pF capacitor at 3.5 MHz and a 79-pF unit at 4 MHz. So any capacitor value within this range will resonate in the 80 meter band with the 20-µH inductor.

Before you go on, study test questions E5A01 through E5A09 and E5A14 through E5A17. Review this section if you have difficulty.

Q AND BANDWIDTH OF RESONANT CIRCUITS

We have talked about ideal resistors, capacitors and inductors, and how they behave in ac circuits. We have shown that resistance in a circuit causes some departure from a circuit of ideal components by dissipating some of the stored energy. But how can we determine how close to the ideal a certain component comes? Or how much of an effect it will have on a designed circuit? We can calculate a value for inductors and capacitors that evaluates the relative merits of that component — the *quality factor* called *Q*. We can also assign a Q value to an entire circuit as a measure of how close to the ideal that circuit performs — at least in terms of its properties at resonance.

One definition of Q is the ratio of reactance to resistance. This is, in effect, the ratio of how much energy is stored to how much energy is dissipated. The lower the component's resistive losses, the higher the Q.

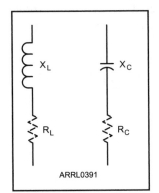

Figure 4-25 — A practical inductor can be considered as an ideal inductor in series with a resistor, and a practical capacitor can be considered as an ideal capacitor in series with a resistor.

Figure 4-25 shows that a capacitor can be thought of as an ideal capacitor in series with a resistor and an inductor can be considered as an ideal inductor in series with a resistor. This parasitic resistance can't actually be separated from the inductor or capacitor, of course, but it acts just the same as if it were in series with an ideal, lossless component. The Q of a real inductor, L, is equal to the inductive reactance divided by the resistance and the Q of a real capacitor, C, is equal to the capacitive reactance divided by the resistance:

$$Q = \frac{X}{R}$$

(Equation 4-16)

If you want to know the Q of a circuit containing both parasitic resistance and actual resistors, both must be added together to find the value of R used in the equation. Since adding a resistor can only raise the total resistance, the Q always goes down when resistance is added in series with an inductor or capacitor. There is no way to raise the Q of an inductor or capacitor except by building a component with less internal resistance.

The internal resistance of a capacitor is usually much less than that for an inductor so we often ignore the resistance of a capacitor and consider only that associated with the inductor when computing Q of a resonant circuit. Stated another way, Q of the inductor is usually the limiting factor on Q of a resonant circuit. **Figure 4-26A** shows a series RLC circuit with a Q of 10. To calculate that value, select either value of reactance and divide it by the resistance.

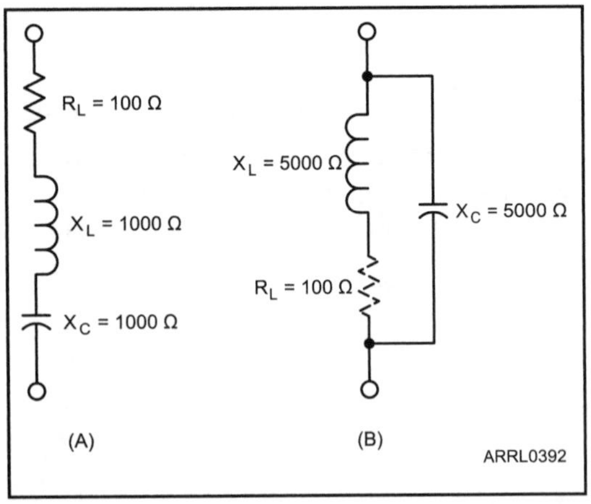

At Figure 4-26B, the frequency has been increased by a factor of 5. The reactance of our inductor has increased 5 times and we have selected a new capacitor to provide a resonant circuit. This time the components are arranged to provide a parallel resonant circuit. The circuit Q is still found using Equation 4-16. Q for this circuit has increased from 10 to 50. Increasing the frequency increases the inductive reactance so if the internal resistance stays the same, the Q increases by the same factor. (See the discussion on Skin Effect below.)

Figure 4-26 — The Q in a series-resonant circuit as shown at A and a parallel-resonant circuit as shown at B is found by dividing the inductive reactance (X_L) by the resistance (R_L).

Resonant-Circuit Bandwidth

Bandwidth refers to the frequency range over which the circuit response in voltage or current is no more than 3 dB below the peak response. The –3 dB points are shown on **Figure 4-27**, and the bandwidths are indicated. (If you are not familiar with the use of decibels,

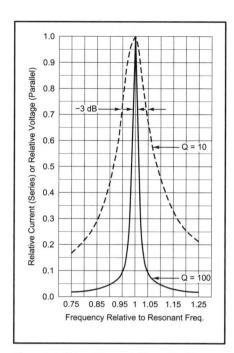

Figure 4-27 — The relative –3 dB bandwidth of two resonant circuits is shown. The circuit with the higher Q has a steeper response, and a narrower bandwidth. The vertical scale represents current for a series circuit and voltage for a parallel circuit.

see the math supplement on the web page for the *Extra Class License Manual*.) Since this 3-dB decrease in signal represents the points where the circuit power is one half of the resonant power, the –3 dB points are also called *half-power points*. At these points, the voltage and current have been reduced to 0.707 times their peak values.

The frequencies at which the half-power points occur are called f1 and f2; Δf is the difference between these two frequencies, and represents the *half-power (or 3-dB) bandwidth*. A circuit with a narrow bandwidth is said to be "sharp" and one with a wider bandwidth "broad." It is possible to calculate the bandwidth of a resonant circuit based on the circuit Q and the resonant frequency:

$$\Delta f = \frac{f_r}{Q} \qquad \text{(Equation 4-17)}$$

where:
Δf = the half-power bandwidth.
f_r = the resonant frequency of the circuit.
Q = the circuit Q

The higher the circuit Q, the sharper the frequency response of a resonant circuit will be, whether it is a series or parallel circuit. [E4B15] Figure 4-27 shows the relative bandwidth of a circuit with two different Q values.

Let's calculate the half-power bandwidth of a parallel circuit that has a resonant frequency of 1.8 MHz and a Q of 95. The half-power bandwidth is found by Equation 4-17: [E5A10]

$$\Delta f = \frac{f_r}{Q} = \frac{1.8 \times 10^6}{95} = 18.9 \times 10^3 \text{ Hz} = 18.9 \text{ kHz}$$

To find the upper and lower half-power frequencies, subtract half the total bandwidth from the center frequency to get the lower half-power frequency and add half the bandwidth to get the upper half-power frequency. The response of this circuit will be at least half of the peak signal power for signals in the range 1.79055 to 1.80945 MHz.

Repeat the calculations for the following combinations of resonant frequency and Q:

f_r = 7.1 MHz and Q = 150: Δf = 47.3 kHz [E5A11]

f_r = 3.7 MHz and Q = 118: Δf = 31.4 kHz [E5A12]

f_r = 14.25 MHz and Q = 187: Δf = 76.2 kHz [E5A13]

Skin Effect and Q

As frequency increases, the electric and magnetic fields of signals do not penetrate as deeply into a conductor such as a wire. At dc, the entire thickness of the wire is used to carry currents. As the frequency increases, the effective area gets smaller and smaller as the current is confined to regions closer and closer to the surface of the wire. [E5D01] This makes the wire less able to carry the electron flow and increases its effective resistance. [E5D02]

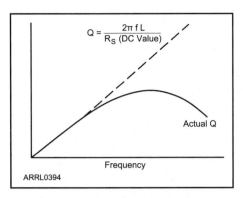

$$Q = \frac{2\pi f L}{R_S \text{ (DC Value)}}$$

Actual Q

Frequency

ARRL0394

Figure 4-28 — At low frequencies, the Q of an inductor is proportional to frequency. At high frequencies, increased losses in the inductor cause Q to be degraded from the expected value.

In the HF range, all current flows in the outer few thousandths of an inch of a conductor. At VHF and UHF, the depth is on the range of a few ten-thousandths of an inch. (This is why many VHF and UHF inductors are silver plated — to provide a low-resistance path for current.) In fact, at VHF and UHF, conductors could be made of metal-plated plastic without any ill effects!

Skin effect is the major cause of why the internal resistance of inductors (due mainly to the resistance of the wire used to wind them) increases somewhat as the frequency increases. Because of the increasing reactance, inductor Q will increase with increasing frequency up to a point but then the internal resistance due to skin effect becomes greater and Q degrades as shown in **Figure 4-28**.

Before you go on, study test questions E4B15, E5A10 through E5A13, E5D01 and E5D02. Review this section if you have difficulty.

MAGNETIC CORES

As you've seen, inductors store magnetic energy, creating reactance. Inductors are usually visualized as the classic winding of wire from one end to the other of a round form — this winding shape is called *solenoidal* — giving rise to the common term "coil." An inductor's *core* is whatever material the wire is wound around, even air. (An inductor whose core consists of air is called *air-wound*.)

Solenoidal coils make great figures in books, but are actually fairly uncommon in electronics. A winding of wire around a hollow form filled with nothing but air is a relatively inefficient way to store magnetic energy. A form made of magnetic material increases the storage of energy because it focuses the magnetic field created by the current in the surrounding winding. The stronger magnetic field increases the inductance of the inductor.

Inductance is determined by the number of turns of wire on the core and on the core material's *permeability*. [E6D06] Permeability refers to the strength of a magnetic field in the core as compared to the strength of the field with a core of air. Cores with higher permeability have more inductance for the same number of turns on the core. In other words, if you make two inductors with 10 turns around different core materials, the core with a higher permeability will have more inductance.

Manufacturers offer a wide variety of materials, or *mixes*, to provide cores that will perform well over a desired frequency range. Powdered-iron cores combine fine iron particles with magnetically-inert binding materials. Combining materials such as nickel-zinc and manganese-zinc compounds with the iron produces ceramic *ferrite* cores. The chemical names for iron compounds are based on the Latin word for iron, *ferrum*, so this is how these materials get the name ferrite. By careful selection of core material, it is possible to produce inductors that can be used from the audio range to UHF. Inductors with magnetic material cores are also called *ferromagnetic inductors*.

The choice of core materials for a particular inductor presents a compromise of features. Powdered-iron cores generally have better temperature stability. [E6D08] Ferrite cores generally have higher permeability values, however, so inductors made with ferrite cores require fewer turns to produce a given inductance value. [E6D16]

Core Shape

The shape of an inductor's core also affects how its magnetic field is contained. For a solenoidal core, the magnetic field exists not only in the core, but in the space around the inductor. This allows the magnetic field to interact with, or *couple* to, other nearby conductors. This coupling often creates unwanted signal paths and interactions between components so external shields or other isolation methods must be used.

To reduce unwanted coupling the donut-shaped *toroid* core is used. When wire is wound on such a core, a toroidal inductor is produced. Nearly all of a toroidal inductor's magnetic field is contained within the toroid core. [E6D10] Toroidal inductors are one of the most popular inductor types in RF circuits because they can be located close to each other on a circuit board with almost no interaction. See **Figure 4-29** for a photo of a variety of toroidal inductors. Toroidal inductors are used in circuits that involve frequencies from below 20 Hz to around 300 MHz. [E6D07]

Figure 4-29 — This photo shows a variety of inductors wound on toroid cores.

Calculating Inductance

Calculating the inductance of a particular toroidal inductor is simple. First, you must know the inductance index value for the particular core you will use. This value, known as A_L, is found in the manufacturer's data. For powdered-iron toroids, A_L values are given in microhenrys per 100 turns. **Table 4-1** gives an example of the data for several core types. (See *The ARRL Handbook* for more complete information about these cores and their applications.)

To calculate the inductance of a powdered-iron toroidal inductor when the number of turns and the core material are known, use Equation 4-18.

Table 4-1

A_L Values for Selected Powdered-Iron and Ferrite Toroids

A_L Values for Powered-Iron Cores (µH per 100 turns)

Size	Mix 2	Mix 3	Mix 6	Mix 10	Mix 12
T-12	20	60	17	12	7.5
T-20	27	76	22	16	10.0
T-30	43	140	36	25	16.0
T-50	49	175	40	31	18.0
T-200	120	425	100	n/a	n/a

A_L Values for Ferrite Cores (mH per 1000 turns)

Size	Mix 43	Mix 61	Mix 63	Mix 77
FT-23	188.0	24.8	7.9	396
FT-37	420.0	55.3	19.7	884
FT-50	523.0	68.0	22.0	1100
FT-114	603.0	79.3	25.4	1270

Data from *The ARRL Handbook*, courtesy of Amidon Associates and Micrometals.

$$L \text{ (for powdered iron cores)} = \frac{A_L \times N^2}{10,000} \qquad \text{(Equation 4-18)}$$

where:

L = inductance in μH.

A_L = inductance index, in μH per 100 turns.

N = number of turns.

For example, suppose you have a size T-50 core made from the number 6 mix which is good for inductors from about 10 to 50 MHz. From Table 4-1 we find that this core has an A_L value of 40. What is the inductance of an inductor that has 10 turns on this core?

$$L = \frac{40 \times 10^2}{10,000} = 0.4 \ \mu H$$

Often you want to know how many turns to wind on the core to produce an inductor with a specific value. In that case, solve Equation 4-18 for N.

$$N = 100 \sqrt{\frac{L}{A_L}} \qquad \text{(Equation 4-19)}$$

Suppose you want to know how many turns to wind on the T-50-6 core used in the previous example to produce a 5 μH inductor? (The A_L value = 40.) [E6D12]

$$N = 100 \sqrt{\frac{L}{A_L}} = 100 \sqrt{\frac{5}{40}} = 100 \sqrt{0.125} = 35 \text{ turns}$$

So we will have to wind 35 turns of wire on this core to produce a 5 μH inductor. Keep in mind that if the wire simply passes through the center of the core, you have a 1-turn inductor. Each time the wire passes through the center of the core it counts as another turn. A common error is to count one complete wrap around the core ring as one turn. That produces a 2-turn inductor, however. This is illustrated in **Figure 4-30**.

The calculations for ferrite toroids are nearly identical but the A_L values are given in millihenrys per 1000 turns instead of microhenrys per 100 turns because the permeability of ferrite is higher. This requires a change of the constant in Equation 4-18 from 10,000 to 1,000,000. Use Equation 4-20 to calculate the inductance of a ferrite toroidal inductor.

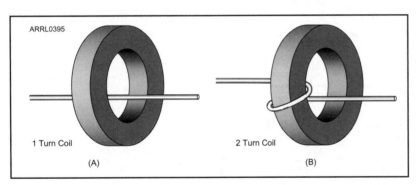

Figure 4-30 — Proper turns counting is important when you wind a toroidal inductor. Each pass through the center of the core must be counted. Part A shows a 1-turn inductor and Part B shows a 2-turn inductor.

$$L \text{ (for ferrite cores)} = \frac{A_L \times N^2}{1,000,000} \qquad \text{(Equation 4-20)}$$

where:

L = inductance in mH.

A_L = inductance index, in mH per 1000 turns.

N = number of turns.

Suppose we have a size FT-50 core made from 43-mix material. What is the inductance of a 10-turn inductor? Table 4-1 shows that A_L = 523 for this core. [E6D11]

$$L = \frac{523 \times 10^2}{1,000,000} = 0.0523 \text{ mH} = 52.3 \text{ }\mu\text{H}$$

Again, it is a simple matter to solve this equation for N, so you can calculate the number of turns required to produce a specific inductance value for a particular ferrite core.

$$N = 1000 \sqrt{\frac{L}{A_L}}$$
(Equation 4-21)

How many turns must we wind on a T-50-43 core to produce a 1-mH inductor? ($A_L = 523$)

$$N = 1000 \sqrt{\frac{L}{A_L}} = 1000 \sqrt{\frac{1}{523}} = 1000 \sqrt{1.91 \times 10^{-3}} = 43.7 \text{ turns}$$

Winding 43 or 44 turns on this core will produce an inductor of about 1 mH.

Toroidal cores are available in a wide variety of sizes. It is important to select a core size large enough to be able to hold the required number of turns to produce a particular inductance value. For a high-current application you will have to use a large wire size so a larger core size is required. To wind an inductor for use in a high-power antenna tuner, for example, you may want to use #10 or 12 AWG wire. If your inductor requires 30 turns of this wire, you would probably select a 200-size core which has an outside diameter of 2 inches and an inside diameter of 1.25 inches. You might even want to select a larger core for this application.

RF transformers are often wound on toroidal cores. If two wires are twisted together and wound on the core as a pair to place two windings on the core, we say it is a *bifilar winding*. It is also possible to wind three, four or more wires on the core simultaneously but the bifilar winding is the most common.

Toroid cores are very useful for solving a variety of radio-frequency interference (RFI) problems. For example, you might select a type-43 mix ferrite core and wind several turns of a telephone wire or speaker leads through the core to produce a *common-mode choke*. Such a choke is designed to suppress any RF energy flowing in common on all of these wires. Audio signals flow through the choke unimpeded but the RF signals are blocked.

A *ferrite bead* is a very small core with a hole designed to slip over a component lead. These are often used as suppressors for VHF and UHF oscillations at the input and output terminals of HF and UHF amplifiers, for example. [E6D09] The use of ferrite beads as parasitic suppressors points out another interesting property of these core materials — their loss changes with frequency. Each mix has a different set of loss characteristics with frequency. While we normally want to select an inductor core material that will have low loss at a particular frequency or over a certain range, at times we want to select a core material that will have high loss to absorb or dissipate energy.

You may wonder how you can tell the difference between, say, a T-50-6 core and a T-50-10 core if you found them both in a piece of surplus equipment or in a grab bag of parts. For that matter, how can you tell if a particular core is powdered-iron or ferrite? Unfortunately, the answer is that you can't! There is no standard way of marking or coding these cores for later identification. So it is important to purchase your cores from a reliable source and store them separately in marked containers.

Before you go on, study test questions E6D06 through E6D12 and E6D16. Review this section if you have difficulty.

Table 4-2

Questions Covered in This Chapter

4.1 Radio Mathematics
E5C11

4.2 Electrical Principles

E4B15	E5B03	E5C10	E5D08
E5A01	E5B04	E5C12	E5D09
E5A02	E5B05	E5C13	E5D10
E5A03	E5B06	E5C14	E5D11
E5A04	E5B07	E5C15	E5D12
E5A05	E5B08	E5C16	E5D13
E5A06	E5B09	E5C17	E5D14
E5A07	E5B10	E5C18	E5D15
E5A08	E5B11	E5C19	E5D16
E5A09	E5B12	E5C20	E5D17
E5A10	E5B13	E5C21	E5D18
E5A11	E5C01	E5C22	E6D06
E5A12	E5C02	E5C23	E6D07
E5A13	E5C03	E5D01	E6D08
E5A14	E5C04	E5D02	E6D09
E5A15	E5C05	E5D03	E6D10
E5A16	E5C06	E5D04	E6D11
E5A17	E5C07	E5D05	E6D12
E5B01	E5C08	E5D06	E6D16
E5B02	E5C09	E5D07	

Chapter 5

Components and Building Blocks

In this chapter, you'll learn about:
- **How semiconductor devices are made**
- **Types of diodes and rectifiers**
- **Bipolar and field effect transistors and RF integrated circuits**
- **Display devices used in ham radio**
- **Digital logic basics and families**
- **Frequency counters, references and markers**
- **Optoelectronics such as solar cells and optocouplers**

The Amateur Radio Extra class license exam presents basic questions about electronic components — diodes, transistors, ICs and other devices. These are contained in question pool Subelement 6 (6 groups of questions), and Subelement 7 (8 groups of questions). You won't have to become a circuit designer to answer the exam questions, but you'll be expected to know what types of devices are used in radio circuits and their important characteristics.

This chapter of the *License Manual* presents the fundamentals of how the devices operate. We'll start from the beginning — with diodes — and work our way up to integrated circuits. As a comprehensive explanation of these electronic building blocks is well beyond the scope of this book, turn to *The ARRL Handbook* or the references on the *Extra Class License Manual* website for more information.

5.1 Semiconductor Devices

Before you can understand the operation of electronic circuits, you must know some basic information about the devices that make up those circuits. This section presents the information about semiconductor and other active devices you need to know to pass your Extra class license exam. You will find descriptions of several types of diodes and transistors, RF and digital integrated circuits (ICs), and various types of display and optoelectronic devices.

MATERIALS

Silicon (Si) and germanium (Ge) are the materials normally used to make semiconductor materials. (The element *silicon* (SIL-i-kahn) is not the same as the household lubricants and rubber-like sealers called *silicone* (sil-i-CONE). Silicon has 14 protons and 14 electrons, while germanium has 32 of each. Silicon and germanium atoms both have four shareable or *valence* electrons in their outer layer of electrons. This arrangement allows these four electrons to be shared with other nearby atoms.

Atoms that arrange themselves into a regular pattern by sharing electrons form *crystals*. **Figure 5-1** shows silicon and germanium crystals. (Different kinds of atoms might arrange themselves into other patterns.) The crystals made by silicon or germanium atoms do not make good electrical conductors or insulators. That's why they are called *semiconductors*.

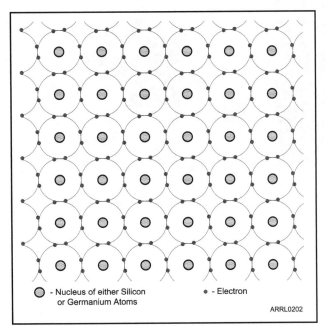

Figure 5-1 — Silicon and germanium atoms arrange themselves into a regular pattern — a crystal. Each atom in this crystal structure is sharing its outermost four electrons with other nearby atoms.

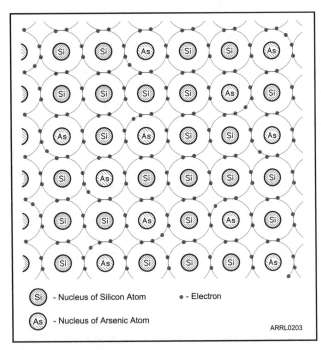

Figure 5-2 — Adding antimony or arsenic atoms to the silicon or germanium crystals results in an extra or free electron in the crystal structure, producing N-type semiconductor material.

Under the right conditions they can act as either conductors or insulators. Semiconductor materials also exhibit properties of both metallic and nonmetallic substances. Semiconductors are solid crystals. They are strong and not easily damaged by vibration or rough handling. We refer to electronic parts made with semiconductor materials as *solid-state devices*.

To control the electrical characteristics of semiconductor material, manufacturers add other atoms to these crystals through a carefully controlled process called *doping*. The atoms added in this way produce a material that is no longer pure silicon or pure germanium. We call the added atoms *impurities*. The impurities are generally chosen for their ability to alter the way in which electrons are shared within the crystal.

As an example, the manufacturer might add some atoms of arsenic (As) or antimony (Sb) to the silicon or germanium while making the crystals. Arsenic and antimony atoms each have five electrons to share — an extra shareable electron compared to the crystal of pure silicon. **Figure 5-2** shows how an atom with five electrons in its outer layer fits into the crystal structure. In such a case, there is an extra or *free* electron in the crystal and we call the semiconductor material made in this way *N-type* material. (This name comes from the extra free electrons in the crystal structure.) [E6A02]

The impurity atoms are electrically neutral, just as the silicon or germanium atoms are. The extra electrons are considered "free" because they are not so strongly shared with adjacent atoms and are more free to move within the crystal structure. Impurity atoms that create (donate) free electrons to the crystal structure are called *donor impurities*.

Now let's suppose the manufacturer adds some gallium or indium atoms instead of arsenic or antimony. Gallium (Ga) and indium (In) atoms only have three electrons that they can share with other nearby atoms. When there are gallium or indium atoms in the crystal there is an extra space where an electron could fit into the structure.

Figure 5-3 shows an example of a crystal structure with spaces where an electron could be present. We call this space for an electron a *hole*. The semiconductor material produced in this way is *P-type* material. Impurity atoms that produce extra holes for electrons in the crystal structure are called *acceptor impurities*. [E6A04, E6A15]

Again, you should realize that the impurity atoms have the same number of electrons as protons. The material is still electrically neutral. The crystal structure is missing an electron in P-type material. Similarly, the crystal structure has an extra electron in N-type material.

Suppose we apply a voltage across a crystal of N-type semiconductor. The side with positive

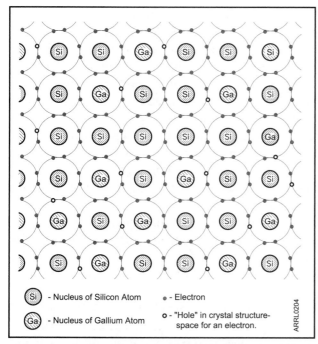

Figure 5-3 — Manufacturers can add gallium or indium atoms to the silicon or germanium crystals. These impurity atoms have only three electronics to share. This leaves a hole or space for another electron in the crystal structure, producing P-type semiconductor material.

Si - Nucleus of Silicon Atom   **•** - Electron

Ga - Nucleus of Gallium Atom   **o** - "Hole" in crystal structure- space for an electron.

voltage attracts electrons. Free electrons in the structure move toward positive voltage through the crystal. Since most of the current through N-type semiconductor material is produced by these free electrons, they are the *majority charge carriers*. [E6A16]

Now suppose we apply a voltage to a crystal of P-type material in which the electrons are all shared between atoms. A shared electron will have a preference to be shared with an atom closer to the positive voltage side. If the shared electron moves to fill a hole created by an acceptor impurity atom it appears as if the hole, attracted by negative voltage, has moved to the position last occupied by the shared electron. In this way, holes in the P-type material move toward the negative side. The majority charge carriers in P-type semiconductor material are holes. [E6A03] Free electrons and holes move in opposite directions through a crystal.

Other materials are used to make semiconductor materials for special-purpose applications. For example, gallium arsenide (GaAs) semiconductor material has performance advantages for use in solid-state devices operating at microwave frequencies. [E6A01]

Before you go on, study test questions E6A01 through E6A04, E6A15, and E6A16. Review this section if you have difficulty.

DIODES

For the Extra class license, you will become acquainted with junction diodes, as well as some specialized types: Schottky, PIN, Zener, varactor, point-contact and hot-carrier diodes are all used in radio circuits in various special applications.

Junction Diodes

The *junction diode*, also called the *PN-junction diode*, is made from two layers of semiconductor material joined together. One layer is made from P-type (positive) material. The other layer is made from N-type (negative) material. The name PN junction comes from the way the P and N layers are joined to form a semiconductor diode. **Figure 5-4** illustrates the basic concept of a junction diode.

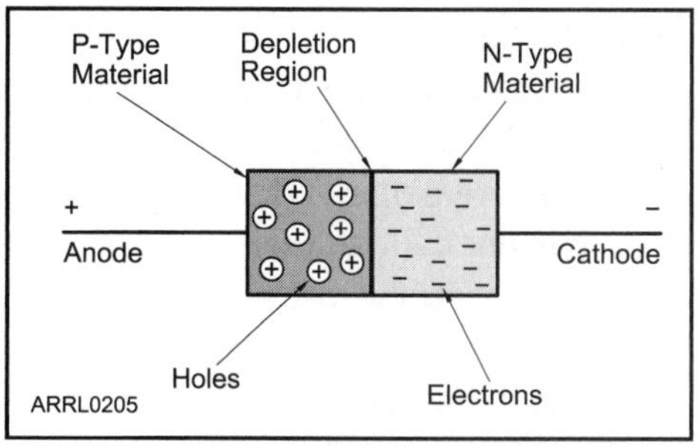

When no voltage is applied to a diode, the junction between the P-type and N-type material acts as a barrier that prevents carriers from flowing between the layers. This happens because the majority carriers (the electrons and holes) combine where the two types of material are in contact, leaving no carriers to support current flow unless a voltage is applied from an external source. This barrier to current flow is called the *depletion region*.

The P-type side of the diode is called the *anode*. The N-type

Figure 5-4 — A PN junction consists of P-type and N-type material separated by a thin depletion region in which the majority charge carriers are not present.

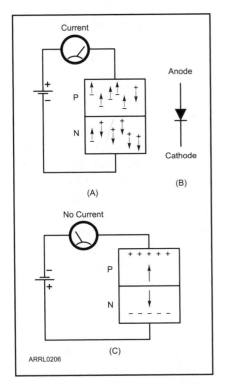

Figure 5-5 — At A, the PN junction is forward biased and conducting. B shows the schematic symbol used to represent a diode, oriented so that its internal structure is the same as in A. Conventional current flows in the direction indicated by the arrowhead in the symbol. At C, the PN junction is reverse biased, so it does not conduct.

side is called the *cathode*. When voltage is applied to a junction diode as shown at A in **Figure 5-5**, charge carriers flow across the barrier and the diode conducts. With the anode positive with respect to the cathode, electrons are attracted across the junction from the N-type material, through the P-type material, and on through the circuit to the positive battery terminal. Holes are attracted in the opposite direction by the negative voltage from the battery. Electrons are supplied to the cathode and removed from the anode by the wires connected to the battery. When the diode is connected in this manner it is said to be *forward biased*. *Conventional current* (which flows from positive to negative) in a diode flows from the anode to the cathode. The electrons flow in the opposite direction.

Figure 5-5B shows the schematic symbol for a diode, drawn as it would be used in the circuit instead of as the pictorial of semiconductor blocks used in part A. The arrow on the schematic symbol points in the direction of conventional current instead of electronic current which is the flow of the actual electrons.

If the battery polarity is reversed, as shown in Figure 5-5C, the excess electrons in the N-type material are attracted away from the junction toward the positive battery terminal. Similarly, the holes in the P-type material are attracted away from the junction toward the negative battery terminal. When this happens, electrons do not flow across the junction to the P-type material and the diode does not conduct. When the anode is connected to a negative voltage source and the cathode is connected to a positive voltage source, the device is said to be *reverse biased*.

The voltage required for carriers to move across the PN junction results in a *forward voltage* across the diode when it is conducting. For silicon diodes, forward voltage is approximately 0.6 to 0.7 V; it is 0.2 to 0.3 V for germanium diodes.

Junction diodes are used as rectifiers to allow current in one direction only. When an ac signal is applied to a diode, it will be forward biased and conduct during one half of the cycle, allowing current to flow to the load. During the other half of the cycle, the diode is reverse biased and current does not flow. The ac current becomes pulses of dc, always flowing in the same direction.

Diode Ratings

Junction diodes have maximum voltage and current ratings that must be observed or damage to the diode could result. The voltage rating is called *peak inverse voltage* (PIV) and the rectified current rating is called *maximum average forward current*. With present technology, diodes are commonly available with voltage ratings of 1000 PIV and current ratings of 100 A.

Peak inverse voltage is the voltage that a diode must withstand when it isn't conducting. Although a diode is normally used in the forward direction it will conduct in the reverse direction if enough voltage is applied. A few hole/electron pairs are generated at the junction when a diode is reverse biased causing a very small reverse current called *leakage current*. Semiconductor diodes can withstand some leakage current. If the reverse voltage reaches a high enough value, however, the leakage current rises abruptly, resulting in a large reverse current. The point where the leakage current rises abruptly is called the *avalanche point*. A large reverse current usually damages or destroys the diode.

When current flows through the junction some power is dissipated in the form of heat just like for a resistor. The amount of power equals the current through the diode multiplied by

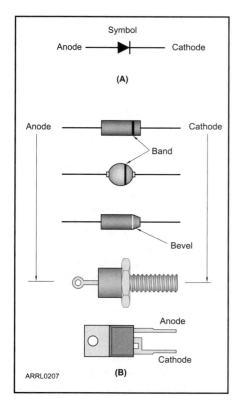

Figure 5-6 — The schematic symbol for a diode is shown at A. Typical diode packages are shown at B.

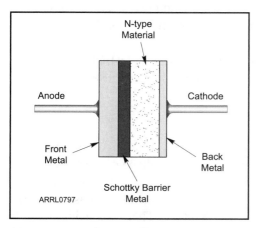

Figure 5-7 — The Schottky barrier diode substitutes a metal layer for the P-type material of a PN-junction. This results in a lower forward voltage drop than for a PN-junction diode.

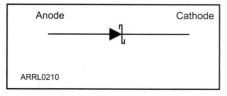

Figure 5-8 — This is the schematic symbol for a Schottky barrier diode.

the forward voltage. For example, approximately 6 W is dissipated by a silicon rectifier with 10 A flowing through it ($P = I \times E$; $P = 10 \text{ A} \times 0.6 \text{ V}$). As the forward current increases, the junction temperature will increase.

The maximum average forward current is the highest average current that can flow through the diode in the forward direction for a specified *maximum allowable junction temperature*. If allowed to get too hot, the diode will be damaged or destroyed. [E6B07]

Diodes designed to safely handle forward currents in excess of a few amps are packaged so they may be mounted on a heat sink. The heat sink helps the diode package dissipate heat more rapidly, keeping the diode junction temperature at a safe level. The metal case or tab of a power diode is usually electrically connected to one of the diode's layers so it must usually be insulated from ground.

Figure 5-6 shows some of the more common diode-case styles, as well as the general schematic symbol for a diode. The line, or spot, on a diode case indicates the cathode lead. Check the case or the manufacturer's data sheet for the correct connections.

Schottky Barrier Diodes

If a PN-junction's P-type material is replaced with a metal layer as in **Figure 5-7** a *Schottky barrier* is created which has similar rectifying properties but with a lower forward voltage than an all-semiconductor junction. [E6B08] (Schottky was the physicist who developed this structure.) For example, the Schottky barrier diode's forward voltage is 0.2 to 0.5 V, compared to the 0.6 to 0.7 V for silicon PN-junction diodes. [E6B02] The lower forward voltage results in lower power dissipation than PN-junction diodes for the same amount of current so Schottky diode rectifiers are widely used in power supply circuits. **Figure 5-8** shows the schematic symbol for a Schottky rectifier.

Point-Contact Diodes

In a junction diode, the P and N layers are separated only by the junction, forming a capacitor: two charged plates separated by a thin dielectric.

Although the internal capacitance of a PN-junction diode may be only a few picofarads this capacitance can cause problems in RF circuits, especially at VHF and above. Junction diodes may be used from dc to the microwave region but the *point-contact diode* has low internal capacitance that is specially designed for RF applications.

Figure 5-9 illustrates the internal structural differences between a junction diode and a point-contact diode. As you can see, the point-contact diode has a much smaller surface area at the junction than does a PN-junction diode. When a point-contact diode is manufactured, the main portion of the device is made from N-type material and a thin aluminum wire, often called a *whisker*, is placed in contact with the semiconductor surface forming a Schottky barrier. The result is a diode that exhibits much less internal capacitance than PN-junction diodes, typically 1 pF or less. This means point-contact diodes are better suited for VHF and UHF applications

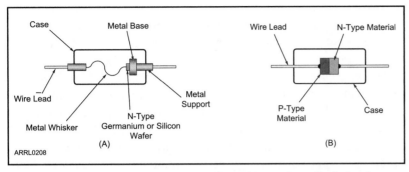

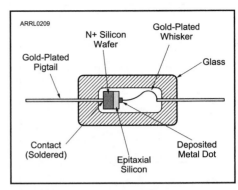

Figure 5-9 — The internal structure of a point-contact diode is shown at A. B shows the internal structure of PN-junction diodes. The schematic symbol for point-contact diodes is the same as junction diodes.

Figure 5-10— This drawing represents the internal structure of a hot-carrier diode. The whisker contact is attached to a metal contact directly on the semiconductor material, improving mechanical and electrical performance over a point-contact diode.

than are PN-junction diodes. Point-contact diodes are generally used as UHF mixers and as RF detectors at VHF and below. [E6B09]

Hot-Carrier Diodes

Another type of Schottky barrier diode with low internal capacitance and good high-frequency characteristics is the *hot-carrier diode*. ("Hot" refers to the diode's higher electron velocities compared to a PN-junction diode.) This device is very similar in construction to the point-contact diode but with an important difference depicted in **Figure 5-10**.

The whisker in a hot-carrier diode is physically attached to a metal dot deposited on the element. The hot-carrier diode is mechanically and electrically superior to the point-contact diode. Some of the advantages of the hot-carrier type are improved power-handling characteristics, lower contact resistance and improved immunity to burnout caused by transient noise pulses.

Hot-carrier diodes are often used in mixers and detectors at VHF and UHF. [E6B06] In this application, hot-carrier diodes are superior to point-contact diodes because they exhibit greater conversion efficiency and lower noise figure.

Zener Diodes

Zener diodes (named for their inventor) are a special class of PN-junction diode used as voltage references and voltage regulators. As discussed earlier, leakage current rises as reverse voltage is applied to a diode. At first, this leakage current is very small and changes very little with increasing reverse voltage. There is a point, however, at which the leakage current rises suddenly. Beyond this point, the current increases very rapidly for a small increase in voltage; this is called the *avalanche point*. The *Zener voltage* is the voltage necessary to cause avalanche. Normal junction diodes would be destroyed if they were operated in this region, but Zener diodes are specially manufactured to safely withstand the avalanche current.

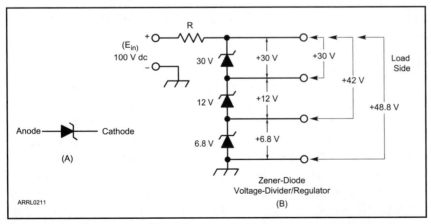

Figure 5-11 — The schematic symbol for a Zener diode is shown at A. B is an example of how Zener diodes are used as voltage regulators.

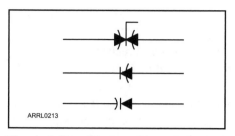

Figure 5-12 — These schematic symbols are commonly used to represent varactor diodes.

Since the current in the avalanche region can change over a wide range while the voltage stays practically constant, this kind of diode can be used as a voltage regulator. [E6B01] The voltage at which avalanche occurs can be controlled precisely in the manufacturing process. Zener diodes are calibrated in terms of avalanche voltage.

Zener diode voltage regulators, shown in **Figure 5-11**, provide a nearly constant dc output voltage, even though there may be large changes in load resistance or input voltage. As voltage references, they exhibit a stable voltage that remains constant over a wide temperature range.

Zener diodes are currently available with voltage ratings between 1.8 and 200 V. Their power ratings range from 250 mW to 50 W. They are packaged in the same case styles as junction diodes. Usually, Zener diodes rated for 10 W dissipation or more are made in stud- or tab-mount cases.

Tunnel Diodes

The tunnel diode is a special type of device that has no rectifying properties. When properly biased, it possesses an unusual characteristic: negative resistance. Negative resistance means that when the voltage across the diode increases, the current decreases. This property makes the tunnel diode capable of amplification and oscillation. [E6B03] At one time, tunnel diodes were expected to dominate in microwave applications, but other devices with better performance soon replaced them. The tunnel diode is obsolete and not used today.

Varactor Diodes

As mentioned above, junction diodes exhibit an appreciable internal capacitance. It is possible to change the internal capacitance of a diode by varying the amount of reverse bias applied to it, changing the separation of the carriers outside the depletion region. *Variable-capacitance diodes* and *varactor diodes* (variable reactance diodes) are designed to take advantage of this property, creating voltage-controlled capacitors. [E6B04] Varicap is a trade name for these diodes.

Varactors provide various capacitance ranges from a few picofarads to more than 100 pF. Each style has a specific minimum and maximum capacitance. The higher the maximum capacitance, the greater will be the minimum capacitance. A typical varactor can provide capacitance changes over a 10:1 range with bias voltages in the 0- to 100-V range.

Common schematic symbols for a varactor diode are given in **Figure 5-12**. These devices are used in frequency multipliers at power levels as great as 25 W, in remotely tuned circuits and in frequency modulator circuits.

PIN Diodes

A PIN (positive/intrinsic/negative) diode is formed by diffusing P-type and N-type layers onto opposite sides of an almost pure silicon layer, called the *I region* because conduction is carried out by the electrons *intrinsic* to a normal silicon crystal. **Figure 5-13** shows the three layers of the PIN diode. This layer is not "doped" with P-type or N-type charge carriers, as are the other layers. Any charge carriers found in this layer are a result of the natural properties of the pure semiconductor material. In the case of silicon, there are relatively few free charge carriers. PIN-diode characteristics are determined primarily by the thickness and area of the I region. The outside layers are designated P+ and N+ to indicate heavier than normal doping

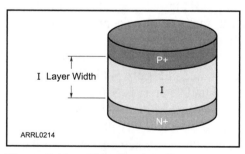

Figure 5-13 — This diagram illustrates the inner structure of a PIN diode. The top and bottom layers are labeled P+ and N+ to indicate very heavy levels of doping impurities are used.

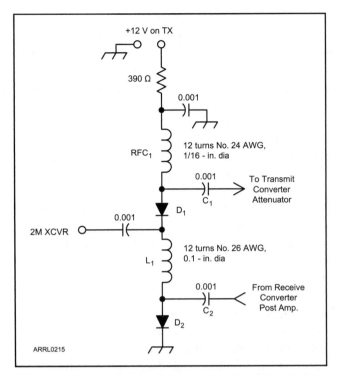

Figure 5-14 — PIN diodes may be used as RF switches. This schematic shows PIN diodes (D1, D2) switching a 2 meter transceiver between a transmit converter and a receive converter. Applying 12 V turns both diodes ON, shorting the receive converter input to ground and connecting the transmit converter to the 2 meter rig. When bias (12 V) is removed, the diodes are both OFF, disconnecting the transmit converter and reconnecting the receive converter.

of these layers. PIN diodes are represented by the same schematic symbol as a PN-junction diode.

PIN diodes have a forward resistance that varies inversely with the amount of forward bias applied. When a PIN diode is at zero or reverse bias, there are essentially no free charge carriers available to conduct, and the intrinsic region can be considered as a low-loss dielectric, conducting RF current like a small capacitor. Under reverse-bias conditions, the charge carriers move very slowly. This slow response time causes the PIN diode to look like a resistor to RF currents, effectively blocking them. The amount of resistivitance that a PIN diode exhibits to RF can be controlled by changing the amount of forward bias applied. Thus, the PIN diode can be voltage-controlled to act as a switch or attenuator. [E6B05, E6B11, E6B12] PIN diodes are faster, smaller, more rugged and more reliable than relays or other electromechanical switching devices.

Figure 5-14 shows a circuit in which PIN diodes are used to build an RF switch. This diagram shows a transmit/receive switch for use between a 2 meter transceiver and a UHF or microwave transverter. With no bias, or with reverse bias applied to the diode, the PIN diode exhibits a high resistance to RF, so no signal will flow from the generator to the load. When forward bias is applied, the diode resistance will decrease, allowing the RF signal to pass.

The amount of insertion loss (resistance to RF current) is determined primarily by the amount of forward bias applied; the greater the forward bias current, the lower the RF resistance.

Before you go on, study test questions E6B01 through E6B09, E6B11, and E6B12. Review this section if you have difficulty.

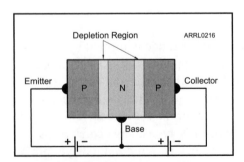

Figure 5-15 — A bipolar junction transistor consists of two layers of N- or P-type material sandwiching a layer of the opposite type of material. This drawing shows the internal structure of a PNP transistor.

BIPOLAR TRANSISTORS

The *bipolar junction transistor* (BJT) is a type of three-terminal, PN-junction device able to use a small current to control a large current — in other words, amplify current. It is made of two layers of N- or P-type material sandwiching a thin layer of the opposite type of material between them as illustrated in **Figure 5-15**. If the outer layers are P-type material and the middle layer is N-type material the device is called a *PNP transistor* because of the layer arrangement. If the outer layers are N-type material the device is called an NPN transistor. A transistor is, in effect, two PN-junction diodes back-to-back. **Figure 5-16** shows the schematic symbols for PNP and NPN bipolar transistors. [E6A07] The three layers of the transistor sandwich are called the *emitter*, *base* and *collector*. A diagram of the construction of a typical PNP transistor is given in Figure 5-15.

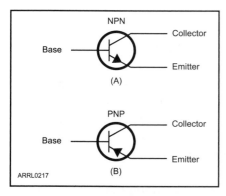

Figure 5-16 — The schematic symbol for an NPN transistor is shown at A and for a PNP transistor at B.

Figure 5-17 — Transistors are packaged in a wide variety of case styles, depending on the application in which they are intended to be used.

In an actual bipolar transistor, the base layer (in this case, N-type material) is much thinner than the outer layers. Just as in the PN-junction diode described in the previous section, a depletion region forms at each junction between the P- and N-type material. These depletion regions form a barrier to current flow until forward bias is applied across the junction between the base and emitter layers.

Forward-bias voltage across the emitter-base section of the sandwich causes electrons to flow through it from the base to the emitter. As the free electrons from the N-type material flow into the P-type material, holes from the P-type material flow the other way into the base. Some of the holes will combine with free electrons in the base, but because the base layer is so thin, most will move right on through into the P-type material of the collector.

As shown, the collector is connected to a negative voltage with respect to the base. Normally, reverse bias would prevent current from flowing across the base-collector junction. The collector, however, now contains an excess of holes because of those from the emitter that overshot the base. Since the voltage source connected to the collector produces a negative voltage, the holes from the emitter will be attracted to that power supply connection, creating current flow from the emitter to the collector.

Bipolar junction transistors are used in a wide variety of applications, including amplifiers (from very low level to very high power), switches, oscillators and power supplies. They are used at all frequency ranges from dc through the UHF and microwave range. Transistors are packaged in a wide variety of case styles. Some of the more common case styles are depicted in **Figure 5-17**.

Transistor Characteristics

Because of the transistor's construction, the current through the collector will be considerably larger than that flowing through the base. When a transistor's base-emitter junction is forward biased, collector current increases in proportion to the amount of bias current applied. The ratio of collector current to base current is called the *current gain*, or *beta*. Beta is expressed by the Greek symbol β. [E6A06] It is calculated from the equation:

$$\beta = \frac{I_c}{I_b}$$

(Equation 5-1)

where:
  I_c = collector current
  I_b = base current

For example, if a 1-mA base current results in a collector current of 100 mA the beta is 100. Typical betas for junction transistors range from as low as 10 to as high as several hundred. Manufacturers' data sheets specify a range of values for β. Individual transistors of a given type can have widely varying betas.

Another important transistor characteristic is *alpha*, expressed by the Greek letter α. [E6A05] Alpha is the ratio of collector current to emitter current, given by the equation:

$$\alpha = \frac{I_c}{I_e}$$

(Equation 5-2)

where:
  I_c = collector current
  I_e = emitter current

The smaller the base current, the closer the collector current comes to being equal to that of the emitter and the closer alpha comes to being 1. For a junction transistor, alpha is usually between 0.92 and 0.98.

The transistor is *saturated* when further increases in base-emitter current do not increase the collector current, and the transistor is said to be *fully on* when the transistor is saturated. At the other end of the scale, when the transistor is reverse-biased, there is no current from the emitter to the collector and the transistor is at *cutoff*. When used to amplify a signal, a transistor operates between these two extremes. By operating at either cutoff or saturation, the transistor can be used as a switch.

Transistors have important frequency characteristics. The *alpha cutoff frequency* is the frequency at which the current gain of a transistor decreases to 0.707 times its gain at 1 kHz. [E6A08] Alpha cutoff frequency is considered to be the practical upper frequency limit of a transistor configured as a common-base amplifier.

Beta cutoff frequency is similar to alpha cutoff frequency, but it applies to transistors connected as common-emitter amplifiers. Beta cutoff frequency is the frequency at which the current gain of a transistor in the common-emitter configuration decreases to 0.707 times its gain at 1 kHz. (These amplifier configurations are explained in the section on amplifier circuits.)

Before you go on, study test questions E6A05 through E6A08. Review this section if you have difficulty.

FIELD EFFECT TRANSISTORS

Field-effect transistors (FETs) are given that name because the current through them is controlled by an electric field or voltage, as opposed to current as for the bipolar junction transistor. There are two types of field-effect transistors in common use today: the *junction FET (JFET)* and the *metal-oxide semiconductor FET (MOSFET)*. The basic characteristic of both FET types is a very high input impedance — typically 1 megohm or greater. This is considerably higher than the input impedance of a bipolar transistor. FETs are made in the same types of packages as bipolar transistors. Some different case styles are shown in **Figure 5-18**.

Figure 5-18 — FETs are packaged in cases much like those used for bipolar junction transistors.

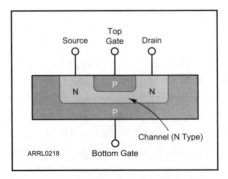

Figure 5-19 — The construction of a junction field-effect transistor (JFET). The top and bottom gate terminals are connected together outside the cross section shown here.

JFETs

The basic JFET construction is shown in **Figure 5-19**. The JFET can be thought of simply as a bar of semiconductor material that acts like a variable resistance. The terminal into which the charge carriers enter is called the *source*. The opposite terminal is called the *drain*. The terminals that control the resistance between source and drain are called *gates*. [E6A17] The mate-

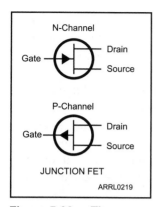

Figure 5-20 — The schematic symbol for an N-channel JFET has an arrow pointing toward the center line that represents the channel. The arrow points away from the channel for a P-channel JFET.

rial connecting the source and drain is called the *channel*. There are two types of JFET, named N-channel and P-channel for the type of material that forms the channel. The schematic symbols for the two JFET types are illustrated in **Figure 5-20**. [E6A11]

Two gate regions, made of the opposite type of semiconductor material used for the channel, are created on opposite sides of the JFET channel and connected together. When a reverse-bias voltage is applied from the top and bottom gate-channel junctions to the source, an electric field is set up across the channel. The electric field controls the normal electron flow through the channel. As the gate voltage changes the electric field varies and that varies source-to-drain current. The gate terminal is always reverse biased, so very little current flows in the gate terminal and the JFET has very high input impedance unlike the bipolar transistor which has much lower input impedance. [E6A14]

Because channel current is controlled by voltage on the gate the gain of a FET is measured as *transconductance* (g_m), the ratio of output current to input voltage. Transconductance is measured in siemens (S), the inverse of ohms.

MOSFETs

The construction of a *metal-oxide semiconductor field-effect transistor (MOSFET)*, sometimes called an *insulated gate field-effect transistor (IGFET),* and its schematic symbol are illustrated in **Figure 5-21**. In the MOSFET, the gate is insulated from the source/drain channel by a thin dielectric layer. Since there is very little current through this dielectric the input impedance is even higher than in the JFET, typically 10 megohms or greater. The schematic symbols for N-channel and P-channel dual-gate MOSFETs are shown in **Figure 5-22**. [E6A10] Some types of MOSFETs have two gates to which different voltages can be applied for special applications, such as mixers.

Nearly all the MOSFETs manufactured today have built-in gate-protective Zener diodes. Without this provision the gate insulation can be punctured easily by small static charges on the user's hand or by the application of excessive voltages to the device. [E6A12] The protective diodes are connected between the gate (or gates) and the source lead of the FET. The diodes are generally not shown on the schematic symbol.

Enhancement and Depletion-Mode FETs

Field-effect transistors are available with two types of channel: *enhancement mode* and

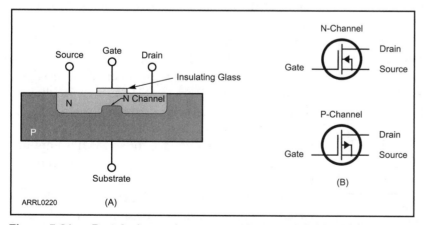

Figure 5-21 — Part A shows the construction of a MOSFET. The schematic symbols (B) for MOSFETs show that the gate terminal is not connected to the channel as in a JFET. As in the JFET symbols, the arrow's direction indicates the type of channel material. These are single-gate MOSFET symbols.

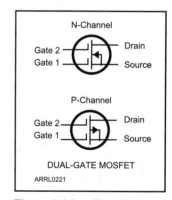

Figure 5-22 — The schematic symbols for N-channel and P-channel dual-gate MOSFETs. Notice that the direction of the arrows again indicates the type of channel material.

depletion mode. A depletion-mode device corresponds to Figure 5-19, where a channel exists without gate voltage applied. [E6A09] The gate of a depletion-mode device is reverse biased in operation. When the reverse bias is applied between the gate and source the channel is *depleted* of charge carriers and current decreases.

Enhancement-mode devices are constructed so there is no channel without voltage applied to the gate. The channel conducts current only when a gate-to-source voltage is applied which causes the channel to be able to conduct current. When the gate of an enhancement-mode device is forward biased, current begins to flow through the source/drain channel. The higher the forward bias on the gate the more current through the channel. JFETs cannot be used as enhancement-mode devices because if the gate is forward biased it will conduct like a forward-biased diode.

The gates of MOSFETs are insulated from the channel region, so they may be used as enhancement-mode devices. Both polarities may be applied to the gate without the gate becoming forward biased and conducting. Some MOSFETs are designed to be used without bias on the gate. In this type of operation the control signal applied to the gate creates forward-bias part of the time and reverse-bias part of the time. The MOSFET operates in the enhancement mode when the gate is forward biased, and in the depletion mode when the gate is reverse biased.

Before you go on, study test questions E6A09 through E6A12, E6A14 and E6A17. Review this section if you have difficulty.

RF INTEGRATED DEVICES

Integrated circuits (ICs) make up most of the internal circuitry of modern electronics. If you open up a new transceiver or a computer you may be hard-pressed to find very many discrete transistors! One of the last types of electronics to convert to ICs has been VHF, UHF and microwave circuits. Even though transistors in an IC are able to handle the high-speed signals, making an IC that would work properly in many different circuits is a tough challenge. Advances in circuit design have finally made it possible to use ICs at these high frequencies just as for lower frequencies. In fact, mobile telephones would be impossible to build without ICs that include UHF and microwave functions!

The most common RF IC used by amateurs is a *monolithic microwave integrated circuit (MMIC)*. It's quite unlike most other ICs you've seen. Most MMICs are quite small, often called "pill packages" because they look like a small pill with four leads coming out of the device at 90° to each other. The most common MMIC has an input lead, an output lead and two ground leads.

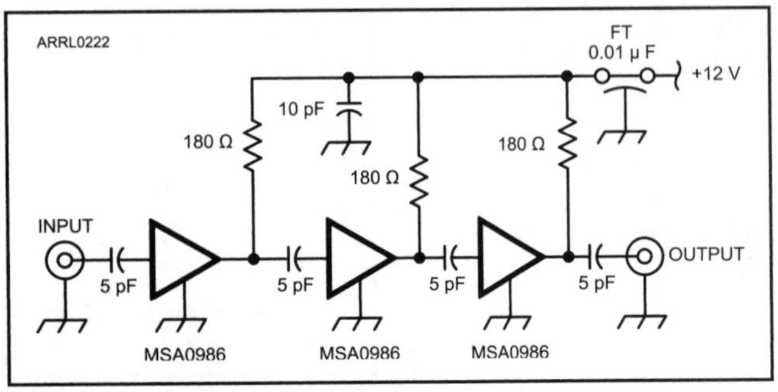

Wait a minute! Two ground leads? Where's the power lead? Many MMICs don't have a separate power lead — dc power to the internal electronics and the RF output from the MMIC both use the same lead! Power is supplied through a resistor and/or RF choke to the output lead. [E6E08] The typical dc operating voltage for an MMIC amplifier is 12 V. A small series *blocking capacitor* keeps the dc voltage from getting to any other circuits as shown in the schematic of **Figure 5-23**. MMICs use this method because of its simplicity and the extra ground lead helps ensure that the amplifier circuit operates properly over the entire frequency range.

MMIC devices have well-controlled operat-

Figure 5-23 — The schematic diagram of the amplifier for UHF and microwave bands shown in Figure 5-24. Three MMIC devices provide a lot of gain without requiring a lot of complex circuitry. Operating power is supplied to the MMICs at their output pins with 5 pF capacitors coupling the RF signal to the next stage while blocking the dc voltage.

Figure 5-24 — This simple utility amplifier is suitable for low-level amplification on the amateur bands from 903 MHz through 5.7 GHz. The schematic is shown in Figure 5-23. The three MMICs are visible as round, black "pills" in a straight line between the input jack at left and the output jack at right.

ing characteristics such as gain, noise figure and input/output impedance, requiring only a few external components for proper operation. [E6E06] As "building blocks," MMICs can greatly simplify an amplifier design for circuits at UHF and microwave frequencies because the circuits and the IC input and output impedances are all close to 50 Ω. [E6E04] MMICs operate well into the microwave range with devices made from gallium nitride reaching the highest frequencies. [E6E11]

As a design example, a MAR-6 MMIC could be used to build a receive preamplifier for a 1296-MHz receiver with just a few external resistors and capacitors. This device provides 16 dB of gain for signals up to 2 GHz, with a noise figure around 3 dB. Many MMIC amplifier devices have noise figures in the range of about 3.5 to 6 dB.

Circuits built using MMICs generally employ *microstrip construction techniques*. [E6E07] **Figure 5-24** is an example of microstrip techniques used to build an amplifier based on the schematic in Figure 5-23. Double-sided circuit board material is used, with one side forming a ground plane. Traces connecting the MMICs or other active devices to the signal input and output connectors form sections of 50-Ω feed line. Components are soldered directly to these feed line sections. The amplifier module in Figure 5-24 includes just three MMICs, three resistors, four chip (surface-mount) capacitors and a feed-through capacitor to bring the supply voltage into the enclosure.

> *Before you go on, study test questions E6E04, E6E06 through E6E08, and E6E11. Review this section if you have difficulty.*

5.2 Display Devices

Visual indicators and imaging systems are very important parts of Amateur Radio, from the power on/off lights to amateur television and test instruments. This section discusses five of the most common devices used for these purposes. Display technology is changing rapidly, with new types of devices announced almost weekly, it seems, but you will encounter everything from the latest gadget to the incandescent bulb in Amateur Radio!

LIGHT-EMITTING DIODES

Light-emitting diodes (LEDs) are designed to emit light when they are forward biased so that current passes through their PN junctions. [E6B13] As a free electron combines with a hole, it gives off light of a specific wavelength or color. LEDs are very efficient light sources.

The color of the LED depends on the material or combination of materials used for the junction. LEDs are available in many colors. By controlling the energy difference between electrons and holes, LED color can also be controlled. The intensity of the light given off is proportional to the amount of current. Red, green and yellow LEDs are typically made from gallium arsenide, gallium phosphide or a combination of these two materials. Blue LEDs use materials such as silicon carbide or zinc selenide. A white LED is really a blue LED with a yellowish phosphor coating on the inside of the package that glows when struck by blue light from the LED. The combination of blue light given off by the LED and yellow light from the

Figure 5-25 — The schematic symbol for an LED is shown at A. B is a drawing of a typical LED case style.

phosphor appears white to the human eye.

LEDs are packaged in plastic cases or in metal cases with a transparent end. LEDs are useful as replacements for incandescent panel and indicator lamps. In this application they offer long life, low current drain and small size. One of their most important electronic applications is as numeric displays in which arrays of tiny LEDs are arranged to provide illuminated segments that form numbers. The schematic symbol and a typical case style for the LED are shown in **Figure 5-25**. [E6B10]

A typical red LED has a forward voltage of 1.6 V. Yellow and green LEDs have higher forward voltages (2 V for yellow and 4 V for green). The forward-bias current for a typical LED ranges between 10 and 20 mA for maximum brilliance. High-intensity LEDs used for lighting use much higher currents. As with other diodes, the current through an LED can be varied with series resistors. Varying the current through an LED will affect its intensity; the voltage across the LED, however, will remain fairly constant.

Before you go on, study test questions E6B10 and E6B13. Review this section if you have difficulty.

LIQUID-CRYSTAL DISPLAYS

In many display applications, *liquid-crystal displays (LCDs)* have become the technology of choice. Liquid crystals are long molecules that form a crystalline liquid suspension at room temperature. If undisturbed, the crystals form a helical structure that affects light passing through the suspension by rotating its polarization. In a display, the molecules are contained between glass plates that have polarizing filter layers attached, similar to Polaroid sunglasses. The combination of filters and the helical crystal structure allows light to pass through the filters with the proper polarization and the display appears transparent. [E6D05] **Figure 5-26** shows the construction of a typical LCD display.

When a voltage is applied across the liquid crystal material, the helical structure is distorted, causing the polarization of light passing through the suspension to be rotated differently. No longer matching the polarity of the front filter, the light is blocked and the display appears dark. By printing transparently thin electrodes on the front filter of the display, the crystal rotation can be turned on and off in different areas to form displays, numbers, and pictures.

A primary advantage of LCDs is that they consume very little power. [E6D15] Small numeric LCD modules are the dominant displays on digital multimeters and other portable devices for this reason. Recent

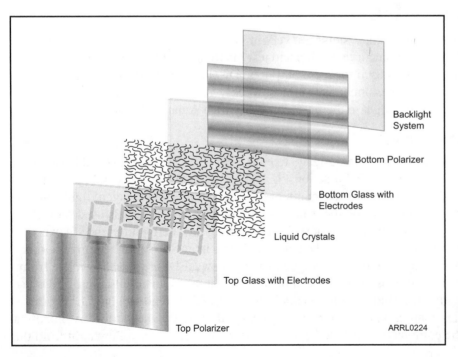

Backlight System

Bottom Polarizer

Bottom Glass with Electrodes

Liquid Crystals

Top Glass with Electrodes

Top Polarizer

ARRL0224

Figure 5-26 — The construction of an LCD display panel. Voltage on the electrodes changes the angle of nearby liquid crystals, causing those areas of the display to appear dark through the polarizing filter.

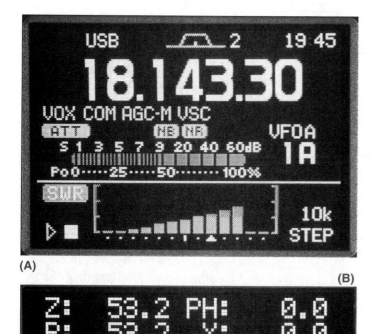

(A)

(B)

Figure 5-27 — LCDs are used in a variety of Amateur Radio equipment. Color LCD panels used in transceivers (A) display many important operating parameters at once. Simple two-line displays like the one shown at B are common on test equipment such as wattmeters and antenna analyzers.

advances in brightness, contrast and the ability to present color images have made them the dominant technology for displaying images, as well. LCDs have much lower operating voltages than the cathode ray tubes discussed below. Since there is no need for an electron gun and deflection circuitry, LCDs require no depth behind the display screen.

Of course, LCDs have their disadvantages, too. Because the mobility of the liquid crystal molecules is proportional to temperature, LCDs are slow to operate at low temperatures, and may be damaged by high temperatures. Some types of LCDs are difficult to read in bright sunlight and have low contrast when viewed from the side of the display. Nevertheless, LCD image quality is sufficient for the technology to have largely displaced CRTs in televisions, computer displays and radio equipment front panels. **Figure 5-27** shows a pair of typical LCDs found in the modern ham shack.

Before you go on, study test questions E6D05 and E6D15. Review this section if you have difficulty.

CATHODE-RAY TUBES

A *cathode-ray tube (CRT)* is a vacuum tube display device found in older TVs, computer displays and oscilloscopes. It displays images by using a beam of electrons to "write" on a surface visible through the tube's front surface. While rapidly being supplanted by LCD displays, CRTs will still be present in many types of equipment for years to come.

The CRTs used for different applications vary in their construction, but all operate on the same basic principles. **Figure 5-28** illustrates the operation of a simple CRT. A heated cathode produces electrons that are formed into a beam and accelerated toward the front of the tube by an anode at a high dc voltage. The beam is directed toward a phosphorescent material or "phosphor" on the inside face of the tube that glows when struck by the electrons. By scanning the electron beam across the tube face, an image is produced. The relative brightness of each part of the image depends on the intensity of the electron beam striking the phosphor at that point.

In a CRT used for TV or computer displays, horizontal and vertical position of the electron beam is controlled by magnetic fields generated by external coils. While sufficient for television purposes, *magnetic deflection* is not suitable for measurement purposes. To precisely display high-frequency signals on a lab-type oscilloscope, *electrostatic deflection* must be used. [E6D13] In this case, a voltage between two *deflection plates* deflects the electron beam. As the voltage between the plates varies, the electron beam's direction is changed to illuminate a different location. One set of plates moves the beam horizontally and the other vertically.

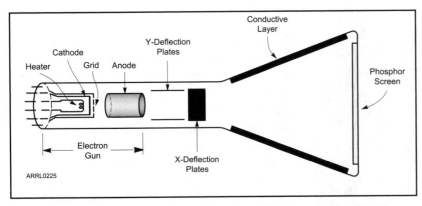

Figure 5-28 — The basic construction of a cathode-ray tube. Electrons are produced by the heated cathode and accelerated toward the front of the tube by the anode. Electrostatic deflection plates move the beam across the phosphor screen at the front of the tube, creating visible images.

Persistence describes the length of time that the tube's phosphor emits light after being illuminated by the electron beam. Or more generally, the length of time an image remains on a CRT screen after the electron beam is turned off. [E6D01] The choice of phosphor used on the inside surface of the display screen determines the tube's persistence and suitability for a particular use.

The anode voltage in a CRT accelerates electrons from the cathode down the length of the tube. Higher anode voltages pull more electrons away from the cathode (increasing beam intensity), and move them at faster speeds, resulting in a brighter image. CRTs should be operated at the specified anode voltage. Operating the tube with excessive anode voltage causes the electrons to strike the phosphor on the front of the tube with sufficient force to produce X-rays. X-rays created in this manner leave the tube through the front of the glass toward the viewer, creating a safety hazard. [E6D02] The increased operating voltage would also make the electrons move at a higher speed, and that would result in less deflection of the electron beam. Therefore, the display image size would be reduced.

Before you go on, study test questions E6D01, E6D02, and E6D13. Review this section if you have difficulty.

CHARGE-COUPLED DEVICES

A *charge-coupled device (CCD)* is a special type of integrated circuit that combines analog and digital signal-handling abilities. In its simplest form, the CCD is made from a string of small capacitors with a MOSFET on its input and output. The first capacitor in the string stores a sample of the input voltage. When a control pulse biases the MOSFETs to conduct, the first capacitor passes its sampled voltage on to the second capacitor and takes another sample. With each successive control pulse, the input samples are passed to the next capacitor in the string. When the MOSFETs are biased off, each capacitor stores its charge. This process is sometimes described as a "bucket brigade," because the analog signal is sampled and then passed in stages through the CCD to the output. [E6D03]

It is easy to imagine that one application of a CCD is as an audio delay line. Each signal sample taken at the input will appear at the output some time later. Delay lines can be used to create audio effects and certain types of audio filters.

A clock signal controls the times at which the CCD samples the input signal voltage, so this is a form of digital sampling. The actual sampled voltage is an analog value, however. The sampled voltage is not transformed into a specific numeric value, as it would be in an analog-to-digital converter, so a CCD cannot be used as an analog-to-digital converter. [E6D14]

Figure 5-29 shows a two-dimensional CCD used for imaging in which an array of sensing elements called *pixels* are coated with light-sensitive material. The light-sensitive material in each pixel *photocharges* a single CCD capacitor. The photogenerated charge on

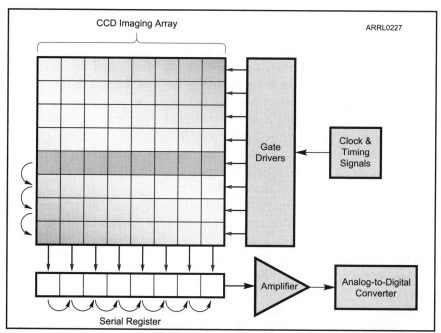

each capacitor is proportional to the amount of light striking the sensing element. The stored charges are then shifted out of the CCD array a line at a time as signals corresponding to each pixel. [E6D04] This CCD array forms the basis for a modern photographic or video camera.

Figure 5-29 — The construction of a two dimensional CCD array that might be used in a digital camera. When a control pulse biases the array to conduct, the first row passes its sampled voltage on to the next row and takes another sample. With each successive control pulse, the input samples are passed to the next row until they reach the edge of the array. There they are transferred to a serial register and digitized.

Before you go on, study test questions E6D03, E6D04, and E6D14. Review this section if you have difficulty.

5.3 Digital Logic

Digital electronics is an important aspect of Amateur Radio. Everything from simple digital circuits to sophisticated microcomputer systems are used in modern Amateur Radio. Even simple equipment often includes a microprocessor. Applications in the radio realm include digital communications, code conversion, signal processing, station control, frequency synthesis, satellite telemetry, message handling and other information-handling operations.

You've already been exposed to digital logic functions in your General license studies. For the Extra exam, we'll examine the fundamentals of digital logic and digital electronics. You'll be introduced to synchronous logic circuits and their applications. Are you ready — 1 or 0? I'll take that as a 1!

LOGIC BASICS

Boolean Algebra

The fundamental principle of digital electronics is that a signal can have only a finite number of discrete values or *states*. In *binary* digital systems signals may have two states, represented in base-2 arithmetic by the numerals 0 and 1. The binary states described as 0 and 1 may represent an OFF and ON condition or as space and mark in a communications transmission such as CW or RTTY. **Figure 5-30** illustrates a typical binary signal.

The simplest digital devices are switches and relays. Electronic digital systems, however, are created using *digital ICs* — integrated circuits that generate, detect or in some way process digital signals. Whether switches or microprocessors, though, all digital systems use common mathematical principles known as *logic*. We'll start with the rules for combining

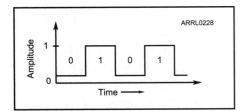

Figure 5-30 — A typical binary signal may have either of two signal levels, shown as 0 when the signal has the amplitude representing a logic value of 0 and 1 when the signal amplitude represents 1. Digital systems use many other combinations of amplitude and timing.

different digital signals, called *combinational logic*. These rules are derived from the mathematics of binary numbers, called *Boolean algebra*, after its creator, George Boole.

In binary digital logic circuits each combination of inputs results in a specific output or combination of outputs. Except during transitions of the input and output signals (called *switching transitions*), the state of the output is determined by the simultaneous state(s) of the input signal(s). A combinational logic function has one and only one output state corresponding to each combination of input states. The output of a combinational logic circuit is determined entirely by the information at the circuit's inputs.

The individual circuits that perform the simplest mathematical functions are called *elements*. Combinational logic elements may perform arithmetic or logical operations. Regardless of their purpose, these operations are usually expressed in arithmetic terms. Digital circuits add, subtract, multiply, and divide but normally do it in binary form using two states that we represent with the numerals 0 and 1.

Binary digital circuit functions are represented by equations using Boolean algebra. The symbols and laws of Boolean algebra are somewhat different from those of ordinary algebra. The symbol for each logical function is shown here in the descriptions of the individual logical elements.

The logical function of a particular element may be described by listing all possible combinations of input and output values in a *truth table*. Such a list of all input combinations and their corresponding outputs characterize, or describe, the function of any digital device. [E7A10]

One-Input Elements

There are two logic elements that have only one input and one output: the *noninverting buffer* and the *inverter* or NOT circuit (**Figure 5-31**). [E6C11] The noninverting buffer simply passes the same state (0 or 1) from its input to its output. In an inverter or NOT circuit, a 1 at the input produces a 0 at the output, and vice versa. NOT indicates inversion, negation or complementation. Notice that the only difference between the noninverting buffer and the inverter is the small circle or triangle on the output lead. This is used to indicate inversion on any digital-logic circuit symbol. The Boolean algebra notation for NOT is a bar over the variable or expression.

The AND Operation

A *gate* is defined as a combinational logic element with two or more inputs and one output state that depends on the state of the inputs. Gates perform simple logical operations and can be combined to form complex switching functions. So as we talk about the logical operations used in Boolean algebra, you should keep in mind that each function is implemented by using a gate with the same name. For example, an AND gate implements the AND operation.

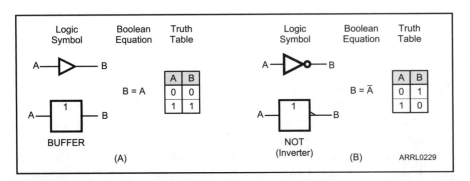

Figure 5-31 — Schematic symbols for a noninverting buffer (A) and inverter or NOT function (B) are shown. The distinctive (triangular) shape is used by ARRL and in most US publications. The square symbol is an alternate. The Boolean equation for the buffer and a truth table for the operation are also given.

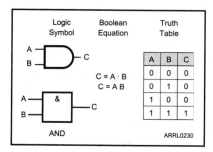

Figure 5-32 — Schematic symbols for a two-input AND gate are shown. The distinctive (round-nosed) shape is used by ARRL and in most US publications. The square symbols in this and following figures are an alternate. The Boolean equation and truth table for the operation are also given.

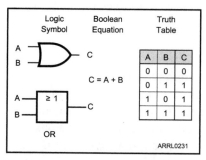

Figure 5-33 — Schematic symbols for a two-input OR gate are shown. The Boolean equation and truth table for the operation are also given.

The AND operation results in a 1 only when all inputs or *operands* are 1. That is, if the inputs are called A and B, the output is 1 only if A and B are both 1. In Boolean notation, the logical operator AND is usually represented by a dot between the variables (•). The AND function may also be signified by no space between the variables. Both forms are shown in **Figure 5-32**, along with the schematic symbol for an AND gate. [E6C07]

The OR Operation

The OR operation results in a 1 at the output if any or all inputs are 1. In Boolean notation, the + symbol is used to indicate the OR function. The OR gate shown in **Figure 5-33** is sometimes called an INCLUSIVE OR. In Boolean algebra notation, a + sign is used between the variables to represent the OR function. [E6C09] Study the truth table for the OR function in Figure 5-33. You should notice that the OR gate will have a 0 output only when all inputs are 0. [E7A08]

The NAND Operation

The NAND operation means NOT AND. A NAND gate (**Figure 5-34**) is an AND gate with an inverted output. A NAND gate produces a 0 at its output only when all inputs are 1. In Boolean notation, NAND is usually represented by a dot between the variables and a bar over the combination, as shown in Figure 5-34. [E6C08, E7A07]

The NOR Operation

The NOR operation means NOT OR. A NOR gate (**Figure 5-35**) produces a 0 output if any or all of its inputs are 1. In Boolean notation, the variables have a + symbol between them and a bar over the entire expression to indicate the NOR function. When you study the truth table shown in Figure 5-35, you will notice that a NOR gate produces a 1 output only when all of the inputs are 0. [E6C10]

The EXCLUSIVE NOR Operation

The EXCLUSIVE OR (XOR) operation results in an output of 1 if only one of the inputs is 1, but if both inputs are 1 then the output is 0. The Boolean expression $\oplus$ represents the EXCLUSIVE OR function. Inverting the XOR function results in the EXCLUSIVE NOR (XNOR) operation. **Figure 5-36** shows the schematic symbol for an EXCLUSIVE NOR gate and its truth table. [E7A09]

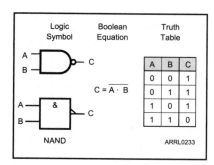

Figure 5-34 — Schematic symbols for a two-input NAND gate are shown. The Boolean equation and truth table for the operation are also given.

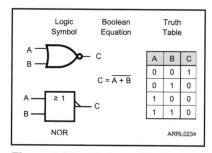

Figure 5-35 — Schematic symbols for a two-input NOR gate are shown. The Boolean equation and truth table for the operation are included.

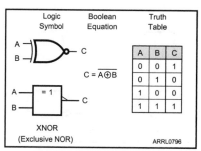

Figure 5-36 — Schematic symbols for a two-input EXCLUSIVE NOR (XNOR) gate are shown. The Boolean equation and truth table for the operation are included.

Positive- and Negative-True Logic

Logic systems can be designed to use two types of *logic polarity*. *Positive* or *positive-true logic* uses the highest voltage level (HIGH) to represent binary 1 and the lowest level (LOW) to represent 0. If the opposite representation is used (HIGH = 0 and LOW = 1), that is *negative* or *negative-true logic*. In the element descriptions to follow, positive logic will be used. [E7A11, E7A12]

Positive and negative logic symbols are compared in **Figure 5-37**. Small circles (*state indicators*) on the input side of a gate signify negative logic. The use of negative logic sometimes simplifies the Boolean algebra associated with logic circuits.

Consider a circuit having two inputs and one output, and suppose you desire a HIGH output only when both inputs are LOW. A search through the truth tables shows the NOR gate has the proper characteristics. The way the problem is posed (the words "only" and "both") suggests the AND (or NAND) function, however. A negative-logic NAND is functionally equivalent to a positive-logic NOR gate. The NAND symbol better expresses the circuit function in the application just described. **Figure 5-38** shows the implementation of a simple function as a NOR or NAND gate, depending on the logic convention chosen. Notice that the truth tables prove the circuits perform identical functions. You should verify this to be true by comparing the lists of input and output conditions.

Tri-State Logic

In digital circuits it is common for many ICs to be connected in parallel on a *data bus* or *address bus* to share data and addressing information. In this configuration, only one IC output may control the signals on the bus at a time and all other IC outputs must "stand by" by changing their outputs to act as a high impedance without attempting to drive the bus connection to a HIGH or LOW state. ICs with this ability are referred to as *tri-state logic* in which an output can be HIGH, LOW, or high-impedance. [E6C03, E6C04]

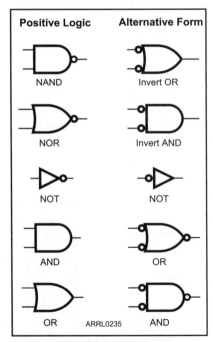

Figure 5-37 — A comparison of positive- and negative-true logic symbols for the common logic elements. The small circles at an input or output indicate negative-true logic.

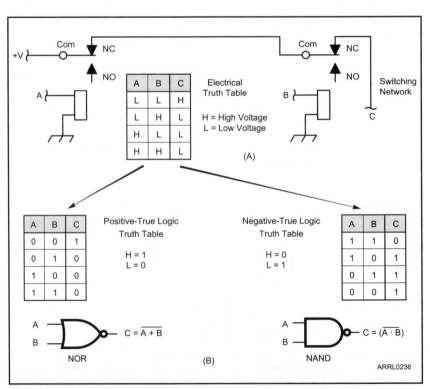

Figure 5-38 — The truth table at A describes a desired logic function. The two truth tables at B describe positive logic (NOR) and negative logic (NAND) implementations of that function.

Before you go on, study test questions E6C03, E6C04, E6C07 through E6C11, and E7A07 through E7A12. Review this section if you have difficulty.

SYNCHRONOUS LOGIC

The output state of a *sequential-logic* circuit is determined by both its present inputs and previous output states. The dependence on previous output states implies that the circuit must have some type of memory.

Flip-Flops

A *flip-flop* (also known as a *bistable multivibrator*) is a binary sequential-logic element with two stable states: the *set* state (1 state) and the *reset* state (0 state). The term *bistable* means that the circuit has two stable states and it can stay in either of them indefinitely. [E7A01] Thus, a flip-flop can store one bit (from *bi*nary dig*it*) of information. A flip-flop used to store information is sometimes called a *latch*. The schematic symbol for a flip-flop is a rectangle containing the letters FF, as shown in **Figure 5-39**. (These letters may be omitted if no ambiguity results.)

Flip-flop inputs and outputs are normally identified by a single letter, as outlined in **Tables 5-1** and **5-2**. A letter followed by a subscripted letter (such as D_C), means *that* input is dependent on the input of the subscripted letter (input D_C is dependent on input C). Note that R and S cannot both be in the 1 state at the same time. The *state table* of Figure 5-39 (a truth table for synchronous logic) shows that you can't be sure what the outputs (Q and $\overline{Q}$) will be if both inputs are high at the same time. There are normally two output signals that are complements of each other, designated Q and $\overline{Q}$ (read as Q NOT). If Q = 1 then $\overline{Q}$ = 0 and vice versa.

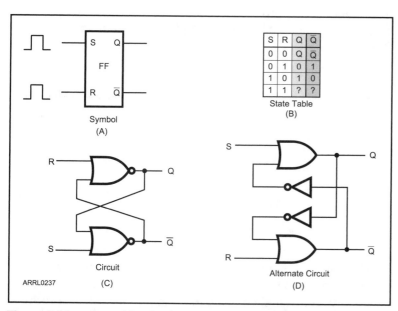

Symbol (A)

S	R	Q	$\overline{Q}$
0	0	Q	$\overline{Q}$
0	1	0	1
1	0	1	0
1	1	?	?

State Table (B)

Circuit (C)

ARRL0237

Alternate Circuit (D)

Figure 5-39 — A positive-logic, unclocked R-S flip-flop is used to illustrate the operation of flip-flops in general. Where Q and $\overline{Q}$ are shown in the state table, the previous output states are retained. A question mark (?) indicates an invalid state in which you can not be sure what the output will be. C shows an R-S flip-flop made from two NOR gates. The circuit shown at D is another implementation using two OR gates and two inverters.

Synchronous and Asynchronous Flip-Flops

The terms synchronous and asynchronous are used to characterize a flip-flop or individual inputs to an IC. In *synchronous flip-flops* (also called *clocked*, *clock-driven* or *gated* flip-flops), the output follows the input only at prescribed times determined by the clock input. *Asynchronous flip-flops* are sometimes called *unclocked* or *data-driven* flip-flops because the output can change whenever the inputs change. Asynchronous inputs are those that can affect the output state independently of the clock. Synchronous inputs affect the output state under control of the clock input.

Dynamic versus Static Inputs

Dynamic (edge-triggered) inputs can affect the outputs only when the clock

Table 5-1
Flip-Flop Input Designations

Input	*Action*	*Restriction*
R (Reset)	1 resets the flip-flop. A return to 0 causes no further action.	Simultaneous 1 states for R and S inputs are not allowed.
S (Set)	1 sets the flip-flop. A return to 0 causes no further action.	
R_D (Direct Reset)	1 causes flip-flop to reset regardless of other inputs.	
S_D (Direct Set)	1 causes flip-flop to set regardless of other inputs.	
J	Similar to S input.	Simultaneous 1 states for J and K inputs cause the flip-flop to change states.
K	Similar to the R input.	
G (Gating)	1 causes the flip-flop to assume the state of G's associated input.	
C (Control)	1 causes the flip-flop to assume the state of the D input.	A return to 0 produces no further action.
T (Toggle)	1 causes the flip-flop to change states.	A return to 0 produces no further action.
D (Data)	A D input is always dependent on another input, usually C. C = 1, D = 1 causes the flip-flop to set. C = 1, D = 0 causes the flip-flop to reset.	A return of C to 0 causes the flip-flop to remain in the existing state (set or reset).

changes state. This type of input is indicated on logic symbols by a small triangle (called a *dynamic indicator*) on the symbol where the input line is attached. Unless there is an inversion or *negation* indicator (a small circle or triangle outside the symbol), the 0-to-1 transition is the recognized transition. This is called *positive-edge triggering*. The negation indicator means that the input is *negative-edge triggered*, and responds to 1-to-0 transitions.

Static (level-triggered) inputs are recognizable by the absence of the dynamic indicator on the logic symbol. Input states (1 or 0) of static inputs are what causes the flip-flop to act.

Many different types of flip-flops exist. These include the clocked and unclocked R-S, D,

Table 5-2
Flip-Flop Output Designations

Output	Action	Restrictions
Q (Set)	Normal output	Only two output states are possible: Q = 1 and Q = 0
Q̄ (Reset)	Inverted output	Output states are the opposite of Q: Q̄ = 0 and Q̄ = 1

Notes
1) Q̄ is the complement of Q.
2) The normal output is normally marked Q or unmarked.
3) The inverted output is normally marked Q̄. If there is a 1 state at Q, there will be a 0 state at Q̄.
4) Alternatively the inverted output may have a (negative) polarity indicated (a small right triangle on the outside of the flip-flop rectangle at the inverted output line). For lines with polarity indicators, be aware that a 1 state in negative logic is the same as a 0 state in positive logic. This is the convention followed by the International Electrotechnical Commission.

T, J-K and master/slave (M/S) types. These names come from the type of input signals of the flip-flop. See Tables 5-1 and 5-2 for a summary of the operation of these signals.

R-S, D, T and J-K Flip-Flops

One simple circuit for storing a bit of information is the R-S (or S-R) flip-flop. The inputs for this circuit are *set* (S) and *reset* (R). Figure 5-39 shows the schematic symbol for this type of flip-flop along with a state table to help you determine the outputs for given input conditions. Two implementations of this circuit using discrete digital logic gates are also shown. When S = 0 and R = 0 the output will stay the same as it was at the last input pulse. This is indicated by a Q in the state table.

If S = 1 and R = 0, the Q output will change to 1. If S = 0 and R = 1, the Q output will change to 0. If both inputs became 1 simultaneously, the output states would be indeterminate, meaning there is no way to predict how they may change. A clocked R-S flip-flop also has a clock input, in which case no change in the output state can occur until a clock pulse is received. [E7A13]

The Q output of a D flip-flop takes the value of the D input when the clock signal *triggers* the flip-flop by changing from 0 to 1. [E7A15] When a D flip-flop is wired with the Q NOT (Q̄) output connected back to the D input, it forms a *toggle* or T flip-flop, also known as a *complementing* flip-flop. The timing diagram of **Figure 5-40** shows that the flip-flop output changes state with each positive clock pulse. So if the output is 0 initially, it will change to a 1 on the leading edge of the first positive clock pulse and it will change back to 0 on the leading edge of the next positive clock pulse. The output of a bistable T flip-flop changes state two times for every two trigger pulses applied to the input. [E7A02] Another way to say this is that the T flip-flop provides one output pulse for every two input pulses. The result is that a bistable multivibrator, such as this flip-flop circuit, electronically divides

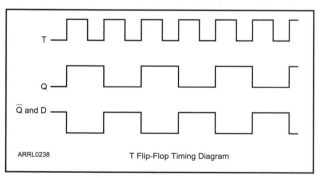

ARRL0238   T Flip-Flop Timing Diagram

Figure 5-40 — The timing diagram for a T flip-flop is shown. Since the output changes at half the rate of the input signal, the T flip-flop acts as a divide-by-two counter.

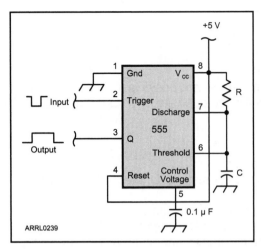

Figure 5-41 — A 555 timer IC can be connected to act as a one-shot multivibrator. See text for the formula to calculate values for R and C.

the input signal by two. [E7A03] Two such flip-flops could be connected to divide the input signal by four, and so on. [E7A04] (There is more about digital frequency-divider circuits later in this chapter.)

The J-K flip-flop has two inputs, J and K. If J and K are in opposite states (1 and 0), the Q output takes the value of J when the clock signal triggers the flip-flop. If J and K are both 0, the Q output does not change. If J and K are both 1, the J-K flip-flop emulates the T type flip-flop by toggling the Q output. [E7A14]

One-Shot or Monostable Multivibrator

A *monostable multivibrator* (or *one-shot*) has one stable state and an unstable (or quasi-stable) state. In the multivibrator circuit the circuit can stay in the unstable state for a time determined by RC circuit components connected to the one-shot. When triggered, it switches to the unstable state and then returns after a set time to its original, stable state. [E7A06] When the time constant has expired the one-shot reverts to its stable state until retriggered. Thus, the one-shot outputs a single pulse when triggered.

In **Figure 5-41**, a 555 timer IC is shown connected as a one-shot multivibrator. The action is started by a negative-going trigger pulse applied between the trigger input and ground. The trigger pulse causes the output (Q) to go positive until capacitor C charges to two-thirds of V_{CC} through resistor R. At the end of the trigger pulse, the capacitor is quickly discharged to ground. The output remains at logic 1 for a time determined by:

$$T = 1.1 \, RC \qquad \qquad \text{(Equation 5-3)}$$

where:

R is resistance in ohms.
C is capacitance in farads.
T is time in seconds.

Astable Multivibrator

An *astable* or *free-running multivibrator* is a circuit that continuously switches between two unstable states. [E7A05] It can often be synchronized with an input signal of a frequency that is slightly higher than the astable multivibrator free-running frequency.

An astable multivibrator circuit using the 555 timer IC is shown in **Figure 5-42**. Capacitor C_1 repeatedly charges to two-thirds V_{CC} through R_1 and R_2, and discharges to one-third V_{CC}

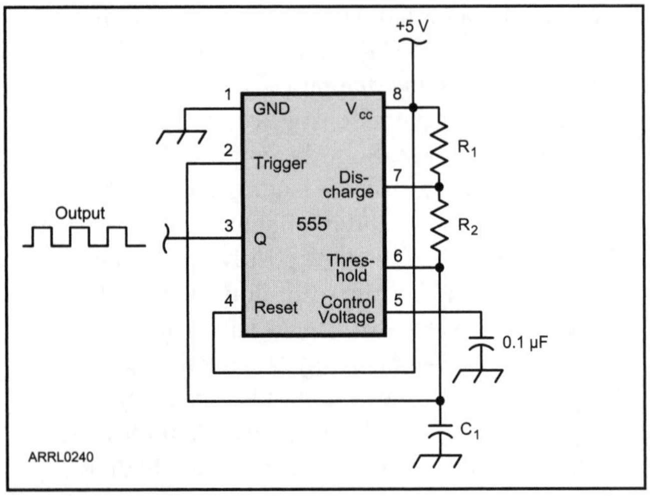

Figure 5-42 — A 555 timer IC can be connected as an astable multivibrator. See text for the formula to calculate values of R_1, R_2 and C_1.

through R_2. The ratio (R_1:R_2) sets the duty cycle. The frequency is determined by:

$$f = \frac{1.46}{C_1\,[R_1 + (2 \times R_2)]}$$

(Equation 5-4)

where:
R is resistance in ohms.
C is capacitance in farads.

Before you go on, study test questions E7A01 through E7A06 and E7A13 through E7A15. Review this section if you have difficulty.

FREQUENCY DIVIDERS AND COUNTERS

A *counter*, *divider* or *divide-by-n* counter is a circuit composed of multiple flip-flops that stores pulses and produces an output pulse after a specified number (n) of input pulses have occurred. In a counter consisting of flip-flops connected in series, when the first stage changes state it affects the second stage and so on. Each input pulse toggles the counter circuit to the next state. The outputs from all of the flip-flops can form a composite output that forms a binary number representing the total pulse count.

A *ripple*, *ripple-carry* or *asynchronous* counter passes the count from stage to stage; each stage is clocked by the preceding stage so that the change in the circuit's state "ripples" through the stages. In a *synchronous* counter, each stage is controlled by a common clock so that the outputs of all stages change at the same time.

Most counters have the ability to clear the count to 0. Some counters can also be preset to a desired count. Some counters may *count up* (increment) and some *count down* (decrement). Up/down counter ICs are available, able to count in either direction, depending on the status of a control input.

Internally, a *decade counter* IC has 10 output states. Some counters have a separate output pin for each of these 10 states while others have only one output connected to the last bit of the counter. The last flip-flop stage produces one output pulse for every 10 input pulses. [E7F03]

Counter or divider circuits find application in various forms. Common uses for these circuits in Amateur Radio include marker generators and frequency counters.

Frequency Counters and References

One of the most accurate means of measuring frequency is the *frequency counter*. This instrument counts pulses of the input signal over a specified time and displays the frequency of the signal on a digital display. For example, if an oscillator operating at 14.230 MHz is connected to the counter input, 14.230 would be displayed. Some counters are usable well up into the gigahertz range. [E7F08, E7F09]

Most counters that are used at such high frequencies make use of a *prescaler* ahead of a lower-frequency counter. A special type of frequency divider circuit, the prescaler reduces a signal's frequency by a factor of 10, 100, 1000 or some other integer divisor so that a low-frequency counter can display the input frequency. [E7F01, E7F02]

Frequency-counter accuracy depends on an internal crystal-controlled reference oscillator, also called the *time base*. The more accurate the crystal reference, the more accurate the counter readings will be. [E7F07] A crystal frequency of 1 MHz has become more or less standard for use in the reference oscillator. The crystal should have excellent temperature stability so the oscillator frequency won't change appreciably with temperature changes.

For very high accuracy counting or for use in microwave operation, a *frequency reference* is required. It may be sufficient to use a signal from a time-and-frequency standard such as signals from a GPS satellite or from terrestrial stations such as WWV and WWVH. Frequency references are also available as separate pieces of equipment, such as a stabilized rubidium reference oscillator locked to the vibrations of rubidium atoms. Other techniques include temperature-controlled oscillators that use quartz crystals or high-Q dielectric resonators. [E7F05]

For counting very low frequencies many frequency counters measure the period of a signal and then compute the frequency. This method is much faster than counting slow pulses over a long period because the counter can use the high-frequency reference signal as its measurement reference, often achieving accurate measurements in a single cycle of the low-frequency signal. [E7F10, E7F11]

A frequency counter will measure the frequency of the strongest signal at its antenna or input connector. Counters can acquire their input signal by an antenna placed close to a transmitter rather than having a direct connection. For low-level signals, however, a probe or other input connection may be used. The measured signal must have a frequency within the measurement range of the frequency counter or prescaler circuit.

Marker Generators

FCC regulations require that your transmitted signal be maintained inside certain frequency limits. The exact frequency need not be known as long as it is not outside the limits. Staying inside the limits is not difficult to do, but you shouldn't rely on your rig's frequency display as the only method of determining your transmit frequency. Checking your radio's frequency calibration requires only a *marker generator* or frequency counter and some care. [E7F06]

A marker generator employs a high-stability crystal-controlled oscillator that generates a series of reference signals at known frequency intervals. The oscillator is followed by one or more divider circuits that generate harmonics at intervals that are fractions of the reference oscillator frequency. For example, you could use a 1-MHz crystal oscillator and a decade counter to produce a 100-kHz signal. You should be able to hear the harmonics of this circuit every 100 kHz with your receiver. Your rig may include such a crystal-controlled marker generator.

For example, dividing a higher frequency signal down to a 25-kHz fundamental frequency will produce marker signals spaced every 25 kHz, provided that the oscillator harmonics are strong enough to be heard throughout the desired range. When these harmonics are detected in a receiver, they mark the exact edges of the amateur frequency assignments. Most US amateur band limits are exact multiples of 25 kHz, whether at the band extremes or at points marking the subdivisions between types of emission and license-class restrictions. But if the receiver calibration is not accurate enough to positively identify which harmonic you are hearing, there may still be a problem in determining how close to the band edge you are operating.

Rather than using a 25-kHz oscillator, an oscillator frequency of 100 kHz might be used. A divider circuit coupled to the oscillator provides markers at increments of other than 100 kHz. In the circuit of **Figure 5-43**, two divide-by-2 D-type flip-flops are switch selected to produce markers at 50 and 25-kHz intervals. [E7F04] A NOR gate and diodes provide the required switching and signal routing. These D-type flip-flops are wired to form T flip-flops by connecting the $\overline{Q}$ output of each to its D input. These flip-flops form two divide-by-two counters to provide 50-kHz and 25-kHz outputs.

Frequency stability is a very important consideration for a crystal-controlled marker generator. Using a 1-MHz crystal also makes it relatively easy to compare harmonics of the crystal-oscillator signal to WWV or WWVH signals at 5, 10 or 15 MHz, for example, and to adjust the oscillator to *zero beat* (that is, the frequencies of the beating signals or their harmonics are equal).

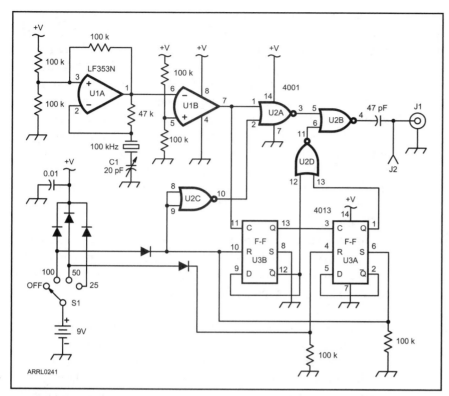

Figure 5-43 — This schematic diagram shows a simple 100, 50 and 25-kHz marker generator. Two switch-selected divide-by-two stages produce the 50 and 25-kHz markers.

Before you go on, study test questions E7F01 through E7F11. Review this section if you have difficulty.

LOGIC FAMILIES

While there may be just one symbol for a NAND gate or a decade counter, there are lots of different types of digital circuits and components that can perform the necessary functions. Digital logic device manufacturers strive for consistency across their product line, creating a whole series of logic ICs with similar characteristics optimized for certain types of applications, such as low power consumption or high switching speed. All of the logic elements are available within that technology so that the circuit's signals are compatible with other similar chips. These groups of similar ICs are called *families*. Within a logic family, all of the devices will have similar input and output signal constraints and will switch at approximately the same speed.

TTL Characteristics

Transistor-transistor logic (TTL) is one of the oldest bipolar logic families, so called because the gates are made entirely of bipolar transistors. Most TTL ICs are identified by 7400/5400 series numbers. For example, the 7490 is a decade counter IC.

All of the logic elements described earlier in this section have TTL IC implementations. Some examples are the 7400 quad NAND gate, the 7432 quad OR gate and the 7408 quad AND gate. (The *quad* in these names refers to the fact that there are four individual gate circuits on the single IC chip.) Other examples of 7400 series ICs are the 7404 hex inverter, and the 7476 dual flip-flop. (*Hex* refers to the six inverters on a single IC.) The 7404 contains six separate inverters, each with one input and one output, in a single 14-pin package. A diagram of the 7404 is shown in **Figure 5-44**. The 7476 includes two J-K flip-flops on one IC.

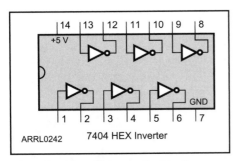

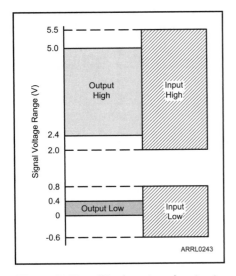

Figure 5-44 — This symbol is the schematic representation of a 7404 TTL hex inverter.

Figure 5-45 — The input and output signal-voltage ranges for the TTL family of logic devices.

TTL ICs require a +5-V power supply. [E6C01] The supply voltage can vary between 4.7 and 5.3 V, but 5 V is optimum. There are also limits on the input-signal voltages. To ensure proper logic operation, a HIGH, or 1 input must be between 2 V and 5 V and a LOW, or 0 input must be no greater than 0.8 V. To prevent permanent damage to a TTL IC, HIGH inputs must be no greater than 5.5 V, and LOW inputs no more negative than –0.6 V. TTL HI outputs will fall somewhere between 2.4 V and 5.0 V, depending on the individual chip and load current. TTL LOW outputs will range from 0 V to 0.4 V. The ranges of input and output levels are shown in **Figure 5-45**. Note that the guaranteed output levels fall conveniently within the input limits. This ensures reliable operation when TTL ICs are interconnected.

TTL inputs that are left open, or allowed to "float," will cause the internal circuitry to assume a HIGH or 1 state, but operation may be unreliable. If an input should be HIGH, it is better to tie the input to the positive supply through a *pull-up resistor* (usually a 1 to 10-kΩ resistor). [E6C02]

There are several variants of the TTL family that provide different characteristics and are identified by letters following the "74" in the part number. For example, a logic device number beginning with "74LS" is from the Low-power Schottky TTL family and a part number beginning with "74HC" is from the High-speed CMOS version of TTL logic. Within a family, it is almost always the case that parts with the same logic function will have the same pin connections. For example, all the inverters in the 7404, 74LS04, 74H04, 74S04, 74HC04, and so on families will have the same pin connections or *pinouts* as in Figure 5-44.

CMOS Characteristics

Complementary metal-oxide semiconductor (CMOS) devices are composed of N-channel and P-channel FETs combined on the same substrate. [E6A13] Because both N and P-channel FETs can be combined on the same substrate, the circuitry can be placed in a smaller-sized area. This also helps reduce the cost of these ICs. CMOS logic has become the most widely used form of digital logic in the world because of its high switching speed, small size of the individual gates and other elements, and far low power consumption than TTL. When a CMOS gate is not switching, it draws very little power, for example. [E6C05]

One of the most popular CMOS families is the parts carrying 4000-series part numbers. For example, a 4001 IC is a quad, two-input NOR gate. The 4001 contains four separate NOR gates, each with two inputs and one output. Some other examples are the 4011 quad NAND gate, the 4081 quad AND gate and the 4069 hex inverter.

Mentioned previously, the 74HC00-series of part numbers are pin-compatible with the 7400 TTL family, offering equivalent switching speed at much lower power. If you come across a device whose part number begins 74C or 74HC device you should be aware that a C in the part number probably indicates that it is a CMOS device.

The 4000-series of CMOS ICs (model numbers between 4000 and 4999) will operate over a much larger power-supply range than TTL ICs. The power-supply voltage can vary from 3 V to as much as 18 V. CMOS output voltages depend on the power-supply voltage. A HIGH output is generally within 0.1 V of the positive supply connection, and a LOW output is within 0.1 V of the negative supply connection (ground in most applications). For example, if you are operating CMOS gates from a 9 V battery, a logic 1 output will be somewhere

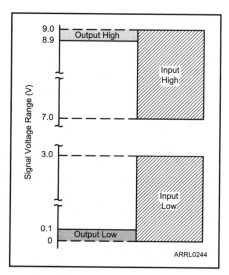

Figure 5-46 — This diagram shows 4000-series CMOS-device input and output signal-voltage ranges with a 9-V supply.

between 8.9 and 9 V, and a logic 0 output will be between 0 and 0.1 V.

The switching threshold for CMOS inputs is approximately half the supply voltage. **Figure 5-46** shows these input and output voltage characteristics. The wide range of input voltages gives the CMOS family great immunity to noise, since noise spikes will generally not cause a transition in the input state. Even the TTL-compatible CMOS families have a slightly higher noise immunity because of their wider HIGH and LOW signal ranges. [E6C06]

All CMOS ICs require special handling because of the thin layer of insulation between the gate and substrate of the MOS transistors. Even small static charges can cause this insulation to be punctured, destroying the gate. CMOS ICs should be stored with their pins pressed into special conductive foam. They should be installed in a socket or else a soldering iron with a grounded tip should be used to solder them on a circuit board. Wear a grounded wrist strap when handling CMOS ICs to ensure that your body is at ground potential. Any static electricity discharge to or through the IC before it installed in a circuit may destroy it.

BiCMOS Logic

Because both bipolar and CMOS technology each have certain performance advantages, combining them in a single IC creates devices that can operate with the speed and low output impedance of bipolar transistors and the high input impedance and reduced power consumption typical of CMOS. This is referred to as BiCMOS technology. [E6C12, E6C13] This allows ICs to combine analog functions such as amplifiers and oscillators, with digital functions such as control and switching circuits.

Before you go on, study test questions E6A13, E6C01, E6C02, E6C05, E6C06, E6C12 and E6C13. Review this section if you have difficulty.

5.4 Optoelectronics

Optics may not seem to have a lot to do with radio, but there are many components that are hybrids of optical and electronic functions. These are called *optoelectronics* and they make use of the optical properties of semiconductors to perform useful functions. The most commonly utilized optical properties are *photoconductivity* in which light interacts with a semiconductor to change its conductivity and the *photovoltaic effect* in which light causes current to flow.

PHOTOCONDUCTIVITY

To understand photoconductivity, we must start with the *photoelectric effect*. In simple terms, this refers to electrons being knocked loose from the atoms of a material when light shines on it. While a complete explanation of light's interaction with semiconductor material is beyond the scope of this book, we will simply describe some of the basic principles behind photoelectricity.

Let's revisit the basic structure of an atom as shown in **Figure 5-47**. The nucleus contains protons (positively charged particles) and neutrons (with no electrical charge). The number of protons in the nucleus determines the atom's element. Carbon has 6 protons, oxygen has 8

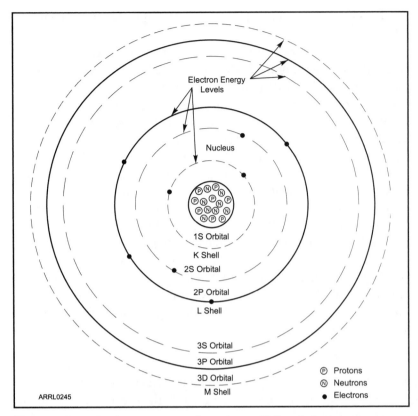

Figure 5-47 — In the structure of an atom, shells of electrons surrounding the nucleus have increasing energy levels with increasing distance from the nucleus. The photoelectric effect is caused by electrons in the outermost shell absorbing a photon of light with sufficient energy to cause them to leave the atom. As free electrons they can flow as current.

and copper has 29, for example. The nucleus of the atom is surrounded by the same number of negatively charged electrons as there are protons in the nucleus so that an atom has zero net electrical charge.

The electrons surrounding the nucleus are found in specific energy levels, as shown in Figure 5-47. The increasing energy levels are shown as larger and larger spheres surrounding the nucleus. (While this picture is not really accurate, it will help you get the idea of the atomic structure.) For an electron to move to a different energy level it must either gain or lose a certain amount of energy. One way that an electron can gain the required energy is by absorbing electromagnetic energy in the form of a photon of light. The electron absorbs the energy from the photon and jumps to a new energy level. An electron that has absorbed energy and jumped to a higher energy level is called *excited*.

If the light photon has enough energy the electron can be freed completely from the atom. In a metallic conductor this *free electron* can now flow as an electric current. The current can then flow through a circuit connected to the material illuminated by the photons. This is the basis of the photoelectric effect.

The Photoconductive Effect

With this simple model of the atom in mind, it is easy to see that an electric current through a wire or other material depends on electrons being pulled away or knocked free from atoms. The rate of electrons moving past a certain point in the wire specifies the current. Every material presents some opposition to this flow of electrons, and that opposition is the *resistivity* of the material. If you include the length and cross-sectional area of a specific object or piece of wire, then you know the resistance of the object:

$$R = \frac{\rho\, l}{A}$$

(Equation 5-5)

where:
  ρ is the lower case Greek letter rho, representing the resistivity of the material.
  l is the length of the object.
  A is the cross-sectional area of the object.
  R is the resistance.

Conductivity is the reciprocal of resistivity, and conductance is the reciprocal of resistance:

$$\sigma = \frac{1}{\rho} \qquad \text{(Equation 5-6)}$$

where σ is the lower case Greek letter sigma, which represents conductivity and

$$G = \frac{1}{R} \qquad \text{(Equation 5-7)}$$

where G is the conductance.

You just learned that according to the photoelectric effect electrons can be knocked loose from atoms when light strikes the surface of the material. With this principle in mind, you can see that those free electrons will make it easier for a current to flow through the material. But even if electrons are not knocked completely free of the atom, excited electrons in the higher-energy-level regions are more easily passed from one atom to another.

All of this discussion leads us to one simple fact: it is easier to produce a current when some of the electrons associated with an atom are excited. The conductivity of the material is increased and the resistivity is decreased. [E6F02] The total conductance of a piece of wire may increase and the resistance decrease when light shines on the surface. That is the nature of the *photoconductive effect*. Materials that respond to the photoconductive effect are said to exhibit *photoconductivity*. [E6F01]

The photoconductive effect is more pronounced and more important in crystalline semiconductor materials than in ordinary metal conductors. [E6F06] With a piece of copper wire, for example, the conductance is normally high so any slight increase because of light striking the wire surface will be almost unnoticeable. The conductivity of semiconductor crystals such as germanium, silicon, cadmium sulfide, cadmium selenide, gallium arsenide, lead sulfide and others is low when they are not illuminated but the increase in conductivity is significant when light shines on their surfaces. (This also means that the resistance decreases.)

Each material shows its biggest change in conductivity over a different range of light frequencies. For example, lead sulfide responds best to frequencies in the infrared region while cadmium sulfide and cadmium selenide are both commonly used in visible light detectors, such as are found in cameras.

Before you go on, study test questions E6F01, E6F02 and E6F06. Review this section if you have difficulty.

OPTOELECTRONIC COMPONENTS

Most semiconductor devices are sealed in plastic or metal cases so that no light will reach the semiconductor junction. Light will not affect the conductivity and hence the operating characteristics of such a transistor or diode. But if the case is made with a window to allow light to pass through and reach the junction, then the device characteristics will depend on how much light is shining on it. Such specially made devices have a number of important applications in Amateur Radio.

A *phototransistor* is a special device designed to allow light to reach the transistor junction. Light, then, acts as the control element for the transistor. In fact, in some phototransistors there is no base lead at all. In others, a base lead is provided, so you can control the output signal in the absence of light. You can also use the base lead to bias the transistor to respond to different light intensities. In general, the gain of the transistor is directly proportional to the amount of light shining on the transistor. A phototransistor can be used as a *photodetector* — a device that detects the presence of light.

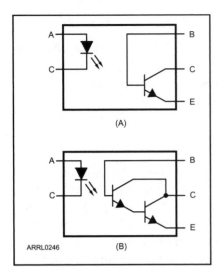

Figure 5-48 — Optocouplers consist of an LED that emits light and a phototransistor to detect light. The device shown at A uses a single-transistor detector while the unit shown at B uses a Darlington phototransistor to increase the current transfer ratio.

Optocouplers and Optoisolators

An *optocoupler* or *optoisolator* is an LED and a phototransistor sharing a single IC package. [E6F03] Applying current to the LED causes it to emit light and the light from the LED causes the phototransistor to turn on. Because they use light instead of a direct electrical connection, optoisolators provide one of the safest ways to transfer signals between circuits using widely differing voltages.

Optoisolators have a very high impedance between the light source (input) and the phototransistor (output). There is no current between the input and output terminals.

The LEDs in most optocouplers are infrared emitters, although some operate in the visible-light portion of the electromagnetic spectrum. For this reason they are often used when 120 V ac circuits are to be switched under the control of low-power digital signals. [E6F08]

Figure 5-48A shows the schematic diagram of a typical optocoupler. In this example, the phototransistor base lead is brought outside the package. The ratio of the output current to input current is called the *current transfer ratio* (CTR). As shown at B, a Darlington phototransistor can be used to improve the CTR of the device.

In an IC optoisolator, the light is transmitted from the LED to the phototransistor detector by means of a plastic light pipe or small gap between the two sections. It is also possible to make an optoisolator by using discrete components.

By combining an optocoupler with power transistors, the functions of an electromechanical relay can be implemented by solid-state components. The resulting *solid-state relay* (SSR) can operate much faster than an electromechanical relay and can be controlled directly by digital circuits. [E6F07]

A separate LED or infrared emitter and matching phototransistor detector can be separated by some small distance to use a reflective path or other external gap. In this case, changing the path length or blocking the light will change the transistor output. This can be used to detect an object passing between the detector and light source, for example.

The Optical Shaft Encoder

An optoelectronic device widely used in radio equipment is an *optical shaft encoder*. It consists of an array of emitters and detectors. A plastic disc with a pattern of alternating clear and black radial bands rotates through a gap between the emitters and detectors as illustrated

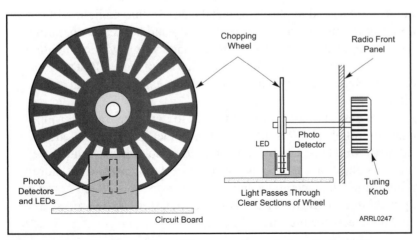

Figure 5-49 — An illustration of the operation of an optical shaft encoder, often used on the tuning and control knobs of transceivers.

in **Figure 5-49**. By using an array of emitters and two detectors, a microprocessor can detect the rotation direction and speed of the wheel. [E6F05] Modern transceivers use a system like this to control the frequency of a synthesized VFO. To the operator, the tuning knob may feel like it is mechanically tuning the VFO, but there is no tuning capacitor or other mechanical linkage connected to the knob and light-chopping wheel. Inexpensive shaft encoders are often used for switches and selector controls, as well.

PHOTOVOLTAIC CELLS

The photoelectric effect can also be put to use to generate electrical energy as well as to change or control electrical properties. At a PN junction, such as in a diode, charge carriers create a depletion region as described earlier in this chapter. For charge carriers to cross the junction, voltage must be applied.

If a PN junction is exposed to light, photons will be absorbed by the electrons in the semiconductor material. [E6F12] If the photons have the correct energy, free electrons in the N-type material can be excited sufficiently to move across the depletion region into the P-type material. Effectively, this is the same as a hole moving the other way. As long as the junction is illuminated, electrons and holes can be made to flow across the junction, creating a voltage difference from one side of the junction to the other.

If a circuit is provided between the two sides of the junction, the voltage caused by the photons being absorbed by the electrons will cause a current to flow. This current represents the conversion of light energy from the photons to electrical energy carried by the electrons in the circuit. This is the *photovoltaic effect*. [E6F04] A PN-junction designed to absorb photons and create electrical energy is called a *photovoltaic* or *PV cell*. The cross section of a photovoltaic cell is shown in **Figure 5-50**.

The voltage developed by a photovoltaic cell depends on the material from which it is made. For

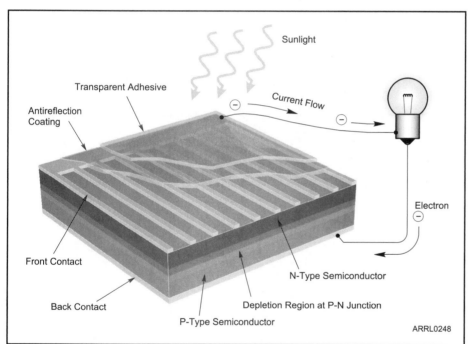

Figure 5-50 — The construction of a photovoltaic cell. As long as the PN junction is illuminated, electrons and holes can absorb energy and flow across the junction causing a current to flow in the external circuit. This current represents the conversion of light energy from the photons to electrical energy carried by the electrons in the circuit.

example, a fully-illuminated cell made of silicon, the most common material used for PV cells, develops an open-circuit voltage of approximately 0.5 V. [E6F10, E6F11] The amount of current such a cell can produce is determined by the degree of illumination and the *conversion efficiency* of the material — the relative fraction of light energy that is converted to electrical energy in the form of current. [E6F09] Almost any semiconductor material can be made into a photovoltaic cell, but the highest efficiency in cells made from a single material is currently found in cells made from gallium arsenide (GaAs).

Photovoltaic energy is rapidly becoming a commercially viable option to generate electrical energy in large quantities. The photo-voltaic or *solar* cell shown in Figure 5-50 is made from semiconductor material but other types of optically-active material such as mixtures of metal and semiconductor material, organic molecules, and nanomaterials may be used.

Before you go on, study test questions E6F04 and E6F09 through E6F12. Review this section if you have difficulty.

Table 5-3

Questions Covered in This Chapter

5.1 Semconductor Devices	*5.2 Display Devices*	*5.3 Digital Logic*	*5.4 Optoelectronics*
E6A01	E6B10	E6A13	E6F01
E6A02	E6B13	E6C01	E6F02
E6A03	E6D01	E6C02	E6F03
E6A04	E6D02	E6C03	E6F04
E6A05	E6D03	E6C04	E6F05
E6A06	E6D04	E6C05	E6F06
E6A07	E6D05	E6C06	E6F07
E6A08	E6D13	E6C07	E6F08
E6A09	E6D14	E6C08	E6F09
E6A10	E6D15	E6C09	E6F10
E6A11		E6C10	E6F11
E6A12		E6C11	E6F12
E6A14		E6C12	
E6A15		E6C13	
E6A16		E7A01	
E6A17		E7A02	
E6B01		E7A03	
E6B02		E7A04	
E6B03		E7A05	
E6B04		E7A06	
E6B05		E7A07	
E6B06		E7A08	
E6B07		E7A09	
E6B08		E7A10	
E6B09		E7A11	
E6B11		E7A12	
E6B12		E7A13	
E6E04		E7A14	
E6E06		E7A15	
E6E07		E7F01	
E6E08		E7F02	
E6E11		E7F03	
		E7F04	
		E7F05	
		E7F06	
		E7F07	
		E7F08	
		E7F09	
		E7F10	
		E7F11	

Components and Building Blocks

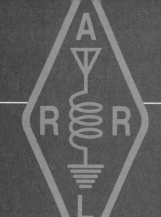

Chapter 6

Electronic Circuits

In this chapter, you'll learn about:
- Amplifier design, adjustment and operation
- Oscillators and frequency synthesis
- Digital signal processing (DSP) basics
- Mixers and modulators
- Detectors and demodulators
- Filter types and characteristics
- Impedance matching
- Power supply regulators

You have studied dc and ac electronics principles and the basic properties of some modern solid-state components. Now you are ready to apply those ideas to practical circuits, the subject of this chapter. Here we present examples and explanations of using those principles and components to create circuits used in Amateur Radio. The circuits include amplifiers (both low- and high-power), signal processing (such as oscillators and modulators), filters, impedance matching circuits and power supplies.

As you might imagine, entire books have been written about every topic covered in this chapter so we're unable to provide a complete treatment of every subject. The goal is to help you understand the circuits well enough to pass your Extra class exam. If you need more background or are interested in the details of the topic, consult the references listed on the *Extra Class License Manual* website (**www.arrl.org/extra-class-license-manual**) or the *ARRL Technical Information Service* web pages at **www.arrl.org/technical-information-service**. *The ARRL Handbook* is an excellent source of information, as well.

At various places throughout your study you will be directed to the related exam questions as a review exercise. Make sure you can answer the questions before going on. Ready for some circuits?

6.1 Amplifiers

When amateurs talk about amplifiers, the subject is often the piece of equipment that amplifies the output of a transceiver to several hundred watts or more. Far more numerous, however, are the much smaller amplifier circuits that increase the voltage and current of small signals in our radios and test instruments. These amplifiers operate from dc to microwaves and at signal levels all the way down to billionths of a watt. Yet all of them have much in common.

In a piece of equipment where several amplifier circuits work together, each separate amplifier circuit is called a *stage*, just as a rocket has stages. A stage whose output signal is the input to another amplifier, particularly in a transmitter, is called a *driver*. The last amplifier in a piece of transmitting equipment is called the *final amplifier*, or simply the *final*. The circuit to which an amplifier delivers its output power is called a *load*. A load may be anything from another circuit to a dummy load to an antenna. Attaching a load to the output of an amplifier is called *loading*.

Amplifier Gain

The *gain* of an amplifier is the ratio of the output signal to the input signal. An amplifier's *voltage gain* is the ratio of its output and input voltages. *Current gain* is the ratio of output and input current, and *power gain* is the ratio of output and input power levels.

This ratio can be a very large number when several amplifier stages are combined or when the gain is very large. The *decibel* is used to express the ratio in terms of a logarithm, making the number smaller and easier to work with. We often state the gain of a stage as a "voltage gain of 16" or a "power gain of 25," both simple ratios. But for very large ratios, such as an IF amplifier gain of 90 dB (1,000,000,000), it is easier to express and work with the smaller numbers. Decibels have been part of your license studies since the Technician exam and won't be covered again in this manual, but if you would like a refresher course in "dee-bees," a primer on the dB is available on the ECLM website.

Input and Output Impedance

An amplifier's *input impedance* is the equivalent impedance that it presents to the preceding or driving stage. There is no single component that creates the input impedance. It is a combined effect of the components making up the circuit and the way the circuit is constructed. Input impedance is measured as the ratio of input voltage to input current at the amplifier's input terminals. Input impedance almost always changes with frequency and may also change with the circuit's operating characteristics.

Output impedance is a bit more difficult to define. It is the equivalent impedance of a signal source representing the amplifier output. Low output impedance implies that the source can maintain a constant voltage over wide ranges of current. High output impedance sources maintain constant current while voltage may vary. Output impedance is highly dependent on the circuit's structure. Amplifiers that are intended to deliver significant output power generally have low output impedances, such as 50 Ω.

DISCRETE DEVICE AMPLIFIERS

The discussion in this section will be limited to amplifier circuits using bipolar transistors. FETs and vacuum tubes are also used in amplifiers, but the techniques and general circuit configurations also apply to bipolar transistor amplifiers. Some topics associated with tube circuits are discussed later in this chapter. To learn more about how FET and tube-type amplifiers function, we recommend that you turn to appropriate sections of *The ARRL Handbook*. At some point in your Amateur Radio career, you may want to learn about these amplifiers, but for the purpose of helping you pass your Extra class exam we will concentrate on bipolar transistor circuits.

Basic Circuits

Amplifier circuits used with bipolar transistors fall into one of three types, known as the *common-emitter, common-base* and *common-collector* circuits. "Common" means that the referenced transistor electrode — base, emitter, collector — serves as a reference terminal for both the input and output connections. The common terminal is usually circuit ground, as shown in the following circuits.

A bipolar transistor amplifier is essentially a current amplifier. Current in the base-emitter circuit controls larger currents in the collector-emitter circuit. To use the transistor as a voltage amplifier, the amplifier's output current flows through a resistive load and the resulting voltage, or change in voltage, is the amplifier's voltage signal output.

Bipolar-transistor diode junctions must be forward biased (see the Components and Building Blocks chapter) in order to act as current amplifiers. (Forward bias will be assumed when the word *bias* is used unless noted otherwise.) In circuits using an NPN transistor the collector and base must be positive with respect to the emitter. Conversely, when using a PNP

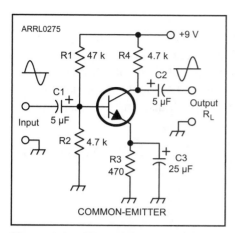

Figure 6-1 — The common-emitter amplifier circuit has a fixed-bias source (R1 and R2), degenerative emitter feedback for bias stabilization (R3), and a resistive collector load (R4). C1 and C2 are dc blocking capacitors for input and output signals. C3 acts as an emitter bypass to increase ac gain while maintaining stable behavior at dc.

transistor, the base and collector must be negative with respect to the emitter. The required bias is provided by a power source that supplies the collector-to-emitter voltage and emitter-to-base voltage. These bias voltages cause two currents to flow: collector-emitter current and base-emitter current. The direction of current flow depends on the type of transistor used.

Either type of transistor, PNP or NPN, can be used with a negative- or positive-ground power supply. Forward bias must still be maintained, however. The combination of bias and collector-emitter current is called the circuit's *operating point*. The operating point with no input signal present is called the circuit's *quiescent* or *Q-point*.

Common-Emitter Circuit

Common-emitter amplifiers are the type of amplifier most often used, so we'll begin by studying one as an example of amplifier operation. The common-emitter circuit is shown in **Figure 6-1**. You can recognize the common-emitter circuit by the value for resistance in the emitter circuit (R3 in Figure 6-1) being much smaller (or even absent) than that in the collector circuit (R4) or the emitter resistor being bypassed with a capacitor (C3 in Figure 6-1). [E7B12]

In bipolar transistors, the emitter current is in phase with the base current. As base-emitter current increases, so does collector-emitter current. As more collector current flows, the voltage drop across R4 increases, lowering the voltage at the circuit's output. When input voltage increases, causing higher base-emitter and collector-emitter current, the output voltage drops. Thus, the input and output signals are out of phase as shown in Figure 6-1.

R1 and R2 form a *voltage divider* to provide a stable operating point. This technique is called *fixed bias*. [E7B10] The amount of forward bias is a compromise between amplifier circuit gain and transistor power dissipation. The lower the forward bias, the less collector current will flow and the smaller the output signal will be.

As forward bias is increased the collector current and gain are higher but so is the temperature of the transistor's junctions. As temperature increases so does the gain of the transistor, causing collector-emitter current to increase even more. Even if the base-emitter bias is kept at a steady value by R1 and R2, for large enough emitter currents or if the circuits get hot enough, the mutually reinforcing conditions of increasing temperature and gain can result in *thermal runaway* in which the transistor junctions are overheated and destroyed. Some kind of negative feedback or *bias stabilization* is required to prevent this from happening.

One solution is to add resistance in the emitter circuit (R3) to create *degenerative emitter feedback* or *self-bias*. [E7B11, E7B15] Here's how it works: As emitter current increases, so does the dc voltage across R3. This increasing voltage reduces the base-emitter forward bias established by R1 and R2, also reducing emitter current. The resulting balancing act stabilizes the transistor's operating point and prevents thermal runaway. Other methods involve the use of temperature-sensitive components in the biasing circuits that reduce base-emitter forward bias at higher temperatures.

The low reactance of C3 at audio frequencies *bypasses* R3 for ac signals to increase ac signal gain as explained below. C1 and C2 are coupling capacitors, used to allow the desired signals to pass into and out of the amplifier, while blocking the dc bias voltages. Their values should be chosen to provide a low reactance at the signal frequency.

The input impedance of the common-emitter amplifier is fairly high — several kΩ is typical. The output impedance depends largely on the collector circuit resistance (R4). The common-emitter circuit has a lower *cutoff frequency* (the frequency at which current gain is reduced by half, or 3 dB) than does the common-base circuit, but it can provide the highest power gain of the three amplifier circuit configurations.

The common emitter amplifier's voltage gain (the letter A represents gain and A_V is voltage gain) is determined by the ratio of two resistances. The first resistance is the parallel combination of an external load, R_L, connected between the output terminals and the sum of R3 and R4 shown in the numerator of the following equation. ("//" means "the parallel combination of") The second resistance is the sum of R3 and the transistor's internal emitter resistance ($26 \text{ mV}/I_e$) in the denominator.

$$A_V \text{ (dc gain)} = \frac{-[(R4+R3)//R_L]}{\left(R3 + \dfrac{26}{I_e}\right)} \qquad \text{(Equation 6-1)}$$

where:

I_e is the emitter current in milliamperes

R_L is the output load.

The minus sign indicates that the input and output signals are out of phase — an increasing input voltage results in a decreasing output voltage.

Equation 6-1 tells us that as R3 gets larger, A_V gets smaller. Wouldn't it be nice if we could get rid of R3 and have higher gain, but retain its stabilizing effect on bias? That is the function of C3 — at dc, R3 provides bias stabilization, while for ac signals, the low reactance of C3 bypasses R3 and changes the equation for voltage gain to:

$$A_V \text{ (ac gain)} = \frac{-(R4//R_L)}{\left(\dfrac{26}{I_e}\right)} \qquad \text{(Equation 6-2)}$$

Thus, the common emitter circuit has a different value of gain at dc than for ac.

For example, if in the circuit of Figure 6-1 the emitter current is 1.3 mA and R_L is omitted:

$$A_V \text{ (dc gain)} = \frac{-(4700+470)}{(470+20)} = -10.55$$

and

$$A_V \text{ (ac gain)} = \frac{-4700}{20} = -235$$

That's quite a difference in gain! To express this voltage-gain ratio in decibels, just take the logarithm of 235 (the minus sign is ignored), and multiply by 20:

$$A_V \text{ (in dB)} = 20 \log (235) = 20 \times 2.37 = 47.4 \text{ dB}$$

In a more complete analysis of the circuit, the effects of C1 and C2 cannot be ignored at low frequencies. The increasing reactance of the coupling capacitors reduces gain at low frequencies until, at dc, input and output signals are completely blocked. The increasing reactance of C3 as frequency decreases also causes it to be less effective as a bypass so gain eventually drops to the dc value. As frequency increases, the transistor's current gain also decreases, reducing gain at high frequencies. For this reason, this value of A_V is said to be *mid-band gain*.

Common-Base Circuit

Figure 6-2 shows the common-base amplifier circuit. It can be recognized as the common-emitter "on its side," with the input signal applied to the emitter instead of the base. When viewed from this perspective, you will notice that the bias resistors, R1 and R2, have much the same configuration as in the common-emitter circuit, and that input and output

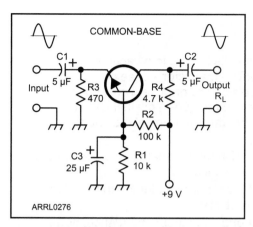

Figure 6-2 — The common-base amplifier is used when a low input impedance and high output impedance are needed. It is often used for RF preamplifiers and input circuits, since its input impedance is near 50 Ω. The circuit has approximately the same output and input current, but can have high values of voltage gain.

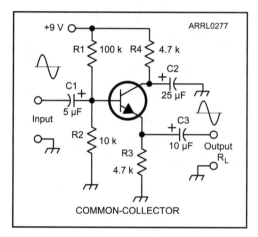

Figure 6-3 — The common-collector amplifier is also known as an emitter follower because the output voltage across the emitter resistor, R3, is in phase with and very nearly equal to that of the input signal.

coupling capacitors are also used. C3 bypasses the base to ground so that a steady dc forward bias current flows in the base-emitter circuit.

By using the transistor in this way the circuit has no current gain because the collector output current is equal to the emitter current less the small base current. This also means that the input and output signals are in-phase. The output impedance of the circuit is high so the collector current is almost independent of whatever R_L is connected to the output terminals. As a result, the circuit can have fairly large voltage gains:

$$A_V = \frac{R_L}{R_3}$$
(Equation 6-3)

The primary use of a common-base amplifier is as an *impedance converter* when signals from a low-impedance source (such as a 50-Ω feed line) must drive a higher impedance load (such as another amplifier circuit's input). Common-base amplifiers are frequently used as receiver preamplifiers.

Common-Collector or Emitter Follower Circuit

The common-collector transistor amplifier in **Figure 6-3**, sometimes called an *emitter follower* amplifier, is another example of a circuit used for impedance conversion. It has high input impedance and low output impedance. The input impedance is approximately equal to the load impedance connected to the output terminals, R_L, divided by $(1 - \alpha)$. (Remember that α is the ratio of collector to emitter current and is close to 1.) Having input impedance depend on load impedance is a disadvantage of this type of amplifier, especially if the load impedance varies with frequency.

The input and output signals are in phase, as shown in Figure 6-3. C2 is a collector bypass capacitor. This amplifier also uses input- and output-coupling capacitors, C1 and C3. [E7B14] R3 is both a bias stabilization resistor and as the *emitter load resistor* also develops the output voltage [E7B13] The common-collector circuit can be recognized by the output voltage being across the emitter resistor and the collector being kept at ac ground with a bypass capacitor. Common-collector amplifiers are often used as *buffer* amplifiers that isolate a low-power or sensitive stage from a heavy output load.

Before you go on, study test questions E7B10 through E7B15. Review this section if you have difficulty.

Each of the transistor's electrodes — emitter, base, collector — has its analog in a vacuum tube electrode. The analog of a transistor's emitter is the tube's cathode. The transistor's base corresponds to a tube's grid and the transistor's collector corresponds to the tube's anode or plate. Thus, each amplifier circuit shown in Figures 6-1, 6-2 and 6-3 has an analog in vacuum tube amplifier circuits. **Figure 6-4** shows the three tube amplifier configurations. And like the transistor amplifiers, the tube circuits have unique characteristics:

✔ Common-cathode: Input signal applied to the grid, relatively high input impedance and power gain, often requires neutralization at VHF, output is out-of-phase with the input signal.

✔ Grounded-grid: Input signal applied to the cathode, no current gain, low input impedance matches well to 50-Ω feed line, grounded grid reduces need for neutralization. [E7B18]

✔ Common-anode (cathode follower): Input signal applied to the grid, high input impedance, no voltage gain, rarely used in amateur equipment.

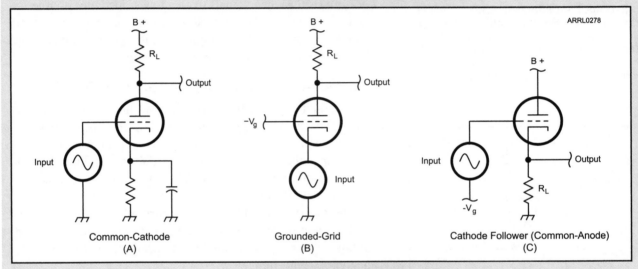

Figure 6-4 — Each of the common tube amplifier configurations — common-cathode (A), grounded-grid (B), common-anode (C) — has its analog in transistor amplifier circuits.

Before you go on, study test question E7B18. Review this section if you have difficulty.

OP AMP AMPLIFIERS

The *operational amplifier* or *op amp* is a high-gain, *direct-coupled*, *differential* amplifier that amplifies dc signals as well as ac signals. [E7G12] Direct-coupling means that the circuit's internal components and stages are connected directly together without blocking, coupling, or bypass capacitors, so that it works with dc and ac signals in exactly the same way. The input to a differential amplifier is the difference between two input signals.

The first op amps were designed for use in analog computers where they performed such mathematical operations as multiplying numbers and extracting square roots; hence the name operational amplifier. By connecting external components between the inputs and the output in different ways, a wide variety of circuit characteristics can be created. Op amps are some of the most versatile ICs available.

Operational amplifiers have two inputs, one *inverting* and one *non-inverting*, as shown in **Figure 6-5**. Signals connected to the inverting (labeled –) and non-inverting (labeled +)

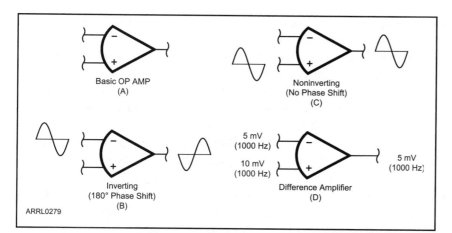

Figure 6-5 — Part A shows the basic schematic symbol for an operational amplifier (op amp). Parts B through D show how the output responds to signals at the op amp's inputs.

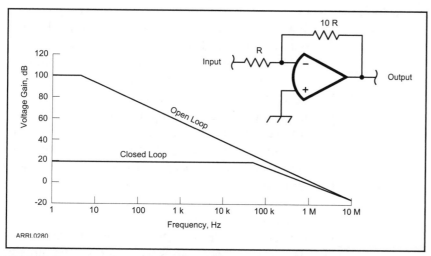

Figure 6-6 — The open-loop gain and closed-loop gain are shown as a function of frequency. The vertical separation between the curves is a measure of the feedback or gain margin.

inputs result in out-of-phase and in-phase output signals, respectively. Because it is a differential amplifier, the op amp amplifies the difference between the signals at its two inputs, regardless of the absolute voltage level at either input — it is only the difference that matters.

Op Amp Characteristics

A theoretically perfect (*ideal*) op amp would have the following characteristics: infinite input impedance, zero output impedance, infinite voltage gain, flat frequency response within its frequency range and zero output when the input is zero. [E7G08, E7G14, E7G15] Because of this, the characteristics of op amp circuits are controlled by components external to the op amp itself. These criteria can be approached in a practical IC op amp, but not realized completely as described in the following paragraphs.

The voltage gain of a practical op amp without feedback (*open-loop gain*) is often as high as 120 dB (1,000,000). Op amps are rarely used as amplifiers in the open-loop configuration, however. Usually, some of the output is fed back to the inverting input, where it acts to reduce and stabilize the circuit gain. The more negative feedback that is applied, the more stable the amplifier circuit will be.

The gain of an ideal op amp does not vary with changing frequency. The open loop gain of a practical op amp does decrease linearly with increasing frequency, however. The *small-signal bandwidth* of an op amp is the frequency range over which the open-loop voltage gain is equal to or greater than 1 (0 dB).

The gain of the circuit with negative feedback is called the *closed-loop gain*. The higher the open-loop gain, the more negative feedback that can be used and still create a circuit with a useful amount of closed-loop gain. Because open-loop gain is many times greater than closed-loop gain, the circuit gain is determined solely by the external feedback network components. By connecting the op amp in a closed-loop circuit as shown in **Figure 6-6,** circuit gain remains constant over a wide frequency range.

The *power bandwidth* of an op amp is a function of its *slew rate*, and is always less than the small-signal bandwidth. Slew rate is a measurement of the maximum output voltage swing per unit time. Values of 1 to 30 volts per microsecond are typical of modern devices.

If the input terminals of an op amp are shorted together, the output voltage should be zero. With most op amps there will be a small output voltage, however. This voltage offset results from imbalances between the op amp's input transistors. The op amp's *input-offset voltage* specifies the voltage between the amplifier inputs that will produce a zero output voltage, assuming the amplifier is in a closed-loop circuit. [E7G13] Offset-voltages range from millivolts in consumer-grade devices down to nanovolts or microvolts in premium op amps.

The temperature coefficient of offset voltage is called *drift*. Drift is usually considered in relation to time as the heat generated by the op amp itself or by associated circuitry causes the offset voltage to change. Specified at the op amp's input terminals, a few microvolts per degree Celsius is a typical drift specification.

Op amps can be assembled from discrete transistors but better thermal stability results from fabricating the circuit on a single silicon chip. IC op amps are manufactured with bipolar, JFET and MOSFET transistors, either exclusively or in combination.

Basic Circuits

The most common application for op amps is in negative-feedback circuits operating from dc to perhaps a few hundred kilohertz. Provided the op amp has sufficient open-loop gain, the circuit's gain and frequency response are determined almost entirely by the components that provide the external feedback connection. The differential inputs on an op amp allow for both inverting and non-inverting circuits.

Op amps make excellent, low-distortion amplifiers. They can be used to make oscillators that generate sine, square and even sawtooth waves. Used with negative feedback, their high input impedance and linear characteristics make them ideal for use as instrumentation amplifiers that amplify signals for precise measurements. There are many books, such as *IC Op-Amp Cookbook* by Walter Jung, that describe useful op-amp circuits.

The op amp's high gain amplifies the difference between the voltages between its inputs. Applying negative feedback causes the op amp to attempt to drive the input difference voltage to zero. The op amp's high input impedance allows current into or out of the inputs to be ignored. The usefulness of these two negative feedback concepts — input differential driven to zero and no input current —

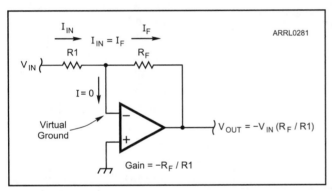

Figure 6-7 — Because of its high gain, negative feedback forces the op amp to keep the inverting and non-inverting inputs at nearly the same voltage. To do so requires balancing input current through R1 and feedback current in R_F, resulting in voltage gain of $-R_F/R1$.

will become apparent as we derive the gain for the simple inverting op amp circuit in **Figure 6-7**. (The circuit is inverting because the input and output signals are out of phase.)

The op amp's high gain forces the voltages at the inverting and non-inverting terminals to be approximately equal. Since the non-inverting input is connected to ground, the voltage at the inverting input will be forced to ground potential, no matter what the value of the circuit input and output voltages (as long as they are within the power supply range). Maintaining one input at ground potential without a direct ground connection is called a *virtual ground*.

Voltage gain for the inverting op-amp circuit in Figure 6-7 is determined solely by R1 (the input resistor) and R_F (the feedback resistor). In order to maintain the inverting input at ground potential, any input current $I_{IN} = V_{IN} / R1$ must be balanced by an equal and opposite feedback current $I_F = -V_{OUT} / R_F$, or:

$$\frac{V_{IN}}{R_1} = \frac{-V_{OUT}}{R_F}$$

so

$$A_V = \frac{V_{OUT}}{V_{IN}} = \frac{-R_F}{R_1}$$ (Equation 6-4)

Op amp circuit gain is generally stated as a magnitude ($|A_V|$) and either as inverting or non-inverting. This dependence only on external components makes computing circuit gain easy as shown in the following examples:

Example 6-1

What is the voltage gain of the circuit in Figure 6-7 if R1 = 1800 Ω and R_F = 68 kΩ? [E7G10]

$$|A_V| = \frac{R_F}{R_1} = \frac{68000}{1800} = 38$$

Example 6-2

What is the voltage gain of the circuit in Figure 6-7 if R1 = 10 Ω and R_F = 470 Ω? [E7G07]

$$|A_V| = \frac{R_F}{R_1} = \frac{470}{10} = 47$$

Example 6-3

What is the voltage gain of the circuit in Figure 6-7 if R1 = 3300 Ω and R_F = 47 kΩ? [E7G11]

$$|A_V| = \frac{R_F}{R_1} = \frac{47000}{3300} = 14$$

Example 6-4

What will be the output voltage of the circuit in Figure 6-7 if R1 = 1000 Ω and R_F = 10 kΩ and the input voltage = 0.23 V? [E7G09]

$$|A_V| = \frac{R_F}{R_1} = \frac{10000}{1000} = 10$$

The circuit is inverting, so $V_{OUT} = -A_V V_{IN} = -10 (0.23) = -2.3$ V

> *Before you go on, study test questions E7G07 through E7G15. Review this section if you have difficulty.*

CLASSES OF OPERATION

Figure 6-8 shows a set of *characteristic curves* for a typical NPN transistor amplifier, such as the circuit in Figure 6-1. The lines on this graph show how the collector current changes as the collector to emitter voltage varies as the base-emitter bias level changes.

- Collector-Emitter Voltage (V_{CE}) — the horizontal axis of the graph shows the voltage from the collector to the emitter
  - Collector Current (I_C) — the vertical axis of the graph shows the collector current
  - Base Current (I_B) — Each of the slanted lines describes the relationship between V_{CE}

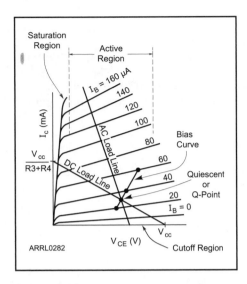

Figure 6-8 — Transistor characteristic curves showing the relationship between bias and operating point for a common-emitter, NPN transistor amplifier. The intersection of the bias curve and load lines determines the Q-point of the circuit. The input signal moves the operating point back and forth along the ac load lines on either side of the circuit's Q-point.

and I_C for a particular value of I_B. Given a value for I_B and V_{CE}, the value of I_C can be determined. These are called *constant-I_B curves*. Note that as V_{CE} is reduced all of the constant-I_B curves merge toward the left-hand side of the graph, dropping steeply to reach the origin where both V_{CE} and I_C are zero.

This set of lines describes the operation of the amplifier's transistor. You can see that for a given V_{CE} (draw a vertical line from the horizontal axis through the set of base current curves) increasing I_B also increases I_C. Current gain (β) can be determined graphically by selecting a point where the vertical line intersects one of the constant-I_B curves and then reading I_C from the vertical axis. Instruments called *curve tracers* generate these curves automatically.

The remaining lines on the graph depend on the values of the components in the amplifier's circuit. They show how the circuit responds to input signals.

- Bias Curve — a sequence of points showing what happens to the circuit's Q-point as R1 and R2 are adjusted to supply different amounts of bias current. A *load line* is a line through the characteristic curves that describes the circuit's output for a certain type and value of load.

- DC Load Line — Given a specific value for R4, the load line represents the relationship between V_{CE} and I_C if a dc input signal is applied to the amplifier. If the dc signal increases I_B, the operating point moves away from the Q-point toward higher I_C and lower V_{CE}. If the dc signal decreases I_B, the operating point moves toward lower I_C and higher V_{CE}. The load line intersects the horizontal axis at $V_{CE} = V_{CC}$ (the power supply voltage). The load line intersects the vertical axis at $I_C = V_{CC} / (R4+R3)$.

- AC Load Line — In the case of the common-emitter amplifier circuit, since R3 is bypassed by C3 the behavior of the operating point for an ac input signal is different than for a dc signal. The circuit has the same Q-point for both ac and dc (the intersection of the ac and dc load lines, but I_C changes more for an ac input signal (higher ac gain). This is reflected in the higher slope of the ac load line.

As a signal is applied to the amplifier, the operating point shifts back and forth along the ac load line. This changes I_C and since the output signal $V_O = V_{CC} - I_CR4$, the output changes, as well. If the input signal increases, the operating point moves to the left (increasing I_B) as I_C increases and the output voltage will drop. If the input signal decreases, the operating point moves to the right (decreasing I_B) as I_C decreases and the output voltage will rise.

As the input signal or *drive level* increases, the operating point will move into the region in which the constant-I_B lines merge and become nearly vertical. This is the *saturation* region in which further increases in input do not result in increasing output. Similarly, as the input signal drops the operating point moves closer to the horizontal axis and no further increase in V_{CE} is possible. The transistor is said to be in the *cutoff* region. For a sine-wave input signal to a common-emitter amplifier, saturation is observed as distortion at the lowest values of output and cutoff as distortion at the highest values of output. On an oscilloscope such a signal would be said to be *flat-topping* as the peaks of the output signal take on a flattened appearance.

The choice of Q-point bias for the amplifier determines the maximum amplitude of an input signal for which the amplifier operates in the *linear region* between cut off and saturation. This also determines the amplifier's *operating class*. This is illustrated by the graph in **Figure 6-9**.

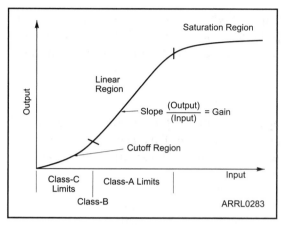

Figure 6-9 — This graph shows the three regions of amplifier operation: cutoff, linear and saturation. Each region of the curve corresponds to a different operating class.

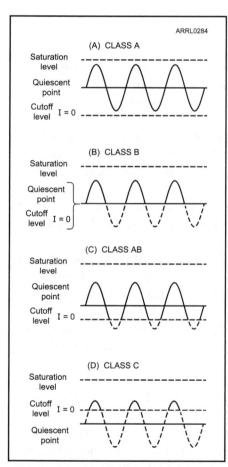

Figure 6-10 — Amplifier output waveforms for various classes of operation. All waveforms assume a sine-wave input signal.

The three basic operating classes are *Class A*, *Class B*, and *Class C*. The class of operation with characteristics intermediate between the A and B classes is called *Class AB* operation. **Figure 6-10** shows the output signal from amplifiers operating in Class A, B, AB and C. The difference between the classes is related to the amplifier's *conduction angle* — the portion of the input waveform cycle during which the amplifying device is conducting current.

Class A

In Class A operation, the bias and drive level are such that all of the output waveform is contained between the saturation and cutoff regions as shown in Figure 6-10A. This means the amplifier is operating in the linear region and the output is a linear (but larger) reproduction of the input signal and the conduction angle is a full 360°. The drive level is never large enough to cause the amplifier device to stop conducting current (cutoff) or to enter the saturation region.

The efficiency of a Class A amplifier is low, because there is always a significant amount of current drawn from the power supply — even with no input signal. The bias sets the Q-point to approximately halfway between saturation and cutoff. [E7B04] The maximum theoretical efficiency of a Class A amplifier is 50%, but in practice it is more like 25 to 30%.

Class B

Class B operation sets the bias at the cutoff level as shown in Figure 6-10B. In this case, the output signal only appears during half the input sine wave. This represents a conduction angle of 180°. The output is not linear as with a Class A amplifier but is still acceptable for many applications and a pair of Class B amplifiers are often connected in *push-pull* to supply both halves of the signal as in **Figure 6-11**. The advantage is increased efficiency. Up to 65% efficiency is theoretically possible with a Class B amplifier, and practical amplifiers often attain 60% efficiency.

Class AB

For a Class AB amplifier the drive level and bias are adjusted so that the conduction angle is between 180° and 360° and the operating efficiency is often more than 50%. Figure 6-10C shows the output signal for a Class AB amplifier. This class of operation is often used for voice signals when a small amount of distortion is acceptable as the price of improved efficiency. [E7B01]

Class C

Class C amplification requires that the bias be well into the amplifier's cutoff region and that the drive level be large enough to cause the amplifying device to conduct current during part of a half-cycle of the input signal. A Class C amplifier has a conduction angle of less than 180°. The output signal will consist of pulses at the signal frequency, as shown in Figure 6-10D. The amplifier is cut off for considerably more than half the cycle, acting much like a switch.

The result is that the operating efficiency can be quite high — up to 80% with proper design. Linearity is very poor, however, so Class C amplifiers are only used for CW and FM signals which do not require linear amplification and a tuned network at the output of the amplifier acts as a filter to reduce harmonics

Class D

The Class D amplifier goes beyond Class C operation with the tube or transistor acting entirely as a switch — either completely ON or completely OFF. This results in very low power dissipation by the tube or transistor and efficiencies of more than 90%. Class D amplifiers are used to amplify audio signals with the switching taking place at many times the highest audio frequency to be amplified. The switching action creates an output waveform that is a series of squared-off pulses very rich in harmonics. A low-pass filter at the amplifier output removes the harmonics while leaving signals in the desired range of frequencies unmodified. [E7B02, E7B03]

Before you go on, study test questions E7B01 through E7B04. Review this section if you have difficulty.

DISTORTION AND INTERMODULATION

The linearity of the amplifier stage is important because it describes how faithfully the input signal will be reproduced at the output. Any nonlinearity results in a distorted output. You can see that a Class A amplifier will have the least amount of distortion while a Class C amplifier produces a severely distorted output.

A consequence of nonlinearity is that the output waveform will contain harmonics of the input signal. Distortion causes a pure sine wave input signal to become a complex combination of sine waves at the output. If the input signal consists of sine waves of more than one frequency, such as a voice signal, the nonlinearity of the amplifier will also create *intermodulation products* or *intermod*. Intermodulation products are created at the sum and difference of all of the harmonics of the input signals and so would be considered spurious signals if transmitted.

The severity of intermodulation distortion depends on the *order* of the products that are created, even or odd. (Intermodulation in receivers is explained in detail in the Radio Signals and Measurements chapter.) *Even-order products* result in spurious signals near harmonics of the input signal and *odd-order products* result in spurious signals near the frequencies of the input signals. [E7B16] The higher the harmonics that are combined, the weaker the product. This means that the lower odd-order intermodulation products, specifically third-order, are more likely to cause interference to signals near the desired transmit frequency. [E7B17]

Tuned Amplifiers

By now you are probably asking, "But why would anyone want to have an amplifier that generates a distorted output signal?" That certainly is a good question. At first thought, it sure doesn't seem like a very good idea. You must remember, however, that every circuit design consists of compromises. We would like to have perfect linearity for our amplifiers, but we would also like them to have 100% efficiency. You have learned that those two ideals are exclusive. The closer you get to one of them, the further you get from the other. So you must compromise to create a useful design. Your particular application and operating conditions help determine which compromises are made.

In the case of trading efficiency against linearity in an RF amplifier, most RF amplifiers use a tuned output circuit to minimize the effects of the distortion. The tuned circuit stores electrical energy like a mechanical flywheel, which is used to store mechanical energy. Once set in motion, a flywheel will keep spinning at the same rate because of its high inertia. Even if energy is applied to the flywheel in pulses, the inertia of the wheel will smooth out the energy in the pulses to keep itself spinning.

The tuned circuit in the RF amplifier's output circuit is the electrical equivalent of a flywheel. Tuned amplifier output circuits are called *tank circuits* because they store RF energy. The energy in a tank current circulates back and forth as current flowing in and out of the inductance (L) and capacitance (C). Like a flywheel, the tank circuit smoothes the energy pulses that occur from turning the amplifying device off for parts of each cycle.

This is especially useful if you are amplifying a pure sine-wave signal, such as for CW, and want to take advantage of the increased efficiency offered by a Class C amplifier. The tuned circuit will reduce the unwanted harmonics generated by a nonlinear amplifier stage.

Selecting Operating Class

If you are amplifying an audio signal, linearity may be the most important consideration. Use a Class A amplifier for audio stages. For an AM or SSB signal, which is an RF signal envelope that varies at an audio rate, you would also want to use a linear amplifier. You may be willing to accept a bit of nonlinearity to obtain increased efficiency, in which case Class AB operation would be indicated with a tank circuit to remove the harmonics created by distortion. Using a Class C amplifier for AM or SSB signals would result in too much distortion and the output signal would occupy excessive bandwidth. [E7B07]

You can take advantage of the nonlinearity of a Class C amplifier by using it as a *frequency multiplier* stage. We mentioned earlier that one consequence of a nonlinear amplifier is that it would generate harmonic signals. If you want to multiply the frequency of a signal for operation on another band, you might do that by using a tank circuit tuned to a desired harmonic of the input frequency generated by the output of a Class C amplifier. This is especially useful for generating an FM signal at VHF or UHF where you may start with a signal in the HF range, apply modulation through a reactance modulator, then multiply the signal frequency to the desired

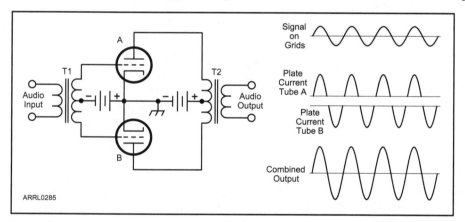

Figure 6-11 — Two triode vacuum tubes operating in Class B with their bias at the cutoff point form a push-pull amplifier. The waveforms show how each tube conducts for half of each cycle. The output transformer T2 combines the signals to form a complete output signal.

range on 2 meters or higher. The exact bias point for the Class C amplifier determines which harmonic will be the strongest in the output.

Class B amplifiers are often used for audio frequencies by connecting two of them back to back in a *push-pull circuit*. Figure 6-11 illustrates a simple vacuum tube push-pull amplifier and the waveforms associated with this type of operation. This circuit is popular in high-fidelity audio amplifiers. A push-pull amplifier can also be used at RF. While one tube is cut off, the other is conducting, so both halves of the signal waveform are present in the output. This reduces the amount of distortion in the output and will reduce even-order harmonics. [E7B06]

Before you go on, study test questions E7B06, E7B07, E7B16 and E7B17. Review this section if you have difficulty.

INSTABILITY AND PARASITIC OSCILLATION

Amplifier Stability

Excessive gain or undesired positive feedback may cause amplifier instability. Oscillation may occur in unstable amplifiers under certain conditions. Damage to the amplifying device from excessive heating is only the most obvious effect of oscillation. Even low-power amplifiers can experience instability — deterioration of noise figure, spurious signals generated by the oscillation and re-radiation of the oscillation through the antenna causing RFI to other services can occur. Negative feedback can stabilize an RF amplifier as described in the following sections on *neutralization* and *parasitic suppression*. [E7B05] Care in terminating the amplifier input and output, attention to proper grounding, and proper shielding of the input from the output can also prevent oscillation.

Neutralization

A certain amount of *interelectrode capacitance* exists between the input and output circuits in any amplifying device. In the bipolar transistor it is the capacitance between the collector or emitter and the base. In an FET it is the capacitance between the drain or source and the gate. In vacuum tubes it is the capacitance between the plate and grid circuit. So far we have simply ignored the effect that this capacitance has on the amplifier operation. In fact, it doesn't have much effect at the lower frequencies. Above 10 MHz or so, however, the capacitive reactance may be low enough that it must be accounted for in the amplifier's design.

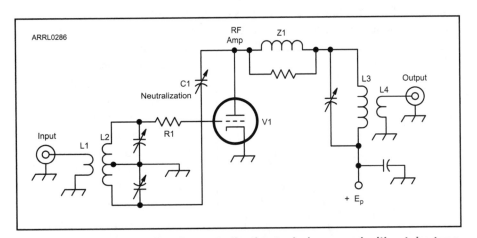

Oscillation can occur when some of the output signal is fed back in phase with the input signal (positive feedback). As the output voltage increases so will the feedback voltage. If the feedback is sufficiently strong, the re-amplified signal can build up to the point where it is self-sustaining and the amplifier is now an oscillator!

Figure 6-12 — An example of a neutralization technique used with a tube-type RF amplifier.

In order to overcome the positive feedback it is necessary to provide an alternate path from output to input that will supply a voltage that is equal to that causing the oscillation, but with opposite phase (negative feedback). [E7B08]

One neutralization technique for vacuum-tube amplifiers is shown in **Figure 6-12**. In this circuit the neutralization capacitor, C1, is adjusted to have the same value of reactance as the plate-to-grid capacitance that is causing the oscillation. By connecting C1 to the tuned input circuit, the phase shift results in the feedback signal being of opposite phase to the unwanted plate-to-grid feedback, canceling it.

With solid-state amplifiers it is more common to include a small value of resistance in either the base or collector lead of a low-power amplifier. Values between 10 and 20 Ω are typical. For higher power levels (above about 0.5 W), one or two ferrite beads are often used on the base or collector leads.

Parasitic Oscillations

Oscillations can occur in an amplifier on frequencies that have no relation to those intended to be amplified. Oscillations of this sort are called *parasitics* because they absorb power from the circuits in which they occur. Parasitics occur because of resonances that exist in the input or output circuits, enabling positive feedback to occur. They can occur above or below the operating frequency.

Parasitics are more likely to occur above the operating frequency as a result of stray capacitance and lead inductance along with interelectrode capacitances. In some cases it is possible to eliminate parasitics by changing lead lengths or the position of leads so as to change their capacitance and inductance and thus the resonant frequency.

An effective method of suppressing parasitics in HF vacuum tube amplifiers is to insert a parallel combination of a small inductor and resistor in series with the grid or plate lead. Such a *parasitic suppressor* is labeled Z1 in Figure 6-12. The coil's reactance is high enough at VHF/UHF that those signals must pass through the resistor while HF signals pass easily through the coil. The resistor value is chosen to load the VHF/UHF circuit heavily enough to prevent oscillation. Values for the coil and resistor are usually found experimentally as different layouts require different suppressor values.

In transistor amplifiers ferrite beads are often used on the device leads or on a short connecting wire placed near the transistor. These beads present high impedance to the VHF or UHF signals, blocking the parasitic currents. In general, the combination of parasitic suppressors and proper neutralization will prevent parasitic oscillations.

Before you go on, study test questions E7B05 and E7B08. Review this section if you have difficulty.

VHF, UHF AND MICROWAVE AMPLIFIERS

At VHF and higher frequencies special techniques are required to design and construct amplifiers. This is because the wavelength at these frequencies is short enough that the amplifying devices and the connecting leads are themselves a significant fraction of a wavelength. In addition, because of limits on the speed at which electrons can respond and interact, transistors and tubes alike begin to lose their ability to amplify signals in these frequency ranges. As a result, new types of devices and circuits are required

As you learned in the Components and Building Blocks chapter, a special type of IC — the MMIC — integrates many devices into a single package for use at microwave frequencies. Making the circuits smaller is one solution that works well at low power. This section presents three examples of other devices used to amplify signals at VHF and above. By no means are these the only solutions available!

Klystrons

While ordinary vacuum tubes work well into the VHF and low UHF ranges, the size of their internal structures makes the time that it takes for electrons to move between the cathode and the plate a limiting factor. Since tubes can only be made so small and still develop the power required for transmitters, different techniques are required to amplify these signals. There are several special tubes designed to work at microwave frequencies and the *klystron* tube is one example.

Instead of the cathode-to-plate transit time of electrons acting as a limitation in a klystron, the tube is constructed to take advantage of it. The electrons are formed into a beam that travels inside a cylindrical cavity called a *drift tube*. The tube has small gaps spaced at precise intervals. RF voltages are applied to the gaps and the resulting electric fields cause the electrons to accelerate or decelerate. As they travel through the drift tube, the repeated acceleration and deceleration causes the density of the electrons in the beam to vary in a regular pattern, like waves on the surface of a lake. When the beam of electrons reaches the anode, the variations in beam density become variations in current in the anode circuit. This technique is called *velocity modulation*. [E7B19]

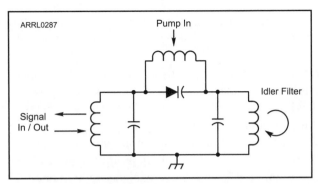

Figure 6-13 — The received signal is input into the parametric amplifier where a varactor diode acts as a variable capacitance. The capacitance of the diode is changed by the pump signal, varying the voltage across it. The idler filter converts the energy to the input signal's frequency.

Klystrons are used primarily at frequencies in and above the UHF range. They vary in size from pocket-sized tubes that develop milliwatts of power to human-sized behemoths that are used as broadcast television transmitters at multi-kilowatt levels. Because of the precise timing relationships involved in velocity modulation, a klystron is only usable over a narrow frequency range.

Parametric Amplifiers

One of Amateur Radio's many contributions to radio technology, the *parametric amplifier* was pioneered by amateurs at a time before transistors had much gain at VHF and higher frequencies. Tubes could amplify but added significant amounts of noise to weak signals. The parametric amplifier circuit shown in **Figure 6-13** amplifies signals by using a varactor diode as a small variable capacitor and a separate oscillator signal that "pumps" the diode to vary its reactance. The resulting amplification is low noise because the only contributor to noise in the output is that of the oscillator. [E7B20]

As the bias voltage on the varactor is changed, the capacitance changes as well. If the total amount of charge in the varactor doesn't change but the capacitance does change, the voltage must increase as capacitance decreases and vice-versa.

The *pump oscillator* and the input signal interact with the changing diode capacitance in such a way that the voltage of a signal at the input signal frequency is increased. An *idler filter* acts in a similar fashion to a tuned amplifier tank circuit, to reinforce signals at the desired frequency. Thus, energy is transferred from the pump oscillator frequency to the input signal frequency and signals at the input frequency are amplified. While VHF and UHF receive amplifiers use transistors today, parametric amplifiers are used at microwave, millimeter-wave and optical frequencies.

Microwave Semiconductor Amplifiers

Once well beyond the frequency at which semiconductor devices could develop useful transmit power, recent advances have made solid-state power amplifiers commonplace at microwave frequencies. This is partly due to the use of higher and higher frequencies for consumer and commercial networking and communications products. The mobile phone industry has a great demand for high-power amplifiers at frequencies of a few GHz. Even inexpensive cordless phone handsets now connect with their base stations at frequencies as high as 5.7 GHz! The vigorous development of microwave devices for high-volume products has led to large improvements in semiconductor materials and devices.

The physics of junction-type transistors and semiconductor materials places the upper frequency limit for these devices at a few GHz. The combination of advanced materials with the structure of an FET, however, results in significant gain and power handling in the microwave region up to 20 to 30 GHz. For example, a single GaAs (gallium arsenide) FET can develop watts of microwave power. [E7B21]

Output power of 100 W or more is not unusual from a power amplifier module that combines several FETs. These modules are available to amateurs both new and surplus and can be used on the 902 MHz bands and higher. They are easy to use, mechanically sturdy, and much less expensive than equipment based on discrete devices.

Before you go on, study test questions E7B19 through E7B21. Review this section if you have difficulty.

6.2 Signal Processing

In this section, we begin to study basic circuits applied to specific functions that are useful in radio: oscillators, modulators and demodulators, detectors, mixers, phase-locked loops and direct digital synthesis. All of these circuits are used to generate or manipulate signals in support of getting information from one point to another via radio. The general name for these functions is *signal processing*. Once exclusively the domain of analog circuits, digital techniques are rapidly becoming the dominant method of performing these functions. Nevertheless, understanding how an analog circuit performs a certain function is an excellent way of preparing to understand the digital method. We begin with the circuit that creates all signals — the oscillator.

OSCILLATOR CIRCUITS AND CHARACTERISTICS

When we discussed amplifiers, it was considered to be a problem if some of the output signal made its way back to the input in a manner that creates positive feedback. That circuit instability turned the amplifier into an *oscillator*. When we want the circuit to behave as an amplifier, this is a problem. But if we want to generate a signal (often at radio frequencies) without any input signal, the instability of an amplifier is just what we need!

To create an oscillator, we need three things: an amplifier with gain at the desired frequency, a circuit that provides positive feedback from the output of the amplifier to its input, and a filter that restricts the feedback to the desired frequency. These are connected in a *feedback loop* as shown in a block diagram by **Figure 6-14**.

The feedback loop is designed so that at the frequency of interest, the product of amplifier gain, A_V, and feedback ratio, β, is equal to or greater than 1. [E7H02] $A_V\beta$ is also known as *loop gain*, the total gain experienced by a signal all the way through the amplifier, the feedback circuit (the LC circuit in Figure 6-14), and back to the input. The requirement that $A_V\beta \geq 1$ for the circuit to oscillate is formally known as the *Barkhausen Criterion*. It's important to note that "1" in this case means, "an amplitude of 1 and a phase difference of zero degrees." If A_V is negative (inverting), then β must also be negative, meaning the feedback must have a 180-degree phase shift.

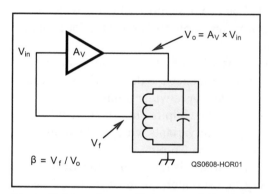

Figure 6-14 — An oscillator consists of an amplifier plus a feedback network, shown here as a parallel LC circuit. The LC circuit acts as a filter, restricting feedback to its resonant frequency.

Signals of the right frequency will be amplified, a portion fed back to the amplifier input, amplified again, and so on, becoming self-sustaining and creating a steady output signal. If the amplitude of the feedback is sufficient but the phase difference is not exactly right, then the returned portion of the output signal becomes progressively farther and farther out of phase on each trip. The result is that the oscillator's output consists mostly of a fundamental frequency plus small amounts of other signals — usually harmonics — for which loop gain is sufficient and the round trip phase difference is some integer multiple of 360 degrees.

Where does the necessary phase shift come from? Some of the phase shift may come from the amplifier itself. For example, the common-emitter amplifier you learned about in the preceding section has a phase shift of 180°. The remaining phase shift must come from the filter and feedback circuitry. Any kind of device or circuit that produces phase shift can be used, including RC or LC circuits and even transmission lines. A non-inverting amplifier such as the common-base or emitter-follower requires 0° or 360° of phase shift in the feedback path. As long as the requirements for total loop gain and phase are satisfied the oscillator will oscillate.

Figure 6-15 shows three different variations of the same oscillator circuit with the gain

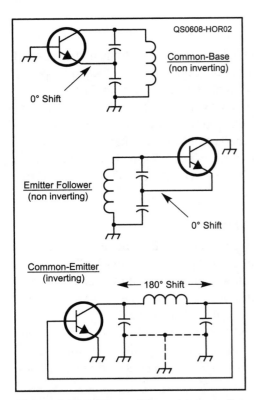

Figure 6-15 — Three different types of oscillator circuits. The type of amplifier circuit determines whether 0 or 180 degrees of phase shift is needed in the feedback path.

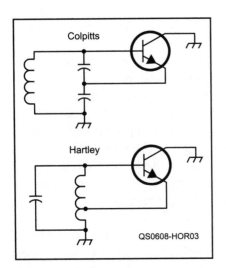

Figure 6-16 — The Colpitts and Hartley oscillators work on the same principle, but use different connections to the LC resonator to provide feedback.

provided by a transistor and a resonant LC tank circuit from which some of the amplifier output signal is extracted to provide feedback.

The resonator serves two purposes. It provides the necessary phase shift at the desired frequency, and it acts as a filter for signals in the amplifier loop so that only the desired signals are amplified. (The dc power and bias connections are omitted for clarity.) To visualize the common-emitter circuit's parallel LC resonator, imagine the capacitors connected together and their connection grounded, as shown by the dashed line.

RF Oscillators

Back in the 1920s, two circuit designers, Mr. Hartley and Mr. Colpitts, came up with the two circuits of **Figure 6-16** that became popular in radio designs. In each, feedback is created by routing part of the emitter circuit through a voltage divider created by two reactances. The connection to the voltage divider is called a *tap* and such a circuit or component is said to be *tapped*.

If the reactive divider is a pair of capacitors, it's a Colpitts oscillator. [E7H04] (All of the circuits in Figure 6-15 are Colpitts oscillators.) If the reactive divider is a pair of inductors or, more frequently, a single tapped inductor, the circuit is a Hartley oscillator. [E7H03] These same circuits are in wide use today at nearly 100 years of age! (You can remember which is which by thinking, "C is for capacitors and Colpitts" and "H is for henrys and Hartley.")

The Hartley and Colpitts oscillator circuits are very similar in behavior but their differences influence the designer's preferred choice. For example, the Hartley has a wider tuning range than the Colpitts and fewer components. The Colpitts, however, avoids the tapped inductor and has several popular variants with good stability — a Hartley oscillator is less stable than the Colpitts.

If the inductor of the Colpitts oscillator is replaced with a quartz crystal, the result is a Pierce crystal oscillator, the most stable of the three major oscillator circuits — Colpitts, Hartley and Pierce. [E7H01] The Pierce circuit controls the frequency of the positive feedback through its use of a quartz crystal. [E7H05] Besides its stability, another reason why the Pierce circuit is popular is that you do not have to build and tune an LC tank circuit. Adjusting the tank circuit for the exact resonant frequency can be touchy, but it is a simple matter to plug a crystal into a circuit and know it will oscillate at the desired frequency!

Variable-Frequency Oscillators

While the quartz crystal oscillator has excellent frequency stability, amateurs need to be able to tune their radios over a frequency range. This requires a *variable-frequency oscillator* (VFO). VFOs are created by using a variable component in the oscillator's resonant circuit. The tradeoff is that the resulting frequency is not as stable as that of a crystal-controlled oscillator. Both Hartley and Colpitts oscillators can

It's Crystal Clear

The use of quartz crystals as a means of controlling frequency and performing filtering functions is nearly as old as radio. That a simple piece of quartz can perform these functions is amazing!

A piezoelectric material has the ability to change mechanical energy (such as pressure or deformation) into an electrical potential (voltage) and vice versa. This property is known as the *piezoelectric effect*. [E6E03] A number of piezoelectric crystalline substances can be found in nature. Piezoelectric crystals were once common in microphones, phonograph pickups and headphones, where mechanical sound vibrations were transformed into ac voltages or vice versa.

Piezoelectric crystals can be sliced into plates with resonant vibration frequencies ranging from a few thousand hertz to tens of megahertz depending on the crystalline material and the dimensions of the plate. What makes the crystal resonator valuable is that it has an extremely high Q, ranging from a minimum of about 20,000 to as high as 1,000,000. The high Q means that the frequency of vibration is very stable and precise. The most suitable material for crystals to be used at radio frequencies is quartz because of its excellent stability with respect to temperature and its mechanical ruggedness.

The mechanical properties of a quartz crystal resonator, or simply *crystal*, are very similar to the electrical properties of a tuned circuit. We therefore have an equivalent circuit for the crystal. The electrical coupling to the crystal is through holder plates or electrodes that sandwich the crystal between them. A small capacitor is formed by the two plates with the crystal plate as then dielectric between them. The crystal itself is equivalent to a series-resonant circuit and, together with the capacitance of the holder, forms the equivalent circuit shown in **Figure 6-17**.

Can the crystal be changed so that it will resonate at a different frequency? Sure. If we want the crystal to vibrate at a higher frequency, we can make it thinner or shorter by grinding away some of the quartz. The resonant frequency can't be made lower, but if we cut a new crystal that is longer or thicker, the resonant frequency will go down. There are two major limitations to the use of crystals. First, it only has two terminals — the electrodes. In other words, a crystal can't be tapped as we might tap an inductor in a circuit. Second, the crystal is an open circuit for direct current, so you can't feed operating voltages through the crystal to a circuit.

The major advantage of a crystal used in an oscillator circuit is its frequency stability. In an LC oscillator, the spacing of coil turns can change with vibration and the plates of a variable capacitor can move. On the other hand, the frequency of a crystal is much less apt to change with thermal or mechanical changes. By controlling the angle at which the crystal plate is cut across the plane of the quartz structure, manufacturers are able to control the crystal's temperature coefficient and other parameters. If better frequency stability is required, the crystal can also be placed in a crystal oven to maintain a constant temperature.

Another way to affect the frequency at which a crystal oscillator operates is by adding capacitance in parallel with the crystal. This changes the resonant frequency of the tuned circuit formed by the crystal and its associated components. In fact, crystals are quite sensitive to the effects of external capacitance. The crystal manufacturer will specify what capacitance must be placed in parallel with the crystal in order for it to resonate at the intended frequency. [E6E09]

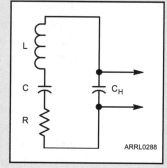

Figure 6-17 — The electrical equivalent circuit of a quartz crystal in a holder. L, C and R are the electrical equivalents of the crystal's mechanical properties and C_H is the capacitance of the holder plates, with the crystal serving as the dielectric.

be used as VFOs. [E7H06] The usual technique for adjustable LC oscillators is to use a Colpitts oscillator in which an adjustable tuning capacitor is placed in parallel with the inductor. Numerous variations on this scheme can be found in the technical references on this book's website.

Microwave Oscillators

Because an oscillator requires amplification, the devices that can be used for oscillators have the same limitations as for amplifiers. At microwave frequencies, in particular, special techniques are required to generate useful signal levels with acceptable frequency stability. Two notable examples are discussed here: the *magnetron* and the *Gunn diode oscillator*.

Magnetron

Like the klystron, the magnetron uses the physical dimensions of a vacuum tube to its advantage and is an efficient oscillator for UHF and microwave frequencies. You probably use a magnetron every day as they are the power source that cooks food in a microwave oven! Magnetrons are also used in high-power radar equipment.

A magnetron tube is a diode. The anode of a magnetron is not a flat plate as in most tubes, but is formed into a set of specially-shaped resonant *cavities* that surround the cathode. A strong external magnet creates a magnetic field inside the tube. [E7H07] When the cathode is heated and raised to a high negative voltage it emits electrons that are attracted to the anode. The strong magnetic field causes the electrons to revolve around the cathode as they travel toward the anode. As they spiral outward, the electrons emit RF energy, creating an oscillating electric field. The resonant cavities act as the filter for the oscillator, reinforcing a single frequency of oscillation. RF power is extracted from the tube through a glass window through which the RF energy is coupled into a *waveguide* (or an oven).

Magnetrons are self-oscillating with the frequency determined by the dimensions of the anode. However, they can be tuned by coupling either inductance or capacitance to the resonant anode. The available tuning range depends on how fast the tuning must be accomplished. A magnetron may be tuned slowly over a range of approximately 10% of the magnetron's natural frequency. If faster tuning is necessary, such as for frequency modulation, the range decreases to about 5%.

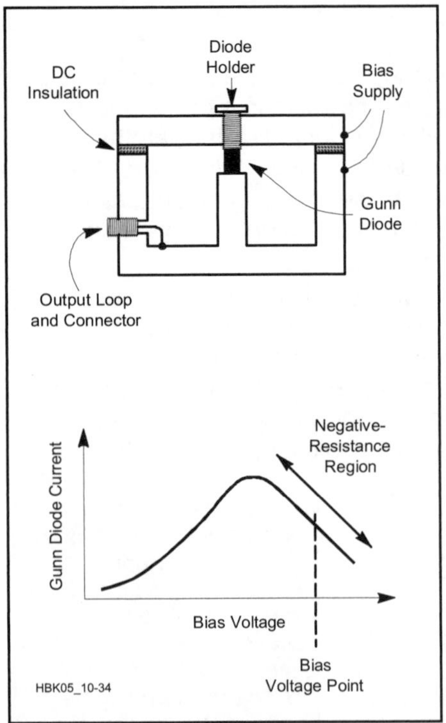

Figure 6-18 — A Gunn diode oscillator uses negative resistance and a cavity resonator to produce RF energy.

Gunn Diode Oscillators

A *Gunn diode* oscillator also makes use of a resonant cavity to control the frequency of the generated signal. It does so by placing an amplifier inside the cavity. Gunn diodes (named for their inventor) and tunnel diodes are semiconductor diodes with special types of doping. As forward bias voltage is increased across these diodes a point is reached at which current *decreases* with increasing voltage. This is *negative resistance* as shown in **Figure 6-18**.

If such a device is placed inside a cavity and forward biased, the negative resistance acts as an amplifier, with the cavity providing the necessary resonant circuit. The cavity acts both as a feedback circuit to the diode and as a resonant circuit to control the frequency. If the negative resistance is strong enough the combination creates an oscillator. [E7H08]

These diodes are capable of operating at extremely high frequencies and were used long before transistors were developed that had gain at microwave frequencies. Efficient at microwave frequencies and inexpensive to produce, the Gunn diode oscillator is the basis of many of the Doppler-radar modules used to detect traffic or intruders. Figure 6-18 shows the cross section of a common Gunn diode oscillator.

Before you go on, study test questions E6E03, E6E09 and E7H01 through E7H08. Review this section if you have difficulty.

DIGITAL SIGNAL PROCESSING (DSP)

Once possible only with the most advanced and esoteric hardware, the manipulation of audio and radio signals as digital data has become commonplace. *Digital signal processing (DSP)* is an integral part of modern electronics, revolutionizing the industry just as the transistor and microprocessor did. Wherever you find signals being processed by analog circuits today, that function is a candidate to be implemented by DSP tomorrow.

To explain the basics of DSP for reference in subsequent sections of this book, we now present edited portions of material originally written for the *ARRL Handbook* by Doug Smith, KF6DX. If you are interested in learning more about DSP in radios, it is strongly suggested that you read the full *Handbook* chapter.

Sequential Sampling

DSP is about rapidly measuring analog signals, recording the measurements as a series of numbers, processing those numbers, then converting the new sequence back to analog signals. How we process the numbers depends on which of many possible functions we are performing.

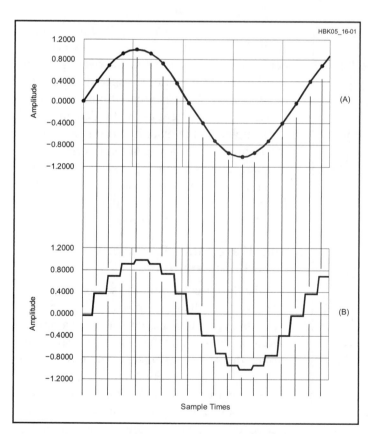

Figure 6-19 — Sine wave of frequency much less than the sampling frequency (A). The sampled sine wave (B).

The process of generating a sequence of numbers that represent periodic measurements of a continuous analog waveform is called *sequential sampling*. Each number in the sequence is a single measurement of the instantaneous amplitude of the waveform at a sampling time. When we make the measurements continually at regular intervals, the result is a sequence of numbers representing the amplitude of the signal at evenly spaced times. This process is illustrated in **Figure 6-19** showing how an analog signal is converted to digital form. [E8A14]

Note that the frequency of the sine wave being sampled is much less than the sampling frequency, f_s. In other words, we are taking many samples during each cycle of the sine wave. The sampled waveform does not contain information about what the analog signal did between samples, but it still roughly resembles the sine wave. Were we to feed the analog sine wave into a spectrum analyzer, we would see a single signal component at the sine wave's frequency. Pretty obviously, the spectrum of the sampled waveform is not the same, since it is a stepwise representation consisting of discrete steps at each value

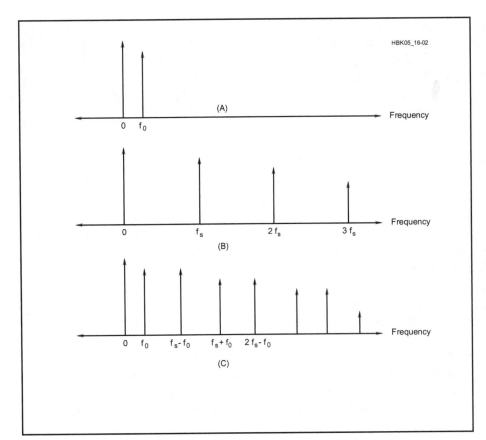

before jumping to the next value.

The sampled signal's spectrum can be predicted and interpreted as shown in **Figure 6-20**. The analog sine wave's spectrum is shown in Figure 6-20A, above the spectrum of the sampling function in Figure 6-20B. The sampled signal is just the product of the two signals; its spectrum is the *convolution* of the two input spectra, as shown in Figure 6-20C.

The sampling process is equivalent to a mixing process: They each perform a multiplication of the two input signals. Note that the sampled spectrum repeats at intervals of f_s. These repetitions are called *aliases* and are as real as the fundamental in the sampled signal data. Each contains all the information necessary to fully describe the original signal. In general, we are only interested in the fundamental, but let's see what happens when the sampling frequency is less than that of the analog input.

Sine Wave, Alias Sine Wave: Harmonic Sampling

Take the case wherein the sampling frequency is less than that of the analog sine wave as shown in **Figure 6-21**. The sampled output no longer matches the input waveform. Notice that the sampled signal retains the general shape of a sine wave, but at a frequency lower than that of the input. Ordinarily, this would not be a happy situation. A downward frequency translation is useful, though, in the design of receivers that use DSP in their IF sections (IF DSP). In addition, lower sampling

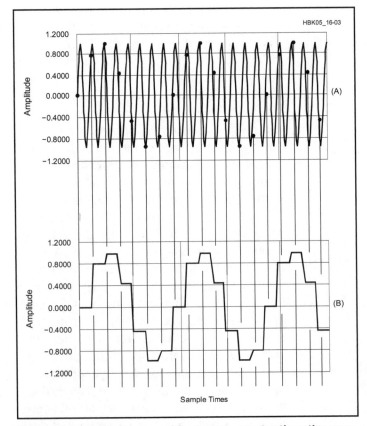

Figure 6-21 — Sine wave of frequency greater than the sampling frequency (A). Harmonically sampled sine wave (B).

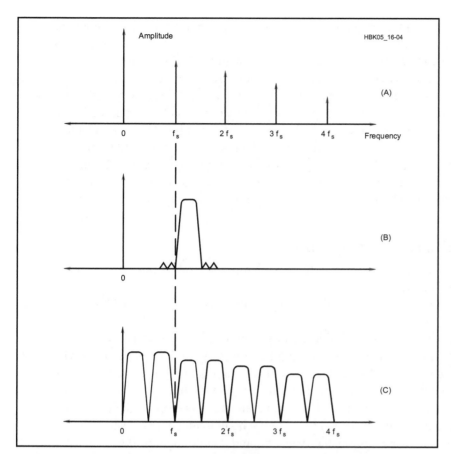

Figure 6-22 — Spectrum of a sampling function (A). Spectrum of a band of real signals (B). Spectrum of a harmonically sampled band of real signals (C).

frequencies are good because they allow more time between samples for signal processing algorithms to do their work; that is, lower sampling rates ease the processing burden.

Caution is required, though: An input signal near twice the sampling frequency would also produce the same output as that of Figure 6-21. To use this technique, then, we must first limit the bandwidth (BW) of the input: A band-pass filter (BPF) is called for. This is known as *harmonic sampling*. The BPF is referred to as an *anti-aliasing filter*. Input signals must fall between the fundamental (or some harmonic) of the sampling frequency, f_s, and the point half way to its next higher harmonic, $2f_s$. A frequency translation will take place, but no information about the shape of the input signal will be lost.

A spectral representation of harmonic sampling is shown in **Figure 6-22**. It reveals the basis for the *Nyquist sampling theorem*: The sampling frequency, f_s, must be at least twice the input BW to avoid aliasing. Such aliasing would destroy information; once incurred, nothing can remedy it.

Data Converters

The device used to perform sampling is called an *analog-to-digital converter* (*ADC*). For each sample, an ADC produces a binary number that is directly proportional to the input voltage. The number of bits in its binary output limits the number of discrete voltage levels that can be represented. An 8-bit ADC, for example, can only give one of 256 values. This means the amplitude reported is not the exact amplitude of the input, but only the closest value from those available. The difference is called the *quantization error*.

A *digital-to-analog converter* (*DAC*) performs the conversion of binary numbers back into analog voltages — the reverse operation of an ADC. Typical DACs are *sample-and-hold* devices. They continue to output the last sample value throughout the sample period. This effect acts as a low-pass filter and also adds quantization error.

Representation of Numbers: Floating-Point vs Fixed-Point

One of the things that makes general-purpose computers so useful is their ability to perform *floating-point* calculations. In this form of numeric representation, similar to scientific notation, numbers are stored in two pieces: a fractional part, or *mantissa*, and an *exponent*. The mantissa is assumed to be a binary number representing an absolute value less than one, and the exponent, a binary integer. This approach allows the computer to handle a large range of numbers, from very small to very large. Some DSP chips support floating-point calculations, but it is not as great an advantage in signal processing as it is in general-purpose computing because the range of values we are dealing with in DSP is limited to the precision of the input ADC. For this reason, *fixed-point* processors are common in DSP.

A fixed-point processor treats numbers as just the mantissa and does away with the exponent. The *radix* point — the separation between the integer and fractional parts of a number — is usually assumed to reside to the left of the most-significant bit. This is convenient, since the product of two fractions less than unity is always another fraction less than unity. The sum of two fractions, though, may be greater than unity; *overflow* would be the result. Overflow is a constant concern for fixed-point DSP programmers and leads to considerations for *scaling* of the input signal data to limit system dynamic range to less than the data converter's capabilities.

Software-Defined Radio (SDR) Systems

What is a software-defined radio? Well, to be as comprehensive as possible, we can state that a software radio is a radio:

1. Whose hardware is able to handle almost any modulation format, signal bandwidth and frequency desired.

2. Whose functionality may be altered at will by downloading new software.

3. That replaces traditional analog circuits and functions by implementing them with DSP.

You may recognize several of these characteristics in radios that you already own! SDR radios, such as the Flexradio (**www.flexradio.com**), are already being produced and sold commercially to amateurs. Groups of interested amateurs are developing their own SDR systems, software and hardware. Your next QSO may be with an amateur using a radio whose only analog signal processing is the front end and the audio amplifier! Yet you may be completely unable to discern that fact from the signal — any function an analog radio can perform, an SDR can perform.

What is important to understand about DSP and SDR systems is that they can perform any mathematically defined signal processing function if hardware is available to adequately sample the signal and perform the required math operations quickly enough. That includes modulation, demodulation, filtering, speech processing and so on — anything found in your radio today.

There are still areas where analog circuits perform better and to be sure, analog circuits are still used to bring the input signal frequencies within range of affordable hardware, but the trends are clear toward tomorrow's fully digital radios.

Before you go on, study test question E8A14. Review this section if you have difficulty.

MIXERS

Mixer circuits are used to change the frequency of a signal. In a superheterodyne receiver, this means converting the received signal to the *intermediate frequency* (IF) so it can be amplified and filtered more efficiently. In this way, the receiver can be optimized for the best signal-handling characteristics such as linearity and selectivity without the need to retune many circuit elements every time you change the received frequency. Mixers are also used to change the frequency of a signal as it progresses through a transmitter. The principles of operation are much the same for mixers, detectors and modulators, so the discussion of how mixers work will prepare you to better understand the remaining topics.

If the mixer uses *passive* components that cannot amplify a signal to perform the mixing function, it is called a *passive mixer*. This also results in *conversion loss* caused by losses in the passive components. If the mixing process is performed by devices that can amplify the

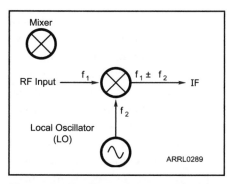

Figure 6-23 — The mixer combines the two input signals (RF and LO). This produces the mixing products that compose the IF signal at the output.

signals, then the circuit is called an *active mixer*.

When two sine waves are combined in a *nonlinear* circuit (one whose output is not a scaled replica of its input) such as a mixer, the output signal is a complex waveform that has principal components at the frequencies of the two original signals and two *product* signals. The product signals are sine waves whose frequencies are the sum and difference of the frequencies of the two original signals. [E7E08] Also included are higher orders of combinations of the harmonics from the input signals, although these are usually weaker than the primary sum and difference frequencies and are ignored in this discussion. The signal of varying frequency (in a receiver, the desired input signal) is usually referred to as the *RF signal*. The signal generated in the radio to mix with the RF signal is called the *local oscillator* (*LO*), and the resulting output signal is called the *IF signal*. **Figure 6-23** shows the symbol for a mixer. In a typical receiver, the LO frequency is varied so that one of the mixer IF output products is at the frequency for which the IF circuits are designed.

One of the products can be selected from the output combination by using a filter. Of course, the better the filter, the lower the level of the unwanted products or the two input signals in the resulting output signal. By using *balanced mixer* techniques the mixer circuit provides isolation of the various signal connections or *ports* so that the RF, LO and IF signals will not appear at any other port. This prevents the two input signals (RF and LO) from reaching the output. In that case the filter needs to remove only the unwanted mixer output products.

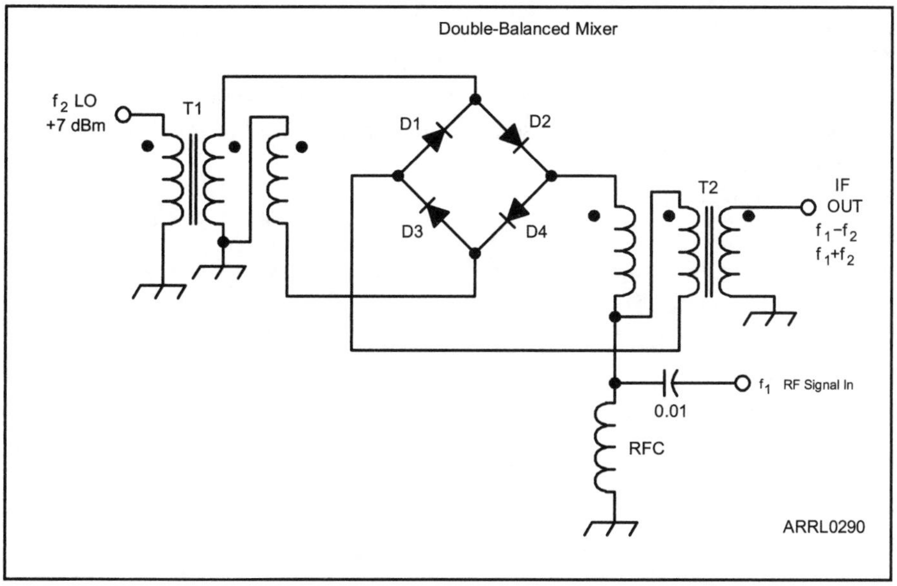

The mixer stages in a high-performance receiver must be given careful consideration because they have a great impact on the ability of the receiver to perform properly in the presence of strong signals. (See the section Receiver Performance in the chapter on Radio Modes and Equipment.) The RF signal should be amplified only enough to overcome mixer losses. Otherwise, strong signals will overload the mixer circuit. This causes desensitization and the higher order combinations of frequencies will appear as highly undesirable spurious

Figure 6-24 — A diode-ring, double-balanced mixer is an example of a passive mixer circuit that has good strong-signal handling characteristics.

signals or *intermodulation distortion* (*IMD*) products at or near the IF frequency, interfering with the desired signal. [E7E09] A mixer should be able to handle strong signals, called *strong-signal performance*, without generating spurious signals.

Passive Mixers

Figure 6-24 shows the schematic of a passive, *double-balanced mixer* (DBM) that uses four diodes in a ring (similar to, but different from a full-wave rectifier) to multiply the RF and LO signals together. Transformers T1 and T2 have an extra *balanced* winding to allow

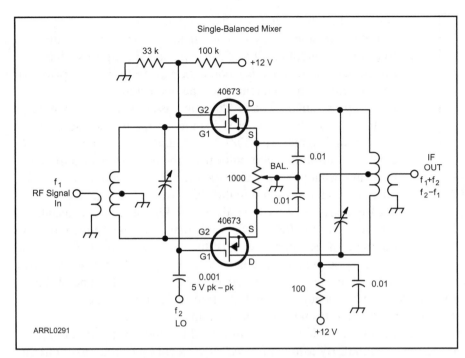

Figure 6-25 — A single-balanced, active mixer that used a pair of dual-gate MOSFETs to both amplify and mix the RF and LO signals.

the LO signal to be applied to the diode ring independently of a ground connection. Commercial DBM modules offer electrical balance at the ports that would not be easy to achieve with homemade transformers. They also use diodes whose characteristics have been carefully matched. Typical loss through a DBM is 6 to 9 dB. The port-to-port isolation is usually on the order of 40 dB.

Active Mixers

While passive mixers have good strong-signal-handling ability, they also have some drawbacks. They require a relatively strong LO signal, and they generate a fair amount of noise. Active mixers can be used to reduce conversion loss, can operate with weaker LO signals, and produce less noise. The drawback is that their strong-signal capabilities are not generally as good as passive mixers.

A JFET or dual-gate MOSFET can be used as a mixer and will provide some gain as well as mixing the signals. Bipolar transistors could be used but they produce higher levels of unwanted mixing products. **Figure 6-25** shows an active mixer circuit. Many variations are possible, and this diagram just shows one arrangement. Integrated-circuit active mixers are available as both single and double balanced types. These devices provide at least several decibels of conversion gain, low noise and good port-to-port isolation.

Before you go on, study test questions E7E08 and E7E09. Review this section if you have difficulty.

MODULATORS

Modulation is the process of adding information to an *unmodulated* radio-frequency (RF) signal, also known as a *carrier* signal. Circuits that perform the modulation process are called *modulators*. Any aspect of the carrier signal can be varied to add the information. Varying the amplitude of the signal is *amplitude modulation (AM)*. Varying the phase or frequency of the carrier are both forms of *angle modulation*, referring to the phase angle of the signal. The two common types of angle modulation are *phase modulation (PM)* and *frequency modulation (FM)*.

Modulation is necessary for transmitting the signal by radio because the *baseband* signals (the frequency components that make up the modulating voice or data signal) cannot be transmitted directly. [E7E07] For any type of signal on the air — amplitude modulated (AM), frequency modulated (FM), phase modulated (PM) and so forth — there are a number of methods of combining the desired information with the carrier. In this chapter, we examine methods for producing AM and FM signals.

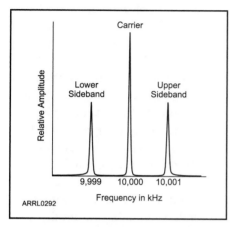

Figure 6-26 — The result of amplitude modulating a 10 MHz carrier with a 1 kHz sine wave shows the upper sideband (USB) at 10 MHz + 1 kHz and \the lower sideband (LSB) at 10 MHz — 1 kHz.

Amplitude Modulation

Double-sideband, full-carrier, amplitude modulation (DSB AM) is not as widely used on the air as the more energy-efficient single-sideband (SSB). However, generating a DSB AM signal is often the first step in creating an SSB signal for transmission.

Generating AM is straightforward and that is the reason it was the first method of voice modulation. AM is produced by applying AF signals to the anode or collector of an RF amplifier. You can also modulate the control element of the amplifier (grid, base or gate). A wide variety of modulator circuits have been used over the years.

In a DSB modulator, an RF *carrier* and an audio (AF) signal are combined into a composite signal in which four principal output signals are found. The carrier and audio signal are present at their original frequencies. The remaining two signals are the *sidebands*: one at a frequency that is the sum of the carrier and audio frequencies (the *upper sideband* or *USB*), the other at their difference (the *lower sideband* or *LSB*). There are a pair of sidebands; thus this is a *double-sideband* or *(DSB)* signal. **Figure 6-26** shows the spectrum of a DSB AM signal generated by modulating a 10 MHz carrier with a 1 kHz sine wave audio signal.

The amplitude of these signals at any given instant depends on the amplitude of the original audio signal at that instant. The greater the audio-signal strength, the greater the amplitude of the sideband signals. As audio frequency increases, both sidebands will move away from the carrier. The upper sideband moves higher in frequency and the lower sideband moves lower.

The result of all this is that the RF *envelope* (the outline of the signal's amplitude) has the general shape of the modulating waveform. The envelope varies in amplitude but the *carrier* amplitude does not change — it is constant and contains no information. What changes is the phase relationship of the three RF signals — carrier, USB and LSB. As the signals add and subtract it is the RF *envelope* that varies in amplitude as a result.

SSB: The Filter Method

One way to generate an SSB signal is to use a filter to remove the carrier and one of the sidebands from an ordinary DSB AM signal. The block diagram shown in **Figure 6-27** shows how this is done. The RF oscillator generates a carrier wave that is injected into a *balanced modulator* where it is combined with audio from the speech amplifier. A bal-

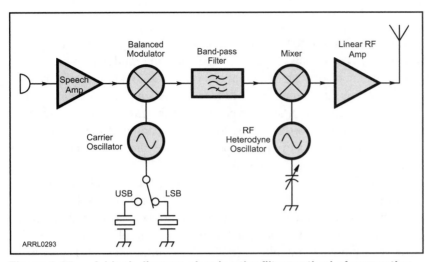

Figure 6-27 — A block diagram showing the filter method of generating an SSB signal.

anced modulator can generate a DSB AM signal, but if properly adjusted, it can remove or *suppress* the carrier signal almost completely. This signal consists only of the USB and LSB signals and is called *double-sideband, suppressed-carrier* or DSB-SC AM.

A balanced modulator is a type of mixer. You'll recall that the output of a mixer consists of the two input signals plus the products that are the sum and difference of the input frequencies — just like a modulator. In the output of a balanced modulator, however, the amplitude of the audio and carrier signals is reduced to very low levels. This leaves only the USB and LSB signals, components of the original DSB AM signal. The first step toward creating a *single-sideband* (*SSB*) signal has been completed by eliminating the carrier.

Both sidebands contain essentially the same information — the amplitude and frequencies of the modulating audio signal. Only one needs to be transmitted to reconstruct the original audio information and so the unwanted sideband can be discarded. In the *filter method* of SSB generation, the unwanted sideband is removed with a band-pass filter (BPF) as shown in Figure 6-27. [E7E04]

The bandwidth of the filter must be narrow enough to pass the desired sideband while removing the unwanted sideband. In our example signal, if the audio signal is that of a human voice with frequency components from 300 to 3000 Hz, the USB will occupy 10000.300 to 10003.000 kHz and the LSB will occupy 9997.000 to 9999.700 kHz. Thus, the filter must have a bandwidth of 2700 Hz, centered on one of those RF frequency ranges. If it is the USB that is desired, the filter's center frequency would be:

$$\frac{10003.000 + 10000.300}{2} = 10001.650 \text{ kHz}$$

If the LSB was desired instead, the center frequency would be 9998.35 MHz. But why use two filters? The diagram in Figure 6-27 shows how a single filter could be used to pass either the USB or LSB signal. The calculations above show that the filter's center frequency must be 1.65 kHz away from the carrier in the appropriate direction to pass the sideband. By shifting the carrier frequency instead of using two filters, either the USB or LSB signal could be centered in a single filter's passband.

Of course 10 MHz isn't a very good choice for our transmitter design because of the potential of interference with WWV's time-and-frequency standard signals, so let's try 9000.000 kHz as the filter's center frequency. If the carrier frequency is chosen to be 9001.65 kHz (above the filter's center frequency) the LSB component will be centered squarely in the filter's passband. Or, by placing the carrier at 8998.35 kHz, the USB signal will pass through the filter instead. Filters are much more expensive than single crystals, so this is a common SSB transmitter design. Once the SSB signal has been generated, another mixer can shift its frequency to whatever amateur band frequency we want and it can then be amplified by the linear amplifier to useful power levels.

SSB: The Quadrature Method

In the first paragraphs of this section, we noted that it was the phase relationships between the three RF components of the DSB AM signal that created the varying envelope. This is a clue that we can create the SSB signal directly by manipulating the phase of the audio and RF signals. This is called the *quadrature method* and the block diagram of such a system is shown in **Figure 6-28**.

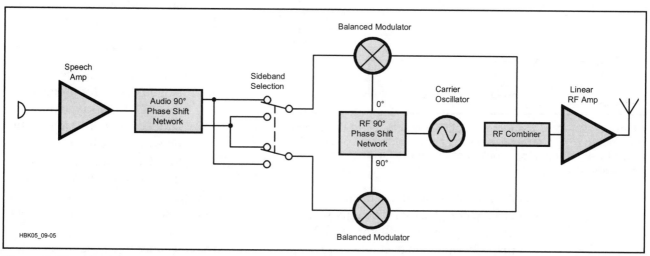

HBK05_09-05

Figure 6-28 — The block diagram of a quadrature SSB generator. By combining the independent DSB signals, one sideband is canceled and an SSB signal results.

The audio and carrier signals are each split into equal components with a 90° phase difference (called *quadrature*) and applied to individual balanced modulators. When the DSB outputs of the modulators are combined one sideband is reinforced and the other is canceled. The figure shows sideband selection by transposing the audio leads but the same result can be achieved by switching the carrier leads.

The phase shift and amplitude balance of the two channels must be very accurate if the unwanted sideband is to be adequately attenuated. To achieve an attenuation of 40 dB, the *phase accuracy* of the 90° audio phase shift must be within 1° across the entire audio band to be transmitted — from 300 to 3000 Hz in a communications voice transmitter. This level of accuracy is very difficult to achieve with an analog circuit.

With DSP (digital signal processing), however, producing a 90° phase shift over a wide frequency range is easily accomplished using a special combination of filters called the *Hilbert transformer*. [E7C09] In Figure 6-28, the phase shift is created by a circuit. A DSP system using the Hilbert transformer creates all of the necessary signals — phase-shifted audio and RF carriers — then performs the balanced modulator function by multiplying the sampled signals together as numbers. The SSB signal is generated by adding the two multiplied sets of data together and using a digital-to-analog converter (DAC) to turn the numbers back into an analog waveform. (For more information on DSP modulators, refer to the "DSP and Software Radio Design" chapter in *The ARRL Handbook*.) This technique makes the quadrature technique the preferred method of SSB generation in DSP systems. [E7E13]

Before you go on, study test questions E7C09, E7E04, E7E07 and E7E13. Review this section if you have difficulty.

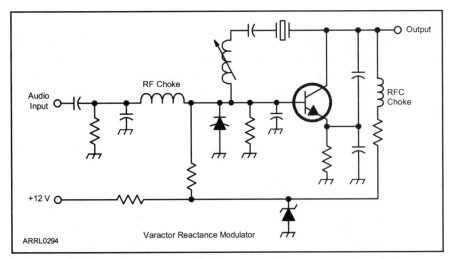

Figure 6-29 — The changing capacitive reactance of the varactor diode connected to the series-LC-crystal circuit of this oscillator changes the output frequency under control of the audio input signal, producing direct FM modulation.

Frequency and Phase Modulation

Most methods of producing FM fall into two general categories: *direct FM* and *indirect FM*. As you might expect, each has its advantages and disadvantages. Let's look at the direct FM method first.

Direct FM

A *reactance modulator* is a simple device for producing direct FM. This type of modulator consists of a varactor diode or a transistor connected to act as a variable reactance in the frequency-determining tank circuit of an oscillator. If the modulating signal controls the variable reactance, the result is a direct FM signal. The only way to produce a true F3E emission is with a reactance modulator acting on an oscillator. [E7E01]

Figure 6-29 shows an example of a reactance modulator. The audio input signal causes the capacitance of the varactor to change and in turn, that changes the LC ratio of the oscillator's tank circuit and its frequency. The modulator's sensitivity (frequency change per unit change in modulating voltage) depends on the varactor diode's change of capacitance per volt of input signal.

For practical reactance modulators, the modulated oscillator is usually operated on a relatively low frequency for high stability of the carrier frequency. Frequency multipliers then increase the signal's frequency to the final desired output frequency as in **Figure 6-30**. It is important to note that when the frequency is multiplied so is the frequency deviation. The amount of deviation produced by the modulator must be adjusted carefully to give the proper deviation at the final output frequency.

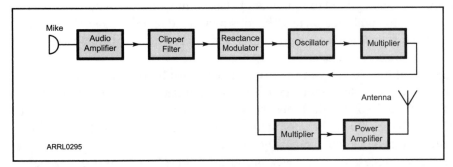

Figure 6-30 — The block diagram of a typical direct-FM transmitter. The amount of frequency deviation produced by the reactance modulator is multiplied along with frequency by the multiplier stages.

Indirect FM

A *phase modulator* varies the tuning of an amplifier tank circuit to produce a PM signal. From a practical view indirect FM is the same as PM. The same type of reactance modulator circuit that is used to vary the oscillator-tank tuning for an FM system can be used to vary reactance in a tuned RF amplifier tank circuit and thus vary the phase of the tank current by acting as an electrically-variable inductance or capacitance. This produces phase modulation (PM). [E7E02] For example, if the varactor diode in Figure 6-29 was connected to an amplifier tank circuit instead of to an oscillator's tank circuit, the result would be PM (emission type G3E). [E7E03]

Reactance modulation of an amplifier stage usually results in simultaneous amplitude modulation because the modulated stage is detuned from resonance as the phase is shifted. The undesirable side effect of AM can be eliminated by feeding the modulated signal through an amplitude limiter; that is, an amplifier that is driven hard enough so that variations in the amplitude of the input signal produce no appreciable variations in the output amplitude and only the frequency changes remain.

Pre-emphasis and De-emphasis

Frequency deviation increases with the modulating audio frequency in PM. (Higher audio frequencies produce greater frequency deviation.) Therefore, it is necessary to use a low-pass filter on the modulating audio to attenuate frequencies above 3000 Hz before modulation takes place. This prevents the generation of unwanted sidebands far from the carrier frequency that cause interference on nearby channels.

In a direct FM system, frequency deviation does not increase with modulating audio frequency so no low-pass filter is needed. However, the lower energy of higher-frequency speech components means that the recovered signal is susceptible to high-frequency noise. To reduce hiss and high-frequency noise, an audio circuit called a *pre-emphasis network* is added to a direct FM modulator to spread the audio signal energy evenly across the audio band. Pre-emphasis applied to an FM transmitter gives the deviation characteristics of PM. [E7E05] The reverse process, called *de-emphasis*, is used at the receiver to restore the audio spectrum to its original relative proportions and reduce noise. [E7E06] A transmitter that uses PM does not need a pre-emphasis network.

> *Before you go on, study test questions E7E01, E7E02, E7E03, E7E05 and E7E06. Review this section if you have difficulty.*

DETECTORS AND DEMODULATORS

Detectors and demodulators have much the same job — to recover the modulating information from a modulated RF signal. A detector circuit extracts the information directly from the signal. A demodulator reverses the modulation process to recover the information. Detectors tend to be simpler circuits than demodulators, but the recovered signal is generally not as accurate a replica of the original modulating signal as when a demodulator is used. Each has a place in radio communications.

Detectors

The simplest type of detector, used in the very first radio receivers, is the *diode detector*. It works by rectifying, then filtering, the received RF signal. [E7E10] A complete, simple receiver is shown in **Figure 6-31**. This circuit only works for strong AM signals so it is not used very much today except for experimentation. It does serve as a good starting point to understand detector operation, however. In early crystal radio sets, a steel "cat's whisker" pressing on a lead crystal created the diode. Sensitive headphones then recovered the audio signal.

The waveforms shown on the diagram illustrate the changes made to the signal as it progresses through the circuit. L1 couples the received RF signal to the tuning circuit of L2-C1. The diode D1 rectifies the RF waveform, passing only the positive half cycles. C2 charges to the peak voltage of each half-cycle, producing a filtered dc waveform, which then goes through the low-pass filter R1 and C3 to remove any residual RF. R2 acts as a volume control and C4 removes the dc offset voltage, leaving only the ac audio signal.

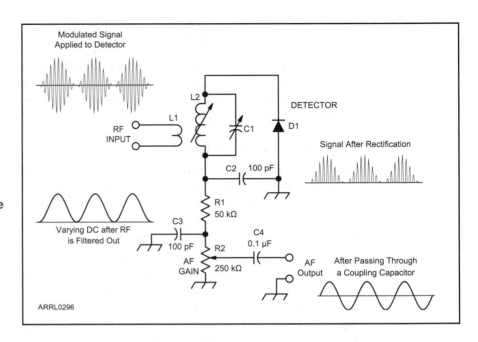

Figure 6-31 — A simple AM receiver circuit using a single diode detector. L2 and C1 are tuned to the desired receive frequency. D1 rectifies the signal, creating a waveform from which the envelope can be easily transformed into an audio signal.

Product Detectors

A *product detector* is a type of mixer that combines an incoming RF signal with a locally generated carrier or *beat-frequency oscillator* (*BFO*). The product detector stage follows the IF stages in a superheterodyne receiver. The BFO frequency is chosen so that one of the sum-and-difference output products (thus the name) is at audio frequencies. Product detectors are used for SSB, CW and RTTY reception. [E7E11]

For example, if the receiver's IF is 455 kHz and the operator prefers to listen to a CW signal with a 700 Hz tone the BFO could be set to 455.7 kHz, creating sum-and-difference products at 700 Hz and 910.7 kHz. An audio filter then removes the higher frequency component. The BFO frequency could also be set to 454.3 kHz, achieving the same result.

The same process can also be used on SSB and RTTY signals, shifting the recovered audio frequencies higher and lower. In the days before high-precision digitally-controlled master oscillators, receivers used a main VFO to convert received signals to the IF frequency and then a "fine tune" BFO to give the signal the proper pitch or tonal balance.

Direct Conversion

In reading about product detectors you might have thought, "Why not do away with the IF? Just recover the signal with the BFO and be done with it!" That is exactly the function of the *direct conversion* receiver. A single mixer stage combines the received signal with the output of an oscillator at or near the same frequency. The resulting sum-and-difference products are at twice the received signal frequency and at audio frequencies.

Direct conversion requires a very stable tuning oscillator because the conversion is done in a single step at the high frequency of the signal itself. The mixing process can also receive *two* signals at once, one above the oscillator frequency and one below. This doubles the potential for interference. Direct conversion or "dc" receivers are also very sensitive to *microphonics* — mechanical vibrations that change the oscillator frequency and thus become part of the recovered signal. Yet the technique adds very little noise and has exceptionally good fidelity.

In a software-defined radio, direct conversion has an additional meaning, that the received signal is converted directly to "baseband" where the cost of the analog-to-digital converter (ADC) is much lower. [E7E14] The sorting out of interference is then performed by signal processing software.

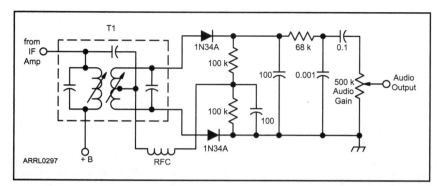

Figure 6-32 — A typical frequency-discriminator circuit used for FM detection. A modulated signal results in an imbalanced output from the transformer that is rectified and turned into audio by the diodes and RC filters.

Detecting FM Signals

There are three common ways to recover the audio information from a frequency-modulated signal. The *frequency-discriminator* circuit of **Figure 6-32** uses a transformer tuned to the receiver's IF to detect FM signals. [E7E12] The primary signal is introduced to the secondary winding's center tap through a capacitor. For an unmodulated input signal, the resulting voltages on either side of the secondary's center tap will cancel. But when the signal frequency changes, there is a phase shift in the two output voltages that varies at the audio frequency of the modulating signal. The two voltages are rectified by a pair of diodes, and the resulting difference in output voltage becomes the audio signal. *Crystal discriminators* use a quartz-crystal resonator instead of the LC tuned circuit in the frequency discriminator which is often difficult to adjust properly.

A *ratio detector* can also be used to receive FM signals. It operates on the output from a tuned transformer similar to that of a frequency discriminator. The rectified output signals are split into two parts by a divider circuit. The ratio of the two voltages is then used to produce the audio signal.

The last method of detecting an FM signal, *slope detection*, can be used if you have an AM or SSB receiver. If you tune the receiver slightly away from the FM signal's carrier frequency, the varying signal frequency moves up and down the slope of the receiver's selectivity curve, producing an AM signal. This AM signal then proceeds through the receiver in the normal fashion, and you will hear the audio in the speaker. Careful tuning will make the FM signal perfectly understandable, although you might have to keep one hand on the tuning knob, and there may be some noise.

Before you go on, study test questions E7E10, E7E11, E7E12 and E7E14. Review this section if you have difficulty.

FREQUENCY SYNTHESIS

Modern radios do not use continuously-tunable oscillator circuits to control signal frequency. Instead, a technique called *frequency synthesis* is used to create signals with precisely controlled frequencies that vary in small steps of 100 Hz or less. Two primary methods of frequency synthesis are used in commercial HF radios: *phase-locked loops* (*PLL*) and *direct digital synthesizers* (*DDS*). Phase-locked loop synthesizers were once universal in commercial radio equipment but they have been largely supplanted by the direct digital synthesizer. Although PLL circuits are simpler, DDS circuits have the advantage of requiring less analog circuitry and are thus easier to integrate into digital ICs and control with a microprocessor. Combinations of PLL and DDS technology are common in commercial amateur equipment.

Phase-Locked Loops (PLL)

In a phase-locked loop, the frequency of a variable oscillator is continuously compared to the phase of a stable, fixed-frequency reference oscillator. If the variable oscillator's fre-

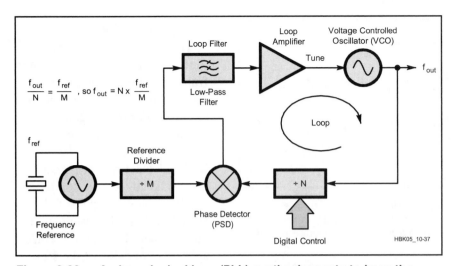

$$\frac{f_{out}}{N} = \frac{f_{ref}}{M} \;,\; \text{so } f_{out} = N \times \frac{f_{ref}}{M}$$

Figure 6-33 — A phase-locked loop (PLL) synthesizer acts to keep the divided-down signal from its voltage-controlled oscillator (VCO) phase-locked to the divided-down signal from its reference oscillator. By changing M and N, fine tuning steps in the VCO frequency can be made with the same frequency stability of the reference oscillator.

quency is too high, its phase will begin to lead that of the reference. This phase difference is used to decrease the frequency of the variable oscillator. Too low a frequency causes an increasing phase lag, which is used to increase the frequency of the oscillator. In this way, the variable oscillator is *phase-locked* to the reference so that their frequencies are kept exactly equal.

To be used as a tunable oscillator, however, the frequency of the variable oscillator must be able to change yet still remain under control of the reference. **Figure 6-33** shows how a PLL works. The *phase detector* outputs a voltage corresponding to the phase difference between the oscillators. This voltage is passed through a low-pass *loop filter*, amplified by a *loop amplifier*, and used to control the frequency of a *voltage-controlled oscillator* (*VCO*). The combination of phase-detector, filter and VCO create an electronic "servo" loop. Using the reference oscillator as the frequency input to the servo loop creates a *phase-locked loop synthesizer*. This combination makes it possible for a radio's VFO to have the same degree of stability as a crystal oscillator. [E7H14, E7H17]

In order to be able to change the frequency of the variable oscillator, the frequency of both oscillators is divided by a fairly large number — N for the VCO and M for the reference. The circuits controlling the PLL select values for N such that the output of the frequency dividers is the same when the VCO has the desired output frequency. Mathematically,

$$\frac{f_{out}}{N} = \frac{f_{ref}}{M} \quad \text{or} \quad f_{out} = N \times \frac{f_{ref}}{M}$$
(Equation 6-5)

where:
    f_{out} = synthesizer output frequency.
    N = the VCO division factor.
    M = the reference oscillator division factor.
    f_{ref} = reference-oscillator frequency.

Since M is fixed, the smallest amount that the VCO's output frequency can change, the synthesizer's *step size*, is f_{ref} / M.

If the loop is "in lock," meaning that the frequency of the VCO is under control, the amount of frequency variation over which the loop can maintain control of the VCO is called the loop's *lock range*. If the loop is not in lock and the divided-down frequencies of the VCO and reference are gradually brought closer to each other the loop will "capture" the VCO at some point. The difference between the maximum and minimum frequencies for which this occurs is the loop's *capture range*. [E7H13] The characteristics of the loop filter and amplifier determine the stability and tuning speed of the PLL.

A PLL feedback loop is constantly "correcting" the output signal frequency. Any variations in the frequency of the reference oscillator will cause variations in the phase of the output

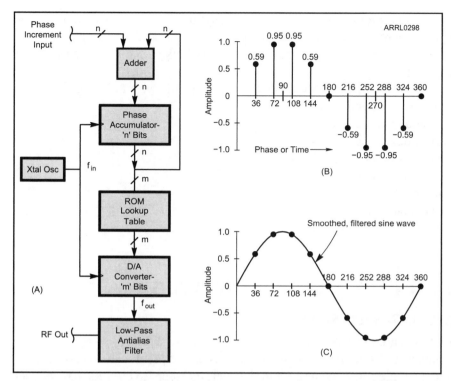

Figure 6-34 — (A) is the block diagram of a direct digital synthesizer (DDS). (B) shows the amplitude values found in the ROM lookup table for a particular sine wave being generated. (C) shows the smoothed output signal from the DDS, after it goes through the low-pass anti-alias filter.

signal from one cycle to the next. [E7H16] These unwanted variations in the oscillator signal phase are called *phase noise*. This is a broadband noise that appears around the desired output frequency and is the major spectral impurity component of a phase-locked loop synthesizer. [E7H18] A very stable reference oscillator must be used for a phase-locked loop frequency synthesizer to minimize this phase noise on the output signal. If you have ever heard a strong hiss from a nearby synthesized radio, that is what phase noise sounds like.

Along with frequency synthesis, a PLL can also be used to perform both FM modulation and demodulation. [E7H15] If the modulating signal is added to the VCO control signal, the output is a direct FM signal. If a divided-down FM signal is input to the phase detector instead of a VCO output, the output of the phase detector will be a replica of the modulating signal.

Direct Digital Synthesizers (DDS)

Figure 6-34 shows the block diagram of a direct digital synthesizer. This type of synthesizer is based on the concept that we can define a sine wave by specifying a series of amplitude values spaced at equal phase angles. The frequency of the sine wave is then determined by the *sampling rate* at which the synthesizer steps through successive values.

The crystal oscillator sets the sampling rate for the amplitude values. The *phase increment* input to the adder block sets the number of samples for one cycle. The oscillator *clock* signal tells the *phase accumulator* to read the data from the adder and then increment the adder value by the phase increment. The phase accumulator value varies between 0 and 360, corresponding to one complete cycle of a sine wave. The *ROM lookup table* contains the amplitude values for the sine (or cosine) of each angle represented by the phase accumulator. A digital-to-analog converter (DAC) changes the digital values from the lookup table to an analog output voltage representing the sine wave. [E7H09, E7H10, E7H12]

The larger the phase increment value, the higher the frequency of the output signal. For example, suppose our synthesizer uses a 10-kHz crystal oscillator. This means there will be one sample every 0.1 ms. If the phase increment is set to 36°, there will be 10 samples in each cycle: 0°, 36°, 72°, 108°, 144°, 180°, 216°, 252°, 288° and 324°. The next sample, at 360° starts the second cycle. The total time for these 10 samples is 1 ms, which means the sine wave defined by these samples has a frequency of 1 kHz. Figure 6-34B shows a representation of the sine values found in the lookup table for these phase angles.

The sample values are fed to the DAC which creates a sine wave output as a series of steps. The DAC output signal goes through the low-pass anti-alias filter so that a smooth sine-wave signal results as in Figure 6-34C.

The frequency of the signal produced by a direct digital synthesizer is controlled by changing the value of the phase increment input to the adder. For example, if the new phase increment is 72° there will be five samples per cycle. Each cycle will take 0.5 ms, so the frequency of this new signal is 2 kHz.

Phase noise is not as much of a problem with DDS circuits as it is with PLL synthesizers. The major spectral impurity components produced by a direct digital synthesizer are *spurs* (unwanted spurious signals) at specific discrete frequencies determined by the clock signal frequency and other digital components of the DDS synthesizer. Careful design can place those spurs outside of the amateur bands. [E7H11]

Before you go on, study test questions E7H09 through E7H18. Review this section if you have difficulty.

6.3 Filters and Impedance Matching

FILTER FAMILIES AND RESPONSE TYPES

In Amateur Radio, filters are used to block, pass or otherwise modify signals within some defined range of frequencies. While the resonant circuits discussed previously also do this and can be considered a simple filter, the term "filter" generally refers to circuits that act over broader ranges of frequencies with well-defined characteristics.

In this section you will learn about *passive* filters and *active* filters. Passive filters are made with unpowered components (R, C or L) and always result in some loss of signal strength. This is called *insertion loss*. Active filters include a powered amplifying device to overcome the filter insertion loss and sometimes even provide signal gain. Some types of filters can only be built using active components.

Passive filters constructed using inductors and capacitors are *LC filters*. There are other types of passive filters, however. For example, *mechanical filters* using internal elements such as disks and rods that vibrate at the frequencies of interest are used as receiver IF filters. *Cavity filters* use the resonant characteristics of a conducting tube or box to act as a filter and are used in repeater duplexers because of their extremely low loss. [E7C10] This chapter won't go into the details of mechanical or cavity filters, but you should be aware of their existence as types of passive filters.

Filter Classification

Filters are classified into the general groups shown in **Figure 6-35**. A *low-pass filter* is one in which all frequencies below the *cutoff frequency* (at which

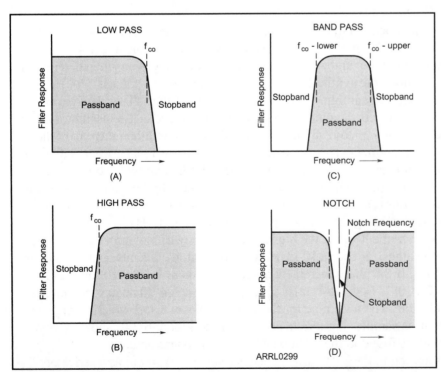

Figure 6-35 — Ideal filter-response curves for low-pass, high-pass, band-pass, and notch filters.

the output signal power is one-half that of the input) are passed with little or no attenuation. Above the cutoff frequency, the attenuation generally increases with frequency. A *high-pass filter* is just the opposite; signals are passed above the cutoff frequency, and attenuated below. The range of frequencies that is passed is the *passband* and the range that is attenuated, the *stopband*.

A *band-pass filter* has both an upper and a lower cutoff frequency. Signals between the cutoff frequencies are passed while those outside the passband are attenuated. The opposite of a band-pass filter is a *band-stop filter*. It attenuates signals at frequencies between the cutoff frequencies. If the stopband is very narrow, that is a *notch filter*.

Filter Design

Filter designs use techniques based on certain types of mathematical equations that describe the filter characteristics. You may have heard of filters referred to as *Butterworth*, or *Chebyshev*, or *elliptical* and these names refer to the family of equations used to design that type of filter. Each type of equation results in filters with different characteristics as described below. These three types of filters are the most common that are used in amateur equipment, but there are many others.

Using these equations, it is possible to build an entire catalog of filters with different characteristics. Tables summarizing these computations can be found in *The ARRL Handbook* and other reference books. From the tables, component values can be determined and the filter constructed with confidence that it will perform as expected. A version of the filter design program *ELSIE* is included with *The ARRL Handbook* to automate filter design.

Before discussing the different types of filters, we should define the terms we use to describe their behavior. **Figure 6-36** shows *response curves* showing the filter's effect on signal amplitude with frequency. The vertical axis has units of dB representing the ratio of output to input signal, so smaller response values correspond to more attenuation of the signal. Frequency increases from left to right, so all of the filter responses in the figure are low-pass filters, attenuating frequencies above the cutoff frequency where the filter response is –3 dB.

Two additional characteristics describe the response curve: *cutoff transition* and *ripple*. *Cutoff* refers to the steepness with which the response curve moves from the passband to the stopband through the *transition region*. A filter with a steep response curve in the transition region is referred to as "sharp." Ripple refers to variations in the response within the passband and the stopband. A "flat" filter response has small amounts of ripple.

A filter's *phase response* describes variations in signal phase from input to output at different frequencies. Typically, as signal attenuation increases, so does the amount by which the signal's phase is delayed as it passes through the filter. A *linear phase response* indicates that the change in phase is smooth and does not exhibit ripple at different frequencies. Ripple in either the passband or stopband means that the phase response is *nonlinear*. A filter response's varia-

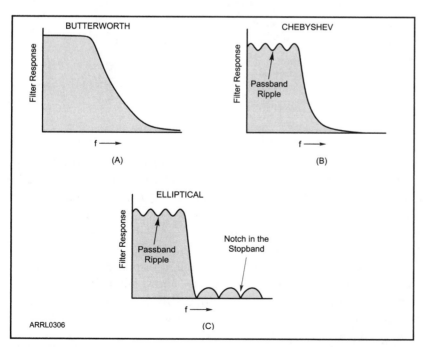

Figure 6-36 — Typical response curves for low-pass Butterworth (A), Chebyshev (B), and elliptical (C) filters.

tions in amplitude and phase can even cause *ringing*, in which certain signal frequencies or oscillations are sustained beyond the duration of the original signal. [E7G02] Nonlinear phase response can result in distortion of complex signals such as those used in digital modes. [E7C14] For such critical applications, DSP techniques may be more appropriate.

The three types of filters have characteristics that complement each other.

• Butterworth: The passband and stopband are both as flat as possible (*maximally flat*) with no ripple at all. The cutoff transition is smooth, but not steep. Butterworth filters are used when smoothly varying phase response is important to minimize signal distortion.

• Chebyshev: The passband has variable amounts of ripple, trading flatness of the passband for a sharper cutoff transition. Chebyshev filters are used when a sharp filter with consistent attenuation in the stopband is more important than maintaining signal phase in the passband. [E7C05]

• Elliptical: Cutoff is the steepest of all three filter types at the expense of ripple in both the passband and stopband. Elliptical filters are used when the most important characteristic is the sharpness of the filter cutoff. Notches in the stopband are positioned at specific frequencies to make cutoff as sharp as possible. [E7C06]

Band-pass, band-stop and notch filters are also characterized by their *bandwidth*, the frequency difference between the filter's cutoff frequencies or other frequencies with specific amounts of attenuation. When selecting a filter for a specific signal type it is important to match the filter and signal bandwidths.

The notch filter, used to remove a single frequency (more accurately, a narrow range of frequencies) and band-stop filters are specified to have a response *depth* in dB. For example, the ability of a notch filter used to remove an interfering carrier from a received SSB signal would be measured in dB with higher numbers indicating more rejection at the notch frequency. [E7C07]

Filters also have a *shape factor* that compares the frequency bandwidth at two levels of attenuation. In amateur equipment, the shape factor is specified between the filter response's –6 dB and –60 dB points. For example, a filter that has a –6 dB bandwidth of 1.8 kHz and a –60 dB bandwidth of 5.4 kHz has a –6/–60 dB shape factor of 1.8/5.4 = 3.0 to 1. The portions of a band-pass filter's response curve outside the passband are called the filter's *skirts*. The closer a filter's shape factor is to 1.0, the sharper its cutoff and the steeper its skirts.

> *Before you go on, study test questions E7C05, E7E06, E7C07, E7C10, E7C14 and E7G02. Review this section if you have difficulty.*

CRYSTAL FILTERS

The IF section of a radio requires very good band-pass filters to provide the narrow bandwidth needed to separate one signal from the many on the band. These filters cannot be built using individual inductors and capacitors. Filters using piezoelectric quartz crystals (discussed previously in the section on oscillators) can provide the high-Q, narrow-bandwidth characteristics required of modern receivers and transmitters.

Although single crystals can be used as filtering devices, the normal practice is to connect two or more together in various configurations to create the desired response. *Crystal-lattice filters* can provide narrow-bandwidth filtering at the frequencies above 500 kHz encountered in a transceiver's signal processing circuits. For example, SSB transmitters often use crystal-lattice filters after the balanced modulator to attenuate only the unwanted sideband from the closely-spaced sideband pair.

Figure 6-37 depicts a filter configuration known as the *half-lattice* filter. In this arrange-

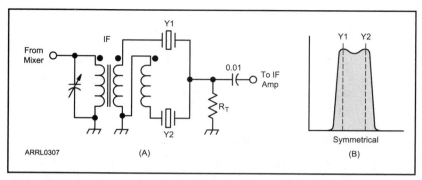

Figure 6-37 — Part A shows a schematic diagram of a half-lattice crystal filter. (B) shows a typical response curve for this type of filter. Note the steep skirts on the response between y₁ and y₂, indicating good rejection of signals outside the passband.

ment, crystals Y1 and Y2 are on different frequencies. The overall bandwidth of the crystal half-lattice filter is equal to approximately 1 to 1.5 times the frequency separation of the crystals. The closer the crystal frequencies, the narrower the bandwidth of the filter.

In general, a crystal lattice filter has narrow bandwidth and steep response skirts, as shown in Figure 6-37B. [E6E01] The bandwidth and response shape of lattice filters depends on the relative frequencies of these the crystals. [E6E02] A special type of crystal lattice filter — the Jones filter — has a variable bandwidth. [E6E12]

A good crystal-lattice filter for double-sideband (DSB) voice signal would have a bandwidth of approximately 6 kHz at the –6 dB points on the response curve so that both of the 3 kHz wide sidebands can pass through it. A crystal filter used for single-sideband (SSB) signals is significantly narrower; typical bandwidth is 2.4 kHz at the –6 dB points. For CW use, crystal filters typically have 250- to 500-Hz bandwidths.

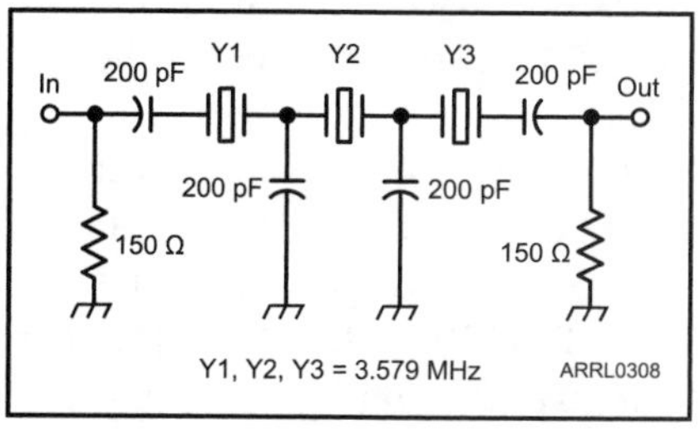

The home construction of crystal filters can be time-consuming but the filters are relatively inexpensive. For home builders, *crystal ladder filters* may be easier to design than the lattice variety. **Figure 6-38** illustrates a simple crystal ladder filter for CW, using three crystals.

To create a crystal ladder filter, start with a collection of crystals that have approximately the same frequency and characteristics. TV color burst oscillator crystals or microprocessor clock crystals are inexpensive, easy to obtain and work well for such filters. Use an oscillator circuit to measure the actual operating frequency of each crystal. Carefully select the crystals such that the frequency difference between the selected crystals is less than 10% of the desired filter bandwidth. In other words, to build a 500-Hz bandwidth filter, select crystals with frequencies all within 50 Hz of each other.

Figure 6-38 — A simple crystal ladder filter that uses three crystals and is suitable for use with CW signals.

Before you go on, study test questions E6E01, E6E02 and E6E12. Review this section if you have difficulty.

ACTIVE FILTERS

An *active filter* is one that uses an amplifier to create its frequency response. In general, passive filters that contain inductive and capacitive elements have a fixed frequency response and exhibit insertion loss. LC filters are usually physically larger and heavier than their active counterparts. Active filters have a number of advantages over LC filters at audio frequencies. They provide gain and good frequency-selection characteristics. [E7G03] They do not require the use of inductors, and they can be accurately tuned to a specific design frequency using a potentiometer.

Op amps are often used to build an active filter because the gain and frequency response of the filter can be controlled by a few resistors and capacitors connected externally to the op amp. [E7G01] There are a few disadvantages to using active filters beyond requiring a

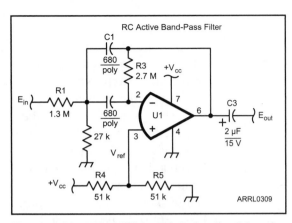

Figure 6-39 — This RC *multiple-feedback* band-pass filter is typical of active filters. This filter has a center frequency of 900 Hz and would be good for use with CW signals. Increasing C1 and C2 would lower the center frequency.

source of power. Low-cost op amps limit the useful upper frequency to a few hundred kilohertz. Their output voltage swing must be less than the dc supply voltage. Strong out-of-band input signals may overload the op amp and distort the output signal. The op amp may add some noise to the signals, resulting in a lower signal-to-noise ratio than you would have with an LC filter.

Active Audio Filters

Figure 6-39 shows a simple RC active band-pass filter suitable for CW use. Filters of this type designed for use at audio frequencies are often called *RC active audio filters*. Their principal use in Amateur Radio is as audio filters used with receivers when more selectivity is needed than is provided by a receiver's IF filters. Not only does a well-designed RC filter help to reduce QRM but it also improves the signal-to-noise ratio in some receivers. [E7G06]

Individual filter sections can be cascaded for greater *selectivity*. One or two sections may be used as band-pass or low-pass filters for improving the audio-channel passband characteristics during SSB or AM reception. Up to four filter sections are frequently *cascaded* (connected in series) to obtain selectivity for CW or RTTY reception. The greater the number of filter sections, up to a practical limit, the steeper the filter skirts will be.

An RC active filter should be inserted in the low-level audio stages. This prevents overloading the filter during strong-signal reception, such as would occur if the filter were placed at the audio output, just before the speaker or phone jack. The receiver AF gain control should be placed between the audio preamplifier and the input of the RC active filter for best results. If audio-derived AGC is used in the receiver, the RC active filter will give its best performance when it is contained within the AGC loop.

An Active Filter Design Example

The use of an op amp IC results in a compact and stable filter. Only five connections are made to the IC. The gain of the filter and its frequency characteristics are determined by the choice of resistors and capacitors external to the IC. What follows is a typical active filter design process:

The filter circuit in Figure 6-39 shows a *single-section* band-pass filter. (A section refers to one circuit that performs a specific filtering function.) To select the component values for a specific filter, you must first state the band-pass characteristics: desired filter Q (the ratio of center frequency to bandwidth), voltage gain (A_V) and center frequency (f_0). The component values for the circuit were calculated based on an $f_0 = 900$ Hz, an A_V of 1 and a Q of 5.

These are typical values of A_V and Q for filters of this type. Both can be increased for a single-section filter, but it is best to restrict the gain to 1 or 2 and limit the Q to no more than 5 to prevent unwanted filter ringing and audio instability. [E7G05]

Next, choose a standard value for C1 and C2, which have equal values in this particular filter circuit. Standard-value 680-pF capacitors were used for this design. For certain combinations of Q, A_V, and f_0 and values of C1 and C2, the design calculations below may result in unwieldy resistance values. If this happens, select a new value for C1 and C2. The capacitors should be high-Q, temperature-stable components. Polystyrene capacitors are excellent for such use. [E7G04] Disc-ceramic capacitors are not recommended. Choose standard capacitor values because capacitors with unusual values are not generally available and it is more difficult to combine capacitors in series or parallel to achieve a desired value than with resistors. C3 should be large enough in value that its reactance is a few ohms at the center frequency of the filter.

Once the capacitor values have been chosen, R1, R2, and R3 are calculated according to the following formulas:

$$R_1 = \frac{Q}{2\pi f_0 A_V C_1}$$

$$R_2 = \frac{Q}{(2Q^2 - A_V)(2\pi f_0 C_1)}$$

$$R_3 = \frac{2Q}{2\pi f_0 C_1}$$

This circuit and these equations create a Butterworth band-pass filter (smooth passband response and cutoff transition). They are called the *Sallen-Key* filter equations after the designers that first derived them. Other sets of equations and circuits can also create band-pass filters but the Sallen-Key equations are the easiest to use.

Once R3 is determined, select values for R4 and R5. They are equal in value and approximately $0.02 \times R3$. R4 and R5 are used to establish the op-amp's reference voltage, $V_{CC}/2$. After the theoretical values for all resistors have been calculated, select the nearest standard-value resistors to build the filter. The actual values can vary slightly from the calculated values, but the main effect of this is a slight alteration of f_0 and A_V. You can also use series and parallel combinations of resistors to achieve the design values, or use variable resistors in the proper range to allow filter adjustment.

Before you go on, study test questions E7G01 and E7G03 through E7G06. Review this section if you have difficulty.

DIGITAL SIGNAL PROCESSING (DSP) FILTERS

There are numerous advantages to filtering signals digitally. Since the processing takes place in a computer or microprocessor, the "circuit" never needs tuning because the computer program doesn't change characteristics with age or temperature. No compromises are required because of standard component values. Changing the filter characteristics is simply a matter of changing the program. In fact, this leads to some interesting applications because the program can respond differently to different types of signals or conditions. This is called *adaptive processing*.

An *adaptive filter* can be useful for removing unwanted noise from a received SSB signal, for example. [E7C08] An adaptive or *automatic notch filter* might automatically identify an interfering tone from a carrier, lock onto that signal and remove it from the received audio. Such a filter can even track the interfering signal as it moves through the receiver passband!

Any type of filter response characteristics that can be created by passive or active components can be implemented in a digital filter. The limits of DSP technique are mainly the upper frequency (sampling rate) at which the input signal can be sampled and the resolution (number of bits) of each sample.

In addition, DSP can create a number of filters that are impractical or impossible to build with physical components. For example, "brick wall" filters with extremely steep cutoffs would require expensive precision components and an impractical number of sections to implement with active filters. These are fairly easy to build using DSP techniques, however. The drawback of DSP filters is that they require the necessary computing hardware to implement them.

Before you go on, study test question E7C08. Review this section if you have difficulty.

IMPEDANCE MATCHING

When most hams talk about impedance matching circuits or *networks*, they are probably thinking of a circuit used between a transmitter or transceiver and an antenna system. Such an *antenna coupler* has two basic purposes: (1) to match the input impedance of the antenna feed line to the output of a transceiver or power amplifier so the amplifier has the proper resistive load, and (2) to reduce unwanted emissions (mainly harmonics) to a very low value. (Antenna couplers are also referred to as *transmatches*, *matchboxes*, *antenna tuners* and *impedance matchers*.) Impedance matching networks are also used inside radio equipment to convert impedances from one value to another.

In the case of an antenna coupler, the impedance matching circuit is usually required to transform a complex load impedance with both resistance and reactance to a purely resistive value, usually 50 Ω. To perform this task, the circuit cancels the reactive part of the impedance and then transforms the remaining resistive portion to the desired value. [E7C04]

An impedance matching network performs the transformation by exchanging energy between the inductor and capacitor in such a way that the ratio of voltage and current (the impedance) is changed between the input and output connections. Aside from small resistive losses, no energy is lost in the transformation. Only the ratio of voltage and current are changed. A mechanical analog is the gearbox in which energy at one combination of speed and torque is changed to a different combination of speed and torque.

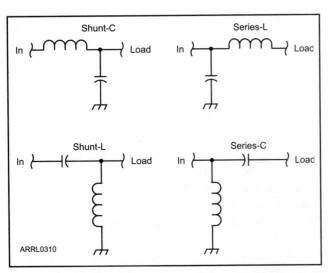

Figure 6-40 — The four variations of the LC impedance-matching L-network. Shunt or series refers to the connection of the component closest to the impedance to be matched.

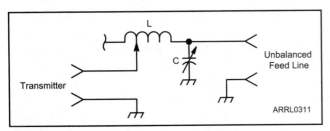

Figure 6-41 — An L-network antenna coupler, useful for an unbalanced feed line, such as coaxial cable. This circuit can transform impedances at the feed line input to the 50-Ω impedances preferred by most transceivers.

L-Networks

The simplest LC impedance matching network is the *L-network*. **Figure 6-40** shows its four variations that have both an inductor and capacitor. (There are four additional variations that either have two inductors or two capacitors, but they are much less common.) The choice of circuit to be used is determined by the ratio of the two impedances to be matched and the practicality of the component values that are required.

The L-network in **Figure 6-41** will transform to 50 Ω any higher impedance presented at the input to the feed line. (At least it will if you have an unlimited choice of values for L and C.) Most antennas and feed lines will present an impedance that can be matched with an L-network.

To adjust this L-network for a proper match, the coil tap is moved one turn at a time, adjusting C for lowest SWR at each step. Eventually a combination should be found that will give an acceptable SWR value. If the impedance at the input to the feed line is lower than 50 Ω, the circuit can be "turned around" to reverse the transformation ratio. Matching networks made entirely of inductors and capacitors work equally well in either direction!

The major limitation of an L-network is that a combination of inductor and capacitor is normally chosen to operate on only one frequency band because a given LC combination has a relatively small impedance-matching range. If the operating frequency varies too greatly, a different set of components will be needed.

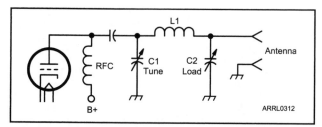

Figure 6-42 — A pi-network output-coupling circuit. C1 adjusts the circuit's tuning to resonance (TUNE) and C2 adjusts the load impedance presented to the tube (LOAD).

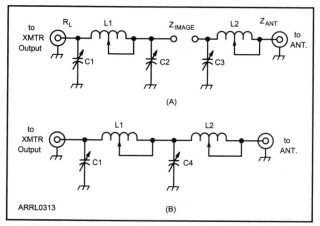

Figure 6-43 — The pi-L-network uses a pi-network to transform the transmitter output impedance (R_L) to an intermediate "image" impedance (Z_{IMAGE}). An L-network then transforms Z_{IMAGE} to the antenna impedance, Z_{ANT}.

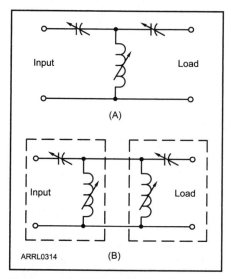

Figure 6-44 — The T-network at (A) can also be thought of as two L-networks back-to-back as shown at (B). This network has low losses and is a widely used circuit in amateur antenna couplers.

Pi and Pi-L Networks

Most tube-type amplifiers use *pi-network output-coupling circuits* as shown in **Figure 6-42**. The most common form of this network consists of one capacitor in parallel with the input and another capacitor in parallel with the output. An inductor is in series between the two capacitors. [E7C01] The circuit is called a pi-network because it resembles the Greek letter pi (π) — if you use your imagination a bit — with the two capacitors drawn down from the ends of the horizontally drawn inductor.

To adjust the pi network in a power amplifier for proper operation, the tuning capacitor (C1) is adjusted for minimum plate current, and the loading capacitor (C2) is adjusted for maximum permissible plate current. The adjustments are interactive, so this procedure is usually performed several times to reach the optimum settings. [E7B09]

You can convert the L-network of Figure 6-41 into a pi-network by adding a variable capacitor to the transmitter side of the inductor. This effectively creates two L-networks back-to-back with each L-network sharing half of the series inductance. [E7C11] Using this circuit, any value of load impedance (greater or less than 50 Ω) can be matched using some values of inductance and capacitance, so it provides a greater impedance-transformation range.

Because of the series coil and parallel capacitors, this circuit acts as a low-pass filter to reduce harmonics as well as acting as an impedance-matching device. (A pi-network with two coils shunted to ground and a series capacitor would make a high-pass filter and is virtually never used as an amateur output-coupling circuit.) Harmonic suppression with a pi-network depends on the impedance- transformation ratio and the circuit Q. While the L-network's Q is fixed as a consequence of the frequency and impedance transformation ratio, the pi-network's Q can be adjusted by selecting different combinations of component values. [E7C13] Circuit design information for pi-networks appears in *The ARRL Handbook*.

If you need more attenuation of the harmonics from your transmitter, you can add an L-network in series with a pi-network, to build a pi-L-network. **Figure 6-43** shows a pi-network and an L-network connected in series. [E7C12] It is common to combine the value of C2 and C3 into a single variable capacitor as shown in Figure 6-43B as C4. The pi-L-network thus consists of two series inductors and two shunt capacitors. The pi-L-network provides the greatest harmonic attenuation of the three most-used matching networks — the L, pi and pi-L-networks. [E7C03]

T-Networks

A T-network as shown in **Figure 6-44** consists of two capacitors in series with the signal lead and a parallel, or shunt-connected inductor between them to ground. The T-network can also be thought of as two L-networks as shown in the figure.

This circuit is commonly used in antenna coupling equipment because the series capacitors and shunt inductor have lower loss than a pi-network. While this type of T-network will transform a wide range of impedances it also acts as a high-pass filter and provides little harmonic rejection. [E7C02]

Before you go on, study test questions E7B09, E7C01 through E7C04 and E7C11 through E7C13. Review this section if you have difficulty.

6.4 Power Supplies

Almost every electronic device requires some type of power supply. The power supply must provide the required voltages when the device is operating and drawing a certain current. The output voltage of most simple power sources, such as batteries or basic rectifier circuits, varies inversely with the load current. If the device starts to draw more current, the supplied voltage will drop. In addition, the operation of most circuits will change as the power-supply voltage changes. For this reason there is a *voltage-regulation* circuit included in the power supply of almost every electronic device. The purpose of this circuit is to stabilize the power-supply output voltage and/or current under changing load conditions.

LINEAR VOLTAGE REGULATORS

Linear voltage regulators make up one major category of voltage regulator. In these circuits, regulation is accomplished by varying the conduction of a *control element* in some proportion to the load current. The control element conduction is varied so as to maintain the output voltage at a constant level. [E7D01]

Shunt and Series Regulators

In Zener diode regulator circuits the control element is a Zener diode that varies the current through a fixed resistor (R1) as shown in **Figure 6-45**. Because the Zener diode's reverse-breakdown voltage is relatively constant, varying load currents do not cause the regulated output voltage to change as long as enough current flows through the Zener diode. Because the Zener diode controls the output voltage by controlling the load on the power source, it is called a *shunt regulator*. Shunt regulators are most useful when a constant load on the input voltage source results in constant output voltage. [E7D05]

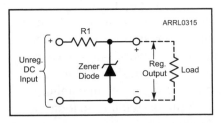

Figure 6-45 — A Zener-diode voltage regulator circuit in which R1 is the control element.

Pass transistors can also be used as the voltage-dropping control element rather than a resistor. By varying the dc current in the base of the transistor its collector-emitter current may be varied as necessary to hold the output voltage constant. This is a *series regulator* in which the current through the control element is also the load current. In effect, series regulators create "smart resistors" whose value is varied to create just the right amount of voltage drop, maintaining a constant output voltage.

Figure 6-46 shows an example of a linear pass-transistor regulator circuit, also shown on the exam in Figure E7-3. [E7D08] The control element is a pass transistor (Q1) whose base current is controlled by the Error Amplifier. The Error Amplifier compares a fraction of the output voltage to that of the voltage reference and adjusts the pass transistor base current until the output voltage has the correct value. The output resistive divider of R1, R2 and the potentiometer both provide a sample of the output voltage and place a small load on the regulator at all times. [E7D12]

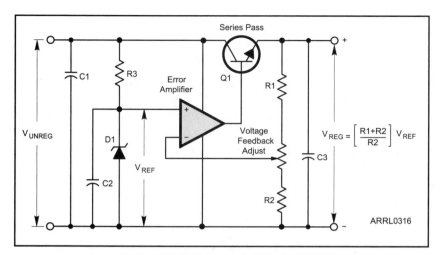

Figure 6-46 — The control element is a pass transistor ("Series Pass") whose base current is varied by the Error Amplifier. The Error Amplifier compares a portion of the output voltage to the voltage reference and adjusts the pass transistor base current until the output voltage has the correct value.

C1 serves to filter the unregulated input supply voltage. [E7D09] This capacitor is often the rectifier output filter capacitor of an unregulated rectifier supply. The *voltage reference*, usually a Zener diode (D1) as shown in the figure, provides the means of comparing output voltage to the desired value or *set point*. [E7D03, E7D13] R3 supplies current to the Zener diode. [E7D11] C2 across the Zener diode serves to filter hum, ripple, and noise from the voltage reference. [E7D07] C3 across the output terminals prevents the regulator from oscillating if the load is removed or is very small. [E7D10]

Power transistors are available that will handle several amperes of current at several hundred volts, but linear regulators of this type are usually operated below 100 V. Regulators using pass transistors have much better current handling capabilities than those of simple Zener shunt regulators. [E7D06]

IC "Three Terminal" Regulators

The modern trend in voltage regulators is toward the use of integrated-circuit devices known as *three-terminal regulators*. Inside these devices are a voltage reference, a high-gain error amplifier, current-sense resistors and transistors, and a series pass element. Some of the more sophisticated units have *thermal shutdown*, *overvoltage protection* and *foldback current limiting*.

Three-terminal regulators have a connection for unregulated dc input, one for regulated dc output and one for ground. They are available in a wide range of voltage and current ratings. It is easy to see why regulators of this sort are so popular when you consider the low price and the number of individual components they can replace. The regulators are available in several different package styles. The package and mounting methods you choose will depend on the amount of current required from your supply. Regulators in the larger all-metal TO-3 or metal tab TO-220 packages mounted on a heat sink will handle quite a bit more current than a plastic DIP IC.

Three-terminal regulators are available with positive or negative voltage outputs. In most cases, a positive regulator is used to regulate a positive voltage and a negative regulator for a negative voltage (with respect to ground). Depending on the system ground requirements, however, each regulator type may be used to regulate the "opposite" voltage.

Efficiency

Both series and shunt regulators dissipate a significant amount of the supply's input power as heat in order to maintain the constant output voltage. Because the series regulators control the load current directly, however, they are more efficient than shunt regulators. [E7D04] Efficiency is calculated as:

$$\text{Efficiency (in \%)} = 100\% \times \frac{\text{Power Out}}{\text{Power In}}$$

Before you go on, study test questions E7D01 and E7D03 through E7D13. Review this section if you have difficulty.

SWITCHING REGULATORS

The second major category of voltage regulators is the *switching regulator* in which the control device is switched on and off electronically. The switching regulator works by storing energy in the magnetic field of an inductor or transformer, then releasing it to an output filter circuit. The duty cycle of the control element controls the rate at which energy is stored and released and is automatically adjusted to maintain a constant average output voltage. [E7D02]

Switching frequencies of tens of kilohertz or more reduce the size of the transformer or energy storage inductor and of the capacitors needed to filter the output voltage. In an inverter-style power supply, the savings in weight and component cost can be substantial. [E7D17] Switching regulators also have a very high efficiency compared to linear regulators, justifying the higher expense and complexities of their design.

Before you go on, study test questions E7D02 and E7D17. Review this section if you have difficulty.

HIGH VOLTAGE TECHNIQUES

The construction of high-voltage supplies poses special considerations in addition to the normal design and construction practices used for lower-voltage supplies. In general, remember that physical spacing between leads, connections, parts and the chassis must be sufficient to prevent arcing. Also, the series connection of components such as capacitor and resistor strings needs to be done with consideration for voltage stresses in the components.

Capacitors

Capacitors are often connected in series strings to form an equivalent capacitor with the capability to withstand the applied voltage. When this is done, equal-value resistors need to be connected across each capacitor in the string in order to distribute the voltage equally across each capacitor. The *equalizing resistors* should have a value low enough to equalize differences in capacitor leakage resistance between the capacitors but high enough not to dissipate excessive power. The equalizing resistors also serve as a bleeder resistor (see below) and place a constant, light load on the supply to prevent excessive voltage with no load connected to the supply. [E7D16]

Capacitor bodies and cases in high-voltage strings need to be insulated from the chassis and from each other by mounting them on insulating panels to prevent arcing to the chassis or other capacitors in the string.

In order to reduce stress on the power supply high-voltage transformer and rectifier circuits when the supply is turned on, a "step-start" function is often used to charge the filter capacitors gradually. This consists of a resistor in the primary circuit of the power transformer that limits the input current to the supply. After a short period of a second or two, the resistor is switched out with a relay and the supply charges to its full output. [E7D15]

For high voltage supplies, oil-filled paper-dielectric capacitors are superior to electrolytics because they have lower internal impedance at high frequencies, lower leakage resistance and are available with higher working voltages. These capacitors are available in values of several microfarads and have working voltage ratings of thousands of volts.

Avoid older oil-filled capacitors. They may contain polychlorinated biphenyls (PCBs), a known cancer-causing agent. Newer capacitors have eliminated PCBs and have a notice on the case to that effect. Should you encounter old oil-filled capacitors, contact your local power utility as they often have the means to safely dispose of them.

Bleeder Resistors

Bleeder resistors provide protection against shock when the power supply is turned off and dangerous wiring is exposed because they "bleed off" the stored charge in the filter capacitors. Bleeder resistors also place a constant load on the supply. Most high-voltage supplies in amateur equipment are not regulated, so a string of bleeder resistors across the filter capacitors improves output regulation of an otherwise unregulated high voltage supply. [E7D14]

A general rule is that the bleeder resistor should be designed to reduce the output voltage to 30 V or less within 2 seconds of turning off the power supply. Take care to ensure that the maximum voltage and power rating of the resistor is not exceeded. The bleeder resistor will probably consist of several individual resistors in series. An additional recommendation is that two separate bleeder strings be used, to provide safety in the event one of the strings fails.

Before you go on, study test questions E7D14 through E7D16. Review this section if you have difficulty.

Table 6-1

Questions Covered in This Chapter

6.1 Amplifiers	*6.2 Signal Processing*	*6.3 Filters and Impedance Matching*	*6.4 Power Supplies*
E7B01	E6E03	E6E01	E7D01
E7B02	E6E09	E6E02	E7D02
E7B03	E6E10	E6E12	E7D03
E7B04	E7C09	E7B09	E7D04
E7B05	E7E01	E7C01	E7D05
E7B06	E7E02	E7C02	E7D06
E7B07	E7E03	E7C03	E7D07
E7B08	E7E04	E7C04	E7D08
E7B10	E7E05	E7C05	E7D09
E7B11	E7E06	E7C06	E7D10
E7B12	E7E07	E7C07	E7D11
E7B13	E7E08	E7C08	E7D12
E7B14	E7E09	E7C10	E7D13
E7B15	E7E10	E7C11	E7D14
E7B16	E7E11	E7C12	E7D15
E7B17	E7E12	E7C13	E7D16
E7B18	E7E13	E7C14	E7D17
E7B19	E7E14	E7G01	
E7B20	E7H01	E7G02	
E7B21	E7H02	E7G03	
E7G07	E7H03	E7G04	
E7G08	E7H04	E7G05	
E7G09	E7H05	E7G06	
E7G10	E7H06		
E7G11	E7H07		
E7G12	E7H08		
E7G13	E7H09		
E7G14	E7H10		
E7G15	E7H11		
	E7H12		
	E7H13		
	E7H14		
	E7H15		
	E7H16		
	E7H17		
	E7H18		
	E8A14		

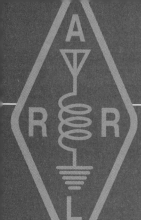

Chapter 7

Radio Signals and Measurements

In this chapter, you'll learn about:
- **Various types of ac waveforms**
- **Measuring the power of modulated RF signals**
- **Test equipment**
- **Oscilloscopes and spectrum analyzers**
- **FM and pulse modulation systems and multiplexing**
- **Transmitter intermodulation**
- **Noise from atmospheric static, power lines and automotive systems**
- **Noise reduction techniques**

This chapter covers radio signals and the instruments used to measure them and other electrical quantities. Like other topics on the Extra class exam, you were introduced to many of these concepts as you studied for your Technician and General class licenses. For the Extra class exam, we'll dive deeper into these topics.

The Extra class license exam includes basic questions about ac waveforms and waves, measurements of waveforms and associated test equipment, FM and pulse modulation, intermodulation, noise and interference. These topics are somewhat related and that is why they are grouped together in this chapter. Questions about these topics are located in several parts of the question pool, so be sure to review each one before moving on.

7.1 AC Waveforms and Measurements

Before we study specific terms and measurements associated with ac signals and waveforms, it's a good idea to review the basic concepts. Students familiar with ac waveforms can skip this section but should check the related exam questions to be sure of their answers.

TYPES OF WAVEFORMS

So far, you have only encountered sine waves as examples of ac waveforms. Sine waves are not only a very common waveform, but they can be added together to make other waveforms. We'll begin by studying the sine wave in depth and then move on to other common types of waveforms.

Sine Waves

The sine wave is the most fundamental of all ac waveforms, consisting of energy at a single frequency. Not only is a sine wave the most fundamental ac waveform, it also describes rotating motion. This equivalence forms a bridge between understanding electrical and mechanical energy. To visualize a sine wave as describing rotation, imagine a rotating wheel. Paint one dot anywhere along the circumference (rim) of the wheel. Spin the wheel at a constant rate and watch that spot!

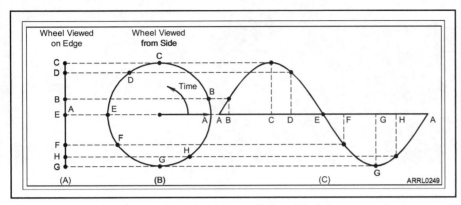

Figure 7-1 — This diagram illustrates the relationship between a sine wave and circular rotation. You can see how various points on the circle correspond to values on the sine wave.

Figure 7-1 illustrates how this works. If you watch the spot from the wheel's edge as in Figure 7-1A, it will just move up and down. If we designate the position at point C as +1 and at point G as –1, then make a table of values as the wheel rotates through 360°, the values will correspond exactly to the values of the sine function (*sin*). When the wheel is in position A corresponding to 0°, sin (0°) = 0, at position C, sin (90°) = 1, and so forth all the way around.

Now move to look directly at the wheel as in from the side as in Figure 7-1B. Plot the dot's position in degrees around its circular path on the horizontal axis. Plot the dot's vertical position using the vertical axis as before. The dot's position will then trace out the waveform in Figure 7-1C. This is called a sine wave because the amplitude of the waveform is equal to the value of the sine of the wheel's position in degrees. The mathematical description of the individual points that make up this waveform is

$$A = \sin (\theta) \hspace{4cm} \text{(Equation 7-1)}$$

where:
A = the instantaneous amplitude of the sine wave in the vertical direction.
θ = the position of the point in degrees along the horizontal axis.

There is one more way to think of the sine wave. In Figure 7-1B, an arrow is drawn from the center of the wheel to point A. This arrow represents a *vector*. Vectors consist of a pair of numbers — amplitude and direction. The vector in Figure 7-1B has an amplitude equal to its length, which we arbitrarily decided would be 1 when constructing the table of sine values. Since the vector is pointing exactly along the horizontal axis, its direction is 0°. Thus, the vector is described as "1 at an angle of 0°." In mathematical notation, this is written $1\angle 0°$. (You've seen this notation before in our discussion of phase angles and impedance.) The rotation of the wheel can then be described as a vector that is rotating (spinning around like the hand of a clock) at the frequency of the sine wave.

Now let's make the connection between angle and time. If the wheel is spinning at a constant rate, every complete rotation takes the same amount of time. That means each degree of rotation represents the same amount of time, as well. The time it takes to make each rotation is the *period*, T, and period is equal to the reciprocal of frequency, f. [E8A08]

$$f = \frac{1}{T} \quad \text{and} \quad T = \frac{1}{f} \hspace{3cm} \text{(Equation 7-2)}$$

Each degree of rotation then takes T/360. For example, a wheel spinning four times per second (T = 0.25 second) has a frequency of 1/T = 1/0.25 = 4 Hz. Each degree of rotation then takes T/360 = 0.25 / 360 = 0.00069 second.

To be able to use Equation 7-1 to calculate the amplitude, A, of the sine wave (the vertical position of the dot) at any point in time, t, we need to be able to convert time to the angle θ. Assume that at time t = 0, the dot is at position A and the wheel is moving 360 degrees in T seconds. The wheel will turn θ degrees in t × (360/T). So our sine wave equation is now:

$$A = \sin\left(360\,\frac{t}{T}\right) \quad = \quad \sin\left(360 \times \frac{1}{T} \times t\right) \quad = \quad \sin(360\,f\,t) \qquad \text{(Equation 7-3)}$$

Example 7-1

What is the amplitude of a 60 Hz sine wave at t = 1 ms? At t = T = 16.7 ms? At t = T/4 = 4.17 ms?

$$A = \sin(360 \times 60 \times 0.001) = \sin(21.6°) = 0.368$$

$$A = \sin(360 \times 60 \times 0.0167) = \sin(360°) = 0$$

$$A = \sin(360 \times 60 \times 0.00417) = \sin(90°) = 1$$

In many technical fields the mathematics works out better if *radians* are used instead of degrees. There are 2π (6.28) radians in a circle, so 1 radian = 360 / 2π = 57.3°. Equation 7-3 is then often written A = sin (2πft). Because this is so common, the symbol ω is used instead, with $\omega = 2\pi$f. ω is called *angular frequency*. The license exam and this book always use f, but you will encounter ω in technical references and articles.

Complex Waveforms

A signal composed of more than one sine wave is called a *complex waveform*. That doesn't necessarily mean "difficult," just that there is more than one sine wave involved. A simple example of a complex waveform is the signaling waveform used by telephones when dialing. This waveform is composed of two different sine wave tones, thus the name "dual-tone multi-frequency" or DTMF for that signaling system. Listen carefully next time you dial and you will hear the two tones of different frequencies.

There are certain well-known and common complex waveforms that are made up of a sine wave and its harmonics. These are termed *regular* waveforms because the harmonic relationship of all the sine waves results in a waveform with a single overall frequency and period. The lowest frequency sine wave in a regular waveform is called the *fundamental*. A waveform that is made of sine waves that are not harmonically related, such as human speech, is an *irregular* waveform. [E8A09] Whether regular or irregular, the sine waves that make up a complex waveform are called its *components*.

To be able to discuss complex waveforms more easily, it is useful to be able to make a graph of all the components by frequency and amplitude as in **Figure 7-2**. This is called a *frequency domain* or *spectrum* graph. Each component of the signal is shown as a vertical line, with the height of the line showing the component's amplitude. Note that each component occupies a single frequency. Which harmonics are combined with the fundamental and the relative amplitude of each determine the final shape of the waveform as you will see in the following two sections. The set of all components that make up a signal is called the signal's *spectrum*.

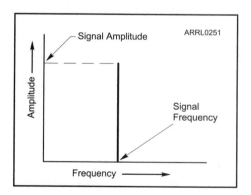

Figure 7-2 — A frequency domain or spectrum graph shows a sine wave as a single vertical line. The horizontal axis represents frequency and the vertical axis represents amplitude. The height of the line representing the sine wave shows its amplitude.

Sawtooth Waves

The shape of a *sawtooth* waveform, as shown in **Figure 7-3**, closely resembles the teeth on a saw blade. It is characterized by having a significantly faster *rise time* (the time it takes for the wave to reach a maximum value) compared to its *fall time* (the time it takes for the wave to reach a minimum value). A sawtooth wave is made up of a sine wave at its fundamental frequency and all of its harmonics. The sawtooth's spectrum is shown in Figure 7-3. The *ramp* waveform is similar to the sawtooth but with a slow rise time

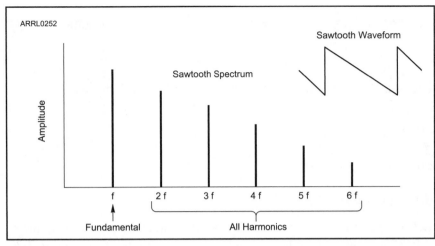

Figure 7-3 — The sawtooth waveform is made up of sine waves at the fundamental frequency and all of its harmonics. The amplitude of the harmonics decreases as their frequency increases.

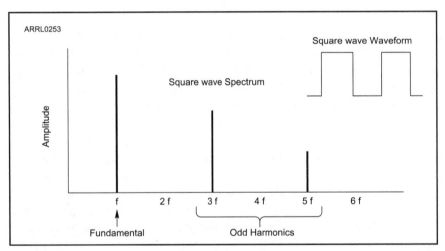

Figure 7-4 — The square wave is made up of sine waves at the fundamental frequency and only the odd harmonics. The amplitude of the harmonics decreases as their frequency increases.

(the ramp) and a fast fall time. [E8A02, E8A03]

Square Waves

A square wave is one that abruptly changes back and forth between two voltage levels and remains an equal time at each level as in **Figure 7-4**. (If the wave spends an unequal time at each level, it is known as a *rectangular wave*.) A square wave is made up of sine waves at the fundamental and all the *odd* harmonic frequencies. [E8A01] The square wave's spectrum is shown in Figure 7-4.

Pulse Waveforms

Information is often represented by *pulse waveforms*. Shown in **Figure 7-5**, these are similar to square and rectangular waveforms, but some characteristic of the pulse is varied (amplitude, rate, width, and so forth) to represent information. Another example of a pulse waveform is found in radar systems that measure the time it takes for a pulse to travel to and from the reflecting surface. In general, pulse waveforms consist of narrow bursts of energy followed by periods of no signal. [E8A10]

Before you go on, study test questions E8A01, E8A02, E8A03, E8A08, E8A09 and E8A10. Review this section if you have difficulty.

AC MEASUREMENTS

Because an ac signal's *instantaneous* voltage and current change from one instant to the next, how do you measure it? Is there a single point at which the measurement is taken? Since a sine wave is positive and negative for exactly the same amount of time, shouldn't the value be zero? Actually, the average dc voltage and current in an ac signal are zero! A dc meter connected to an ac voltage will read zero, too. Obviously, ac signals do deliver power and so other measurements than averaging are used to measure the value of ac voltage and current.

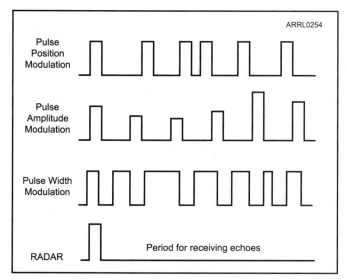

Figure 7-5 — A pulse waveform consists of varying durations of energy with no signal between the pulses.

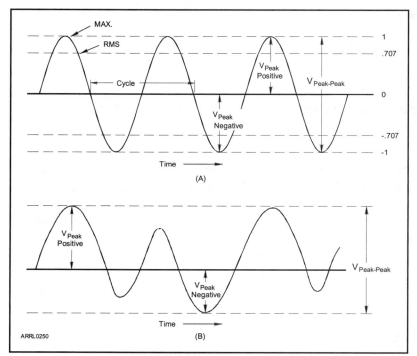

Figure 7-6 — The various parameters of a sine wave are shown at A. At B, an asymmetric, complex waveform is shown whose positive peak voltage and negative peak voltage are different.

If you look at the sine wave of Figure 7-1, the easiest dimension to measure is the vertical height of the waveform. The maximum positive or negative voltage is called the *peak voltage*, V_{Peak} or V_P. The voltage from the maximum positive to the maximum negative peak is called the *peak-to-peak* value, V_{P-P} or V_{Pk-Pk}. (From here on, we'll assume the ac waveforms represent voltage, unless specifically noted otherwise.) In a symmetrical waveform, peak voltage has half the value of the peak-to-peak voltage. [E8D02] **Figure 7-6** illustrates the various voltage parameters of a sine wave.

Peak-to-peak voltage measurements are particularly useful when working with devices such as linear amplifiers. These devices can only reproduce input signals below a certain level. Input signals above this level will produce distortion or spurious outputs. Because it is important to avoid such nonlinear operation, the maximum peak-to-peak input amplitude is an important parameter for evaluating linear (Class A) amplifiers. [E8D03]

When an ac voltage is applied to a resistor, the resistor will dissipate energy in the form of heat, just as if the voltage were dc. The dc voltage that would cause an identical amount of heating to the ac voltage is called the *root-mean-square* (*RMS*) or *effective value* of the ac voltage. [E8A04] The phrase "root-mean-square" describes the mathematical process of actually calculating the effective value. The method involves squaring the instantaneous values for a large number of points along the waveform, then finding the average (or mean) of the squared values of those points, and taking the square root of that average.

The RMS value of any waveform, voltage or current, can also be determined by measuring the heating effect in a resistor. This is actually the process used for precise measurements of complicated waveforms when a computer is unavailable to perform the point-by-point calculation. Because equivalent heating defines RMS value, this is the most accurate method of measuring the RMS value of a waveform of any type, ac or dc. [E8A05]

This sounds like a very difficult measurement but for common symmetric ac waveforms the conversions between peak, peak-to-peak, average and RMS are simple. **Table 7-1** shows how to convert between peak, peak-to-peak, average and RMS waveforms of sine and square

Table 7-1

AC Measurements for Sine and Square Waves

	Sine Wave	Square Wave
Peak-to-Peak	2 × Peak	2 × Peak
Peak	0.5 × Peak-to-Peak	0.5 × Peak-to-Peak
RMS	0.707 × Peak	Peak
Peak	1.414 × RMS	RMS
Average	0 (full cycle)	0 (full cycle)
	0.637 × Peak (half cycle)	0.5 × Peak (half cycle)

waves. You will make frequent use of the sine wave conversions, in particular.

Example 7-2

What is the RMS value of a sine wave with a 10 V peak value?

$$V_{RMS} = 0.707 \times 10 = 7.07 \text{ V}$$

Example 7-3

What is the peak value of a sine wave with an RMS value of 120 V, the common household voltage? [E8D13]

$$V_P = 120 \times 1.414 = 170 \text{ V}$$

Example 7-4

What is the peak-to-peak value of a sine wave with an RMS value of 65 V? [E8D05]

$$V_P = 1.414 \times 65 = 91.9 \text{ V}$$

$$V_{P-P} = 2 \times V_P = 2 \times 91.9 = 184 \text{ V}$$

Example 7-5

What is the peak value of a sine wave with an RMS value of 34 V? [E8D12]

$$V_P = 1.414 \times 34 = 48 \text{ V}$$

Unless otherwise specified or if it is obvious from the context, ac voltage is given as an RMS value. For example, the nominal value of household ac voltage in the U.S. 120 V, is an RMS value. [E8D15] The voltage at your household outlets varies with the amount of load that the power company must supply and is sometimes even specified as 110 V. This also means that the peak and P-P values also vary.

Example 7-6

What is the peak-to-peak value of a sine wave with an RMS value of 120 V, the common household voltage? [E8D14]

$$V_P = 120 \times 1.414 = 170 \text{ V}$$

$$V_{P-P} = 2 \times V_P = 340 \text{ V}$$

Example 7-7

What is the RMS value of a 340-V_{P-P} sine wave? [E8D16]

$$V_P = 0.5 \times 340 = 170 \text{ V}$$

$$V_{RMS} = 0.707 \times V_P = 120 \text{ V}$$

As you can tell from Examples 7-3 and 7-6, it is important to know the peak value of the voltage when picking an electronic component to be used with that waveform. For example, suppose you wanted to install a capacitor from the ac power input connections of a piece of equipment to ground to get rid of some interference. If you ordered a component rated for 120 V, or even 150 V, it would likely fail immediately because the peak voltage it would experience is actually 170 V!

AC Power

The terms RMS, average and peak have different meanings when referring to ac power. When the sine waves for voltage and current are in phase such as in a resistor (see the discussion on power in the Electrical Principles chapter), power is the *product* of RMS voltage and current. In this case, RMS and average power are the same, so only average power, P_{AVG}, is used.

$$V_{RMS} \times I_{RMS} = P_{AVG}$$ (Equation 7-4)

For continuous sine wave signals with voltage and current in phase:

$$V_{PEAK} \times I_{PEAK} = P_{PEAK} = 2 \times P_{AVG}$$ (Equation 7-5)

Before you go on, study test questions E8A04, E8A05, E8D02, E8D03, E8D05, and E8D12 through E8D16. Review this section if you have difficulty.

POWER OF MODULATED RF SIGNALS

In the case of an unmodulated radio signal, average power is calculated as in Equation 7-4. If the load impedance (Z) is known, it is easier to use Equation 7-6 than to measure the RF current.

$$P_{VAG} = \frac{V_{RMS}^2}{Z}$$ (Equation 7-6)

Example 7-8

What is the average power in a 50-Ω load during one complete RF cycle if the RMS voltage is 70 V?

$$P_{AVG} = \frac{70^2}{50} = 98 \text{ W}$$

Example 7-9

What is the average power in a 50-Ω load during one complete RF cycle if the peak voltage is 35 V? [E8D11]

$$V_{RMS} = V_P \times 0.707 = 35 \times 0.707 = 24.7 \text{ V}$$

$$P_{AVG} = \frac{24.7^2}{50} = 12.2 \text{ W}$$

Only when using CW is a steady sine-wave signal being produced by a transmitter. For AM signals, the waveform varies with time in order to carry information. The peak power output of an AM or SSB transmitter is defined to be the power averaged over a single, complete RF cycle having the greatest amplitude. This is *peak envelope power* or *PEP*. For an unmodulated signal, PEP is equal to average power.

Voice-modulated AM signals are not purely sinusoidal because they are composites of RF

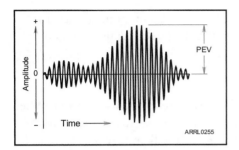

Figure 7-7 — An amplitude-modulated signal as an example of a complex signal. Peak envelope voltage (PEV) is an important parameter for determining the power of a complex waveform.

signals and the various frequencies in the modulating voice signal. The cycle-to-cycle variation of the RF waveform is small enough, compared to the modulating audio frequencies, that sine-wave measurement techniques produce accurate results.

For determining PEP of such a waveform, the most important parameter is the *peak envelope voltage* (PEV), shown in **Figure 7-7**. PEV is used in calculating the power in a modulated signal, such as that from an amateur SSB transmitter. To compute the PEP of a modulated waveform, multiply the PEV by 0.707 to obtain the RMS value, square the result and divide by the load resistance.

$$PEP = \frac{(PEV \times 0.707)^2}{R_{LOAD}}$$ (Equation 7-6A)

Example 7-10

What is the PEP output power of a transmitter that has a PEV of 100 V across a resistive load of 50 Ω?

$$PEP = \frac{(100 \times 0.707)^2}{50} = 100 \text{ W}$$

Example 7-11

What is the PEP output power of a transmitter that has a PEV of 30 V across a resistive load of 50 Ω? [E8D04]

$$PEP = \frac{(30 \times 0.707)^2}{50} = 9 \text{ W}$$

Envelope peaks occur only sporadically during voice transmission. They occur too quickly for meter readings to accurately represent their values. *Peak-hold* or *peak-reading wattmeters* are special meters used to display the peak value of a waveform. [E8D06]

If you use a power amplifier in your amateur station, you might want to use a peak-reading wattmeter in the feed line to your antenna to insure that you do not exceed the maximum allowable output power. [E8D10] Such a meter will indicate the peak envelope power output (PEP) of a modulated signal, as long as the transmitter and amplifier are properly adjusted to eliminate spurious outputs. You can also measure the peak voltage and calculate PEP from Equation 7-6A. Remember that wattmeters are generally calibrated for use in a system with a particular characteristic impedance, such as 50 Ω, and will give inaccurate readings otherwise.

Most meters and displays respond to the amplitude (current or voltage) of the signal, averaged over many cycles of the modulation envelope. When amateurs refer to *average output power* in this context, they are referring to a long-term average of power, not P_{AVG} of Equation 7-4.

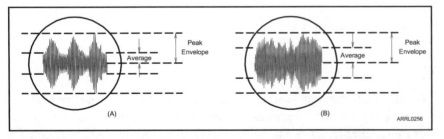

Figure 7-8 — Two modulation envelope patterns that show the difference between average and peak power levels. In each case, the RF amplitude is plotted as a function of time. In B, the average power level has been increased compared to the peak value.

The ratio of peak-to-average power in SSB voice signals varies widely with voices of different characteristics. [E8A07] **Figure 7-8** shows two typical envelope patterns. In the case shown in Figure 7-8A, the average power (estimated graphically) is such that the peak-to-average ratio is almost 3:1. By raising the minimum modulated signal levels, the average output power can be increased as in Figure 7-8B. Typical ratios of PEP to average power are about 2.5:1. [E8A06] Depending on the type of voice and manner of speaking the ratio may be more than 10:1. The PEP of an AM or SSB signal will be several times greater than the power output averaged over many cycles of the modulating signal.

Before you go on, study test questions E8A06, E8A07, E8D04, E8D06, E8D10, and E8D11. Review this section if you have difficulty.

ELECTROMAGNETIC FIELDS

You learned about electric and magnetic fields while studying capacitors and inductors. These fields store energy in and around those components. The fields do not move. *Electromagnetic waves* (what we fondly call *radio waves*) carry energy as well, but they move through space independently of any component or conductor.

An electromagnetic wave, as the name implies, is composed of both an electric field and a magnetic field. Both fields must be changing with time — most commonly in a sinusoidal pattern. [E8D08] The fields are also oriented at right angles to each other as shown by **Figure 7-9**. [E8D07] The term "lines of force" in the figure means the direction in which a force would be felt by an electron (from the electric field) or by a magnet (from the magnetic field). The direction of the right angle between the electric and magnetic fields determines the direction the wave travels, as illustrated by Figure 7-9.

To an observer staying in one place, such as a fixed station's receiving antenna, the electric and magnetic fields of the wave appear to oscillate as the wave passes. That is, the fields create forces on electrons in the antenna that increase and decrease in a sine wave pattern. Some of the energy in the propagating wave is transferred to the electrons as the forces from the changing fields cause them to move. This creates a sine wave current in the antenna whose frequency is determined by the rate at which the field strength changes as the wave passes.

If the observer is moving in the same direction as the wave and at the same speed, however, the strength of the fields will not change. To that observer, the electric and magnetic field strengths are fixed, as in a photograph. This is a *wavefront* of the electromagnetic wave — a flat surface or plane moving through space on which the electric and magnetic fields have a constant value as illustrated in Figure 7-9.

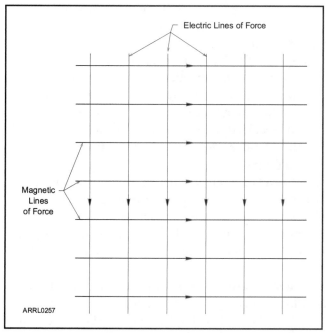

Figure 7-9 — Representation of electric and magnetic lines of force in an electromagnetic wavefront. Arrows indicate the instantaneous directions of the fields for a wavefront in a wave traveling toward you, out of the page. Reversing the direction of either of the fields would also reverse the direction of the wave.

Just as an ac voltage is made up of an infinite sequence of instantaneous voltages, each slightly larger or smaller than the next, an infinite number of wavefronts make up an electromagnetic wave, one behind another like a deck of cards. The direction of the wave is the direction in which the wavefronts move. The fields on each successive wavefront have a slightly different strength, so as they pass a fixed location the detected field strength changes as well. The fixed observer "sees" fields with strengths that vary as a sine wave.

Figure 7-10 is a drawing of what would happen if we could suddenly freeze all of the wavefronts in the wave and take measurements of the electric and magnetic field strengths

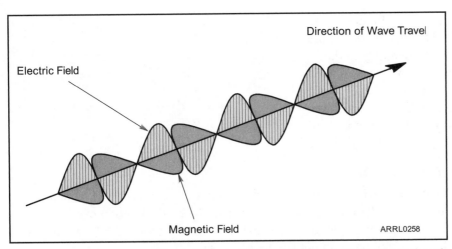

Figure 7-10 — Representation of the magnetic and electric field strengths of a vertically polarized electromagnetic wave. In the diagram, the electric field is oriented vertically and the magnetic field horizontally.

in each. In this example, the electric field is oriented vertically and the magnetic field horizontally. (Each of the vertical lines in the electric field can be thought of as representing an individual wavefront.) All of the wavefronts are moving in the direction indicated — the whole set of them moves together at the same speed. As the wave — the set of wavefronts — moves past the receive antenna, the varying field strengths of the different wavefronts is perceived as a continuously changing wave. What we call a "wave" is really this entire group of wavefronts moving through space.

In free space, the wavefronts move at the speed of light, approximately 300 million meters per second (3×10^8 m/s). (Light is also an electromagnetic wave of extremely high frequency — many thousands of GHz.) If the receive antenna is moving toward or away from the wave's source, the antenna will encounter the wavefronts more or less frequently than they were emitted, raising or lowering the received frequency. The same effect occurs if the transmitting antenna is moving. This is the *Doppler effect* and it can be observed in the frequencies of signals from rapidly moving satellites.

The speed at which electromagnetic waves travel or *propagate* depends on the characteristics of the medium through which they travel. In general, the speed of light is highest in the vacuum of free space and only slightly lower in air. In materials such as glass or plastic, however, velocity can be quite a bit lower. For example, in polyethylene (commonly used as a center insulator in coaxial cable) the *velocity of propagation* is about two-thirds (67%) of that in free space.

One more important note about electromagnetic waves: The electric and magnetic fields are *coupled*, that is they are both aspects of the same entity — the electromagnetic wave. They are not perpendicular electric and magnetic fields that simply happen to be in the same place at the same time! The fields cannot be separated although the energy in the wave can be detected as either electric or magnetic force. The fields are created as a single entity — an electromagnetic wave — by the motion of electrons in the transmitting antenna.

Polarization

The orientation of the pair of fields in an electromagnetic wave can have any orientation with respect to the surface of the Earth, but the electric and magnetic fields will always be at right angles to each other. The orientation of the wave's electric field determines the *polarization* of the wave. If the electric field's lines of force are parallel to the surface of the Earth (meaning those of the magnetic field are perpendicular to the Earth), the wave is *horizontally polarized*. Conversely, if the magnetic field's lines of force are parallel to the surface of the Earth (and those of the electric field are perpendicular to the Earth), the wave is *vertically polarized*. Knowing the polarization of the wave allows the receiving antenna to be oriented so that the passing wave will exert the maximum force on the electrons in the antenna, maximizing received energy and thus signal strength.

For the most part, the wave's polarization is determined by the type of transmitting antenna and its orientation. For example, a Yagi antenna with its elements parallel to the Earth's surface transmits a horizontally polarized wave. On the other hand, an amateur mobile whip antenna, mounted vertically on an automobile, radiates a vertically polarized wave. If a vertically polarized antenna is used to receive a horizontally polarized radio wave (or vice versa), received signal strength can be reduced by more than 20 dB as compared to using an antenna with the same polarization as the wave. This is called *cross-polarization*.

It is also possible to generate electromagnetic waves in which the orientation of successive wavefronts rotates around the direction of travel — both the electric and magnetic fields. This is called *circular polarization*. [E8D09] Imagine the wave of Figure 7-10 being twisted so at one point the direction of the electric field is horizontal and a bit further along the wave it is vertical. As the twisted, circularly polarized wave passes the receiving antenna, the polarization of its fields will appear to rotate. The rate at which the polarization changes and the direction of the rotation — *right-handed* or *left-handed* — is determined by the construction of the transmitting antenna. Note that the electric and magnetic fields rotate together so the right-angle between them remains fixed. Polarization that does not rotate is called *linear polarization* or *plane polarization*. Horizontal and vertical polarization are examples of linear polarization.

To best receive a circularly polarized wave, the structure of the receiving antenna should match that of the transmitting antenna. It is particularly helpful to use circular polarization in satellite communication, where polarization tends to shift with the orientation of the satellite and the path of its signal through the atmosphere. Circular polarization is usable with linearly polarized antennas at one end of the signal's path. There will be some small loss in this case, however.

Before you go on, study test questions E8D07, E8D08 and E8D09. Review this section if you have difficulty.

7.2 Test Equipment

The following types of test equipment are just a few of the many different instruments used in radio. You have been introduced to the multimeter in studies for your previous license exams and the Extra class license goes a bit further. The other types of test equipment covered in this section may be new to you, but are often encountered in building and testing amateur equipment and antennas.

INSTRUMENTS AND ACCURACY

Multimeters

One of the simplest and most useful pieces of test equipment is a *multimeter*, also known as a *volt-ohm-meter*, *volt-ohm-milliammeter*, *VOM*, or most frequently, just plain *voltmeter*. This basic piece of test equipment is used to display a variety of measurements, including voltage, current and resistance. Everyone who works with electronics will find many uses for meters. A multimeter is one of the first pieces of test equipment that most amateurs will own.

Where analog, needle-swinging meters were once the norm, digital meters are the most common today, with microprocessor-controlled electronics inside and a liquid crystal display (LCD). These meters are called *digital multimeters* or *DMMs*. Regardless of whether the meter is digital or analog, their basic specifications are similar.

The accuracy of most meters is specified as a percentage of full scale. If the specification states that the meter accuracy is within 2% of full scale, the possible error anywhere on a scale of 0 to 10 V is 2% of 10 V, or 0.2 V. The actual value read anywhere within that range could be as much as 0.2 V above or below the indicated meter reading. If you are using a

0 to 100-mA scale, the possible error would be ± 2 mA anywhere on that scale. Many digital multimeters are *autoranging* and can automatically choose the right range to display volts, amps or ohms.

The resolution of almost any multi-meter sold today is sufficient for general measurements in radio equipment. Most offer "3½ digit" displays, meaning that the left-most of four digits is 1 or blank. Such a meter has a resolution of 0.05% at full scale, plenty good for amateur use! (This would be a good time to read the sidebar "Accuracy, Precision and Resolution.")

Another useful specification is the meter's sensitivity. A sensitive meter draws very little current from the circuit being tested. Sensitivity of analog meters is often specified in ohms-per-volt (Ω/V), the input impedance in ohms divided by the full-scale reading in volts. Or the input impedance can be obtained by multiplying the full-scale meter reading by the sensitivity in Ω/V. [E4B12] Digital meters just specify their input impedance directly. Higher values in Ω/V or of input impedance is generally good for both ac and dc meters. [E4B08]

Dip Meters

Once called a grid-dip meter because it employed a vacuum tube with a meter to indicate grid current, this handy device is actually an RF oscillator. Modern solid-state *dip meters* use a field-effect transistor (FET) in the oscillator circuit. The fixed-value inductor of the tuned LC circuit that determines the oscillator's frequency is exposed so that it can be placed near circuits to be tested. A variable capacitor inside the dip meter determines the oscillator's frequency. A meter shows the voltage output from the oscillator.

The principle of operation of a dip meter is that when the meter is brought near a circuit and its frequency of oscillation adjusted to match that of the circuit, some power will be transferred from the dip meter to the circuit. This causes a drop in the meter reading, and the frequency can be read from a calibrated dial on the variable capacitor. There are usually several plug-in inductors so that the dip meter can operate over a wide frequency range. This is generally not a highly accurate frequency measurement, but is often sufficient for general tuning purposes.

One common amateur use of a dip meter is to determine the resonant frequency of an antenna or antenna traps. Dip meters also can be used to determine the resonant frequency of other circuits, such as an LC tuned circuit. The circuit under test should have no signal or power applied to it while you couple the dip meter to the circuit. Another common use is to provide a small signal at some specific frequency to inject into a circuit. The tunable RF oscillator of a dip meter is ideal for this task. Dip meters have even been used as very low power CW transmitters!

Dip meters are usually coupled to a circuit by allowing the field of the dip meter's inductor to interact with an inductor in the circuit under test. This is called *inductive coupling*. The energy is transferred through the magnetic fields of the inductors. Sometimes it is not possible to couple the meter to an inductor, so *capacitive coupling* is used. In this case,

Accuracy, Precision and Resolution

The terms *accuracy*, *precision* and *resolution* are often confused and used interchangeably, when they have very different meanings. When dealing with measurements and test instrumentation, it's important to keep them straight.

✔ Accuracy is the ability of an instrument to make a measurement that reflects the actual value of the parameter being measured. An instrument's accuracy is usually specified in percent or decibels referenced to some known standard.

✔ Precision refers to the smallest division of measurement that an instrument can make repeatedly. For example, a metric ruler divided into mm is more precise than one divided into cm.

✔ Resolution is the ability of an instrument to distinguish between two different quantities. If the smallest difference a meter can distinguish between two currents is 0.1 mA, that is the meter's resolution.

It is important to note that the three qualities are not necessarily mutually guaranteed. That is, a precise meter may not be accurate, or the resolution of an accurate meter may not be very high, or the precision of a meter may be greater than its resolution. It is important to understand the difference between the three when selecting and using test instruments.

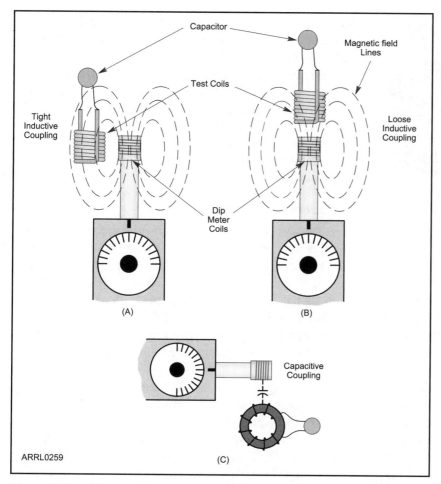

Figure 7-11 — The methods of coupling a dip meter to a tuned circuit. Tight inductive coupling is shown at A, and loose inductive coupling is shown at B. Capacitive coupling is shown at C. The choice of which method to use depends on the type of circuit being tested.

the dip meter inductor is simply brought close to any component of the circuit being tested and the capacitance between the components couples a signal to the circuit. **Figure 7-11** indicates the methods of coupling a dip meter to a circuit.

The coupling should be as light ("loose") as possible and still provide a definite, but small, dip in the meter reading when coupled to a circuit resonant at the dip meter's oscillator frequency. Coupling that is too loose will not give a dip sufficient to be a positive indication of resonance. Coupling that is too strong ("tight"), however, almost certainly will change the tuning of the dip meter's oscillator, reducing the accuracy of the frequency measurement. [E4B14]

The procedure for using a dip meter is to bring the dip meter inductor within a few inches of the circuit to be tested and then sweep the oscillator through the frequency band until the meter reading drops ("dips") sharply. This indicates that energy from the dip meter oscillator is being coupled to the circuit being tested. Erratic readings, such as a very abrupt dip or a sharp change above or below the frequency of the dip, may indicate that coupling is too tight. Nearby transmitters can also cause jumps in the meter reading. If no dip appears during the frequency sweep, try another inductor on the dip meter. The actual resonant frequency of the circuit being tested might be far from that expected.

Impedance Bridges

Another test instrument that uses a meter as its main indication is the *impedance bridge*. A bridge circuit, shown in **Figure 7-12**, works by balancing the two voltage dividers (called *arms*) — Z1-ZS and Z2-ZX. Z1 and Z2 have equal values. A bridge circuit works by connecting the unknown impedance (ZX) in one arm of the bridge and adjusting the other arm (ZS) until it has the same value as ZX. When this occurs, the bridge is said to be *balanced* and an equal voltage appears at each of the arm midpoints. With equal voltages on either side, the voltmeter reading is then zero or *nulled*. Bridges are excellent instruments for measuring impedance because obtaining the null of the meter reading can be done precisely. [E4B02]

The actual circuits of ac bridges take many forms, depending on the ranges of impedance and frequency to be covered. As the fre-

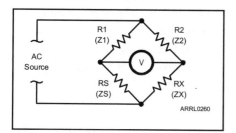

Figure 7-12 — This generalized bridge circuit works by balancing the two voltage dividers, Z1-ZS and Z2-ZX

Figure 7-13 — Typical frequency counter used by hams for troubleshooting and construction. Many lab-grade counters are available as surplus or used equipment. Frequency counters are also available in kit form and as new equipment.

quency increases, stray effects (unwanted capacitances and inductances) become more pronounced. At RF, it takes special attention to minimize them.

Most amateur-built bridges are used for RF measurements, especially SWR measurements on transmission lines, as well as lumped constant components. *Antenna analyzers* have become the preferred instrument used by many hams for measuring impedance, but the analyzer circuit uses an internal calibrated bridge circuit to measure impedance.

Frequency Counters

A *frequency counter*, once considered almost a luxury item, is now often included as a feature on multimeters. The name describes its function completely: A frequency counter makes frequency measurements. It counts the number of cycles per second (hertz) of a signal, and displays that number on a digital readout. Advanced frequency counters can also measure pulse widths and signal periods, count pulses, and measure the time intervals. **Figure 7-13** shows a typical frequency counter. Lab-grade counters are widely available as surplus or used equipment and as inexpensive new equipment and kits.

You can use a frequency counter to measure signal frequencies throughout a piece of equipment and to make fine adjustments to tuned circuits and oscillators. In this case, a frequency counter becomes a valuable piece of test equipment.

A frequency counter can only measure time as accurately as its reference crystal oscillator, or *time base*. The more accurate this crystal is, the more accurate the readings will be. [E4B01] Close-tolerance crystals are used, and there is usually a trimmer capacitor across the crystal so the frequency can be set exactly once it is in the circuit. One way to increase the accuracy of a frequency counter is to increase the accuracy of the time base oscillator. The internal circuits of a typical counter are illustrated as the block diagram in **Figure 7-14**.

Frequency counters that operate well into the gigahertz range are available. Counters that operate at VHF and higher frequencies, and sometimes those that operate near the top of the HF range, employ a *prescaler*. The prescaler is a circuit that divides the input frequency by a known factor (usually 10 or 100), greatly extending the useful range of the frequency counter.

Although usually quite accurate, a frequency counter should be checked regularly against WWV, WWVH or some other *frequency standard*. The accuracy of frequency counters is often expressed in parts per million

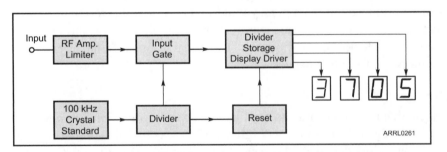

Figure 7-14 — This block diagram shows the basic parts of a frequency counter.

(ppm). Even after checking the counter against WWV, you must take this possible error into account. The counter error can be as much as:

$$\text{Error in Hz} = \frac{\text{f (in Hz)} \times \text{counter error in ppm}}{1,000,000}$$

(Equation 7-7)

Suppose you are using a frequency counter with a time base accuracy of 10 ppm to measure the operating frequency of a 146.52 MHz transmitter. How much could the actual frequency differ from the reading? Use Equation 7-7 to calculate the maximum frequency-readout error:

$$\text{Error in Hz} = \frac{146,520,000 \times 10 \text{ ppm}}{1,000,000} = 1465.20 \text{ Hz} \quad [\text{E4B05}]$$

This means the actual operating frequency of this radio could be as high as 146.5214652 MHz or as low as 146.5185348 MHz. You should always take the maximum frequency-readout error into consideration when deciding how close to a band edge you should operate.

Example 7-12

If a frequency counter with a time base accuracy of ±0.1 ppm reads 146,520,000 Hz, what is the most the actual frequency being measured could differ from the reading?

$$\text{Error in Hz} = \frac{146,520,000 \times 0.1 \text{ ppm}}{1,000,000} = 14.652 \text{ Hz} \quad [\text{E4B04}]$$

If a frequency counter with a time base accuracy of ±1.0 ppm reads 146,520,000 Hz, what is the most the actual frequency being measured could differ from the reading?

$$\text{Error in Hz} = \frac{146,520,000 \times 1 \text{ ppm}}{1,000,000} = 146.52 \text{ Hz} \quad [\text{E4B03}]$$

Any variation of the time base oscillator frequency affects the counter accuracy and precision. The counter keeps track of the number of RF cycles that occur in a given time interval (called the *gate interval*). Even a slight increase in the gate interval could result in a significant increase in the displayed frequency. As a simple example, suppose you are using a frequency counter that counts RF cycles for 1 second, and displays the frequency as a result of that count. If the time-base oscillator slows down, RF cycles will be counted for a longer period, so the displayed frequency will increase. If the oscillator speeds up, RF cycles will be counted for a shorter period, so the displayed frequency decreases. The displayed frequency on this counter will change even if the signal it is counting does not vary!

Some transceivers use a built-in frequency counter to measure and display the operating frequency, although not all radios with a digital display use a frequency counter *per se*. Rigs that use direct digital synthesizer (DDS) technology use the controlling microprocessor to calculate the operating frequency for display. If a counter if used, though, then the frequency readout will be as accurate as the counter itself.

Before you go on, study test questions E4B01 through E4B05, E4B08, E4B12, and E4B14. Review this section if you have difficulty.

THE OSCILLOSCOPE

Direct observation of high-speed signals and waveforms is not possible using any kind of meter or numeric instrument. There is just too much information to be conveyed at too high a rate. Enter the *oscilloscope*, or *scope* — the amateur's electronic eyes. A scope is used to display a signal's amplitude versus time so that the shape and other characteristics of the waveform can be seen and measured, even if the signal is changing very quickly.

The *oscilloscope* is built around a *cathode-ray tube (CRT)*, discussed in the Components and Building Blocks chapter. A sawtooth-type *ramp* waveform with a slow rise time and a sudden fall time applied to the horizontal deflection plates causes the spot to move from left to right, creating a narrow line of light (called the *trace*) on the face of the CRT as shown in **Figure 7-15**. Because the ramp moves the beam across the face of the tube at a known rate, it is called the scope's *time base*. The rate at which the trace moves is called the *sweep speed*, and it is selected by the operator.

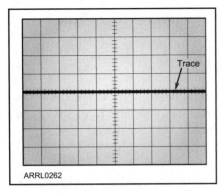

Figure 7-15 — Without any voltage applied to the vertical channel of the oscilloscope, the electron beam creates a flat trace across the face of the CRT. The grid markings (graticule) allow the operator to make amplitude and time measurements.

The signal to be analyzed is applied to the CRT's vertical deflection plates through the scope's *vertical channel* amplifier. Positive voltage moves the trace upward. If a sine wave signal is applied to the vertical amplifier and the appropriate sweep speed is selected, the trace will form a sine wave as the trace is moved up and down simultaneously with its movement across the tube. An important limitation to the accuracy, frequency response and stability of an oscilloscope is the bandwidth (frequency response) of the scope's vertical amplifiers. Scopes are often specified in terms of this bandwidth. Scopes are also specified by how many vertical amplifier channels they have (special circuits can make it appear as if there are separate traces for each channel), so you might see a particular model listed as a "20 MHz dual-channel scope."

Another important performance limitation is the accuracy and linearity of the scope's time base. Unless the sweep oscillator is stable and the sweep speed can be set accurately, frequency measurements made with the scope will not be very accurate. Scopes with good performance at RF through the HF range are widely available both as new and used instruments.

The grid of marks on the face of the tube is called a *graticule* and each line is called a *division*. The graticule's vertical axis is calibrated in *volts/division* or *V/div* and various scales are selectable by the operator. The horizontal axis is calibrated in *time/division* or seconds/division (*s/div*). Vertical scales are often available from mV/div to tens of V/div. Most scopes offer sweep speeds of a few s/div to ns/div. This allows the display of signals with frequencies of less than 1 Hz to hundreds of MHz and with amplitudes ranging from mV to tens of V.

By using a positioning control, the amplitude and frequency of a signal can be compared to the fine divisions on the graticule's central axes. The easiest amplitude measurement to make with a scope is an ac signal's peak-to-peak voltage by using the graticule lines as shown in **Figure 7-16**. [E8D01] Similarly, by using the points at which the waveform crosses a horizontal line, its period (and thus frequency) is easy to determine, as well. Some scopes even have built-in circuits to make these and other measurements automatically!

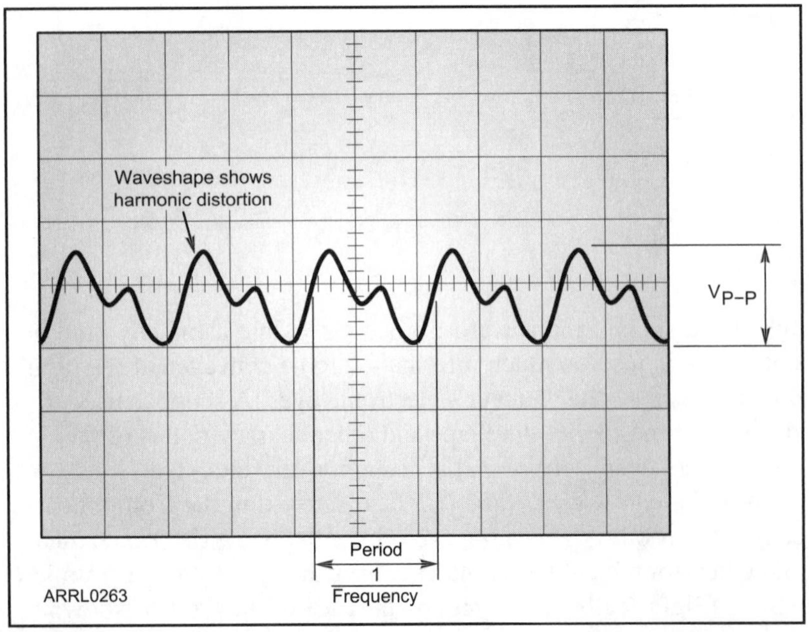

There are many uses for an oscilloscope in an amateur station. This instrument is often used to display the output waveform of a transmitter during a two-tone test. Such a test can help you determine if the amplifier stages in your rig are operating in a linear manner. Complex digital signals require an oscilloscope for detailed analysis, as well. [E4A11] An oscilloscope can also be used to display signal waveforms during troubleshooting procedures. For example, consider the waveform display of Figure 7-16. The fundamental frequency sine wave can be seen to have some significant distortion due to the presence of harmonics.

Oscilloscopes are connected to the circuits being tested using a special *oscilloscope probe* or *scope probe*.

Figure 7-16 — This display shows how the peak-to-peak voltage and period of a complex waveform are measured using an oscilloscope. It can also be seen that the waveform contains enough harmonic energy to cause significant distortion.

Such probes are specially constructed so that they convey the signal to the vertical input with its frequency and timing characteristics as little-changed as possible. Each probe has its own ground lead that is connected to the circuit being tested. For the most accurate measurements at high frequencies, it is important to keep the ground connection as short as possible. [E4B07]

The probe incorporates a high-impedance voltage divider that loads the circuit as little as possible. The signal's amplitude at the output of the probe (the input to the scope) is typically divided by 10 — this is a *times-10* (*×10*) probe. (There are also ×1 and ×100 probes, but they are used much less frequently.) Most scopes assume a ×10 probe is used when displaying the vertical scale in V/div.

Because the frequency response of the probe can affect the frequency response of the signal being displayed on the scope, it is important to ensure that the probe is adjusted properly. This is referred to as the probe's *compensation*. Most scopes provide a square wave *calibration* signal of a few kHz and with an amplitude of 0.1 to 1 V. The probe is connected to this signal and its compensation control adjusted until the square wave's horizontal portions are flat and the corners are sharp. (The scope's user manual will describe this procedure in detail.) [E4B13]

Before you go on, study test questions E4A11, E4B07, E4B13 and E8D01. Review this section if you have difficulty.

THE SPECTRUM ANALYZER

In the previous section on AC Waveforms and Measurements, we discussed how the spectrum of signals can be graphed with amplitude on the vertical axis and frequency on the horizontal axis. This is called the *frequency domain*. Just as an oscilloscope displays a signal's amplitude versus time (called the *time domain*), the *spectrum analyzer* displays a signal in the frequency domain. This type of display is useful when measuring amplifiers, oscillators, detectors, modulators, mixers and filters — the frequency content of their outputs being of primary importance. [E4A01]

Time and Frequency Domains

To better understand the concepts of time and frequency domains, refer to **Figure 7-17**. In Figure 7-17A, the three-dimensional coordinates show time (as the line sloping toward the bottom right), frequency (as the line sloping toward the top right), and amplitude (as the vertical axis). The two frequencies shown are harmonically related (f_1 and $2f_1$). The time domain is represented in Figure 7-17B, in which all frequency components are added together. If the two frequencies were applied to the input of an oscilloscope, we would see the bold line that represents the amplitudes of the signals added together. The frequency domain contains information not found in the time domain, and vice versa. Hence, the spectrum analyzer offers advantages over the oscilloscope for certain measurements, but for measurements in the time

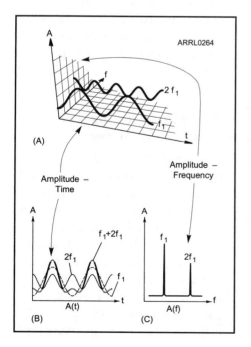

Figure 7-17 — This diagram shows how a complex waveform may be displayed in either the time domain or frequency domain. Part A is a three-dimensional display of amplitude, time, and frequency. At B, this information is shown in the time domain as on an oscilloscope. At C, the signal's frequency domain information is shown as it would be displayed on a spectrum analyzer.

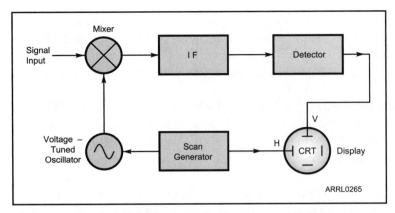

Figure 7-18 — This simplified block diagram illustrates the operation of swept-superheterodyne spectrum analyzer. It is basically an electronically tuned, narrow-band receiver with a CRT display of signal amplitude and frequency.

domain, the oscilloscope is an invaluable instrument.

The display shown in Figure 7-17C is typical of a spectrum analyzer presentation of a complex waveform. Here the signal is separated into its individual frequency components, and a measurement made of the amplitude of each signal component. A signal's amplitude can be represented on a spectrum analyzer's vertical scale as its voltage or as its power.

Spectrum Analyzer Basics

A simplified block diagram of the common *swept superheterodyne* analyzer is shown in **Figure 7-18**. The analyzer is basically a narrow-band receiver that is electronically tuned in frequency. Tuning is accomplished by applying a linear ramp voltage to a *voltage-controlled oscillator* (*VCO*). The same ramp voltage is simultaneously applied to the horizontal deflection plates of the cathode ray tube (CRT). This means the horizontal axis of the spectrum analyzer displays frequency. The receiver output is applied simultaneously to the vertical deflection plates of the CRT, which means the vertical axis of the spectrum analyzer displays signal amplitude. The resulting spectrum analyzer display shows amplitude versus frequency. [E4A02, E4A03]

Spectrum analyzers are calibrated in both frequency and amplitude for relative and absolute measurements. The frequency range, controlled by the *scan width* control, is calibrated in hertz, kilohertz or megahertz per division on the graticule. Within the limits of the voltage-controlled oscillator, you can test a transmitter over any frequency range. Whether you are testing an HF or a VHF transmitter, the spectrum analyzer displays all frequency components of the transmitted signal.

The vertical axis of the display is commonly calibrated as 1 dB, 2 dB or 10 dB per division. (Linear scales in V/division are also available, but not used as frequently.) For transmitter testing, the 10 dB/div range is commonly used because it allows you to view a wide range of signal strengths, such as those of the fundamental signal, harmonics and spurious signals.

Transmitter Testing with a Spectrum Analyzer

Among other practical uses, the spectrum analyzer is ideally suited for checking the output from a transmitter or amplifier for spectral quality. You can easily see any spurious signals from the transmitter on a spectrum analyzer display. [E4A04] **Figure 7-19** shows two test setups commonly used for transmitter testing. The setup at B is the more accurate approach for

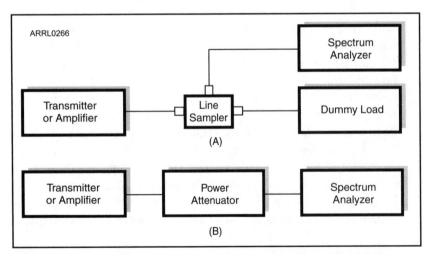

Figure 7-19 — These diagrams show two commonly used test setups to observe the output of a transmitter or amplifier on a spectrum analyzer. The system at A uses a transmission line power sampler to obtain a small amount of the transmitter or amplifier output power. At B, the majority of the transmitter output power is dissipated by the power attenuator, leaving a small amount to be measured by the spectrum analyzer.

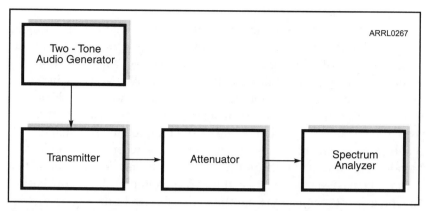

Figure 7-20 — This diagram illustrates the test setup used in the ARRL Laboratory for measuring the IMD performance of SSB transmitters and amplifiers.

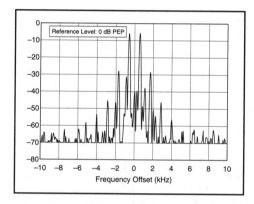

Figure 7-21 — This is the spectrum analyzer display showing the result of a two-tone intermodulation distortion test of an SSB transmitter. Each horizontal division is equal to 2 kHz and each vertical division is 10 dB. The transmitter's PEP output for a single tone is represented by the top line of the display. The sideband's components representing the two audio tones are the large peaks at the center of the display. IMD products can be observed 28, 47 and 52 dB below the PEP output level.

broadband measurements because most transmission line power sampling devices do not have a constant-amplitude output across a broad frequency spectrum. The ARRL Headquarters Laboratory staff uses the setup shown in B. The power attenuator should be able to attenuate the transmitter output power to a safe level for the sensitive spectrum analyzer input, typically 10 mW or less. [E4A12]

Another area of concern in the realm of transmitter spectral purity has to do with the *intermodulation distortion (IMD)* levels of SSB transmitters and amplifiers. The test setup in **Figure 7-20** shows how a spectrum analyzer is used for SSB transmitter IMD testing. [E4A05] The transmitter is first adjusted to produce full PEP output with a single audio tone input. Two equal-amplitude, but not harmonically related, audio tones are then input to the transmitter. (The ARRL Lab uses 700 and 1900-Hz tones.) **Figure 7-21** shows a typical display during a transmitter IMD test. Signals other than from the two individual tones (the large peaks at the center of the display) are distortion products. In this example, IMD products are observed at approximately 28, 47 and 52 dB below the transmitter's single-tone PEP output level, which is at the very top line of the display. The two individual tones are 6 dB below PEP output because they are displayed as two discrete frequencies. [E4B10]

As mentioned earlier, you can make many types of measurements with a spectrum analyzer. In addition to checking a transmitter's spectral output and measuring the two-tone IMD levels, you can determine whether a crystal is operating on its fundamental or overtone frequency by building a simple oscillator circuit and measuring the output frequency. You can even measure the degree of isolation between the input and output ports of a 2 meter duplexer. [E4A06] Because the spectrum analyzer does not display time, however, you cannot measure the speed at which a transceiver switches from transmit to receive, for example. You can find a more detailed discussion of spectrum analyzer measurement techniques in *The ARRL Handbook*.

Before you go on, study test questions E4A01 through E4A06, E4A12 and E4B10. Review this section if you have difficulty.

TRANSISTOR CIRCUIT PARAMETERS

As long as we're discussing test equipment, here's a useful topic — testing transistors and diodes without removing them from their circuits. This is a part of troubleshooting that can save you a lot of time trying to track down a problem. As you go through a circuit, a quick check of voltages with a multimeter can find a problem fairly quickly and keep you from poking around where there isn't any problem!

Start by checking the power supply voltages at the circuit. If the power supplies aren't right, then nothing else is going to work right. If you encounter a power supply voltage that is lower than it should be, but isn't completely off, that's a clue to look in the power supply section.

Assuming your power supply voltages are reasonable, it's time to look at voltages in the circuit itself. If you are lucky, the designer made notes right on the schematic about what voltages are expected in various places. You can often isolate a problem to a particular section of the circuit by finding where the voltages differ significantly from those on the schematic.

To check out individual transistors and diodes, you'll need to know whether the devices are supposed to be on (conducting current) or off (not conducting current). We'll assume that all of the devices are silicon. The device electrode listed first (for example, "base" in "base-to-emitter") is the one to which you'll connect the multimeter's positive lead. V_{CC} or V_{DD} is the power supply voltage.

PN Junction Diode or Rectifier
ON — 0.6 to 0.75 V from anode to cathode
OFF — 0 to the diode's PIV rating from cathode to anode

Schottky Rectifier
ON — 0.1 to 0.5 V from anode to cathode
OFF — 0 to the diode's PIV rating from cathode to anode

NPN Transistor
ON — V_{BE} of 0.6 to 0.75 V from base to emitter, between 0.3 V and V_{CC} from collector to emitter [E4A10]
LINEAR REGION — Same V_{BE} as ON, ⅓ to ⅔ V_{CC} from collector to emitter
SATURATED — Same V_{BE} as ON, but 0.3 V or less from collector to emitter
OFF (CUTOFF) — V_{BE} of less than 0.5 V from base to emitter, V_{CC} from collector to emitter

PNP Transistor
ON — V_{EB} of 0.6 to 0.75 V from emitter to base, between 0.3 V and V_{CC} from emitter to collector
LINEAR REGION — Same V_{EB} as ON, ⅓ to ⅔ V_{CC} from emitter to collector
SATURATED — Same V_{EB} as ON, but 0.3 V or less from emitter to collector
OFF (CUTOFF) — V_{EB} of less than 0.5 V from emitter to base, V_{CC} from collector to emitter

FET transistors are less easy to troubleshoot because there are many varieties (N-channel and P-channel, both enhancement and depletion mode). Here is the information for the two most common varieties:

N-Channel Enhancement MOSFET
ON — V_{GS} of 2 V to V_{DD} from gate to source; between 0.1 V and V_{DD} from drain to source
LINEAR REGION — Same V_{GS} as ON, ⅓ to ⅔ V_{DD} from drain to source
SATURATED — Same V_{GS} as ON, but 0.3 V or less from drain to source
OFF (CUTOFF) — V_{GS} of less than 0.5 V from gate to source, V_{DD} from drain to source

N-Channel Enhancement JFET
ON — V_{GS} between pinch-off voltage (usually –2 to –5 V) from gate to source; between 0.1 V and V_{DD}— Same V_{GS} as ON, ⅓ to ⅔ V_{DD} from drain to source
SATURATED — Same V_{GS} as ON, but 0.3 V or less from drain to source
OFF (PINCH OFF) — V_{GS} of pinch-off voltage or less from gate to source, V_{DD} from drain to source

By learning the basic parameters of many devices you can quickly determine if they are seriously damaged or if their supporting circuit has a problem. Of course, if they are cracked and charred, that might be a tip-off, too — don't forget the visual inspection! You'll become an ace troubleshooter by being organized, keeping records, and trying to understand the circuit before diving in. Good luck!

Before you go on, study test question E4A10. Review this section if you have difficulty.

7.3 Modulation Systems

The process of adding information to and recovering information from radio frequency signals is what radio is all about! You've already studied AM techniques for the General class exam and both modulators and demodulators for AM signals were covered in the chapter on Electronic Circuits. In this chapter, we'll cover two important signal definitions for FM, the most popular mode of all, including commercial and industrial signals. We'll also take a look at pulse modulation and multiplexing (a method of combining more than one flow of information in a single signal). Very multimedia!

FCC EMISSION DESIGNATIONS AND TERMS

Although the question pool does not include any direct questions on the system of emission identifiers used by the FCC, it is a good idea to be familiar with them, since they are frequently used in spectrum allocations and regulations.

The International Telecommunication Union (ITU) has developed a special system of identifiers to specify the types of signals (emissions) permitted to amateurs and other users of the radio spectrum. This system designates emissions according to their necessary bandwidth and their classification. While a complete emission designator might include up to five characters, generally only three of them are used.

The designators begin with a letter that tells what type of modulation is being used. The second character is a number that describes the signal used to modulate the carrier. The third character specifies the type of information being transmitted.

Table 7-2 summarizes the most common characters for each of the three symbols that make up an emission designator. Some of the more common combinations encountered in Amateur Radio are:

N0N — Unmodulated carrier
A1A — Morse code telegraphy using amplitude modulation
A3E — Double-sideband, full-carrier, amplitude-modulated telephony
J3E — Amplitude-modulated, single-sideband, suppressed-carrier telephony
J3F — Amplitude-modulated, single-sideband, suppressed-carrier television
F3E — Frequency-modulated telephony
G3E — Phase-modulated telephony
F1B — Telegraphy using frequency-shift keying without a modulating audio tone (FSK RTTY). F1B is designed for automatic reception.
F2B — Telegraphy produced by modulating an FM transmitter with audio tones (AFSK RTTY). F2B is also designed for automatic reception.
F1D — FM data transmission, such as packet radio

An emission designator is assembled by selecting one character from each of the three sets, based on knowledge of the transmission system. For example, suppose you know that a certain signal is produced by an AM (double-sideband, full carrier) transmitter. That is represented by the letter A for the first character. If the transmitter is modulated with a single-channel signal containing digital information without the use of a modulating subcarrier, you would select the number 1 as the second character. Finally, suppose the resulting signal is a telegraphy signal primarily intended for aural (by ear) reception rather than for machine or

Table 7-2

Partial List of Emissions Designators

(1) First Symbol — Modulation Type

Unmodulated carrier	N
Double sideband full carrier	A
Single sideband reduced carrier	R
Single sideband suppressed carrier	J
Vestigial sidebands	C
Frequency modulation	F
Phase modulation	G
Various forms of pulse modulation	P, K, L, M, Q, V, W, X

(2) Second Symbol — Nature of Modulating Signals

No modulating signal	0
A single channel containing quantized or digital information without the use of a modulating subcarrier	1
A single channel containing quantized or digital information with the use of a modulating subcarrier	2
A single channel containing analog information	3
Two or more channels containing quantized or digital information	7
Two or more channels containing analog information	8

(3) Third Symbol — Type of Transmitted Information

No information transmitted	N
Telegraphy — for aural reception	A
Telegraphy — for automatic reception	B
Facsimile	C
Data transmission, telemetry, telecommand	D
Telephony	E
Television	F

computer (automatic) copy. Then the third character of our emission symbol will be A. So we have just completely described an A1A signal, which is Morse code telegraphy!

Emission Types

Part 97, the FCC rules governing Amateur Radio, refers to emission types rather than emission designators. The emission types are CW, phone, RTTY, data, image, MCW (modulated continuous wave), SS (spread spectrum), pulse and test. Any signal may be described by either an emission designator or an emission type. Emission types are what amateurs refer to as "modes."

While emission types are fewer in number and easier to remember, they are a somewhat less descriptive means of identifying a signal. There is still a need for emission designators in Amateur Radio and they are used in Part 97. In fact, the official FCC definitions for the emission types include references to the symbols that make up acceptable emission designators for each emission type.

The US Code of Federal Regulations, Title 47, consists of telecommunications rules numbered Parts 0 through 200. These Parts contain specific rules for many telecommunications services the FCC administers. Part 2, Section 2.201, Emission, modulation and transmission characteristics, spells out the details of the ITU emission designators, as the FCC applies them in the US.

FM/PM MODULATION AND MODULATORS

Frequency modulation (emission F3E) operates on an entirely different principle from amplitude modulation. You learned about direct and indirect FM modulators in the chapter on Electronic Circuits. With FM, the signal frequency is varied above and below the carrier frequency at a rate equal to the modulating-signal frequency. (Carrier frequency refers to the frequency of the FM signal with no modulation applied.) For example, if a 1000-Hz tone is used to modulate a transmitter, the modulated signal's frequency will vary above and below the carrier frequency 1000 times per second.

The amount of frequency change, however, is proportional to the modulating signal amplitude. This frequency change is called *deviation*. Let's say that a certain signal produces a 5-kHz deviation. If another signal, with only half the amplitude of the first, were used to modulate the same transmitter, it would produce a 2.5-kHz deviation.

To more completely describe an FM signal, you will need to understand two terms that refer to FM systems and operation: *deviation ratio* and *modulation index*. They may seem to be almost the same — indeed, they are closely related.

Deviation Ratio

In an FM system, the ratio of the maximum frequency deviation to the highest modulating frequency is called the *deviation ratio*. [E8B09] It is a constant value for a given modulator and transmitter, and is calculated as:

$$\text{Deviation Ratio} = \frac{D_{MAX}}{M} \qquad \text{(Equation 7-8)}$$

where:

D_{MAX} = peak deviation in hertz
M = maximum modulating frequency in hertz.

Peak deviation is defined as half the difference between the maximum and minimum signal frequencies. That is, a sine wave modulating signal will cause the signal frequency to move higher and lower symmetrically about the carrier frequency. If maximum deviation is specified as ±5 kHz, a total difference of 10 kHz between maximum and minimum frequency, the peak deviation is one-half that value, or 5 kHz.

Peak deviation is usually controlled by setting an audio gain control in the FM modulator's circuit. Because it is fixed for that transmitter, there is no microphone gain control on an FM transmitter's front panel.

Example 7-13

In the case of narrow-band FM (the type used in amateur analog FM voice communications), peak deviation at 100% modulation is defined as 5 kHz (or less). What is the deviation ratio if the maximum modulating frequency is 3 kHz? [E8B05]

$$\text{Deviation Ratio} = \frac{D_{MAX}}{M} = \frac{5\,\text{kHz}}{3\,\text{kHz}} = 1.67$$

Example 7-14

If the maximum deviation of an FM transmitter is 7.5 kHz and the maximum modulating frequency is 3.5 kHz, what is the deviation ratio? [E8B06]

$$\text{Deviation Ratio} = \frac{D_{MAX}}{M} = \frac{7.5\,\text{kHz}}{3.5\,\text{kHz}} = 2.14$$

Notice that since both frequencies were given in kilohertz we did not have to change them to hertz before doing the calculation. The important thing is that they both be in the same units.

Modulation Index

The ratio of the maximum signal frequency deviation to the instantaneous modulating frequency is called the *modulation index*. [E8B01] Modulation index is a measure of the relationship between deviation and the modulating signal's frequency. That is:

$$\text{Modulation Index} = \frac{D_{MAX}}{m}$$ (Equation 7-9)

where:

D_{MAX} = peak deviation in hertz.

m = modulating frequency in hertz at any given instant.

Example 7-15

If the peak deviation of an FM transmitter is 3000 Hz, what is the modulation index when the carrier is modulated by a 1000-Hz sine wave? [E8B03]

$$\text{Modulation Index} = \frac{D_{MAX}}{m} = \frac{3000 \text{ Hz}}{1000 \text{ Hz}} = 3$$

When the same transmitter is modulated with a 3000-Hz sine wave that results in the same peak deviation (3000 Hz), the index would be 1; with a 100-Hz modulating wave and the same 3000-Hz peak deviation, the index would be 30, and so on.

Example 7-16

If the peak deviation of an FM transmitter is 6 kHz, what is the modulation index when the carrier is modulated by a 2 kHz sine wave? [E8B04]

$$\text{Modulation Index} = \frac{D_{MAX}}{m} = \frac{6 \text{ kHz}}{2 \text{ kHz}} = 3$$

The modulation index varies inversely with the modulating frequency in a frequency modulator if the peak deviation is kept the same, as Equation 7-9 shows. A higher modulating frequency results in a lower modulation index. In a frequency modulator, the actual deviation depends only on the amplitude of the modulating signal and is independent of frequency. Thus, a 2-kHz tone will produce the same deviation as a 1-kHz tone if the amplitudes of the tones are equal. The modulation index in the case of the 2-kHz tone is half that for the 1-kHz tone.

By contrast, in a phase modulator the modulation index is constant regardless of the modulating frequency, as long as the amplitude of the modulating signal remains constant. [E8B02] In other words, a 2-kHz tone will produce twice as much deviation as a 1-kHz tone if the amplitudes of the tones are equal.

Notice that with either an FM or a PM system, the deviation ratio and modulation index are independent of the frequency of the modulated RF carrier. It doesn't matter if the transmitter is a 10 meter FM rig or a 2 meter FM rig.

What are deviation ratio and modulation index used for? Since deviation ratio is fixed, it is used to describe and specify an FM or PM modulator. Once the transmitter is built, deviation ratio is adjusted and not changed.

Modulation index, on the other hand, varies with the input signal's amplitude (because it changes D_{MAX}) and frequency. The actual spectrum of an FM signal is quite complex and modulation index provides a way to describe how the energy is distributed within its spectrum. Just as with an AM signal, modulation index is a way to describe a modulated signal's bandwidth. Modulation indexes that are too high produce signal components farther and farther from the carrier frequency. This can cause interference to signals on adjacent channels, indicating that the modulating signal amplitude should be reduced, particularly at low frequencies. Controlling modulation index is another reason why pre-emphasis and de-emphasis are used in FM modulators and demodulators. The maximum modulation index allowed by FCC rules is 1.0. [E1B12]

Before you go on, study test questions E1B12, E8B01 through E8B06 and E8B09. Review this section if you have difficulty.

PULSE MODULATION SYSTEMS

There are several methods used for transmitting analog information as a series of brief RF pulses. These methods are known as pulse modulation and they vary in how the modulating signal's information is conveyed by the pulses. These systems are different from those that transmit data as digital codes that you will study in the chapter on Radio Modes and Equipment. Nevertheless, because the data is transmitted as a series of on/off pulses, it is considered to be a digital data system. [E8A11]

Pulse modulation systems represent the modulating signal's information directly by varying the pulse's characteristics — amplitude, repetition rate or frequency, or width. The modulating signal's information can be recovered directly from the pulse-modulated signal, without conversion to an intermediate form in which the data is encoded as characters or numbers.

In *pulse-width modulation* (*PWM*), also known as *pulse-duration modulation* (*PDM*), the duration (width) of the transmitted pulses varies with the modulating signal's amplitude. In *pulse-position modulation* (*PPM*) the modulated signal is produced by varying the position of each pulse with respect to some fixed time base. This means the modulating signal changes the timing of the transmitted pulses. [E8B08]

Both systems start with a sample of the information signal. This sampling can be done by using a sawtooth waveform with a rapid rise time and a frequency that is several times higher than the information-signal frequency. **Figure 7-22** shows how this sawtooth waveform can be used to trigger a sampling circuit and measure the difference between the information signal amplitude and the sampling wave.

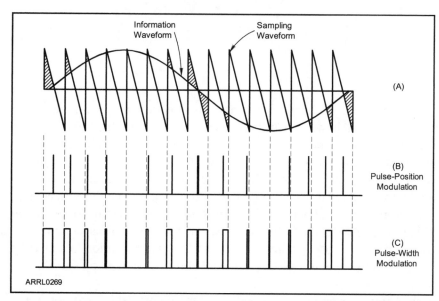

ARRL0269

Figure 7-22 — An information signal and a sampling waveform are shown at A. B shows how the sampled information signal can produce pulse-position modulation. C shows how the same sampling waveform can be used to produce pulse-width modulation.

While both PWM and PPM use pulses of uniform amplitude, the third common form of pulse modulation, *pulse-amplitude modulation* (*PAM*), changes the amplitude of each pulse to reflect the modulating signal amplitude. **Figure 7-23** illustrates how PAM signals are generated. PAM is less immune to noise than are either of the other methods, and so it is seldom considered for communication systems, although it may be the easiest method to visualize and understand.

Notice that in all three cases, the duty cycle of the transmission is very low. A pulse of relatively short duration is transmitted, with a relatively long

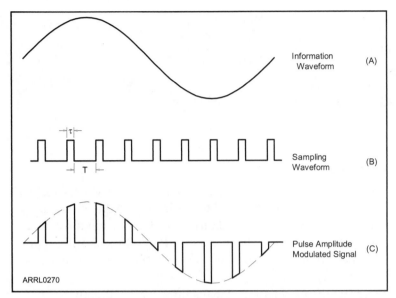

Figure 7-23 — The information signal at A is sampled during the pulses of the sampling waveform (B) to produce a pulse-amplitude modulated signal at C.

period of time separating each pulse. This causes the peak power of a pulse-modulated signal to be much greater than its average power. [E8B07]

Our examples have used a sine-wave-like information signal. If the information signal that is being sampled to produce pulse modulation is a voice, then the transmitted signal will be a representation of that speech pattern. When the signal is demodulated in the receiver, the voice audio will be reproduced. A voice signal can be made to vary the duration, position, amplitude or code of a standard pulse to produce a pulse-modulated signal. For example, in a pulse-width modulation system, the standard pulse is varied in duration by an amount that depends on the voice waveform at a particular instant.

Amateurs use pulse-position modulation and pulse-width modulation systems for radio control purposes. The transmitted signal contains coded information about the position of a control device and the receiver decodes the information to control the position of a servo motor.

Pulse-modulation techniques find applications in some other electronics circuits as well. Switch-mode power supplies and switching regulators commonly use a pulse-width modulator IC. In a switching regulator for a power supply the output voltage is usually higher than the desired regulated output voltage. The pulse-width modulator IC turns the switching transistor on and off at the proper time to ensure smooth regulation. The switching usually occurs rapidly, up to several hundred kilohertz.

Before you go on, study test questions E8A11, E8B07 and E8B08. Review this section if you have difficulty.

MULTIPLEXING

Multiplexing means to combine more than one stream of information into one modulated signal. This allows one RF transmitter and transmitted signal to carry more than one information stream. There are two common methods of multiplexing, *frequency division multiplexing* (*FDM*) and *time division multiplexing* (*TDM*). [E8B10]

FDM uses more than one *subcarrier*, each modulated by a separate analog signal. The subcarriers are combined into a single *baseband* signal and used to modulate the RF carrier. [E8B11] FDM was used for multiplexed telephone systems but is rarely used in Amateur Radio. An example of FDM in use today is the SCA (Sub-Carrier Authorization) system used by broadcast FM stations to distribute background music and digital information to subscribers. The subcarriers are ignored by regular FM receivers, but are demodulated by special receivers.

TDM is the transmission of two or more signals over a common channel by interleaving so that the signals occur in different, discrete time slots of a digital transmission. [E8B12] The Global System for Mobile Communications (GSM) mobile phone system uses TDM with each phone sharing the common channel and transmitting bursts of data at a rate of around 200 Hz. In Amateur Radio, TDM is used mostly for telemetry, such as from amateur satellites and remote repeaters.

Before you go on, study test questions E8B10, E8B11 and E8B12. Review this section if you have difficulty.

7.4 Interference and Noise

Interference and noise are the bane of receivers everywhere. Interference, or QRM, is the term given to unwanted signals that have the characteristics of transmitted signals, whether they are actually a transmitter output or not. Noise, or QRN, is either randomly generated by natural processes or is the unintentional output of non-transmitting equipment. Confronted with a mix of both when trying to receive a weak signal, the distinction can seem a bit strained. Different techniques can be applied to reduce, filter or eliminate each type. This section touches on a few of the different types of interference and noise and how to manage them.

TRANSMITTER INTERMODULATION

Intermodulation (IMD) has been and will be discussed at several points in this manual — so far, in the section on mixer circuits (see the Electronic Circuits chapter) and in this chapter's discussion of spectrum analyzers. IMD is a serious problem because it generates interfering signals both internally and externally to equipment. To a receiver, these signals are no different than signals received over the air and so cannot be filtered out. It is important that IMD be reduced or eliminated as much as possible.

Nonlinear circuits or devices can cause intermodulation distortion in just about any electronic circuit. IMD (also called *cross-modulation*) often occurs when signals from several transmitters, each operating on a different frequency, are mixed in a nonlinear manner, either by an active electronics device or a passive conductor that happens to have nonlinear characteristics. [E4D08] The mixing, just like in a mixer circuit, produces mixing products that may cause severe interference in a nearby receiver. Harmonics can also be generated and those frequencies will add to the possible mixing combinations. The *intermod*, as it is called, is radiated and received just like the transmitted signal. [E4D06] It is a clue that an interfering signal is intermod because the modulation from the off-frequency signal is combined with that of the desired signal in the interfering product signal. [E4D07]

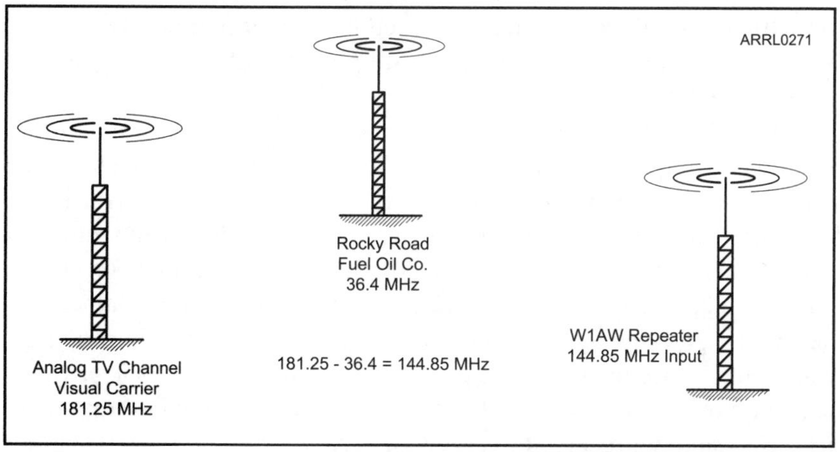

For example, suppose an amateur repeater receives on 144.85 MHz. Nearby, are relatively powerful non-amateur transmitters operating on 181.25 MHz and on 36.4 MHz (see **Figure 7-24**). Neither of these frequencies is

Figure 7-24 — A potential intermod situation. It is possible for the 36.4 MHz and 181.25 MHz signals to mix together in the output stages of either transmitter. While the sum of the two frequencies (36.4 + 181.25 = 217.65 MHz) is not in an amateur band, the difference frequency is (181.25 −36.4 = 144.85 MHz) and such a signal is on the repeater's input frequency. The repeater would treat this signal as a desired signal, attempt to demodulate it, and retransmit the audio on its output channel. An isolator or circulator should be installed at the outputs of the transmitters to prevent the creation of intermodulation products.

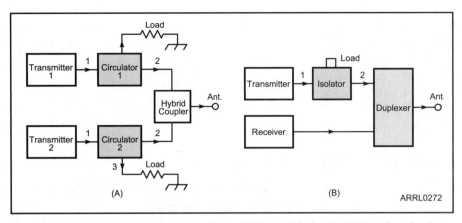

Figure 7-25 — This block diagram shows the use of circulators and an isolator. Circulators may be used to share one antenna with two transmitters, as in A, particularly at multi-transmitter sites. Duplexers are also used along with the circulators. In B, an isolator is placed between the transmitter and duplexer to reduce intermodulation.

harmonically related to 144.85. The difference between the frequencies of the two non-amateur transmitters, however, is 144.85 MHz. If the signals from these transmitters are somehow mixed, an inter-mod mixing product at the difference frequency could be received by the amateur repeater, demodulated as if it were a desired signal, and retransmitted on the repeater's output frequency. A listening station would hear a signal with the modulation of both non-amateur transmitters.

Intermodulation interference can be produced when two transmitted signals mix in the final amplifiers of one or both transmitters and unwanted signals at the sum and difference frequencies of the original signals are generated. [E4D03] In this example, the two signals could be perfectly clean, but if they are mixing in one of the transmitters, the intermod signal may actually be transmitted along with the desired signal from that transmitter.

Two devices that are highly effective in eliminating this type of intermod are *isolators* and *circulators*. Circulators and isolators are ferrite components that function like a one-way valve to radio signals. Very little transmitter energy is lost as RF travels to the antenna, but a considerable loss is imposed on any energy coming back down the feed line to the transmitter. Circulators can also be used to allow two or more transmitters to use a single antenna. Thus, circulators and isolators effectively reduce intermod problems. [E4D04] Another advantage of a circulator is that it provides a matched load to the transmitter output, regardless of what the antenna-system SWR might be, by routing reflected power to a dummy load. **Figure 7-25** illustrates how circulators or isolators may be included in a repeater system.

Low-pass and band-pass filters usually are ineffective in reducing intermod problems, because at VHF and UHF they are seldom sharp enough to suppress the offending signal without also weakening the wanted one.

Intermod, of course, is not limited to repeaters. Anywhere two relatively powerful and close-by transmitter fundamental-frequency outputs or their harmonics can combine to create a sum or a difference signal at the frequency on which any other transmitter or receiver is operating, an intermod problem can develop. Any nonlinear device or conductor in which energy from the two transmitters can combine will generate both harmonics and IMD products from the two signals. For example, corroded metal joints are very nonlinear. If two strong AM broadcast stations are nearby, the signals can mix in the joint and generate intermodulation products across a wide range of frequencies, some in the ham bands. [E4E11] Tracking down such a problem can be quite time-consuming, but fortunately, the harmonics and other products are usually fairly weak and not received over a wide area.

Another IMD topic mentioned earlier has to do with transmitter spectral-output purity. When several audio signals are mixed with the carrier signal to generate the modulated signal, spurious signals will also be produced. These are normally reduced by filtering after the mixer, but their strength will depend on the level of the signals being mixed, among other things, and they will be present in the transmitter output to some extent. You can test your transmitter's output signal purity by performing a transmitter two-tone test as described earlier in this section. This is important because excessive intermodulation distortion of an SSB transmitter output signal results in *splatter* being transmitted over a wide bandwidth. The transmitted signal will be distorted, with spurious (unwanted) signals on adjacent frequencies. This is not a good way to make friends on the air!

Before you go on, study test questions E4D03, E4D04, E4D06, E4D07, E4D08 and E4E11. Review this section if you have difficulty.

ATMOSPHERIC STATIC

When a charge of static electricity builds up on some object and then discharges to ground, it can produce a popping, crackling noise in receivers. The most common source of static noise is caused by the buildup and discharge of static electricity in the atmosphere, mostly during thunderstorms. [E4E06] Lightning is the most visible manifestation of this process but charge is continuously being transferred between the atmosphere and the Earth.

When there is a thunderstorm in your area you will definitely hear this type of noise on the lower-frequency HF bands, such as 160, 80 and 40 meters! The noise from a thunderstorm can be heard in radio receivers for up to several hundred miles from the storm center. (In fact, when propagation is right on the higher HF bands such as 15 meters, static from afternoon thunderstorms in East Africa can be heard in North America!) Thunderstorm static is louder on the lower HF frequencies, and the noise is more frequent in summer. Also, storms are more frequent in the lower latitudes than in the higher ones. Most static electricity noise in a receiver is produced by a thunderstorm somewhere in the area.

You can be bothered by static noise even when there is no thunderstorm. Any form of precipitation picks up some static charge from friction with the air as it falls, so there can be high noise levels during any storm in your area. Noise produced by the falling precipitation is often called *snow static* or *rain static*, depending on which type of precipitation is falling. Both mobile and fixed stations can be bothered by atmospheric static noise.

Before you go on, study test question E4E06. Review this section if you have difficulty.

AC LINE NOISE

Electrical line noise can be particularly troublesome to operators working from a fixed location. The loud buzz or crackling sound of ac line noise can cover all but the strongest signals. Most of this man-made interference is produced by some type of electrical arc. An electrical arc generates varying amounts of RF energy across the radio spectrum. (In the early days of radio, amateurs used spark gap transmitters to generate their radio signals.)

When an electric current jumps a gap between two conductors as in **Figure 7-26**, an arc is produced as the current travels through the air. To produce such an arc, the voltage must be large enough to ionize the air between the conductors. Once an ionized path is established, there is a current through the gap. The electron flow through this gap is highly irregular compared to the rather smooth flow through a conductor. The resistance of the ionized air varies constantly, so the instantaneous current is also changing. This causes radio-frequency energy to be radiated and the noise can be conducted along the power wires that act as an excellent antenna. The longer the width of the gap and the higher the voltage, the greater the interference the arc causes. Because of the high voltages and power available, poor connections or defective insulators in the power distribution system are frequent sources of potentially severe line noise.

Small arcs are created in a variety of electrical appliances, especially those using brush-

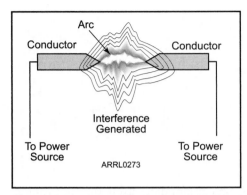

Figure 7-26 — An electrical arc can form through the air gap between conductors. Arcs radiate noise across a wide RF spectrum.

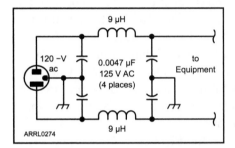

Figure 7-27 — A "brute force" ac line filter may reduce or eliminate line noise when installed in the power leads to an ac brush-type motor.

type motors. Electric shavers, sewing machines and vacuum cleaners are just a few examples. In addition, devices that control voltage or current by opening or closing a circuit can produce momentary arcs that cause interference as pops or clicks. Light dimmers, heater elements, and blinking advertising displays are a few examples of devices like this. Defective or broken appliances and wiring inside the home can also generate line noise. Doorbells or doorbell transformers and other types of devices that are powered continuously often generate low-level arcs (and fire hazards!) when they begin to fail. [E4E13]

One effective way to reduce electrical noise produced by an electric motor is to use a "brute force" ac line filter in series with its power leads. This filter will block the noise from being conducted along the power wiring away from the motor. [E4E05] **Figure 7-27** shows the schematic diagram for such a filter. All components must be ac line rated, and able to carry the current required by the motor or appliance connected to the filter. UL-listed commercial filters are recommended for this application.

Before you go on, study test questions E4E05 and E4E13. Review this section if you have difficulty.

LOCATING NOISE AND INTERFERENCE SOURCES

Perhaps one of the most frustrating facts about line noise is its intermittent nature. The noise will come and go without warning, usually with no apparent pattern or timing (except that it occurs when you are trying to operate). This can make it extremely difficult to locate the source of the interference. There are two ways that noise interference can find its way into a receiver. If the interference source is located in the same building as the receiver, it's likely that the noise will flow along the house wiring. For example, if a furnace's blower motor is at fault, the interference may be carried by the ac wiring from the furnace to wherever the receiver is located. Interference from outside the home is usually picked up through an antenna and feed line.

The first step in tracking down the interference is to determine if it is being generated in your own house. Check this by opening the mains circuit breaker, powering down your entire house. You will need a battery-powered receiver for this test, but a portable AM receiver should work fine. Check to be sure you can hear the noise on your portable receiver. Tune to a clear frequency and listen for the interference. [E4E07] An FM receiver, with or without a directional antenna, will not be helpful in tracking down the noise source because FM receivers are not affected by line noise.

If the noise goes away when you turn off all power to your house, check to see if it comes back when you turn the power on again. It's possible that the offending device only produces the noise after it has been operating for a while, so the interference may not come back immediately when you restore power. If it doesn't return, continue your investigation later.

After you are reasonably confident that the noise is being produced in your house, proceed with your investigation by removing power from one circuit at a time. When you narrow it down to a particular circuit, unplug the appliances or other electrical devices one by one. Be persistent. You may have to continue your investigation over a long time until you discover the culprit device.

If the interference is not being produced in your house, the search will be a bit more difficult. The problem may be an arc in the utility company's power distribution system, a neighbor's appliance or any of a number of other items. Your portable AM receiver can be used to "sniff" along the power lines, looking for stronger interference. Driving around with your car's AM radio tuned between stations is sometimes effective. A directional antenna may help you locate the noise source. When you get close to the noise source you may find that the null (direction of weakest signal in the antenna pattern) is more helpful in pinpointing the direction to the source.

When you think you've located the source, contact the power company and explain the problem to them. Be as specific as possible about where you believe the interference is originating. Note the identification numbers on the utility pole, for example. The power company may send a technician with even more sophisticated equipment to help pinpoint the location and source of the interference. The technician may have a handheld "RF sniffer" that will pick up the radio frequency noise. They may also use an ultrasonic transducer that uses a parabolic reflector antenna and an amplifier to listen for the sound of an arc at frequencies just above the audio spectrum.

In addition to receiving interference from various noise sources, your transmitted signal may also cause interference to devices such as a TV, radio or telephone. Such interference is often caused by *common-mode* signals which means the ac and telephone wiring in your house may pick up your signals and conduct them to the device. [E4E08] Common-mode means the that signal flows in the same direction on all of wires in the power or phone line rather than in opposite directions along the wires as it would on a transmission line. You'll need a common-mode choke to cure this type of interference. Wind several turns of the power cord or phone line around a ferrite toroid core. Ferrites made of type -31,-73 material are a good choice for most HF problems. Type -43 material works best on the upper HF bands and at VHF and UHF.

Interference may also be generated by computer and network equipment and switching power supplies. The characteristics of noise from this type of device are unstable modulated or unmodulated signals at specific frequencies. [E4E14] The signal may change as the device performs different functions. Interference from switching power supplies is usually a series of signals spaced at regular intervals over a wide spectrum.

"Touch controlled" devices such as lamps generate similar signals plus other signals that sound like ac hum on an SSB or CW receiver, drift slowly across the band, and can be several kHz wide because the oscillator that generates them is unstable. [E4E10] All of these devices are "unintentional radiators" covered by Part 15 of the FCC's rules.

For more information about electrical power line noise and other types of interference, see *The ARRL RFI Book*. That book contains detailed information about how noise and interference are generated, how to locate the source, and how to cure it.

Before you go on, study test questions E4E07, E4E08, E4E10 and E4E14. Review this section if you have difficulty.

AUTOMOTIVE NOISE

One of the most significant deterrents to effective signal reception during mobile or portable operation is electrical *impulse noise* from the automotive ignition system. The problem also arises during use of gasoline-powered portable generators. This form of interference can completely mask a weak signal. Other sources of noise include conducted interference from the vehicle battery-charging system, instrument-caused interference, static and corona discharge from the mobile antenna. Atmospheric static and electrical line noise also present some difficult problems for home-station operation. This section offers a short description of the sources and some suggestions about how to solve these noise problems.

Vehicular System Noise

Most electrical noise can be eliminated by taking logical steps to suppress it. The first step is to clean up the noise source itself, then to use the receiver's built-in noise-reducing circuit as a last measure to minimize any noise impulses from passing cars or other man-made sources. In general, you can suppress electrical noise in a mobile transceiver by applying shielding and filtering. The exact method of applying these fixes will depend on the type of ignition system your car uses.

Most vehicles manufactured prior to 1975 were equipped with inductive-discharge ignition systems. A variety of noise-suppression methods were devised for these systems, such as resistor spark plugs, clip-on suppressors, resistive high-voltage spark-plug cable, and even complete shielding. Resistor spark plugs and resistive high-voltage cable provide the most effective noise reduction for inductive-discharge systems at the least cost and effort.

Modern automobiles use sophisticated, high-energy, electronic-ignition systems (sometimes called capacitive-discharge ignition) to reduce exhaust pollution and increase fuel mileage. Solutions to noise problems that were effective for inductive-discharge systems cannot uniformly be applied to the modern electronic systems. Such fixes may be ineffective at best and at worst may impair engine performance. One significant feature of capacitive-discharge systems is extremely rapid voltage rise, which combats misfiring caused by fouled plugs. Rapid voltage rise depends on a low RC time constant being presented to the output transformer. High-voltage cable designed for capacitive-discharge systems exhibits a distributed resistance of about 600 Ω per foot, compared with 10 kΩ per foot for cable used with the old inductive-discharge systems. Increasing the RC time constant by shielding or installing improper spark-plug cable could seriously affect the capacitive-discharge-circuit operation.

The first rule when installing a mobile transceiver in a modern vehicle is to follow the manufacturer's recommended procedures. Some manufacturers provide detailed installation guidelines for installing transmitting equipment. Others may recommend against installing any transmitters, or provide little instruction. If possible, you should check with the manufacturer before buying a vehicle. There is a lot of information about mobile installations from the ARRL's Technical Information Service at **www.arrl.org**.

Ferrite beads and cores are a possible means for RFI reduction in modern vehicles. Both primary and secondary ignition leads are candidates for beads. Install them liberally then test the engine under load to ensure adequate spark-plug performance.

Electrical bonding of all vehicle body parts was an effective way to reduce noise in older vehicles. This may not be desirable on modern vehicles, however.

Charging-System Noise

Noise from the vehicle battery-charging system can interfere with both reception and transmission of radio signals. The charging system of a modern automobile consists of a belt-driven, three-phase alternator and a solid-state voltage regulator. Interference from the charging system can affect receiver performance in two ways. Radiated noise can be picked up directly by the antenna and noise can be conducted directly into the radio through its power leads.

Alternator whine is a common form of conducted interference and can affect both transmitting and receiving. This noise is characterized by a high-pitched buzz on the transmitted or received signal. The tone changes frequency as the engine speed changes.

VHF FM communications are the most affected by alternator whine, since synthesized carrier generators and local oscillators are easily frequency modulated by power-supply voltage fluctuation. Alternator whine is most noticeable when transmitting because the alternator is most heavily loaded in that condition.

Conducted noise and radiated noise can be minimized by connecting the radio power leads directly to the battery, as this point has the best voltage quality and regulation in the system. Connect both the positive and negative leads directly to the battery, with a fuse rated to carry the transmit current installed in each lead. Coaxial capacitors in series with the alternator leads may also help. [E4E04]

Instrument Noise

Some automotive instruments can create noise. Among these gauges and senders are the engine-heat and fuel-level indicators. (Fuel pumps are common noise producers, as well.) Ordinarily the installation of a 0.5-µF coaxial capacitor at the sender element will cure this problem.

Other noise-generating accessories include turn signals, window-opener motors, heating-fan motors and electric wiper motors. The installation of a 0.25-µF capacitor across the motor winding will usually eliminate this type of interference.

Before you go on, study test question E4E04. Review this section if you have difficulty.

NOISE REDUCTION

Once inside the receiver, noise is very difficult to eliminate. Two basic types of noise reduction techniques are widely used. The first, *noise blanking*, works by detecting noise pulses and muting the receiver when they are present. The second, *noise reduction*, uses special DSP techniques to separate noise from the desired signal.

Noise Blankers

Special IF circuits detect the presence of a noise impulse and open or mute the receive signal path just long enough to prevent the impulse from getting through to the audio output stages where it is heard as a "pop" or "tick." This technique, called *gating*, is particularly effective on power line and mobile ignition noise. [E4E01]

A diode or transistor is used as a switch to control the signal path. An important requirement is that the IF signal must be delayed slightly, ahead of the switch, so that the switch is activated precisely when the noise arrives at the switch. The circuitry that detects the impulse and operates the switch has a certain time delay, so the signal in the mainline IF path must be delayed also.

To detect the sharp noise pulses, the noise blanker must detect signals that appear across a wide bandwidth. [E4E03] This usually means that the noise blanker cannot be protected by the narrow receive filters. As a consequence, the noise blanker can be fooled by strong signals into shutting down the receiver as if they were noise pulses. This can cause severe distortion of desired signals, even if no noise is present. It might sound as if the strong signal is very "wide" with lots of spurious signals. Before getting upset at the station with the strong signal, make sure your noise blanker is turned off. [E4E09]

DSP Noise Reduction

DSP noise reduction filters operate by using *adaptive filter* techniques in which software programs search for signals with the desired characteristics of speech or CW or data and remove everything else, such as impulse noise and static. These techniques work particularly well at removing broadband audio "white" noise. DSP noise reduction also works on impulse noise such as ignition and power line noise. [E4E02]

Automatic notch filters are a particularly useful feature of these systems. Modern DSP auto-notch features can track and remove several interfering tones from an audio channel! This is very useful under crowded band conditions or on shared allocations where carriers from shortwave broadcast stations can be quite strong. One drawback of these systems is that they sometimes confuse CW or low-rate digital signals with an interfering tone and attempt to remove them, as well! [E4E12]

> Before you go on, study test questions E4E01, E4E02, E4E03, E4E09 and E4E12. Review this section if you have difficulty.

Table 7-3

Questions Covered in This Chapter

7.1 AC Waveforms and Measurements
E8A01
E8A02
E8A03
E8A04
E8A05
E8A06
E8A07
E8A08
E8A09
E8A10
E8D02
E8D03
E8D04
E8D05
E8D06
E8D07
E8D08
E8D09
E8D10
E8D11
E8D12
E8D13
E8D14
E8D15
E8D16

7.2 Test Equipment
E4A01
E4A02
E4A03
E4A04
E4A05
E4A06
E4A10
E4A11
E4A12
E4B01
E4B02
E4B03
E4B04
E4B05
E4B07
E4B08
E4B10
E4B12
E4B13
E4B14
E8D01

7.3 Modulation Systems
E1B12
E8A11
E8B01
E8B02
E8B03
E8B04
E8B05
E8B06
E8B07
E8B08
E8B09
E8B10
E8B11
E8B12

7.4 Interference and Noise
E4D03
E4D04
E4D06
E4D07
E4D08
E4E01
E4E02
E4E03
E4E04
E4E05
E4E06
E4E07
E4E08
E4E09
E4E10
E4E11
E4E12
E4E13
E4E14

Chapter 8

Radio Modes and Equipment

In this chapter, you'll learn about:
- **Digital codes, protocols and modes**
- **Spread spectrum techniques**
- **Fast-scan television**
- **Slow-scan television**
- **Receiver sensitivity and noise**
- **Dynamic range and intercept point**
- **Phase noise**

Building on your growing knowledge of radio terminology, circuits and signals, this chapter deals with topics that tie them all together. In the discussion of digital protocols and modes, you'll meet the fastest-growing part of Amateur Radio. Image transmissions using fast-scan and slow-scan amateur TV are the next topic. You're also ready to take a close look at receivers, using the information about modulation and mixing and filters from earlier chapters to understand how performance measurements describe how receivers actually perform.

The Amateur Radio Extra class license exam has groups of questions on each topic — digital communications, image transmission and receiver performance. There is also a sprinkling of questions from other parts of the question pool that are easier to discuss in the context of these topics. As in all of the sections, it's a good idea to review all of the listed questions before moving on. If you have trouble with any group of questions, review material from earlier chapters and make use of the online references at **www.arrl.org/extra-class-license-manual**.

8.1 Digital Protocols and Modes

Before plunging into details (and there are many) of the digital transmissions becoming so popular with amateurs, we'll pause to clearly define some digital terminology. It's very easy to misapply a term or to confuse the meaning of one term with another, leading to certain confusion.

SYMBOL RATE, DATA RATE AND BANDWIDTH

Two of the most important and useful characteristics of a digital communications system concern the speed with which data is transferred. "Speed" actually means the rate of information transfer, and so it is measured in units of data per second as described below. It also depends on where the rate is measured in the total communications system — from the generation of data to the delivery of data.

There are two ways of defining digital signal speed, depending on whether you are referring to the *air link* — meaning the actual transmitted signal — or the *data stream* that occurs within the computer equipment that handles the digital data. For example, when listening to an RTTY signal's characteristic two-tone warble on the air by ear, you are listening to the

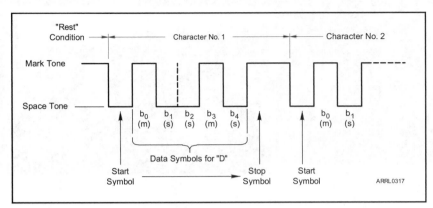

Figure 8-1 — The pattern of symbols transmitted for the letter "D" in Baudot code. Each "bit" of the code is transmitted as one symbol.

RTTY communication system's air link. If you are reading the received characters from an RTTY decoder box or a software program, that is the data stream. The speed of the data stream is often called *data throughput*, meaning the overall speed with which the entire communications system transfers data.

When discussing the speed of the air link, the unit of speed is *baud* (*Bd*) or *bauds*. Baud, like hertz, refers to a quantity of events per second. Just as frequency counts the cycles of an ac waveform, baud counts *signaling events* that refer to changes in the transmitted signal representing information. Each signaling event transfers one *symbol* across the air link to the receiving station. Thus, baud refers to the number of data symbols transmitted per second. [E2D02] A rate of one baud means that one symbol is transmitted every second. (Baud is also referred to as *baud rate* but that is redundant because baud is already a rate! Just say "baud" or "bauds.")

In an RTTY signal's air link, for example, (see **Figure 8-1**) the symbols *mark* and *space* are pulses of transmitted tones with separate frequencies. The tone frequency during the pulse is the signaling event. The duration of each tone pulse is always the same, so the receiving system measures the tone frequency at the time a symbol's tone is expected to be present. Sometimes the tone changes from one symbol to the next (such as from b_0 to b_1 in the figure) and sometimes it doesn't (such as from b_1 to b_2). The signal's baud is the number of those events — the transmitted symbols — per second.

Figure 8-1 also shows how a typical Baudot code waveform might look as a digital data stream displayed on an oscilloscope as the voltage of an electrical signal inside computing equipment processing the letter D. The mark symbol is represented here as a high voltage and the space symbol as a low voltage. Older teleprinters using *current loops* to send data represented mark as current on and space as current off.

Regardless of the mode in use, each symbol in the waveform of a digital data stream can be thought of as being represented as a pulse (or the absence of a pulse) with the varying data creating varying patterns of pulses in time. [E8A15]

Data Rate vs Symbol Rate

Figure 8-1 serves two purposes. It describes how the air link would sound if you listened to an RTTY signal and it also shows a possible data stream waveform. The figure shows how the five bits (b_0 through b_4) are encoded as mark or space symbols, one after the other. Each symbol on the air link corresponds one-to-one with a bit in the data stream. In many simple transmission systems — such as RTTY and 1200-baud packet — the system *data rate*, which is measured in *bits per second* or *bps*, is exactly the same as the *symbol rate* in bauds.

This is not always the case. In some transmission systems — such as 9600-baud packet or D-STAR's digital voice and data systems — each data symbol can encode more than one bit at a time. This is done by simultaneously transmitting more than one signal or by varying more than one attribute of the signal over the air link. Without delving into the details of modern communications technology, imagine that you could send the both the RTTY mark and space tones independently with either tone being on or off during each symbol period. That creates four possible symbols, with the presence or absence of the two tones representing the bit patterns 00, 01, 10 and 11. Because each symbol represents two bits of information, this is a *dibit* (pronounced "dye-bit") system. The data rate in a dibit system is two times the symbol rate.

So you can see that you must use the correct term for data transfer when discussing the air link or data stream portions of the transmission system. For example, a 9600 bps modem used to send high-speed packet data actually sends symbols across the air link at 4800 baud. Each symbol encodes 2 bits, so the data rate is 9600 bps.

Digital Signal Bandwidth

It's also important to know the bandwidth of a digital signal when transmitted over the air link. *Shannon's Information Theorems*, fundamental laws of information transmission, link the symbol rate of a signal to its bandwidth. The higher the signal's baud, the wider the air link signal will be. In Amateur Radio, certain simplifications can be made so that the equation for signal bandwidth generally takes this form:

$$BW = B \times K \hspace{4cm} \text{(Equation 8-1)}$$

where:

    BW is the necessary bandwidth of the signal.

    B is the speed of the transmission in bauds.

    K is a factor relating to the shape of the keying envelope.

Equation 8-1 and its variations are useful in making estimates of signal bandwidth and predicting the effects of changing the various aspects of the signal.

Before you go on, study test questions E2D02 and E8A15. Review this section if you have difficulty.

PROTOCOLS AND CODES

Before studying the characteristics of digital mode communications, some definitions are in order. A *protocol* is the set of rules that controls the encoding, packaging, exchanging and decoding of digital data. For example, packet radio uses the AX.25 protocol. The AX.25 standard specifies how each packet is constructed, how the packets are exchanged, what characters are allowed, and so forth.

The protocol standard doesn't say what kind of modulation to use or what the signal will sound like on the air. The method of modulation, such as SSB packet (on HF) or FM packet (at VHF and above), is determined by conventional operating practice and regulation.

Codes

A *code* is the method by which information is converted to and from digital data. The individual symbols that make up a specific code are its *elements*. The elements may be numbers, bits, tones or even images (think of the code "one if by land and two if by sea" used by Paul Revere).

A digital code doesn't specify how the data is transmitted, the rules for its transmission or the method of modulation. The code doesn't care about those things — it's only a set of rules for changing information from one form to another. Certain types of codes are more suitable for different applications, a matter of preference by the communications system designer. Amateur Radio uses three common types of codes: varicodes (Morse and PSK31's Varicode), Baudot, and ASCII.

Morse and Varicode

While most types of digital codes use a fixed number of identical length bits to make up each character, *variable-length codes* or *varicodes* can vary both the length of the bits and the number of bits in each character or symbol. The length of the codes for a character varies with its frequency of use to save transmission time. Maximizing data rate within a signal's bandwidth also maximizes *bandwidth efficiency* or *spectral efficiency.*

Morse is a varicode because the elements and characters are of different lengths. Morse code is constructed from the dot and its absence, the inter-element space. The dash and longer spaces are made up of multiple dots and inter-element spaces. This creates the elements of Morse: dot, dash, and three lengths of spaces between elements, characters and words. The elements of Morse are unusual in that they are of different lengths, with the dash being three (usually) times longer than the dot. The inter-character (three dots long) and inter-word spaces (seven dots long) are made up of multiple inter-element spaces. [E8C01]

PSK31 uses a type of varicode that is also named Varicode, invented by Peter Martinez, G3PLX. The elements of the PSK31 Varicode are the same length but its characters have different lengths with the most-common text character E being the shortest, similarly to Morse code. [E2E09]

Baudot

The Baudot code is used by RTTY systems and has two elements — mark and space — each the same length. The code is made up of different combinations of five mark and space elements as illustrated in Figure 8-1. (The mark tone can also be transmitted continuously when the system is idle, but no information is being sent during that period.) Each combination of elements always has the same length.

The mark and space elements of Baudot are formed into groups of five, with each element representing one data bit. These five data bits can make up 32 different combinations (2^5), so there are only 32 characters possible with the Baudot code. Some of these combinations are used twice, forming two sets of characters, with the characters LTRS (letters) and FIGS (figures) signaling the change from one set to the other. This is how the 26 letters of the alphabet, 10 numbers and several punctuation characters can be represented using the Baudot code. The Baudot code only allows upper-case letters, however. These sets of codes and characters are known as International Telegraph Alphabet Number 2 (or ITA2).

Figure 8-1 also shows additional elements called *start* and *stop* bits at the beginning and end of the group of five that represents the character. These are called *framing* bits and allow the receiving system to synchronize itself with the transmitted codes. A complete received character, including the framing bits, is called a *frame.*

ASCII

ASCII stands for American National Standard Code for Information Interchange and is the most commonly used code in computer systems. Basic ASCII code uses seven information bits, so 128 characters are possible (2^7). This makes it possible for the ASCII character set to include upper- and lower-case letters, numbers, punctuation and special *control* characters. [E8C03] The ASCII code does not need a shift character, like Baudot, to change between the letters and figures characters.

An eighth bit is often included with the ASCII characters. This is called a *parity bit*, and is used to detect transmission errors. The parity bit is set to 1 or 0 to maintain either an even number of 1s (even parity) in the character, or an odd number of 1s (odd parity). The receiving system checks the number of 1s in the received character. If it does not match the system convention of odd or even, the receiving system knows an error has occurred during transmission and it can reject the character. [E8C12] Some systems use the eighth bit for another data bit, providing 256 possible characters with each representing a full byte of data. [E8C02]

ASCII codes also have framing bits — one start bit at the beginning of the code and one or two stop bits at the end. A full ASCII character is thus 10 or 11 bits long: 7 data bits, one parity bit (or extra data bit), one start bit, and one or two stop bits. It is necessary for both the transmitting and receiving systems to be preset to agree on these conventions or the recovered data will be garbled.

When transferring ASCII data, remember to account for the difference between bits per second and bytes per second. If each ASCII character includes two framing bits, the *byte rate* of data transfer will be about one-tenth the *bit rate*.

Before you go on, study test questions E2E09, E8C01, E8C02, E8C03 and E8C12. Review this section if you have difficulty.

DIGITAL MODES

A digital *mode* consists of both a protocol and a method of modulation. Since digital protocols can be used to convey speech, video or data files, each specific use forms a different *emission* or mode, as referred to in Amateur Radio. The FCC includes the type of data being transmitted — voice, text, data or image — in the definition of each emission, assigning a different emission designator to each. [E8A12] (FCC emission designators were presented in the chapter on Radio Signals and Measurements.)

The advantage of using digital modes to transfer information is that the signal and its information can be copied and retransmitted multiple times without necessarily introducing errors. [E8A13] Thus, digital modes have become very popular in all types of communications. The following sections present the most common amateur digital modes and compare some of their basic characteristics. Digital protocol development is a hotbed of innovation in the amateur community with new modes and variations on existing modes appearing all the time. If you are interested in digital mode technology in Amateur Radio, the TAPR group (**www.tapr.org**) should be your first stop on the Internet.

CW

A CW signal produced by turning an AM transmitter on and off is described by the emission designator A1A. The bandwidth of a CW signal is determined by two factors: the speed of the CW being sent and the shape of the keying envelope as described in Equation 8-1 on page 8-3. To solve this equation, we must find values for B and K.

Morse code speed is usually expressed in words per minute (WPM), so we need to convert WPM to bauds. Since one baud equals one signal symbol per second, and the symbols in Morse are the dot and inter-element space, we must determine the number of dots and inter-element spaces in a Morse code word. The standard word used for this calculation is "PARIS," which contains 50 symbols (including the seven-symbol inter-word space before the next word). 50 symbols per minute means that

$$1 \text{ WPM} = 50 \text{ symbols}/60 \text{ sec} = 0.83 \text{ baud} \qquad \text{(Equation 8-2)}$$

or

$$\text{baud} = \frac{\text{WPM}}{0.83} = 1.2 \times \text{WPM}$$

(Equation 8-3)

For CW signals, substituting 1.2 WPM for B in Equation 8-1:

$$\text{BW} = \text{WPM} \times 1.2 \times \text{K} \qquad \text{(Equation 8-4)}$$

The second variable, K, reflects the abruptness of the keying waveform with typical values of 3 to 5 for amateur signals. As CW rise and fall times get shorter (more abrupt, harder keying), K gets larger. This is because signals with short rise and fall times contain more harmonics than longer, softer envelopes. (Remember that a square wave contains an infinite number of odd harmonics.) The more harmonics that are required to construct the keying envelope, the greater the bandwidth of the resulting CW signal must be. This is why too-sharp keying waveforms cause *key clicks*. The burst of harmonics modulating the signal (CW is, after all, an AM signal) appear as signals on nearby frequencies.

Suppose you are sending Morse code at a speed of 13 WPM with your transmitter adjusted so that K = 4.8. The bandwidth of the transmitted signal is:

BW = WPM × 1.2 × 4.8 = 13 × 4 = 52 Hz [E8C05]

FSK/AFSK

Most amateur data transmissions (and all of those on the HF bands) employ *frequency shift keying* (FSK). [E2E01] In FSK systems, the transmitter uses one frequency to represent one state and another frequency to represent the other binary state. By shifting between these two frequencies (called the mark and space frequencies), the transmitter creates data symbols. The difference between the mark frequency and the space frequency is called the *shift*.

FSK signals can be generated in two ways. *Direct FSK* is created by shifting a transmitter oscillator's frequency with a digital signal. *Audio FSK* or *AFSK* is created by injecting two audio tones, separated by the correct shift, into the microphone input of a single-sideband transmitter. [E2E11] The FCC emission designators for RTTY are F1B if FSK is used and J2B for AFSK. FSK data emissions are F1D, and AFSK data emissions are J2D. A J2B or J2D signal generated by a correctly adjusted SSB transmitter appears identical to an F1B or F1D signal. The necessary bandwidth of that signal is determined by the frequency shift used and the speed at which data is transmitted. The bandwidth is not affected by the type of data being transmitted or its code. The equation relating necessary bandwidth to shift and data rate is:

BW = (K × Shift) + B   (Equation 8-5)

where:

BW is the necessary bandwidth in hertz.

K is a constant that depends on the allowable signal distortion and transmission path. For most practical Amateur Radio communications, K = 1.2.

Shift is the frequency shift in hertz.

B is the symbol rate in baud.

Example 8-1

What is the bandwidth of a 170-Hz shift, 300-baud ASCII signal transmitted as a J2D emission? [E8C06]

BW = (1.2 × 170 Hz) + 300 = 504 Hz

This is a necessary bandwidth of about 0.5 kHz.

Example 8-2

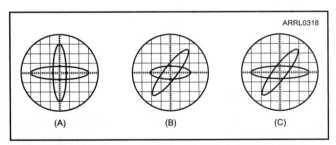

Figure 8-2 — The two tones of an FSK or AFSK signal are represented as a pair of ellipses on a crossed-ellipse display. For the best copy, the signal should be tuned so that the ellipses are of equal size and at right angles as in (A). Displays such as at (B) and (C) indicate a mistuned signal.

What is the bandwidth of a 4800-Hz shift, 9600-baud ASCII signal transmitted as an F1D emission? [E8C07]

BW = (1.2 × 4800 Hz) + 9600 = 15360 Hz = 15.36 kHz

RTTY and other FSK/AFSK modes require careful tuning of the SSB transceiver so that the tones of the signals are as close to exactly right as possible. Errors in tuning will result in poor copy and garbled characters, particularly when using RTTY. To assist in tuning, several "cross-style" indicators have been developed. One of the most popular is the crossed-ellipse indicator seen in **Figure 8-2**. The signal should be tuned in so that the ellipses (representing the amplitude of the two FSK tones) are of equal size. This display shows *selective fading* very clearly, in which one or both of the tones is severely attenuated for a short period as the ellipse shrinks dramatically. [E2E04] It is instructive to observe the effect on the received data stream in the presence of fading.

PSK

Peter Martinez, G3PLX, developed PSK31 for real-time keyboard-to-keyboard QSOs. The name derives from the modulation type — *phase-shift keying* (*PSK*) — and data rate, which is actually 31.25 bauds. PSK31 uses the 128-character ASCII code and the full 256 ANSI character set. (ANSI stands for the American National Standards Institute.) The emission designator for PSK31 is J2B. Variations of PSK31, such as the faster PSK63, are available with more under development.

PSK31 uses Varicode to encode the characters and as with the Baudot and ASCII codes, there is a need to indicate the gaps between characters. Varicode does this by using "00" to represent a gap. Varicode is structured so that two zeros never appear together in any of the combinations of 1s and 0s that make up the characters.

With Varicode, a typing speed of about 50 words per minute requires a 32 bit/s transmission rate. Martinez chose a rate of 31.25 bps because it is easily derived from the 8-kHz sample rate used in many DSP systems.

The bandwidth of PSK31 signals is minimized by the special sinusoidal shaping of the transmitted data symbols. [E8C04] This reduces the harmonic content of each symbol. Using Equation 8-1 and K = 1.2, the bandwidth of a PSK31 signal is approximately BW = 31.25 × 1.2 = 37.5 Hz, narrowest of all HF digital modes used by amateurs, including CW. [E2E10] The mode's narrow bandwidth and phase-shift keying require precise tuning and transmitter adjustment.

HF Packet

Packet radio on HF uses the same AX.25 protocol as on VHF, but is limited to 300 baud by regulation to control the signal's bandwidth. Most HF packet transmissions use FSK at 300 baud compared to the 1200-baud AFSK more common on VHF FM packet systems. [E2E06] The emission designator for HF packet is J2D.

The length of the AX.25 packets (typically 40 bytes) and the hostile environment of HF propagation combine to make HF packet a niche mode. Nevertheless, when conditions are good and fading is mild, HF packet has a significantly higher data rate than RTTY, AMTOR or PSK31. [E2D09]

PACTOR

The original PACTOR mode (referred to as PACTOR-I) is an HF digital mode developed by German amateurs Hans-Peter Helfert, DL6MAA, and Ulrich Strate, DF4KV. It was designed to overcome the shortcomings of AMTOR and HF packet. It performs well under both weak-signal and high-noise conditions. The protocol also supports the transfer of binary files, making it quite useful in today's data-intensive world. [E2E08]

The most popular PACTOR modes in use today are PACTOR-II and PACTOR-III, particularly for using email over HF via the Winlink system. PACTOR systems use *ARQ* (*Automatic Repeat Request*) to correct errors. When a packet is received with errors, the receiving system requests a retransmission until all the data in the packet is received correctly. All three PACTOR modes have the emission designator J2D.

PACTOR systems automatically evaluate the conditions between the receive and transmit stations and *train* to the highest speed supported by the path. PACTOR-III systems running at better than 5 kbps offer the highest data rate of any amateur HF digital mode.

Note that Winlink is not a mode, it is a system of modes and protocols and Internet services that allow email to be exchanged using Amateur Radio. Winlink can only be used to exchange email and attached files. It does not support keyboard-to-keyboard or "chat" communications. [E2E12]

Multi-tone Protocols

In the explanation of data rate and symbol rate at the beginning of this chapter, it was suggested that two tones could be sent at once to increase data rate. In fact, several digital modes do exactly that. Two of the more common as this manual is being prepared in early 2012 are MFSK16 and MT63.

MFSK16 uses frequency-shift keying to modulate 16 different tones. The resulting sound on the air is quite musical. An MFSK16 signal occupies a bandwidth of 316 Hz and has a typical data rate of about 63 bps. [E2E07] MFSK16 includes error correction (as described below) and uses a variation of PSK31's Varicode. MFSK16 can be generated and received using an ordinary computer sound card as the modem.

MT63 has a wider bandwidth (1 kHz, typically) than MFSK16 and uses 64 different tones. The faster keying and more tones make the signal sound like noise to the listener. MT63 makes heavy use of error correction techniques to get the data through. MT63, although it uses more tones, is more tolerant of tuning errors than MFSK16.

WSJT Protocol

Developed by Joe Taylor, K1JT, the WSJT family of digital protocols was developed for demanding VHF/UHF weak signal communication such as Earth-Moon-Earth (EME). The WSJT software has five different operating modes: FSK441 for meteor scatter, JT65 for EME (moonbounce), JT6M for meteor scatter on 6 meters, EME Echo for measuring your own station's signals off the moon, and CW for 15 WPM EME QSOs. [E2D01, E2D03]

These protocols use sophisticated codes to recover signals buried deep in noise, undetectable by ear. In fact, the JT65 mode decodes signals virtually perfectly that have a very low signal-to-noise ratio, even signals that are well below the noise level! [E2D12, E8C13] The protocol uses precise timing between the transmitting and receiving stations and extensive codes to allow the received signal to gradually be separated from noise. For more information about the WSJT protocols, read K1JT's website at **physics.princeton.edu/pulsar/K1JT**.

Transmitting Digital Mode Signals

For all of these digital modes to perform well it is important to pay attention to transmitted signal quality. This is even important with the most basic digital mode, CW, because the on/off characteristics of the signal can cause spurious signals on nearby frequencies (*key clicks*) or make the signal difficult to copy if the waveshape is too "soft", meaning turns on and off too slowly.

For protocols that require very precise phase and amplitude control, it is important to avoid introducing distortion anywhere in the transmit path. For example, when using AFSK modulation such as PSK31, excessive signal levels into the transmitter's audio input can overdrive the modulator or RF amplifiers, creating intermodulation distortion and other spurious signals. Not only do these signals cause interference but they also make it harder for the receiver to recover the data. To make sure your digital signal is clean, follow the manufacturer's instructions on setting the transmitter controls, especially ALC and microphone gain.

Once you have the transmitter and your computer or TNC operating properly, do an "on-air" check yourself or with a nearby station. Measuring your own digital signal can be done in your own shack by transmitting into a dummy load, receiving the signal on a second receiver, and feeding the received audio into the sound card of a computer running a demodulation program for that mode. [E4A09] You'll quickly see whether you have the audio and RF levels set properly on the transmitting system.

Before you go on, study test questions E2D01, E2D03, E2D09,
E2D12, E2E01, E2E04, E2E06, E2E07, E2E08, E2E10, E2E11,
E2E12, E4A09, E8A12, E8A13, E8C04 through E8C07 and E8C13.
Review this section if you have difficulty.

SPREAD SPECTRUM TECHNIQUES

The usual measure of efficiency for a modulation scheme is to examine how tightly it concentrates the signal for a given rate of information — less bandwidth for equivalent data rate is good. While compactness of the signal appeals to the conventional wisdom, spread-spectrum modulation techniques take the exact opposite approach. They spread the signal over a very wide bandwidth by rapidly varying the carrier frequency of the signal in a predefined sequence. [E8C08]

Communications signal bandwidth is increased (called *spreading*) by factors of 10 to 10,000 by using a sequence of bits (the *spreading code*) to vary the signal's frequency. The exact techniques are discussed below.

Spreading has two beneficial effects. The first effect is the dilution of the signal energy on a given frequency, so that while occupying a very large bandwidth, the power density present at any point within the spread signal is very low (see **Figure 8-3**).

Dilution of the signal across many frequencies causes spread-spectrum signals to appear as wideband noise to a conventional receiver. [E8C09] Spectrum spreading may result in the digital signal being below the noise floor of a conventional receiver, and thus invisible to it, while the signal can still be received with a spread-spectrum receiver.

The second beneficial effect of spectrum spreading is that a spread-spectrum receiver can reject strong undesired signals — even those much stronger than the desired spread-spectrum signal's power density. This is because the receiver uses the spreading code to "de-spread" the signal, a process akin to following the signal as it changes frequency. Non-spread signals are then suppressed in the processing because they are not consistent with the spreading code. [E2C08] The effectiveness of this interference-rejection property has made spread-spectrum a popular communications security technique.

Conventional signals such as narrowband FM, SSB and CW are rejected by a spread-spectrum receiver, as are other spread-spectrum signals not bearing the desired coding sequence. The result is a type of private channel, one where only spread-spectrum signals using the correct spreading code are received. A two-party conversation can take place, or if the spreading code is known to a number of people (as it is in amateur applications), net-type operations are possible.

The use of different spreading codes allows several spread-spectrum systems to operate independently while using the same frequencies. This is a form of frequency sharing called *code-division multiple access* or *CDMA* and it is used by most mobile phone systems. If the spreading parameters are chosen judiciously for the propagation conditions that exist on the selected frequencies, conventional users in the same amateur band will experience very little interference from

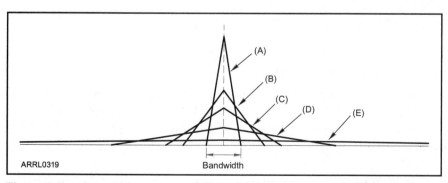

Figure 8-3 — A graphic representation of the distribution of power as the signal bandwidth is increased. The unspread signal (A) concentrates most of its energy near a center frequency. As the bandwidth increases (B), the power near the center frequency falls. At C and D, the energy is distributed wider and wider across the spread signal's wider bandwidth. At E, the signal energy is spread over a very wide bandwidth and there is little power at any one frequency.

spread-spectrum users. This allows more signals to be packed into a band, but each additional signal (conventional or spread spectrum) will add some interference for all users by raising the received noise level.

Types of Spread Spectrum

There are many ways to cause a signal to spread, but all spread-spectrum systems can be viewed as a combination of two modulation processes. First, a conventional form of modulation, either analog or digital, is used to add the information to the carrier. Next, the modulated carrier is again modulated by the spreading code, causing it to spread out over a large bandwidth. Four spreading techniques are commonly used in military and space communications, but amateurs are currently only authorized to use *frequency hopping* and *direct sequence*.

Frequency Hopping

Frequency hopping (FH) is a form of spreading in which the center frequency of a conventional carrier is altered many times per second in accordance with a *pseudo-random* list of channels. [E2C09, E8C10] (Pseudo-random means that the list is not truly random, but is a very long list of numbers that appears random before it repeats.) The same channel list must also be used by the receiving station. The amount of time the signal is present on any single channel is called the *dwell time*. To avoid interference both to and from conventional frequency users, the dwell time must be very short, typically less than 10 milliseconds.

Direct Sequence

In *direct sequence* (DS) spread spectrum, a very fast binary bit stream is used to shift the phase of the modulated carrier. [E8C11] DS spread spectrum is typically used to transmit digital information.

Like the pseudo-random frequency list of FH systems, the sequence of the bits created by a digital circuit is designed to appear random. This binary sequence can be duplicated and synchronized at the transmitter and receiver. Such sequences are called *pseudo-noise* or *PN*.

Each bit of the PN code is called a *chip* and the rate at which the chips shift carrier phase is called the *chip rate*. If the RF carrier's phase is shifted 0 or 180 degrees, it is called *binary phase-shift keying (BPSK)*. Other types of phase-shift keying are also used. For example, *quadrature phase-shift keying (QPSK)* shifts between four different phases (0, 90, 180 and 270 degrees).

Before you go on, study test questions E2C08, E2C09 and E8C08 through E8C11. Review this section if you have difficulty.

ERROR DETECTION AND CORRECTION

Even the best transmitter and receiver systems cannot guarantee 100% accuracy of data transmitted across an air link. There are just too many ways that Mother Nature can disrupt the signal; noise, multipath and fading are just a few causes of errors. To get an idea of what can happen to a data signal, try to copy RTTY signals that are weak or fluttery from ionospheric variations!

In recognition of the realities of radio propagation, data communications engineers have devised a number of strategies. The first challenge is to find out when an error has occurred! This is called *error detection*. Without some clue about what the data should have been, however, there is no way to detect errors. To be able to discern transmission errors, information describing the data is sent along with the original data.

Error detection data can be as simple as the parity bit of ASCII data discussed earlier. The addition of the single parity bit is a great improvement over no error detection at all, at the cost of a small decrease in data rate. Parity cannot be used to detect any type of error, however. Parity can only detect errors that cause the number of even or odd bits to differ from that which the system has been configured to expect. You can probably imagine how changing one bit would cause the *parity check* to fail, but what if two bits are changed so that the net number of even or odd bits is unchanged? Parity checking can only detect single-bit errors.

Another popular technique of error detection, used by packet radio's AX.25 protocol, TCP/IP networking systems, and Ethernet (among many others) is *checksums*. Originally, this was just the sum of all the data values in a packet, sent as a single byte at the end of the packet. It was compared to the sum computed by the receiver. If the sum matched, the entire packet was judged to be good. The simple checksum has evolved into a more sophisticated technique called the *cyclical redundancy check* or (*CRC*). The two-byte CRC is a lot like a checksum and is evaluated by the receiving system in the same way as a checksum. Using a CRC detects most errors.

Once the system has detected an error, what it decides to do about it is another matter. This moves the process from error detection to *error correction*. The simplest form of error correction is ARQ (Automatic Repeat Request), introduced in the section on PACTOR earlier in this chapter. If the receiving system detects an error, it requests a retransmission of the corrupted packet or message by sending a NAK (Not Acknowledge) message to transmitting station. [E2E05] The information is retransmitted until the receiver responds with an ACK (Acknowledge) message. If the errors persist long enough, the system gives up and drops the connection.

Another popular error correction technique is to send some extra data about the information in the packet or message so that the receiving station can actually correct some types of errors. [E2E03] This technique is called *Forward Error Correction* or (*FEC*). [E2E02] The term "forward" stems from sending extra error correction data "ahead" with the original information. The combination of the FEC data and the algorithm by which errors are detected and corrected is called an *FEC code*.

There are many types of FEC codes. Reed-Solomon, Hamming, BCH and Golay codes are all used in consumer electronics. FEC data is sometimes spread out over several data packets to account for fading. FEC is used with digitized voice to help preserve the quality of the received speech. This is why digital voice systems (such as mobile phones) tend to have good quality up to a certain error threshold and then become completely garbled — their FEC code fails at that point.

One type of FEC builds the error correction into the structure of the transmitted data itself. Instead of allowing any possible sequence of symbols to be sent over the air link, the system called *Viterbi encoding* restricts the sequence to a smaller number. That way, the receiving system has fewer possibilities to choose from when deciding what symbol was received. The receiver then reports the most likely sequence of symbols to have created the signal received — the *Viterbi path*.

> Before you go on, study test questions E2E02, E2E03, and E2E05. Review this section if you have difficulty.

8.2 Amateur Television

Many new hams are surprised to find that amateurs can (and do!) communicate using television signals "just like broadcast stations." They could be equally surprised to find that they may have some of the necessary equipment already in their home. Nor does it take a broadcast-sized installation to become active.

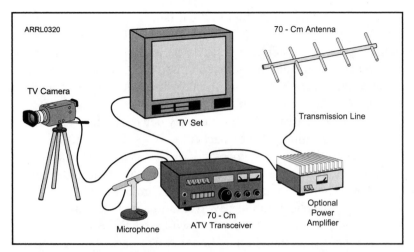

Figure 8.4 — A basic ATV station consists of a camera to produce video, an AM or FM amateur television transceiver (or transmitter and receive converter), an optional amplifier, a directional antenna with gain, and a commercial TV receiver (the receive converter is built into the transceiver).

Slow-scan television also has an enduring presence on the amateur bands, taking advantage of computer sound card and signal processing technology to do away with special imaging tubes and displays. Again, you may already have quite a bit of the necessary equipment already within arm's reach. Generating a television signal does require a bit of know-how, but nothing beyond the understanding of an Extra class ham, as you will see, so to speak!

FAST-SCAN TELEVISION

Fast-scan TV (*FSTV*) can be used by any amateur holding a Technician or higher-class license. FSTV or *Amateur TV* (*ATV*) closely resembles broadcast-quality television, because it normally uses the same technical standards. It is called "fast-scan" because the images are transmitted quickly enough to support full-motion video. Amateurs typically use commercial transmission standards for TV signals, but are not limited to commercial standards. Nevertheless, due to the wide availability of equipment that conforms to those standards, most amateur television is compatible with commercial broadcast equipment for analog signals.

Aside from the sheer enjoyment of sending and receiving "real" television signals, amateurs have put ATV to work in a number of interesting and useful ways. Public service events can be monitored and actual video provided to organizers and sponsors. (The rules prohibit one-way transmissions, so it's not OK to set up an ATV station and broadcast live coverage of a sporting event or parade.) Amateurs have put ATV to work in support of emergency and disaster relief efforts — damage assessment and status reporting benefit greatly from live video. With video equipment getting smaller every day, ATV has also found its way into remote-control planes, balloons and other mobile platforms. Video footage of such ATV adventures found on popular video websites gives you an idea of the quality of such systems.

Fast-Scan System Components

A basic ATV station is constructed as shown in **Figure 8-4**. This is where you may recognize some common home entertainment gear. Any camera that produces a standard video signal can be used, color or black-and-white. Older cameras use a *vidicon* tube while newer models use *charge-coupled device* (*CCD*) imaging ICs. (CCDs are described in Chapter 5.)

Image monitors for "off-air" use can be any TV or computer display that accepts a video signal, such as from a camera or video player. Regular broadcast TV sets can be used with a *receive converter* to pull in UHF ATV signals that use the broadcast transmission standards. The converter shifts the ATV signal to an unused UHF TV channel for display.

An ATV transmitter is one piece of equipment not found in most ham shacks. Nevertheless, low-power ATV transmitters with a few watts of output power are available for purchase. They take a standard baseband video signal as input and produce the UHF ATV signal ready for amplification or connection directly to the antenna. A linear amplifier can be used to boost the output signal as long as it can tune to the portions of the UHF bands where ATV signals are found and has sufficient bandwidth for the ATV signal as described below.

Antennas for ATV need to have fairly high gain (to boost the strength of the weak, wide bandwidth signal) and a consistent radiation pattern over the wide bandwidth consumed by an ATV signal. Yagis and corner reflectors are popular for ATV stations. At microwave frequencies, both Yagi and dish antennas are used.

Image Signal Definitions

Understanding television signals begins with understanding the image itself — how it is converted to an electronic signal, and how the electronic signal represents that image. The process of acquiring an image is called *scanning*. To scan an image, some type of light sensor separates the image into horizontal lines, called *scan lines*. The set of video and control signals that define the image is called the *raster*.

The method by which the raster's video and control signals are combined to transmit and reassemble the image is defined by a video standard. US amateurs use the *NTSC* (National Television Standard Committee) standard. [E2B16] NTSC was the North American broadcast television standard for many years, before the switch to digital TV. The PAL (Phase Alternating Line) and SECAM (Séquentiel Couleur Avec Mémoire — sequential color with memory) standards are used in other countries. These standards are similar but are not compatible with each other. Equipment that complies with the PAL standard, for example, cannot display an NTSC video image.

Table 8-1 lists the primary elements of the NTSC standard used for ATV. In the NTSC standard, a total of 525 horizontal scan lines comprise a *frame* to form one complete image. [E2B02] Thirty frames are generated each second. [E2B01] Each frame consists of two *fields*, each field containing 262½ lines, so 60 fields are generated each second. **Figure 8-5** illustrates the movement of an electron beam (in the scanning or receiving device) across and down, producing the television image. (CCD cameras and solid-state displays work differently, but use the same video signals in this description.)

Scan lines from one field fall between lines from the next field. This is called *interlacing* and is done to reduce flicker, improve the smoothness of motion from frame to frame, and reduce bandwidth while maintaining adequate image quality. If all 525 scan lines are numbered from top to bottom, then one field contains the even-numbered lines and the alternate field contains the odd-numbered lines. [E2B03] This process of slicing up an image is reversed to display the image line-by-line on the surface of a picture tube or CRT. (Computer displays are not interlaced and create the image in a single field, one field per frame.)

Field-one scanning begins in the upper left corner of the image area. The electron beam sweeps across the image to the right side. At the end of the line, the beam is turned off, or *blanked*, and returned to the left side where the process repeats. In the meantime, the beam also has been moved slightly downward to scan the next part of the image. [E2B04]

Table 8-1
NTSC Standards for ATV

Line rate	15,750 Hz
Field rate	60 Hz
Frame rate	30 Hz
Horizontal Lines	262 ½ / field
	525 / frame
Sound subcarrier	4.5 MHz
Channel bandwidth (VSB-C3F)	6 MHz

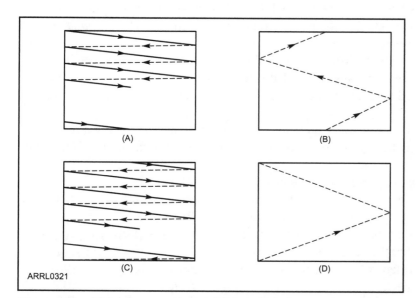

ARRL0321

Figure 8-5 — This diagram shows the interlaced scanning used in TV. In field one, 262½ lines are scanned (A). At the end of field one, the electron scanning beam is returned to the top of the picture area (B). Scanning lines in field two (C) fall between the lines of field one. At the end of field two, the scanning beam is again returned to the top, where scanning continues with field one (D).

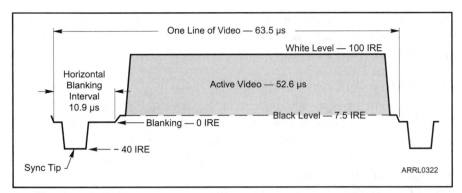

Figure 8-6 — RS-170 monochrome waveform for one video line. A full NTSC video frame consists of 525 video lines organized as two interlaced fields. Each field takes 1/60th of a second to transmit, so that 30 frames are transmitted every second. IRE units are used to measure the relative amplitude of the different parts of the video signal

At the end of 262½ horizontal scan lines, the beam is blanked again and rapidly returned to the top of the image area. At that point, scanning of field two begins. Notice that this time the beam starts scanning from the center of the image line. For that reason, the scanning lines of field two will fall halfway between the lines of field one and not directly on top of them. At the end of 262½ lines the beam is rapidly returned to the top of the image and scanning continues — this time with field one of the next frame.

For every scan line of the two fields to interlace properly, the horizontal and vertical oscillators that control the scanning motion must be precisely and permanently synchronized. Otherwise, the proper interlacing of lines will be lost and vertical resolution or detail will be degraded. The video signal's sync pulses serve to keep the oscillator frequencies locked together.

Video Signal Definitions

The electronic signal that carries all of the image and display coordination information is called *baseband video* or *composite video* and is described in the standard ANSI RS-170. **Figure 8-6** shows the basic structure of one frame of an RS-170 video signal. Sync signals have a negative voltage and the video portion of the signal a positive voltage. The standard voltage level between white video and the *sync tip* is 1-V peak to peak.

Vertical sync pulses tell the display electronics when a new field is about to begin. There are two vertical sync pulses per frame, one for each field. Note that information in the vertical sync pulse tells the display whether the field is field-one or field-two. *Horizontal sync* pulses occur between each horizontal scan line.

Sync pulses are required even for display devices that do not use a swept electron beam, such as LCD or plasma displays, because the video signal is separated into lines sent one after another in the interlaced fields. Sync pulses are markers to keep the image's information aligned properly.

During the sync pulses, the swept electron beam is turned off to avoid creating a line across the image as it returns from right to left or from bottom to top of the image. These are called *blanking intervals*.

Between horizontal blanking pulses, the voltage of the signal represents the brightness or *luminance* of the image. In monochrome black-and-white video, higher video voltages create whiter areas on the image. Lower voltages create blacker areas. Note that the sync pulses have lower voltages than the blackest video voltage. Sync pulses are thus termed "blacker than black."

Television engineers measure video levels in IRE or IEEE Units as shown in **Table 8-2** and **Figure 8-7**. (Percent of Peak Envelope Voltage, or %PEV, is sometimes used by amateurs but it is an obsolete terminology.)

Table 8-2
Standard Video Levels

	IRE Units
Zero carrier	120
White	100
Black	7.5
Blanking	0
Sync tip	−40

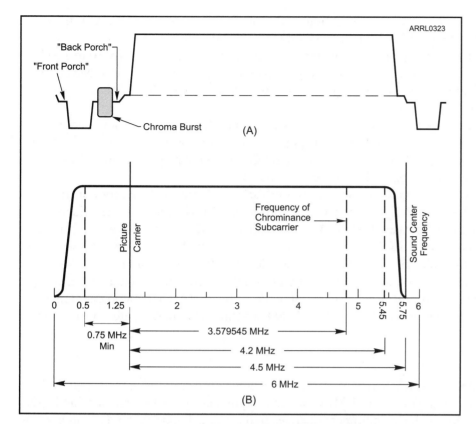

Figure 8-7 — Composite video (A) adds the color burst at the start of each line to synchronize the chroma circuits. (B) shows the spectrum of a standard broadcast TV signal. ATV may or may not use the FM subcarrier for sound.

Composite and RGB Video

In a composite color video signal as shown in Figure 8-7A, all of the information about the image is contained in a single waveform. The color (called *chroma*) information is combined with the luminance information through the use of a separate *chrominance subcarrier* signal. [E2B07]

In each scan line, between the end of the horizontal sync pulse and the start of the video (just before the portion of the waveform called "the back porch") is found a short interval of a 3.5789 MHz signal. This is the chroma subcarrier frequency. The short *chroma burst* helps keep the display's chrominance subcarrier oscillator synchronized with that of the electronics generating the image. If the two are not precisely locked, the hues of the reproduced colors will be wrong.

Color video can also be sent as three separate monochrome RS-170 signals, one for red video, one for blue video and one for green video. These signals are applied separately to the red, green and blue electron guns of the display CRT or video processing electronics. This set of three signals is called *RGB video*. To avoid the three sets of sync pulses causing confusion, sync is only used from one of the three video signals — typically the green channel. RGB video creates a cleaner, sharper image than composite video because there is no need to combine and then re-separate the luminance and chroma information.

Modulated Television Signals

In an RS-170 video signal from a TV camera, positive voltage creates a whiter image and the sync pulses are at ground level. For transmission as an AM signal over the air, however, the polarity or sense of the video is inverted — the sync pulses are at the maximum signal envelope level. The video signal is inverted again in the receiver to restore it to the RS-170 standard's conventions.

Modulating the AM carrier with the inverted video signal means the sync tip will be at peak power output from the transmitter. That means TV receivers are better able to recover a stable image when the received signal is noisy or weak. It is far better to have a bit more noise in the image than for it to be unstable.

RF ATV Signal Characteristics

A fast-scan color ATV video signal has a bandwidth of about 4 MHz. (Satisfactory black-and-white signals require somewhat less bandwidth.) The wide bandwidth is necessary to send the information required for full-motion, real-time images. Consequently, ATV is permitted only in the 420 to 450-MHz band and at higher frequencies.

Because the power density of the ATV signal is comparatively low, typically 10 to 100 W spread across 6 MHz, reliable amateur coverage is only on the order of 20 miles. Nevertheless, you might find yourself exchanging images with stations up to 200 miles away when tropospheric conditions are good.

Most FSTV activity occurs in the 420 to 450 MHz band. The exact frequency used depends on local custom and band plans. Some population centers have ATV repeaters. ATVers should avoid interfering with the weak-signal operation (moonbounce, for example) near 432 MHz and with repeater operation above 442 MHz.

Commercial TV stations and most amateurs use *vestigial sideband* (VSB) for transmission. VSB is like SSB plus full carrier, except a portion (vestige) of the unwanted sideband is retained. [E2B06] In the case of VSB TV, approximately 1 MHz of the lower sideband and all of the upper sideband plus full carrier comprise the transmitted image signal. **Figure 8-7B** shows the spectrum of a color TV signal. VSB uses less bandwidth than a full DSB AM signal but can still be demodulated satisfactorily by simple video detector circuits. [E2B05] The FCC uses emission designators A3F or C3F to describe FSTV. Under an old system of emission designators this was known as A5. Many long-time ATVers still refer to Amateur TV as A5.

FM Television

It is also possible to use frequency modulation systems to transmit the video signal. FM systems have a bandwidth of 17 to 21 MHz, which is significantly greater than the bandwidth for AM systems. Most FM TV equipment operates in the 1.2, 2.4 and 10.25 GHz bands. [E2B18] FM TV receiving systems are more complex than AM TV systems. AM systems give better weak-signal performance, while FM systems give better image quality for stronger signals. The FM systems do not provide immunity from signal fading, however.

The Audio Channel

There are at least three ways to transmit voice information with a TV signal. The most popular method is by talking on another band, often 2 meter FM. This has the advantage of letting other local hams listen in on what you are doing — a good way to pick up some ATV converts! Rather than tie up a repeater for this, it is best to use a simplex frequency.

Commercial TV includes an FM voice subcarrier 4.5 MHz above the TV image carrier (see Figure 8-7B). If your ATV transmitter provides FM audio via a subcarrier 4.5 MHz above the video carrier, the audio can be received easily in the usual way on an analog TV set. The AM carrier of the video signal can also be FM modulated with the audio channel and a modified FM receiver used to demodulate the audio information. [E2B08]

> Before you go on, study test questions E2B01 through E2B08, E2B16 and E2B18. Review this section if you have difficulty.

SLOW-SCAN TELEVISION

Since fast-scan TV systems take several megahertz of bandwidth that must mean there are no images sent on the HF bands, right? Certainly images can't be sent via HF as fast-scan video, but there are two image modes that do work on HF: *facsimile* and *slow-scan television*. You are probably familiar with facsimile, or *fax*, and, yes, amateurs can use the same fax protocols over the airwaves. Amateur facsimile is rarely heard these days because slow-scan television (*SSTV*) is used instead, even though it offers somewhat lower resolution than fax.

Only still images are suitable for transmission by SSTV. Computer sound cards and soft-

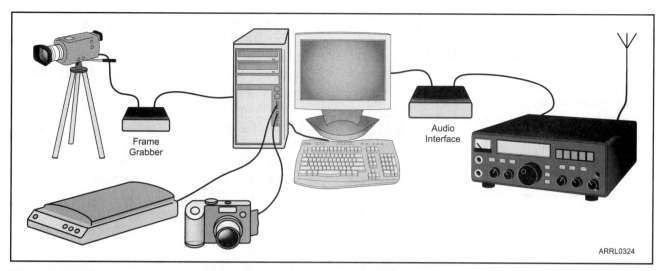

Figure 8-8 — The most common SSTV station uses a video or digital camera, or an image scanner, to generate still images. Software then processes the image and converts it to audio tones that are input to an SSB transmitter.

ware can be used to convert digital images to and from the audio signals used to transmit the SSTV images. The audio tones that encode the SSTV image are then transmitted using SSB.

A typical SSTV system shown in **Figure 8-8** uses a computer and software to generate the required audio tones to represent an image for transmission and to decode the received audio tones and display the image on the computer monitor. Computer graphics open up many possibilities for creating your own SSTV images. An image frame from a video camera can be captured and transmitted via slow-scan by using a *frame-grabber* or images from a digital camera or scanner can be used. You can store received images on disk for later retransmission, as well. **Figure 8-9** shows a screen shot of a program used to handle SSTV images.

An important transceiver consideration is that SSTV is a 100%-duty-cycle transmission mode. This means your transmitter will be producing full power for the entire image-transmission time. Most transmitters and amplifiers will have to be run at reduced power output to avoid overheating.

Figure 8-9 — This screen shot shows an SSTV program that uses the computer sound card to send and receive images.

Analog SSTV Signal Basics

Like fast-scan television, SSTV divides the image into scan lines and frames. (Basic television image terminology is defined in the preceding section on fast-scan TV.) Between each scan line is a horizontal sync pulse to establish the beginning of each line. The beginning of a frame is established by the vertical sync pulse.

Instead of using voltage levels as video and sync signals, SSTV uses frequencies with specific frequencies encoding specific functions. This allows SSTV signals to be transmitted using SSB modulation. [E2B12] For example, sync pulses signifying new lines are sent as

Table 8-3
Black-and-White SSTV Standards

Frame time	8 seconds
Lines per frame	120
Time to send one line	67 ms
Duration of horizontal sync pulse	5 ms
Duration of vertical sync pulse	30 ms
Horizontal and vertical sync frequency	1200 Hz
Black frequency	1500 Hz
White frequency	2300 Hz

bursts of 1200-Hz tones. [E2B15] The image's luminance or brightness is transmitted as a tone of varying frequency. [E2B14] For monochrome black-and-white SSTV signals, a 1500-Hz signal produces black and a 2300-Hz signal produces white. Frequencies between these limits represent shades of gray. The 1200 Hz of the sync pulses are thus "blacker than black" like their fast-scan counterparts and do not show up on the display.

Table 8-3 summarizes the standard parameters for an analog SSTV signal. The horizontal sync pulse is included in the time required to send one line, but the vertical sync pulse adds 30 ms of "overhead." So, it takes a bit more than 8 seconds to actually transmit a black-and-white image frame.

A basic monochrome black-and-white SSTV image takes 8 seconds for one frame and has only 120 scan lines. This works out to 15 scan lines per second. The bandwidth of a monochrome SSTV signal is about 2 kHz. Unlike fast-scan images, the scan lines of SSTV are not interlaced and so each image frame is composed of a single field.

Most SSTV operators today send color images. There are several modes of formatting color images, transmitting 120, 128, 240 or 256 scan lines. These images take from 12 seconds to more than 4 minutes to transmit a single frame. The 128-line and 256-line formats are the most popular. [E2B13] **Table 8-4** summarizes the characteristics of some of the popular color SSTV formats. The bandwidth of the color signals is higher than for monochrome, about 3 kHz. The extra bandwidth is needed for the extra information included to encode the image's colors. [E2B17]

For receiving equipment and software to discern the mode of the SSTV image, a code is transmitted with each image frame. The code is sent during the vertical sync pulse and is called *Vertical Interval Signaling* or (*VIS*). Receiving software reads the code and adjusts its decoding settings to properly capture and display the image. Similarly, the operator can select the mode for transmission and the appropriate code will be incorporated into the image. [E2B11]

The various SSTV modes allow the operator to select different resolutions and control image transmission time. Low to moderate resolution images can be transmitted when crowded band conditions favor short transmission times. Higher resolution images can be transmitted when longer transmissions are acceptable.

Table 8-4
Color SSTV Standards

Format	Name	Time (sec)	Lines
Wraase SC-1	24	24	128
	48	48	256
	96	96	256
Martin	M1	114	256
	M2	58	256
	M3	57	128
	M4	29	128
Scottie	S1	110	256
	S2	71	256
	S3	55	128
	S4	36	128

Digital SSTV

Why not simply send the digitized image as a computer file? A good question! Until recently, file transfer was difficult over HF because digital data transfer protocols were unsuitable. That all changed when shortwave broadcasters developed and released the *Digital Radio Mondiale* (*DRM*) protocol, meaning "Digital World Radio." DRM signals are increasingly used on the shortwave bands to broadcast digitally encoded programs with a much higher received audio quality. Amateurs adapted DRM's file transfer capabilities to send digitized images instead and now digital SSTV is a rapidly growing mode. Since DRM signals can be generated and decoded with software on a PC, no additional special equipment beyond a receiver is required to use DRM for SSTV communications. [E2B09] While broadcast DRM signals have bandwidths of 4 kHz or more, amateur DRM signals on HF are restricted to a normal SSB signal's bandwidth of 3 kHz. [E2B10]

SSTV Operating

SSTV is permitted in the phone segments of all bands (which excludes 30 meters, since phone operation is not allowed there). SSTV signal bandwidth must be no greater than that of a phone signal using the same modulation. [E2B19] SSTV is usually transmitted on HF because fast-scan is available on UHF and higher bands. Standard HF calling frequencies for SSTV are 3.845, 7.171, 14.230, 21.340 and 28.680 MHz. The most popular bands for SSTV are 20 and 75 meters.

An analog SSTV signal must be tuned in properly so that the image will be displayed with the proper brightness and colors and so that the 1200-Hz sync pulses are detected properly. If the signal is mistuned, the image will appear wildly skewed. With some experience, you may find that you are able to zero in on an SSTV signal by listening to the sync pulses and watching for proper synchronization on the display. Most SSTV software provides a visual tuning aid for this purpose.

In general, SSTV operating procedures are quite similar to those used on SSB. The FCC requires you to identify your station every 10 minutes during a transmission or series of transmissions and at the end of the communication. You may use SSTV images to identify your station in meeting this requirement. It is recommended to identify by voice for the benefit of stations that may be listening but unable to copy your SSTV images.

> Before you go on, study test questions E2B09 through E2B15, E2B17, and E2B19. Review this section if you have difficulty.

8.3 Receiver Performance

This is the last section on radio equipment and, given the amount of time you will spend receiving during your ham career, it may be the most important! Your enjoyment of radio depends on quality reception of signals, as free of noise and distortion as possible, able to ignore unwanted signals, and sensitive enough to hear the weakest. That is why it's important to be able to measure and evaluate receiver quality.

When you read product reviews of new radios, you'll notice that a great deal of energy is spent evaluating the receiver. Its performance will be measured using the parameters you're about to study. By learning more about each one, you'll be able to compare receivers and make informed decisions about which radio design is better. We'll begin with the receiver's ability to hear and then move on to how well it hears a particular signal.

SENSITIVITY AND NOISE

One of the basic receiver specifications is *sensitivity* or *minimum discernible signal* (*MDS*). The MDS of a receiver is the strength of the smallest discernable input signal. [E4C07] The sensitivity or MDS of a receiver depends on two factors: noise figure and the bandwidth of the system.

The MDS is also called the receiver's *noise floor*, because it represents the signal strength that produces the same audio output as the receiver noise. You can measure receiver MDS by measuring its audio output power when the antenna input is connected to a dummy load of the proper impedance. After making that measurement, feed a signal from a calibrated signal generator into the receiver antenna input. When the audio output power is twice what it was without an input signal (a 3-dB increase), then the input signal is just strong enough to produce an audio output equal to the receiver's internal noise. The strength of that signal is equal to the MDS for that receiver. The lower the MDS, the more sensitive the receiver.

MDS and other receiver performance specifications are often given in dBm. This abbreviation means "decibels with respect to one milliwatt." 0 dBm is the same as 1 mW, +10 dBm is 10 mW, −20 dBm is 0.01 mW (or 10 μW), and so forth. Using dBm allows us to discuss an extremely wide range of signal power levels.

MDS may also be given in μV, such as 0.5 μV. This can be converted to power if the receiver input impedance is known — usually 50 Ω. Power, $P = V^2/50$ in this case. The equivalent in dBm = 10 log (P/0.001). For example, an MDS of 0.5 μV equals an MDS of −113 dBm. This is a practical MDS on the HF bands.

It is useful to know that the theoretical noise power at the input of an ideal receiver, with an input-filter bandwidth of 1 Hz, is −174 dBm at room temperature. (*The ARRL Handbook* contains more detailed information about how to calculate this number and why it depends on temperature.) This is considered to be the theoretical best (lowest) noise floor any receiver can have. In other words, for this ideal receiver, the strength of any received signal would have to be at least −174 dBm to be detected. [E4C05] Because the noise power increases linearly with bandwidth, the theoretical MDS with a 1 Hz bandwidth filter is specified as −174 dBm/Hz.

Of course −174 dBm is an incredibly small power level — only four billionths of a billionth of a milliwatt! With an antenna attached, an HF receiver's noise floor is actually determined by atmospheric noise which is far higher than the theoretical noise floor. Atmospheric noise, therefore, is the limiting factor for sensitivity of receivers on the HF bands. [E4C15]

A receiver bandwidth of 1 Hz is impractical, but is used as a reference for when wider filters are used. An actual receiver might have a 500-Hz bandwidth for CW operation, or even wider for SSB or FM voice. As the filter gets wider, more noise will be received by a factor equal to the ratio of the filter used to 1 Hz. For example, a 500-Hz bandwidth increases the received noise power by a factor of 500 over the 1 Hz bandwidth, a ratio of 27 dB = 10 log 500. That increase in filter width also increases the noise floor of this theoretical receiver to −174 dBm + 27 dB = − 147 dBm.

For whatever filter width the receiver uses, you can calculate the theoretical MDS for that receiver by calculating the log of the bandwidth and multiplying that value by ten. Add the result to the 1-Hz bandwidth value for MDS of −174 dBm.

Example 8-3

What is the MDS for a receiver with a −174 dBm/Hz noise floor if a 400 Hz filter bandwidth is used? [E4C06]

Step 1 — Calculate the bandwidth ratio in dB = 10 log (400) = 26 dB

Step 2 — To get MDS, add that figure to −174 dBm = −174 + 26 = −148 dBm

Noise Floor and Signal-to-Noise Ratio

Noise figure is a "figure of merit" for the receiver — it is the ratio in dB of the noise generated by the receiver itself to the theoretical MDS. [E4C04] Noise figure evaluates how much noise the receiver's internal circuits contribute. Noise figure is measured in dB.

The higher a receiver's noise figure, the more noise that is generated in the receiver itself. This also means the receiver will have a higher noise floor. Lower noise figures are more desirable. For example, a low-noise UHF preamplifier might have a noise figure of 2 dB or less. [E6E05]

The receiver's internal noise degrades the noise floor, or raises the power that actual signals must have to be heard. You can calculate the actual noise floor of a receiver by adding the noise figure (expressed in dB) to the theoretical best MDS value.

Actual Noise Floor = Theoretical MDS + noise figure   (Equation 8-6)

For example, suppose our 500-Hz-bandwidth receiver has a noise figure of 8 dB. We can use Equation 8-6 to calculate the actual noise floor of this receiver.

Actual Noise Floor = –147 dBm + 8 dB = –139 dBm

The noise figure of a receiver is related to the *signal-to-noise ratio* (*SNR*) of the input and output signals. SNR is defined as signal power divided by noise power and is expressed in dB. Lowering a receiver's noise figure lowers its actual noise floor and improves weak signal sensitivity. By lowering noise without changing the input signal level, the SNR ratio will be increased, meaning that the signal is easier to copy. [E4C08]

Another type of signal-to-noise ratio is *signal-to-noise-and-distortion* (*SINAD*). This figure includes the ability of the receiver to accurately detect or demodulate the input signal. Any distortion of the signal is added to the noise in the sense that it degrades the ratio of the desired signal's energy to that of the undesired energy.

Before you go on, study test questions E4C04 through E4C08, E4C15, and E6E05. Review this section if you have difficulty.

SELECTIVITY

The perfect receiver is one that can tune any frequency and reject every signal except the one you want to receive. That is one definition of *selectivity* — the ability to select a specific signal. That broad definition of selectivity has more specific meanings in the different parts of the receiver. For example, selectivity in a receiver's front end may apply to rejection of strong out-of-band signals, such as shortwave broadcasts. Thus, selectivity is a relative term that depends on where it is applied.

The degree of selectivity is determined by the bandwidth of the receiver's entire filter chain, from the front end to the audio output. (Filters are discussed in the Electronic Circuits chapter.) In superheterodyne receivers, there are several filters in the signal path. As the signal passes through the receiver, it encounters progressively narrower filters that remove more and more of the unwanted signals.

Band-pass *front-end filters* that pass an entire amateur band (or a significant portion of it) are used at the receiver's antenna input. They provide *front-end selectivity* and their purpose is to keep strong out-of-band signals of nearby transmitters or broadcast stations from overloading the sensitive input circuits. A *preselector* is a tunable input filter adjusted to pass signals at the desired frequency and increase rejection of out-of-band unwanted signals. [E4D09] Both improve receiver performance by rejecting signals that can cause image responses in the receiver. [E4C02]

Farther along the signal's path, receivers often use relatively wide filters in each IF amplifier circuit. These may be LC filters, quartz crystal filters, or ceramic resonator filters with characteristics similar to crystal filters. These are used to reject unwanted mixing products and to prevent spurious signals from slipping into the receiver's signal path. These filters pass many signals on or near the desired frequency.

Increasing a superheterodyne receiver's IF improves selectivity. As the IF is increased, the frequency at which image responses occur becomes farther from the desired signal and easier to filter out. This filtering can be performed in the receiver's front-end circuitry and at each conversion stage in the receiver. [E4C09] For example, if a low IF such as 455 kHz is used to receive a signal on 14.300 MHz the BFO could be tuned to 14.3 + 0.455 = 14.755 MHz. (The BFO could also be tuned to 14.3 – 0.455 = 13.845 MHz.) The receiver would also receive an image signal on 15.210 MHz because 15.210 – 14.755 = 0.455 MHz. [E4C14] If the IF was raised to 9 MHz, the BFO would be set to 14.3 + 9 = 23.3 MHz and the image

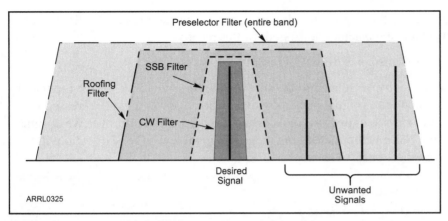

Figure 8-10 — A typical multiple-conversion superheterodyne receiver has several stages of filtering. Preselector filters reject out-of-band signals. Roofing filters at the input to each IF further restrict receiver bandwidth, attenuating strong in-band signals that might overload the IF amplifiers. In the final IF stage, single-signal filters are used to select just the desired signal.

frequency would be 23.3 + 9 = 32.3 MHz which is much farther away and easier to filter out.

At the input to each IF stage where the most strenuous filtering is performed, a *roofing filter* is often used. (Most superheterodyne receivers have two or more IF stages.) Roofing filters are high-performance filters (generally crystal filters) that have a bandwidth wider than that of the widest signal that will be received. Their purpose is to reject as many as possible of the strong signals on adjacent frequencies without affecting the desired signal. Such signals can cause amplifiers to overload or affect the AGC system so as to distort the desired signals. **Figure 8-10** illustrates why the filters are called roofing filters — their broad response acts as a "roof" over the narrower filters intended to pass just a single signal. [E4C13]

In the final IF circuit, narrow filters are used to select only one signal from the many that may be present. Crystal filters are generally used for this purpose, although special *mechanical resonator filters* are sometimes used. The bandwidth of these filters is selectable for the type of desired signal.

It is important to match the filter bandwidth with that of the desired signal. If the filter is too wide, unwanted signals will be received. If the filter is too narrow, the desired signal will be distorted. For example, using a 500-Hz CW filter to receive an SSB signal would render it nearly unintelligible. In general, the final IF filter bandwidth should be slightly greater than the bandwidth of the signal you want to receive.

Table 8-5 shows the customary filter bandwidths for common amateur signals. [E4C10, E4C11] Narrower filters are used when the band is crowded, but they limit signal fidelity and the ability to detect the presence of nearby signals. Wider filters are more comfortable to listen to, but allow undesired signals on nearby frequencies to be heard, as well. [E4C12] Very narrow filters also have a tendency to *ring*, meaning they output a sustained response after the original signal has stopped. This comes from energy stored in the filter's circuits or crystals.

DSP filters are now replacing the crystal and mechanical resonator filters traditionally used in the final IF stages. The technology for these filters has reached equivalent performance with the best of the traditional filters. In addition, these filters have adjustable bandwidths and shape factors. DSP filters can be made very narrow with little ringing. In the coming years, DSP filters will be applied closer and closer to the receiver's front end.

Table 8-5

IF Filter Bandwidths by Signal Type

RTTY or digital	300 Hz
CW	200 to 500 Hz
SSB	1.5 to 2.7 kHz
AM	6 kHz
VHF FM	15 kHz

Before you go on, study test questions E4C02, E4C09 through E4C14, and E4D09. Review this section if you have difficulty.

DYNAMIC RANGE AND INTERMODULATION

Dynamic range is an important receiver parameter. This refers to the ability of the receiver to tolerate strong signals outside the normal passband and still operate properly. We can state a general definition of dynamic range as the ratio between the MDS and the largest input signal that does not cause audible distortion products. There are several types of dynamic range measurements used to describe receiver performance, based on input signal levels in dBm. Dynamic range measurements are stated in dB.

Blocking Dynamic Range

An input signal can be strong enough that the receiver no longer responds linearly and its gain begins to drop. This reduction in gain due to the strong signal causes weaker signals to appear to fade. This reduction in gain is called *gain compression* or *blocking*. Blocking may be observed as *desensitization* or *desense* — the reduction in apparent strength of a desired signal caused by a nearby strong interfering signal. [E4D12, E4D13]

A receiver's *blocking level* is the power of an input signal that causes 1 dB of gain compression. *Blocking dynamic range* (BDR) as illustrated in **Figure 8-11** is the difference between the level of the receiver's MDS and the blocking level. When the blocking dynamic range is exceeded, the receiver begins to lose the ability to amplify weak signals. [E4D01] If the interfering signal is far enough from the desired signal, it may be possible to reduce desensitization by using IF filters to reduce the receiver's RF bandwidth and reject the strong signals. [E4D14]

Intermodulation (IMD)

A perfectly linear receiver will produce an output signal with a strength that changes exactly the same as the input signal. If the input signal changes by 1 dB, the output signal will also change by 1 dB. This is called the *first order* response and is shown as the dashed line in Figure 8-11. No receiver is perfectly linear, however.

As the input signal strength increases, the receiver's response becomes nonlinear and IMD products are created as discussed in the previous sections on mixers and transmitter intermodulation. (This discussion touches on several key points regarding IMD products and receiver linearity but is not meant to be complete. For additional detailed information about receiver performance, see *The ARRL Handbook*.)

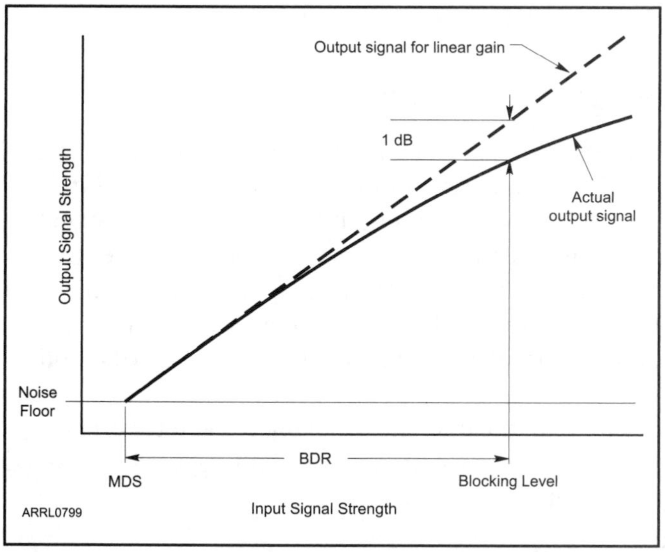

The frequency and amplitude of the IMD products depends upon the *order* of the IMD response. IMD products are created at frequencies which are the sum and difference of the input signals and their harmonics.

$$f_{IMD} = nf_1 \pm mf_2 \qquad \text{(Equation 8-7)}$$

where:
  f_1 and f_2 are the input signal frequencies
  n and m are positive integers; 1, 2, 3, etc

Second-order IMD products are created for n + m = 2 (both n and m equal to 1). *Third-order* IMD products are created if n + m = 3.

The frequencies of second-order IMD products caused by signals that are close together are far from the frequency of either input signal and so are generally not a problem if caused by

Figure 8-11 — Gain compression occurs when the input signal is too strong for the receiver to develop full gain. Blocking Dynamic Range (BDR) is measured in dB from MDS to a level at which the input signal strength causes a 1 dB drop in receiver gain, called the blocking level.

signals within an amateur band. Second-order products can be created in an amateur band, however, by strong out-of-band signals such as from shortwave broadcast stations. Preselectors and front-end band-pass filters can reduce or eliminate second-order IMD products caused by those signals.

There are four third-order IMD product frequencies. Two are additive (f_{IMD1} and f_{IMD3}) and two are subtractive (f_{IMD2} and f_{IMD4}):

$$f_{IMD1} = 2f_1 + f_2 \qquad \text{(Equation 8-8)}$$
$$f_{IMD2} = 2f_1 - f_2 \qquad \text{(Equation 8-9)}$$
$$f_{IMD3} = 2f_2 + f_1 \qquad \text{(Equation 8-10)}$$
$$f_{IMD4} = 2f_2 - f_1 \qquad \text{(Equation 8-11)}$$

where:

f_{IMD} is the frequency of the IMD product.
f_1 and f_2 are the input signals

If the frequencies of the signals causing the IMD products are close together, such as in the same amateur band as the desired signal, the subtractive IMD products (f_{IMD2} and f_{IMD4}) could possibly be very close to the desired signal frequency. Therefore, the third-order IMD performance of a receiver is an important receiver specification. [E4D11]

Here's an example of third-order IMD performance being important. Let's say your receiver is tuned to 146.70 MHz. Whenever a nearby station is transmitting on 146.52 MHz, you receive intermittent bursts of garbled speech. This is likely to be a third-order intermodulation product generated in your receiver which is very sensitive but becomes nonlinear for very strong input signals.

What are the likely frequencies for a second strong signal that could combine with the one on 146.52 MHz to produce the IMD product you hear on 146.70 MHz? [E4D05] You know that the subtractive products are the likely source of the interfering signal because one of the signals causing the interference is close to the desired frequency. If the frequency of the IMD product is 146.70 MHz and you know one of the strong signal frequencies, $f_1 = 146.52$ MHz, you can solve for f_2 using equation 8-9:

$$f_{IMD2} = 2f_1 - f_2$$

$$f_2 = 2f_1 - f_{IMD2} = 2 \times 146.52 \text{ MHz} - 146.70 \text{ MHz} = 146.34 \text{ MHz}$$

This is a common repeater input frequency! Solving Equation 8-11 for f_1 using the same intermod and strong signal frequency, you'll find the other possible frequency to be (146.70 + 146.52) / 2 = 146.61 MHz. It would not be practical to filter out these strong input signals because they are *in-band signals*, close to your operating frequency. It would be better to use a receiver with a high enough dynamic range to accommodate these signals linearly and not produce the IMD products. (If the input signals are simply too strong, an attenuator at the receiver input may reduce the signal levels to a level at which they do not create IMD products.)

Another example from the HF bands will help illustrate the problem. If the interfering IMD product occurs at 14.020 MHz whenever a strong station is transmitting at 14.035 MHz, you can expect to find the other strong signal at:

$$f_2 = 2 \times 14.035 - 14.020 = 14.050 \text{ MHz}$$

or

$$f_2 = (14.035 - 14.020) / 2 = 14.0275 \text{ MHz}$$

With many strong signals closely spaced on a typical amateur band, IMD products can be a real problem! Reducing intermodulation is another reason to use roofing filters. A 6-kHz-wide roof-

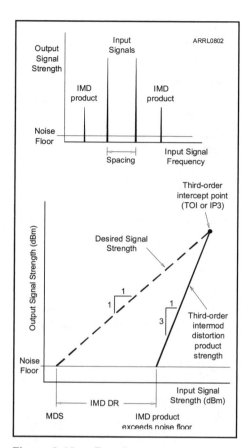

Figure 8-12 — Receiver output power for a desired signal and for third-order distortion products varies with changes of input signal power. The input signal consists of two equal-power sine-wave signals. Higher intercept points represent better receiver IMD performance.

ing filter would significantly reduce the level of any signal at all three of these frequencies at which the IMD product could be generated. Other remedies include adding attenuation as mentioned previously or reducing RF Gain. By eliminating (or at least reducing) strong in-band signals near the desired signal, the receiver's dynamic range is improved and IMD is reduced.

Intercept Points

Second-order IMD products occur at the sum and the difference of the input signal frequencies and their amplitude changes 2 dB for every 1 dB of input signal change. Third-order IMD product amplitudes change 3 dB for every 1 dB of input-signal change. (This assumes that the input signals have equal amplitudes.) **Figure 8-12** shows the output power of the desired signal versus the output power of the *third-order* distortion products at different input signal power levels. (The graph of second-order product strength would be similar but have a slope of 2 instead of 3.)

The input signal power at which the level of the distortion products equals the output level for the desired signal (where the lines cross) is the receiver's *intercept point*. There is a separate intercept point for each order of IMD product; second-order intercept point (SOI or IP2), third-order intercept point (TOI or IP3), and so on. By measuring signal and IMD product amplitudes, the intercept points can be calculated or estimated graphically. Specifically, the output power level at which the curves intersect is the *output intercept*. Similarly, the input power level corresponding to the point of intersection is called the *input intercept*.

For example, a 40 dBm third-order intercept point means that a pair of 40 dBm signals would produce an IMD product of the same 40 dBm level. [E4D10] A level of 40 dBm is equal to 10 watts so calculating the intercept point level is only a method of evaluating receiver performance and not a specification of how much signal the receiver can actually accept.

Although signals on the air are not strong enough to reach the intercept point levels, the intercept point values are useful for assessing receiver linearity. The higher the intercept point, the lower the amplitude of the IMD products generated by the receiver due to nonlinearities at actual received signal levels. Third-order intercept performance of a receiver usually gets worse as the frequencies of the strong signals (f_1 and f_2 in the equations above) get closer together. That is why the IMD performance of a receiver is typically given for several spacings of the input signals.

Intermodulation distortion dynamic range measures the ability of the receiver to avoid generating IMD products. When input signal levels exceed the IMD dynamic range, IMD products will begin to appear along with the desired signal. We can also calculate the third-order IMD dynamic range using the third-order intercept point and the receiver noise floor or MDS value.

$$\text{IMD DR}_3 = (2/3)\ (\text{IP3} - \text{MDS})$$

(Equation 8-12)

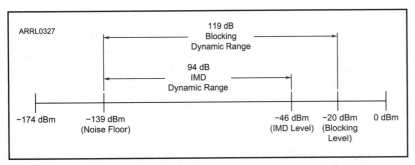

Figure 8-13 — The dynamic-range performance of a hypothetical (though typical) receiver. The noise floor is –139 dBm, blocking level is –20 dBm and the IMD level is –46 dBm. This corresponds to a receiver blocking dynamic range of 119 dB and an IMD dynamic range of 94 dB.

where:
IMD DR$_3$ is the third-order inter modulation distortion dynamic range in dB.
IP3 is the third-order input intercept point in dBm.
MDS is the noise floor or MDS of the receiver in dBm.

Figure 8-13 illustrates the relationship between the input signal levels, noise floor, blocking dynamic range and IMD dynamic range. With modern receiver designs, blocking dynamic ranges of more than 100 dB are possible. This is an acceptable dynamic range figure. So the receiver we have used in this example could be any typical modern receiver. If a receiver has poor dynamic range, cross-modulation or IMD products will be generated and desensitization (blocking) from strong adjacent signals will occur. [E4D02]

Before you go on, study test questions E4D01, E4D02, E4D05, and E4D10 through E4D14. Review this section if you have difficulty.

PHASE NOISE

Phase noise is a problem that has become more apparent as receiver improvements have reduced the noise floor and increased dynamic range. Most modern commercial transceivers use phase-locked loop (PLL) or direct digital synthesis (DDS) frequency synthesizers to create VFOs and other signal sources in a transceiver. PLL synthesizers are continually adjusting their frequency compared to a reference oscillator. DDS synthesizers exhibit small variations in frequency based on the clock signal and artifacts from creating the output waveform as small discrete steps. Both result in the phase of the output signal continuously and randomly shifting back and forth a slight amount, creating *phase noise*.

Phase noise creates a random collection of low-level sidebands that are increasingly stronger close to the desired signal frequency. On a transmitted signal, phase noise sounds like a strong hiss that can often be heard across an entire band to a nearby receiver, even on other bands. This can be a serious problem for other stations on the band if the transmitted signal is strong.

On receive, the phase noise mixes with the received signals just as the desired signal does but the result is random noise, not the desired mixing product. Thus, as you tune toward a strong signal, the receiver noise floor appears to increase. In other words, you hear an increasing amount of noise in an otherwise quiet receiver as you tune toward the strong signal.

Excessive phase noise in a receiver local oscillator allows strong signals on nearby frequencies to interfere with the reception of a weak desired signal. [E4C01] This increased receiver noise can cover a weak desired signal, or at least make copying it more difficult. **Figure 8-14** illustrates how phase noise can cover a weak signal.

A transmitter with excessive phase noise on its output signal also causes noise to be received up and down the band for some range around the desired transmit frequency. This additional noise can fall within the passband of a receiver tuned to some weak signal. So

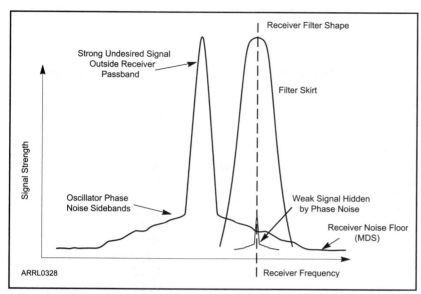

Figure 8-14 — In a receiver with excessive phase noise, a strong signal near the receiver passband can raise the apparent receiver noise floor in the passband. This increased noise can cover a weak signal that you are trying to receive.

even if you have a receiver with very low phase noise, you can be bothered by this type of interference!

Phase noise can be a serious problem when trying to operate two stations close together, such as during Field Day. The phase noise from one transmitter can cause severe interference to nearby receivers, even if they are operating on a different band! In this case, a band-pass filter at the transmitter is required to eliminate the interference to receivers on other bands.

Before you go on, study test question E4C01. Review this section if you have difficulty.

CAPTURE EFFECT

One of the most notable differences between an amplitude-modulated (AM) receiver and a frequency-modulated (FM) receiver is how noise and interference affect an incoming signal. The limiter and discriminator stages in an FM receiver can eliminate most of the impulse-type noise, unless the noise has frequency-modulation characteristics.

FM receivers perform quite differently from AM, SSB and CW receivers when QRM is present, exhibiting a characteristic known as the *capture effect*. The loudest signal received, even if it is only a few dB stronger than other signals on the same frequency, will be the only signal demodulated, blocking all weaker signals. [E4C03]

Capture effect can be an advantage if you are trying to receive a strong station and there are weaker stations on the same frequency. At the same time, this phenomenon will prevent you from receiving one of the weaker signals. When two stations transmit an FM signal on the same frequency, the receiving station will hear only the stronger signal, with buzzing or clicks that are all that's left of the weaker signal. This can be a problem on busy repeaters, particularly during nets with multiple stations attempting to check in or pass messages.

Before you go on, study test questions E4C03. Review this section if you have difficulty.

Table 8-6
Questions Covered in This Chapter

8.1 Digital Protocols and Modes	8.2 Amateur Television	8.3 Receiver Performance
E2C08	E2B01	E4C01
E2C09	E2B02	E4C02
E2D01	E2B03	E4C03
E2D02	E2B04	E4C04
E2D03	E2B05	E4C05
E2D09	E2B06	E4C06
E2D12	E2B07	E4C07
E2E01	E2B08	E4C08
E2E02	E2B09	E4C09
E2E03	E2B10	E4C10
E2E04	E2B11	E4C11
E2E05	E2B12	E4C12
E2E06	E2B13	E4C13
E2E07	E2B14	E4C14
E2E08	E2B15	E4C15
E2E09	E2B16	E4D01
E2E10	E2B17	E4D02
E2E11	E2B18	E4D05
E2E12	E2B19	E4D09
E4A09		E4D10
E8A12		E4D11
E8A13		E4D12
E8A15		E4D13
E8C01		E4D14
E8C02		E6E05
E8C03		
E8C04		
E8C05		
E8C06		
E8C07		
E8C08		
E8C09		
E8C10		
E8C11		
E8C12		
E8C13		

Chapter 9

Antennas and Feed Lines

In this chapter, you'll learn about:
- **Antenna basics —** radiation patterns, gain and beamwidth, polarization and bandwidth
- **The effects of ground on antennas**
- **Shortened, multiband and satellite antennas**
- **Traveling wave antennas and phased arrays**
- **Loops and direction finding antennas**
- **Impedance matching techniques**
- **Transmission line mechanics and the Smith Chart**
- **Antenna measurements and analyzers**
- **Antenna modeling and design**

Antennas and antenna systems are of primary importance to amateurs. Equipment available to amateurs is of the best quality in the history of radio but even the best radios require an antenna to communicate effectively. It is up to the station owner to select and install antennas and antenna system components that get the most out of the radio equipment.

The topics covered by the license exam touch on many important subjects for Amateur Radio antenna systems. Basic antenna concepts are explored in more detail than for other license classes. Practical issues affecting antennas, such as ground systems, shortened antennas and popular multiband designs are covered. We'll also look into the functions of transmission lines as they affect antenna systems. The section concludes with some discussion of using computers to model antennas. The set of exam questions for each subject is listed at the end of each section so that you can review your understanding and build a solid background in antennas. All of these topics are covered in great detail in *The ARRL Antenna Book* if you would like more information.

9.1 Basics of Antennas

We'll explore another level of detail beyond that which you learned for your General class license, but before plunging in, here is a short refresher course on antenna radiation patterns.

ANTENNA RADIATION PATTERNS

An antenna radiation pattern contains a wealth of information about the antenna and its expected performance. By learning to recognize certain types of patterns and what they represent, you will be able to identify many important antenna characteristics and to compare design variations. **Figure 9-1** is an example of a dipole's radiation pattern, including a three-dimensional view (Figure 9-1A).

Figure 9-1B shows one "slice" through the three-dimensional pattern of Figure 9-1A. The orientation of the antenna is shown and the directions around the antenna are shown on the outer circle. The strength scale from the center to the edge of the pattern is usually shown in decibels, but the step size of each ring can be adjusted to show desired pattern details. Make sure the strength scales are the same when comparing the patterns of different antennas.

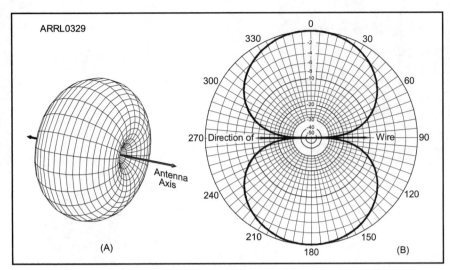

Figure 9-1 — The three-dimensional radiation pattern of a dipole appears at A. B shows a plane radiation pattern of the same antenna — a slice through the pattern at A. The solid line in B represents the relative amount of power radiated at the angles on the outer ring.

The radiation pattern is usually drawn so it just touches the outer circle at the point of its maximum strength. The scale then shows the relative strength of signals radiated in any direction with the maximum point representing 0 dB with respect to all other directions. A legend on the chart often gives the value of gain at the outer circle in absolute terms. To compare antennas, use patterns with the same value of gain at the outer circle. Or, if you're modeling the antennas, plot the radiation patterns for all antennas on a common chart.

Antenna radiation patterns describe the antenna's radiated signal in the antenna's *far field* which begins several wavelengths from the antenna and extends to infinity. Antennas also have a *near field* region which is too close to the antenna for the final pattern to emerge. In the far field, the pattern shape is independent of distance. [E9B12]

ANTENNA GAIN

In many applications, the antenna's most important property is its ability to concentrate its radiated power in useful directions. This property, however, only has meaning with respect to other antennas, so a reference must be established.

The Isotropic Radiator

An *isotropic radiator* is a theoretical, point-sized antenna that is assumed to radiate equally in all directions. This hypothetical antenna has a radiation pattern that is omnidirectional, because the signal is equal in all directions. No such antenna actually exists, but it serves as a useful theoretical reference for comparison with real antennas. The isotropic antenna also provides a useful reference for comparing the differences among real antennas. [E9A01]

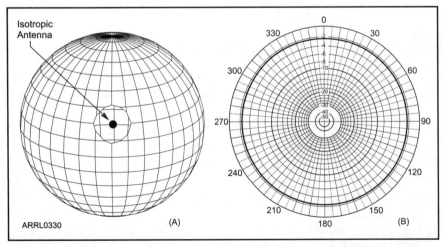

Figure 9-2 — The three-dimensional radiation pattern of an isotropic radiator is shown at A. B shows the radiation pattern of the isotropic antenna in any plane. This plot has been reduced by 2.15 dB so that it can be compared with the radiation pattern of the dipole in Figure 9-1.

As you might expect, the three-dimensional radiation pattern of an isotropic radiator (see **Figure 9-2**) is a sphere, since the same amount of power is radiated in all directions. If the spherical pattern is sliced such that the slice contains the isotropic antenna at the center, the resulting pattern is a circle.

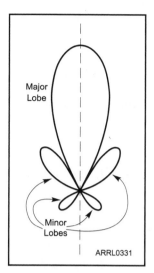

Figure 9-3 — The radiation pattern typical of a VHF beam antenna, illustrating major and minor radiation lobes.

Directional Antennas

The radiation from a practical antenna never has the same intensity in all directions. The intensity may even be near zero in some directions from the antenna; in others it will probably be greater than an isotropic antenna. *Directional antennas* are designed specifically to concentrate their radiated power in one (or more) directions. In a perfect directional antenna, the radio energy would be concentrated in one direction only, called the *forward* direction. This is known as the *major lobe* or *main lobe* of radiation. (Most beams also have *minor lobes* in the *back* and *side* directions.) **Figure 9-3** is an example of a radiation pattern for a typical VHF beam antenna, illustrating major and minor lobes. The directions of minimum radiation between the lobes are the pattern's *nulls*. By reducing radiation in the side and back directions and concentrating it instead in the forward direction, a beam antenna can transmit or receive a stronger signal than an isotropic antenna would in that direction.

An antenna's *gain* is the ratio (expressed in decibels) between the signal radiated from an antenna in the direction of its main lobe and the signal radiated from a reference antenna (usually a dipole) in the same direction and with the same power. [E9A08] A typical beam might have 6 dB of gain compared to a dipole which means that it makes your signal sound four times (6 dB) louder than if you were using a dipole with the same transmitter. An isotropic radiator has no directivity at all, because the radiated signal strength is the same in all directions. We can also say the isotropic radiator has zero or no gain in any direction for this same reason. [E9A03]

The gain of directional antennas is the result of concentrating the radio wave in one direction at the expense of radiation in other directions. There is no difference in the total amount of power radiated. [E9B07] Directional antennas just concentrate it in preferred directions. Since practical antennas are not perfect, there is always some radiation in undesired directions.

In contrast to an isotropic antenna, the solid radiation pattern of an ideal dipole antenna in *free space* (without any ground or conducting surfaces nearby to affect the pattern) resembles a doughnut (Figure 9-1). Radiation in the main lobe of a dipole is 2.15 dB greater than would be expected from an isotropic radiator (assuming the same power to the antennas). [E9A02]

While the isotropic radiator is a handy mathematical tool, it can't be physically constructed. On the other hand, the dipole is a simple antenna that can be constructed and easily tested on an antenna range. For that reason, the dipole is also used as a reference of comparison.

So we have a situation where there are two reference antennas used to compare the radiation patterns of other antennas. Fortunately, it's simple to convert from the isotropic reference to the dipole because the dipole has 2.15 dB of gain over an isotropic radiator. That makes it possible to convert the gain of an antenna that is specified using one reference antenna to a gain specified using the other reference.

If the gain of an antenna is specified as being compared to that of an isotropic radiator, you can find the gain compared to a dipole by:

Gain in dBd = Gain in dBi − 2.15 dB   (Equation 9-1)

where:

dBd is antenna gain compared to a reference dipole.
dBi is antenna gain compared to an isotropic radiator.

Gain in dBd is less than gain in dBi because the antenna being measured has less gain compared to a dipole than it does compared to an isotropic antenna.

If antenna gain is specified compared to a dipole, you can find the gain compared to an isotropic radiator by:

Gain in dBi = Gain in dBd + 2.15 dB (Equation 9-2)

Gain in dBi is greater than gain in dBd because the antenna being measured has more gain compared to an isotropic radiator than if compared to a dipole.

Example 9-1

If an antenna has 6 dB more gain than an isotropic radiator, how much gain does it have compared to a dipole? [E9A13]

Gain in dBd = Gain in dBi – 2.15 dB = 6 dBi – 2.15 dB = 3.85 dBd

Example 9-2

If an antenna has 12 dB more gain than an isotropic radiator, how much gain does it have compared to a dipole? [E9A14]

Gain in dBd = Gain in dBi – 2.15 dB = 12 dBi – 2.15 dB = 9.85 dBd

These gains are free-space gains, meaning that there are no reflecting surfaces near the antenna, such as the ground. Nearby reflecting surfaces can dramatically increase or decrease an antenna's gain. When you compare specifications for several antennas, be sure that they all use the same reference antenna for comparison or convert the gains from one reference to another.

Before you go on, study test questions E9A01, E9A02, E9A03, E9A08, E9A13, E9A14, E9B07 and E9B12. Review this section if you have difficulty.

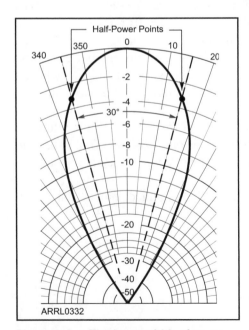

Figure 9-4 — The beamwidth of an antenna is the angular distance between the directions at which the antenna gain is one-half (–3 dB) its maximum value.

BEAMWIDTH AND PATTERN RATIOS

In comparing antennas, it is useful to know their *beamwidth*. Beamwidth is the angular distance between the points on either side of the major lobe at which the gain is 3 dB below the maximum. This is also sometimes called the *–3dB beamwidth*. [E9B08] **Figure 9-4** illustrates the idea. The antenna in the figure has a beamwidth of 30°. This means if you turn your beam plus or minus 15° from the optimum heading, the signal you receive (and the signal received from your transmitter) will drop by 3 dB. As gain is increased, beamwidth decreases. [E9D03] For example, a high-gain antenna, such as is often used at VHF/UHF/microwave frequencies, can have a very narrow beam. Stations significantly to the side of such a beam antenna may not be heard at all!

Figure 9-5 illustrates how to determine beamwidth from a radiation pattern. The major lobe of radiation from the antenna in the figure points to the right, and is centered along the 0° axis. You can make a pretty good estimate of the beamwidth of this antenna by carefully reading the graph. Notice that angles are marked off every 15° and the –3 dB circle is the first one inside the outer circle. The pattern crosses the –3 dB circle at points about 25°

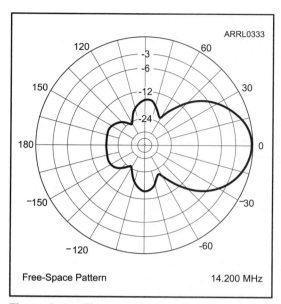

ARRL0333

Free-Space Pattern   14.200 MHz

Figure 9-5 — The radiation pattern of a typical HF beam antenna. The text shows how to read beamwidth and other pattern ratios from the graph.

either side of 0°. So we can estimate the beamwidth of this antenna as 50°. [E9B01]

There are other useful measurements to be taken from an antenna's radiation pattern. Gain is not everything! The ability to reject received signals from unwanted directions is also important. This ability, called a *pattern ratio*, is measured with respect to the directivity of the antenna in the forward or "front" direction. There are three primary pattern ratios:

• *Front-to-back (F/B):* the difference in gain in the direction of the major lobe to the gain in the exact opposite (back) direction.

• *Front-to-rear (F/R):* the difference in gain in the direction of the major lobe to the average gain over a specified angle centered on the back direction.

• *Front-to-side (F/S):* the difference in gain in the direction of the major lobe to the gain at 90° to the front direction. (F/S usually assumes a symmetrical pattern to either side of the major lobe.)

In the example shown in Figure 9-5, find the front-to-back ratio by reading the maximum value of the minor lobe at 180°. This maximum appears to be halfway between the −12 dB and −24 dB circles, so estimate it to be about 18 dB below the major lobe. [E9B02]

The pattern in Figure 9-5 has a minor lobe to each side of the antenna whose maximum strength is a bit more than 12 dB below the main-lobe maximum. A front-to-side ratio of 14 dB looks like a pretty good estimate for this pattern. [E9B03]

If an angle of 30° is specified for the front-to-rear ratio, the values of gain at several points between ±15° from 180° would be measured and averaged. In the figure, the average is just slightly more than 20 dB below the major lobe's maximum, so 19 dB is a reasonably good estimate of front-to-rear ratio. While the difference between front-to-back and front-to-rear is not large in this example, some antennas have very sharp pattern nulls in the back direction that would make the difference much larger.

Before you go on, study test questions E9B01, E9B02, E9B03, E9B08 and E9D03. Review this section if you have difficulty.

RADIATION AND OHMIC RESISTANCE

The power supplied to an antenna is dissipated in the form of radio waves and in heat losses in the wire and materials nearby that absorb the waves, such as foliage or buildings. The radiated energy is the useful part, of course. The power being radiated away by the antenna can be represented as if the energy was dissipated in a resistance, R. There is a pair of resistances, one for the radiated power and one for the power dissipated as heat.

In the case of the heat losses, R is a *real* or *ohmic* resistance that gets warm. In the case of the radiated power, R is an assumed resistance, that if actually present, would dissipate the power actually radiated by the antenna. This assumed resistance is called the *radiation resistance*, R_R. [E9A15] The total power dissipated by the antenna is therefore equal to $I^2 (R_R + R)$. These two resistances form the total resistance of an antenna system, R_T. [E9A06]

Taking as an example an ordinary half-wave dipole antenna for the HF bands, the power lost as heat in the antenna does not exceed a few percent of the total power supplied to the antenna. This is because the RF resistance of copper wire even as small as #14 AWG is very low compared with the radiation resistance if the antenna is reasonably clear of surrounding objects and not too close to the ground. It is reasonable to assume that the ohmic (heat) loss in a reasonably located antenna is negligible and that all of the resistance shown by the antenna is radiation resistance. Such an antenna is a highly efficient radiator of radio waves!

FEED POINT IMPEDANCE

An important characteristic of an antenna is the *feed point impedance* presented to the transmission line. Feed point impedance is simply the ratio of RF voltage to current wherever the transmission line is attached to the antenna. If the voltage and current are in-phase, the feed point impedance is purely resistive and the antenna is resonant, regardless of the value of the resistance. If the voltage and current are not in-phase, the impedance will have some reactance as well and may be inductive or capacitive. Feed point impedance consists of the antenna's radiation resistance, ohmic losses including ground losses, and any reactance caused by the antenna being nonresonant.

Feed point impedance also changes with position on the antenna. Consider a resonant, half-wavelength dipole with maximum current at the mid-point and minimum current at the end. The feed point impedance is lowest in the middle of the antenna where the ratio of voltage to current is lowest. As the feed point is moved toward either end of the dipole, however, the voltage increases and current decreases, causing the feed point impedance to increase. Near the center, feed point impedance is less than 100 ohms but at the ends, feed point impedance can be several thousand ohms. If the same amount of power is transferred to the dipole at both locations for the feed point, the radiated signal will be the same.

The value of an antenna's feed point impedance is also affected by a number of other factors. One is the location of the antenna with respect to other objects, particularly the Earth. For example, in free space with nothing else near it, the radiation resistance of a resonant ½-wavelength dipole antenna made of thin wire is approximately 73 Ω at its center. As the antenna is lowered closer to the ground, the radiation resistance drops as well, lowering the feed point impedance. Other nearby conducting surfaces such as buildings and other antennas can also affect the antenna's feed point impedance.

Another factor is the *length/diameter ratio* of the conductor that makes up the antenna. As the conductor is made thicker, radiation resistance decreases. For most practical wire sizes, the half-wave dipole's radiation resistance is close to 65 Ω. At VHF and above, the dipole's radiation resistance will be from 55 to 60 Ω for antennas constructed of rod or tubing. [E9A05]

ANTENNA EFFICIENCY

Antenna efficiency — the ratio of power radiated as radio waves to the total power input to the antenna — is given by:

$$\text{Efficiency} = \frac{R_R}{R_T} \times 100\% \tag{Equation 9-3}$$

where:
   R_R = radiation resistance.
   R_T = total resistance. [E9A10]

Example 9-3

If a half-wave dipole antenna has a radiation resistance of 70 Ω and a total resistance of 75 Ω, what is its efficiency?

Efficiency = (70 / 75) × 100% = 93.3%

The actual value of the radiation resistance has little effect on the radiation efficiency of a practical antenna. This is because the ohmic resistance is only on the order of 1 Ω with the conductors used for thick antennas. The ohmic resistance does not become important until the radiation resistance drops to very low values — say less than 10 Ω — as may be the case when several antenna elements are very close together or for antennas such as mobile whips that are very short in terms of wavelength.

Nearby materials that dissipate radio frequency energy also increase total resistance. For antennas that have a ground connection or use a ground system as part of the antenna, the *ground resistance* can be a significant contributor to antenna system losses — even if the antenna itself is constructed from low-loss materials.

A half-wave dipole operates at very high efficiency because the conductor resistance is negligible compared with the radiation resistance. In the case of a ¼ wavelength vertical antenna, with one side of the feed line connected to ground, the ground resistance usually is not negligible. If the antenna is short compared to one-quarter wavelength, such as a mobile antenna, the resistance of the necessary loading coil may become appreciable.

To be effective, a ¼-wavelength ground-mounted vertical antenna requires a ground system of *radial* wires, laid out as spokes on a wheel with the antenna element in the center to reduce losses that would otherwise result from current flowing in the lossy soil. [E9A11]

Before you go on, study test questions E9A05, E9A06, E9A10, E9A11 and E9A15. Review this section if you have difficulty.

ANTENNA POLARIZATION

When we consider real antennas near the Earth, we usually refer to the antenna polarization relative to the ground. For example, if the electric field of the radiation from an antenna is oriented parallel to the surface of the Earth, we say the antenna is *horizontally polarized*. If the electric field is oriented perpendicular to the surface of the Earth, the antenna is *vertically polarized*.

The simplest way to determine the polarization of an antenna is by looking at the orientation of the elements. The electric field will be oriented in the same direction as the antenna elements. So if a Yagi antenna has elements oriented parallel to the ground, it will produce horizontally polarized radiation, and if the antenna elements are oriented perpendicular to the ground, it will produce vertically polarized radiation.

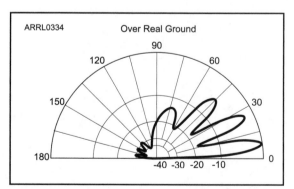

Figure 9-6 — This radiation pattern for a beam antenna mounted above ground is an elevation pattern. The major lobe in the forward direction is about 7.5° above the horizon. There are four radiation lobes in the forward direction and three to the back, all at different vertical angles.

ANTENNA PATTERN TYPES

E and H Planes

Two types of radiation patterns are often used to picture the overall, three-dimensional radiation pattern — the *E-plane* and the *H-plane* radiation patterns. The E-plane pattern is taken in the plane of the electric field and the H-plane pattern in the plane of the magnetic field. In general, the E-plane pattern is in the plane of the antenna's elements and the H-plane pattern is perpendicular to them.

Azimuthal and Elevation Patterns

For a horizontally polarized antenna, the E-plane pattern is parallel to the surface of the Earth and shows the antenna's radiation pattern in directions around the antenna. This is called an *azimuthal pattern*. The H-plane pattern of the same antenna is called the *elevation pattern* and shows the antenna's radiation pattern at different angles above the Earth. Figure 9-5 in the section on Beamwidth and Pattern Ratios shows a typical azimuthal pattern for a beam antenna.

Figure 9-6 shows a typical elevation pattern with multiple lobes in the forward and back directions. [E9C07] The antenna pattern shows four forward lobes (those on the same half of the graph as the largest or main lobe) and three rear lobes. [E9C10] For antennas mounted over ground, an elevation pattern shows only the half of the radiation pattern at positive angles above ground. The angles on the graph represent the vertical angle above ground at which radiated power is measured. The vertical angle at which the antenna's major lobe has its maximum radiation is called the *takeoff angle*. In Figure 9-6, the takeoff angle is 7.5 degrees. [E9C08]

Just as with an azimuthal pattern, an antenna's front-to-back ratio can be determined from the maximum radiation in the forward direction to that in the opposite direction. In an elevation pattern, however, it is necessary to specify the elevation angle at which the pattern is to be measured — usually the takeoff angle. In the case of Figure 9-6, toward the back of the pattern at the takeoff angle antenna gain is about 28 dB below that of the main lobe so front-to-back ratio for this antenna would be 28 dB. [E9C09] At other elevation angles, the front-to-back ratio would be quite different.

Before you go on, study test questions E9C07 through, E9C10. Review this section if you have difficulty.

BANDWIDTH

Along with the characteristics of the antenna's radiation pattern, feed point impedance, and so forth, it's also important to know the range of frequencies over which those characteristics exist.

It's also important to know the range of frequencies over which the antenna will behave as expected. As frequency changes, the electrical size of the antenna and all of its components changes, too. This means the antenna's gain, feed point impedance, radiation pattern, and so forth will also change. The antenna is expected to perform to some specified level — these are the *performance requirements* of the antenna. In general, the bandwidth of an antenna is the frequency range over which it satisfies a performance requirement. [E9A09]

For example, an antenna may be specified to have an SWR of less than 1.5:1, but without an associated bandwidth, the specification is incomplete. Is the SWR going to be less than 1.5:1 at just a few frequencies or over the whole band? If that antenna's *SWR bandwidth* is specified to be 200 kHz, then you can expect to see an SWR of 1.5:1 or less over a range of 200 kHz. Notice that the frequency of minimum SWR is not known, so the range may shift depending on factors relating to assembly or installation. Other common bandwidth specifications are *gain bandwidth* and *front-to-back bandwidth*.

Before you go on, study test question E9A09. Review this section if you have difficulty.

9.2 Practical Antennas

Armed with a better theoretical understanding of what affects antenna system efficiency, you are now ready to consider practical antenna systems. Rarely does an antenna get installed in a location without compromises. What happens if the number of radials available for an HF vertical antenna can't be as many as you'd like? What would be the effect of a beam antenna being close to the roof? How will that nearby hill or building affect your signal? These are all questions that amateurs have to deal with every time an antenna is erected.

EFFECTS OF GROUND AND GROUND SYSTEMS

By far, the biggest effect on antenna system efficiency is the losses in nearby ground, grounded structures, or the antenna's ground system. The radiation pattern of an antenna over real ground is always affected by the electrical conductivity and dielectric constant of the soil, and most importantly by the height of a horizontally polarized antenna over ground. Signals reflected from the ground combine with the signals radiated directly from the antenna. If the signals are in phase when they combine, the signal strength will be increased but if they are out of phase, the strength will be decreased. These ground reflections affect the radiation pattern for many wavelengths from the antenna.

This is especially true of the *far-field pattern* measured many wavelengths from of a vertically polarized antenna operating at HF frequencies. In theory, this type of antenna should produce a major lobe at a low vertical angle — good for DX! Losses caused by low conductivity in the soil near the antenna dramatically reduce signal strength at low angles, however. [E9A12, E9C13] The low-angle radiation from a vertically polarized antenna mounted over seawater will be much stronger than for a similar antenna mounted over rocky soil, for example. [E9C11] The far-field, low-angle radiation pattern for a horizontally polarized antenna is not as significantly affected, however.

There are several things you can do to reduce the near-field ground losses of a vertically polarized antenna system. Adding more radials is one common technique. If you can only manage a modest on-ground radial system under a ⅛-wavelength long, inductively loaded vertical antenna, then a wire-mesh screen about ⅛-wavelength square is a good compromise.

There is little you can do to improve the far-field, low-angle radiation pattern of a vertically polarized antenna if the ground under the antenna has poor conductivity, the most important factor affecting ground losses. Most measures that people might try will only be applied close to the antenna and will only affect the near-field pattern. For example, you can add water to the ground under the antenna but unless you watered the ground for 100 or more wavelengths, it won't improve the ground conductivity for these distant ground reflections. (Even then, it would only be a temporary fix, or would have to be repeated often enough to keep the ground wet.) Adding more radials or extending the radials more than ¼-wavelength won't help either, unless you are able to build an extensive ground screen under the antenna out to 10 or more wavelengths!

Height Above Ground

The height at which an antenna is mounted above ground can also have a large effect on an antenna's radiation pattern. First, by moving the antenna away from ground, the amount of power dissipated in the Earth will be reduced. Second, reflections of the radiated signal from the ground will reinforce the direct radiation from the antenna at lower vertical angles, resulting in a stronger low-angle signal. In general, raising an antenna lowers the vertical takeoff angle of peak radiation.

Horizontally polarized antennas, such as dipoles and Yagis, have less ground loss than ground-mounted vertical antennas. As they are raised, ground losses drop to negligible levels. As such an antenna is raised, the vertical angle of maximum radiation drops until a height of ½ wavelength is reached. Raising the antenna farther causes additional lobes to appear in the elevation pattern above the main lobe. Figure 9-6 shows an example of these additional lobes. In general, mounting horizontally polarized antennas as high as possible is a good rule of thumb. [E3C07]

Vertical antennas, however, may depend on the ground or ground system to create an electrical image of the antenna to complete the overall half-wavelength antenna system. By moving the antenna higher, ground losses may be reduced but the ground system is still needed. Unless elevated radial wires or rods are used, the antenna support and outside of the antenna feed line shield will make up the ground system, with unpredictable results. ("Ground independent" vertical antennas are ⅜ or ½-wavelength long electrically and are fed at a point of high impedance on the antenna. Only a small system of radial-like counterpoise rods is required to reduce ground currents on the feed line shield.)

Terrain

The terrain on which an antenna is mounted affects both the azimuthal and elevation pattern of an antenna. Over flat ground and without nearby obstructions the radiation patterns for the antenna will resemble those in the antenna design books. Once buildings and uneven terrain enter the picture however, all bets are off!

Nearby buildings can serve as "passive" reflectors (or absorbers) of radio waves, most strongly at VHF and higher frequencies. This can be used to advantage, for example, by aiming an antenna to reflect a signal off a building toward a distant station. (Take care, of course, to avoid exposing the inhabitants to excessive levels of RF.) At HF, small buildings are less of a problem, but large buildings can have the same effect as on the higher bands.

Hills and slopes have an effect on both the azimuthal and elevation patterns. A hilltop is highly sought after for radio work because the reflections from the ground's surface are either reduced or are more likely to reinforce the signal at low takeoff angles. This is particularly true for horizontally-polarized antennas; the major lobe's takeoff angle will typically be lower in the direction of a slope. [E3C10]

Ground Systems

RF grounding of antennas and station equipment has many similar concerns. The object is to create a path to ground or earth potential that has as little impedance as possible. To accomplish this, the connection has several basic requirements. It must be electrically short, it must be as straight as possible, and it must be a type of conductor that has low impedance to RF.

The connection must be short in terms of wavelength because once the length exceeds one-tenth of a wavelength, the connection becomes more like a transmission line (discussed later in this chapter). Long ground connections can present significant impedance at frequencies where they are not close to ½-wavelength long. Long connections also begin to act like an antenna, radiating and receiving signals.

Conductors, even straight ones, have a significant amount of inductance. For example, a piece of #10 AWG wire that is one foot long has an inductance well over 100 nH! Obviously, this can be a problem at high frequencies for the typical ground connection.

The final reason pertains to how RF current flows on the surface of a conductor. Because of skin effect (see the discussion of Skin Effect and Q in the Electrical Principles chapter), RF current flows mainly on the surface of any conductor. The larger the amount of surface area, the lower the resistance of the conductor at RF. This is why wide, flat copper straps up to several inches wide, are the best for minimizing losses and impedance of a station's RF ground system. [E9D14]

Once the ground connection reaches the Earth, an RF connection is different than a power system's safety ground. A single ground rod at RF doesn't offer enough surface area to guarantee a low-impedance connection to the Earth. Several interconnected ground rods (three or four is a good compromise) make a much better RF connection. [E9D15] Because RF does not penetrate far into the ground, an RF ground does not require the full eight-foot rods required for electrical safety. Shorter rods or pipes will do. (See *The ARRL Handbook* for more information about safety and RF grounding.)

Before you go on, study test questions E3C07, E3C10, E9A12, E9C11, E9C13, E9D14 and E9D15. Review this section if you have difficulty.

SHORTENED AND MULTIBAND ANTENNAS

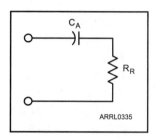

Figure 9-7 — At frequencies below its resonant frequency, the feed point impedance of a whip antenna can be represented as a capacitive reactance, C_A, in series with the antenna's radiation resistance, R_R.

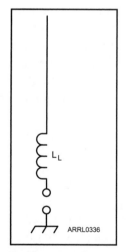

Figure 9-8 — The capacitive reactance of a whip antenna can be cancelled by adding an equivalent amount of inductive reactance as a loading coil in series with the antenna.

It would be just terrific if we could all have a full-sized antenna on every band, far above the ground, but this is rarely possible. Life is full of compromises and a typical antenna system is a good example! Two of the most common antenna aspects requiring compromise, especially on the HF bands, are the number of separate antennas that can be erected and their length. To address the length issue, amateurs use a number of techniques to shorten the antennas while still maintaining acceptable electrical performance. There are also a number of ways to make a single physical structure work well on a number of different frequencies. The next two sections give a few examples of common techniques and the tradeoffs they require.

Loaded Whips

The most difficult place to achieve effective HF antenna performance is a mobile station. Not only is the antenna system exposed to the mobile environment's vibration, temperature extremes and corrosion, but the antenna's size is quite limited by the constraints on vehicle size and maneuverability. The 10 and 12 meter bands are the only bands on which "full-sized" ¼-wavelength ground-plane antennas can realistically be used for a mobile station.

Practically, mobile antennas for HF are almost all some variation on the *whip* — a flexible, vertical conductor attached to the vehicle with a threaded or magnetic mount. The vehicle acts as a ground-plane for the antenna. Whips are usually 8 feet or less in length. At 21 MHz and lower frequencies, the antenna is "electrically short," meaning less than ¼-wavelength long. As the operating frequency is lowered, the feed point impedance of such an antenna is a decreasing radiation resistance in series with an increasing capacitive reactance as shown by the equivalent circuit in **Figure 9-7**. [E9D13] A full-size ¼-wavelength whip's radiation resistance is approximately 36 Ω.

To tune out the capacitive reactance and resonate the antenna, a series inductive reactance, or *loading coil* is used. (Remember that resonance occurs when the feed point impedance is entirely resistive.) [E9D11] The amount of inductance required is determined by the desired operating frequency and where the coil is placed along the antenna. **Figure 9-8** shows the loading coil as an inductance in series with the whip. The tradeoff of using loading coils in a shortened antenna is that the SWR bandwidth of the antenna is reduced. [E9D08]

Base loading (placing the loading coil at the feed point, assumed to be at the base of the antenna) requires the lowest value of inductance for a given antenna length. As the coil is moved along the whip farther from the feed point, the required amount of inductance increases. This is because the amount of capacitive reactance increases as the feed point moves closer to the end of the whip. The addition of the resonating inductance has one drawback in that SWR bandwidth of the antenna system is reduced from that of a full-size, ¼-wavelength antenna. The reduction in bandwidth occurs because reactance of the tuned system increases more rapidly away from the resonant

Loading Coil Construction

Constructing suitable loading coils for typical antenna lengths becomes more difficult at lower frequencies. Since the required resonating inductance gets larger and the antenna's radiation resistance decreases at lower frequencies, most of the power may be dissipated in the coil resistance and other ohmic losses. For example, once the antenna is tuned to resonance, the radiation resistance for an 8-foot whip drops from nearly 15 Ω on 15 meters to 0.1 Ω on 160 meters! When loading coil resistance and other losses are included, the feed point impedance increases to approximately 20 Ω on 160 meters and 16 Ω on 15 meters. And these values are for relatively high-efficiency systems! From this, you can see that the antenna's efficiency is much poorer on 160 meters than on 15 meters under typical conditions.

This is one reason why it is advisable to build or buy the best quality loading coil with the highest power rating possible, even though you may only be considering low-power operation. Percentage-wise, the coil losses in the higher power loading coils are usually less, with subsequent improvement in radiating efficiency, regardless of the power level used.

The primary goal is to provide a coil with the highest possible Q, meaning a high ratio of reactance to resistance, so that heating losses will be minimized. High-Q coils require a large conductor, "air-wound" construction, large spacing between turns, the best insulating material available, a diameter not less than half the length of the coil (this is not always mechanically feasible) and a minimum of metal structures or surfaces near the coil.

frequency than does the feed point reactance of a non-tuned antenna.

One advantage of placing the coil at least part way up the whip, however, is that the current distribution along the antenna is improved, and that increases the radiation resistance. The major disadvantage is that the requirement for a larger loading coil means that the coil losses will be greater, although this is offset somewhat by lower current through the larger coil and is the reason loading coils should have a high Q (ratio of reactance to resistance). [E9D06] Assuming a high-Q coil, center loading offers the best compromise for minimizing losses in an electrically-short vertical antenna. [E9D05]

Figure 9-9 shows a typical bumper-mounted, center-loaded whip antenna suitable for operation in the HF range. The antenna could also be mounted directly on the car body (such as a fender or trunk lid). The base spring acts as a shock absorber for the base of the whip, since continual flexing would weaken the antenna. A short, heavy, mast section is mounted between the base spring and loading coil. Some models have a mechanism that allows the antenna to be tipped over for adjustment or for fastening to the roof of the car when not in use. Optional guy lines can be used to stabilize the antenna while in motion.

A word of caution — the bumper mount technique assumes that the bumper (or its mounting brackets) are metal and bonded to the rest of the car's body. With modern cars increasingly made from nonconductive materials, it is important to be sure that whatever mounting technique is used makes good connection to as much of the vehicle's metal frame and surfaces as possible!

Other common types of mobile HF antennas include a magnetic-mounted, two-section tubular fiberglass base helically wound with the antenna conductor and topped with a short metal whip. The entire base becomes the loading coil. These inexpensive antennas work on a single band, requiring multiple antennas to be carried for operation on multiple bands. They give good performance for their modest price.

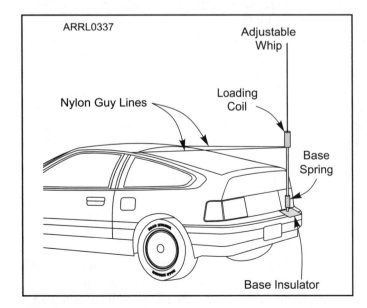

Figure 9-9 — A drawing showing a typical bumper-mounted HF-mobile antenna. Optional guy lines help stabilize the antenna while the vehicle is in motion.

At the other end of the price scale are the tunable "screwdriver" antennas similar to that in Figure 9-9 but for which the loading coil inductance is adjusted from inside the vehicle. (The name derives from the small dc motor used to tune the coil, similar to those found in electric screwdrivers.)

Losses in the loading coil can be reduced if the required loading coil inductance is reduced, allowing a smaller coil. To use a smaller coil, the capacitive reactance that must be tuned out must also be reduced. One method of decreasing capacitive reactance is to increase capacitance. The *top loading* method is one such technique.

Top loading adds a "capacitive hat" above the loading coil, either just above the coil or near the top of the whip. The "hat" usually consists of short wires perpendicular to the whip, often with the ends of the wires connected by a metal ring for additional strength. The added capacitance reduces the resonating value of inductance and the size of the loading coil. Using a smaller loading inductor reduces the loading coil's resistive loss and improves the antenna radiation efficiency. [E9D09]

Trap Antennas

Loading coils resonate an electrically short antenna but this only works on a single band. Using a single antenna on multiple bands requires a different approach. By using tuned circuits called *traps* strategically placed in a dipole, the antenna can be made resonant and used as a multiband antenna at a number of different frequencies. [E9D12] The general principle is illustrated by **Figure 9-10** for a trap antenna that acts like a half-wavelength dipole on three bands — 20, 15 and 10 meters.

The rectangles in the figure represent the traps, which are parallel-LC circuits. At their resonant frequency, f_0, they act as a very high impedance, effectively an open circuit. Above f_0 the traps are capacitive, and below f_0 they are inductive. The traps labeled 1 are identical as are the traps labeled 2. (The name "trap" stems from their appearance. They are often constructed from large-diameter open coils with the capacitor and a strengthening insulator inside, appearing to be "trapped" inside the coil.)

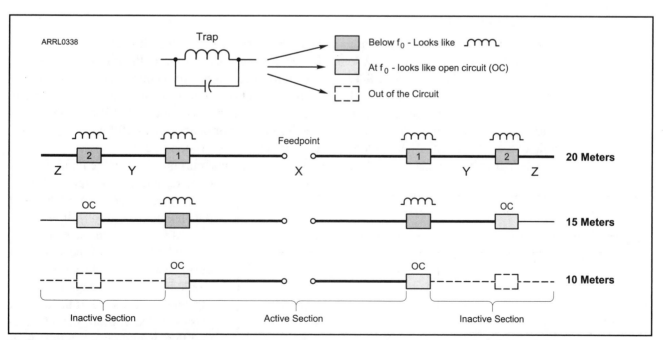

Figure 9-10 — A tri-band antenna for 20, 15 and 10 meters can be constructed using traps. On the resonant band for each trap, it acts like an open-circuit, isolating the outer portions of the antenna. Below resonance, the trap is inductive, acting as a loading coil.

The inner length of wire, X, acts like a resonant ½-λ dipole on the highest frequency of operation, 10 meters. The traps labeled 1 are resonant at 10 meters and electrically disconnect the remainder of the antenna. On 10 meters the only active portion of the antenna is the inner section.

On the next lowest frequency band, 15 meters, trap 1 is now below its resonant frequency and presents inductive reactance, acting like a loading coil in a whip antenna. Trap 2 is resonant on 15 meters and disconnects the outer length of wire, Z, from the antenna. The combination of wire X, the inductive reactance of trap 1, and wire Y are active as a resonant ½-λ dipole on 15 meters.

Finally, on 20 meters, both traps 1 and 2 are below resonance and act as loading coils. All three wire sections of the antenna — X, Y and Z — are active, creating a ½-λ antenna on 20 meters. The loading effect of the traps also means that the overall antenna length will be somewhat shorter than a full-size ½-λ antenna on the 20 and 15 meter bands, an added bonus of the trap technique.

Since the tuned circuits have some inherent losses, the efficiency of this antenna depends on the Q of the traps. Low-loss (high-Q) coils should be used, and the capacitor losses likewise should be kept as low as possible.

Trap dipoles and beam antennas have two major disadvantages. Because they are a multiband antenna, they will do a good job of radiating any harmonics present in the transmitter output. [E9D07] Further, during operation on the lower frequency bands, the series inductance (loading) from the traps raises the antenna Q. Just as for a loaded whip antenna, that means their SWR bandwidth is lower than for a full-size antenna.

Quarter-wavelength vertical antennas can also use trap construction. Only one trap is needed for each band in that case, since only the vertical portion of the antenna uses traps. The ground system under the antenna does not use traps.

Before you go on, study test questions E9D05 through E9D09 and E9D11 through E9D13. Review this section if you have difficulty.

FOLDED DIPOLE

A *folded dipole* antenna is a wire antenna made from a 1-wavelength long wire that is formed into a very thin loop, ½-λ long. [E9A07] The antenna is fed in the middle of one side of the loop as shown in **Figure 9-11**. The antenna is ½-λ long from end to end, thus the name "folded dipole." The folded dipole has the same directional characteristics as a regular dipole, but its feed point impedance is four times that of a regular dipole and its SWR bandwidth is wider than for the single-conductor antenna. The higher impedance is useful when it is desirable to feed the antenna with ladder line or twin lead transmission line.

The higher impedance results from the division of antenna current between the multiple conductors. Without the upper conductor between B and C in Figure 9-11, the antenna is a simple dipole with current flowing in the same direction in both halves of the antenna during each half cycle. Current would be zero at the end points of the antenna, B and C. If the antenna was extended beyond B and C, current flow would reverse and flow in the opposite direction because RF is ac current.

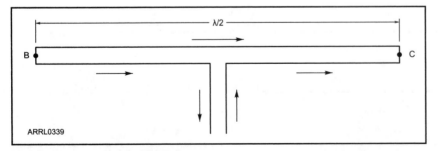

Figure 9-11 — The current distribution on the multiple conductors of a folded dipole. The arrows indicate the direction of current flow during one half-cycle of applied current.

Now imagine that instead of the antenna being extended beyond B and C in a straight line, the extensions are instead folded at B and C and brought together to make an additional conductor between B and C. Presto! The folded dipole! Now, during each half cycle, the direction of current flow is shown by the arrows in the figure. Just remember that current flow reverses in the conductor at B and C.

Why does the antenna impedance change? First, the power input to the folded dipole is just the same as to the single-conductor dipole so no extra power is created or available. That means the total current flowing in the antenna's two conductors must be the same as the current in the regular dipole's single conductor. One-half of the current flows in each conductor.

Comparing the single-conductor and folded dipoles, the power, P, going into the feed point of either antenna is the same. Ignoring the small ohmic loss resistance, for the single-conductor dipole:

$$P = I^2 R_R$$

where:
    I is the feed point current.
    R_R is the radiation resistance of the single-conductor dipole, about 73 Ω.

For the folded dipole, where feed point current is half that of the single-conductor dipole:

$$P = \left(\frac{I}{2}\right)^2 \times R_{FD} = \frac{I^2}{4} \times R_{FD}$$

where R_{FD} is the folded dipole's radiation resistance.

Because P is the same for both antennas, $I^2 R_R = (I^2 / 4) \times R_{FD}$ and that means $4R_R = R_{FD}$. The folded dipole's feed point impedance must be four times that of the single-conductor dipole and $R_{FD} = 4 \times 73 = 292 \ \Omega$, a fairly close match to 300-Ω twin lead or open-wire transmission line. [E9D10]

Feed point impedance (which is mostly radiation resistance) can be increased further by adding additional conductors. The resulting $R_{FD} = 73 \times N^2$, where N is the number of parallel conductors. A three-conductor folded dipole's feed point impedance will be close to $9 \times 73 = 657 \ \Omega$ and so forth. Feed point impedances this high are rarely used for center-fed dipoles in Amateur Radio.

> Before you go on, study test questions E9A07 and E9D10. Review this section if you have difficulty.

TRAVELING WAVE ANTENNAS

So far, the antennas you've studied for your license exams have been resonant antennas, based on ½-λ or ¼-λ elements. Another class of antennas lets the power flow along elements up to several wavelengths long, being dissipated as it goes. These are *traveling wave* antennas.

The traveling wave antenna is the *long-wire* antenna. Long wires, as the name implies, are just, well, long wires — one wavelength long or longer. They can be fed anywhere along their length, but typically ¼-λ from one end as a dipole with one end extended. The radiation pattern for such an antenna is shown at the left in **Figure 9-12**. The feed point impedance of the long-wire antenna varies dramatically as a consequence of the changing current patterns with frequency.

The long wire has four major lobes (and several minor lobes). The longer the wire, the closer to the direction of the wire the lobes become. If two long wires are combined as shown in the figure, and fed out-of-phase (as with a single transmission line where the wires

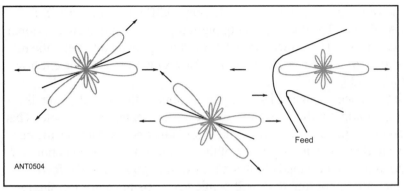

Figure 9-12 — Two long wires and their radiation patterns are shown at the left. If the two are combined to form a V with the major lobes aligned and fed out of phase, the resulting patterns reinforce the major lobes of each wire along the axis of the antenna.

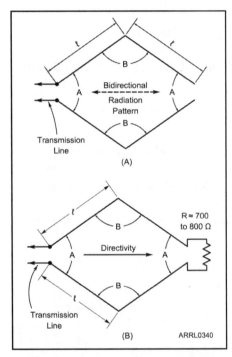

Figure 9-13 — The resonant rhombic at A is a diamond-shaped antenna. All legs are the same length and opposite angles of the diamond are equal. The radiation pattern is bidirectional as shown. Adding a terminating resistor as shown at B makes the pattern unidirectional with no loss of gain in the preferred direction.

come together) their major lobes will coincide in two directions, creating the pattern shown at the right of the figure. This is the antenna known as the *Vee beam*. The open wire ends reflect any power that hasn't been dissipated back toward the feed point, creating the "backward" major lobe of the radiation pattern.

Rhombic Antennas

Vee beams can be combined, as well, creating the *rhombic antenna*, shown in **Figure 9-13**. The diamond-shaped rhombic antenna can be considered as two Vee beams placed end-to-end; it has four equal-length legs and the opposite angles are equal so the antenna is symmetrical. Each leg is at least one wavelength long. Rhombic antennas are installed horizontally with supports at the four corners.

Rhombics are usually characterized as high-gain antennas. They may be used over a wide frequency range, and the directional pattern is essentially the same over the entire range. This is because a change in frequency causes the major lobe from one leg to shift in one direction, while the lobe from the opposite leg shifts the other way, keeping their sum more or less the same. There are two major types of rhombic antennas: *resonant* and *nonresonant*.

The radiation pattern of a resonant rhombic is bidirectional as shown by the arrows in Figure 9-13. The antenna is bidirectional, although the pattern is not symmetrical with slightly higher gain in the direction of the end opposite the feed point. There are minor lobes in other directions; their number and strength depend on the length of the legs. Notice that the wires at the end opposite the feed point are open. Just as for Vee beams, the open wire ends reflect power that's not dissipated back toward the feed point and create the second major lobe. [E9C04]

This antenna has the advantage of simplicity as compared to other rhombics. Input impedance varies considerably with input frequency, as it would with any long-wire antenna. Resonant rhombic antennas are not widely used.

As you can see in Figure 9-13B, a terminating resistor has been added to the resonant rhombic to form the *terminated rhombic* or *nonresonant rhombic*. While the unterminated or resonant rhombic is always bidirectional, the main effect of adding a terminating resistance is to change the pattern to unidirectional by absorbing the power that would have been reflected to create the unwanted second major lobe. [E9C06] The loss of power as heat (about a third of the input power) does not result in a loss in the desired direction, however.

Although there is no marked difference in the gain attainable with resonant and terminated rhombics of comparable design, the terminated antenna has the advantage that it presents an essentially resistive and constant load over a wide frequency range. In addition, the antenna's unidirectional pattern has a good front-to-back ratio.

There aren't many amateurs using rhombic antennas. The primary disadvantage of these antennas is that they require a very large area, since each leg should be a minimum of one wavelength. Four tall, sturdy supports are required for a single antenna, and the direction of radiation from it can't be changed readily. [E9C05] For those lucky enough to be able to install a rhombic, the performance is exceptional in the preferred direction. It is not for nothing that the ARRL symbol is the rhombic's diamond!

Beverage Antennas

Nearly every antenna installed by radio amateurs is used for both receiving and transmitting. One common exception is the traveling wave receive antenna invented by H.H. Beverage and shown in **Figure 9-14**. The Beverage antenna acts like a long transmission line, with one lossy conductor (the Earth) and one good conductor (the wire). As for the nonresonant rhombic antenna, a Beverage antenna has a terminating resistor to ground at the end farthest from the radio. Like the rhombic, the terminating resistor absorbs the power of signals arriving from the unwanted direction instead of allowing them to be reflected back toward the feed point.

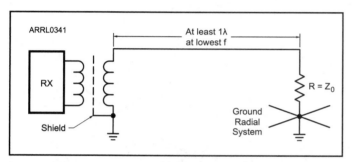

Figure 9-14 — The one-wire Beverage antenna forms a transmission line with the ground and is terminated at the far end. The antenna's preferred direction is to the right in this drawing.

Beverage antennas are effective directional antennas for 160 meters and 80 meters. They are less effective at higher frequencies, however, and are seldom used on 40 meters and shorter wavelength bands. Beverage antennas should be at least one wavelength long at the lowest operating frequency. [E9C12] Longer antennas provide increased gain and directivity. Beverage antennas are installed at relatively low heights, normally 8 to 10 feet above ground. They should form a relatively straight line extending from the feed point toward the preferred direction.

The Beverage's chief benefit is in rejecting noise from unwanted directions, not in having high gain. In fact, the Beverage is rather lossy! The improvement in signal-to-noise ratio, however, can be dramatic. Vertically-polarized signals arriving at low vertical angles along the antenna cause current to flow as they pass along the antenna. Signals arriving from other directions do not create as much signal in the antenna. Horizontally polarized signals from any direction also fail to build up much signal. Thus, the Beverage is able to reject atmospheric static, which is fierce at low frequencies!

Before you go on, study test questions E9C04, E9C05, E9C06 and E9C12. Review this section if you have difficulty.

PHASED ARRAYS

Various pattern shapes can be obtained using an antenna system that consists of two vertical antennas fed with various phase relationships. These are examples of *phased arrays* in which the phase differences of the signals that the antennas receive or transmit create the desired radiation pattern.

Figure 9-15 illustrates the basics of how a phased array pattern is created. In Figure 9-15A, a single antenna (oriented so that the current, I, is flowing perpendicularly to the page) is radiating a signal as shown by the different circles, representing the positive and negative peaks of the signal. When a second antenna is added at Figure 9-15B, the spacing of the antennas

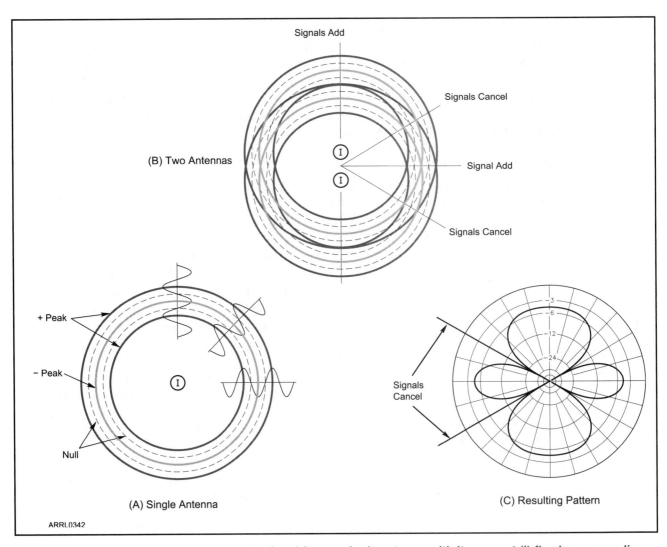

Figure 9-15 — (A) shows the signal being radiated from a single antenna with its current (I) flowing perpendicularly to the page. At (B) a second antenna has been added and the radiated signals reinforce and cancel along different directions, leading to the radiation pattern for the array in (C).

causes the signals to reinforce at some angles to the array and cancel at others. The resulting radiation pattern is shown at Figure 9-15C.

The relative phase between the antennas can be varied by changing their physical spacing or by changing the electrical phase of the currents that feed the antenna. **Figure 9-16** shows patterns for a number of common spacings and current phases. The patterns assume that the antennas are identical ¼-wavelength verticals and are fed with equal magnitudes of current. As in Figure 9-15, the currents in both antennas are perpendicular to the page. The antennas themselves are aligned along a vertical line from top to bottom of the figure. The antenna toward the top of the pattern has the lagging current phase.

By studying the patterns shown in Figure 9-16 you can see that when the two antennas are fed in phase, a pattern that is *broadside* to the elements always results. If the antennas are ¼ wavelength apart and fed in phase, the pattern is elliptical, like a slightly flattened circle. This system is substantially omnidirectional. If the antennas are ½ wavelength apart and fed in phase the pattern is a figure-8 that is broadside to the antennas. [E9C03]

At spacings of less than ⅝ wavelength, with the elements fed 180° out of phase, the maximum radiation lobe is in line with the antennas. This is an *end-fire array*. For example, if the antennas are ⅛ wavelength, ¼ wavelength or ½ wavelength apart and fed 180° out of phase

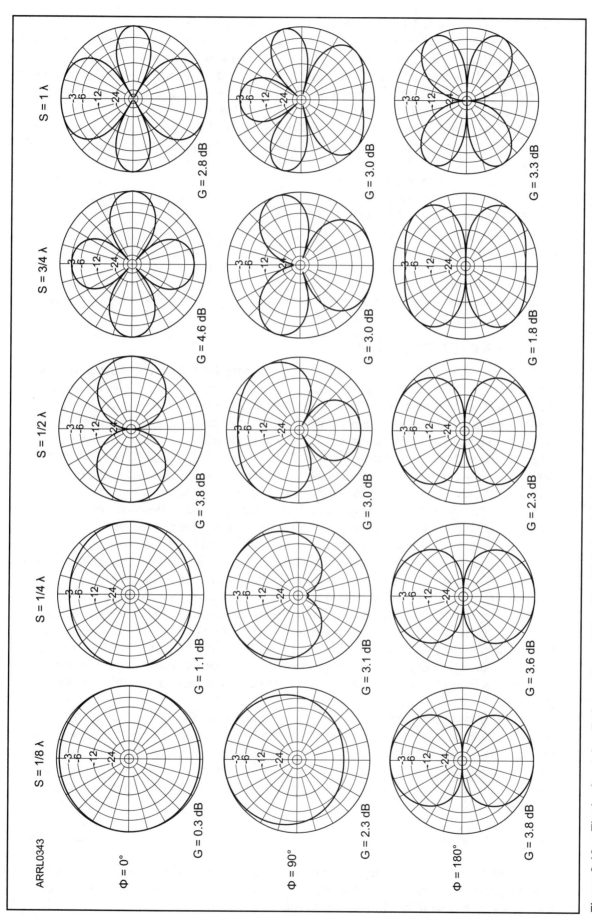

ARRL0343

S = 1/8 λ   S = 1/4 λ   S = 1/2 λ   S = 3/4 λ   S = 1 λ

Φ = 0°
G = 0.3 dB   G = 1.1 dB   G = 3.8 dB   G = 4.6 dB   G = 2.8 dB

Φ = 90°
G = 2.3 dB   G = 3.1 dB   G = 3.0 dB   G = 3.0 dB   G = 3.0 dB

Φ = 180°
G = 3.8 dB   G = 3.6 dB   G = 2.3 dB   G = 1.8 dB   G = 3.3 dB

Figure 9-16 — The horizontal radiation patterns of two identical ½-λ dipole or ¼-λ ground plane antennas, physically spaced as indicated by S and with relative phases of their identical feed point currents indicated by φ. The two elements are aligned along a vertical line from top to bottom of the figure and the antenna toward the top of the pattern has the lagging current phase. The gain values, G, are with respect to that of a single antenna.

the pattern is a figure-8 that is in line with the antennas. [E9C01]

With intermediate amounts of phase difference, the results cannot be stated so simply. Patterns evolve that are not symmetrical in all four quadrants. If the antennas are ¼-wavelength apart and fed 90° out of phase, an interesting pattern results. (See the middle pattern of the second row from the left in Figure 9-16.) This is a *unidirectional cardioid* pattern. [E9C02] A more complete table of patterns is given in *The ARRL Antenna Book*.

These arrays are constructed by using physically identical antennas that are fed with *phasing lines* that create the necessary phase differences between them. This ensures that each element radiates a signal with the necessary phase to create the desired antenna pattern. [E9E12] When the antennas are identical and being fed with identical currents, such as any of the "in-phase" designs with $\phi = 0°$, a *Wilkinson power divider* can be used to split the power from the transmitter into equal portions while preventing changes in the loads from affecting power flow to the other loads. [E9E13]

Before you go on, study test questions E9C01, E9C02, E9C03, E9E12 and E9E13. Review this section if you have difficulty.

EFFECTIVE RADIATED POWER

When evaluating total station performance, accounting for the effects of the entire system is important, including antenna gain. This allows you to evaluate the effects of changes to the station. Transmitting performance is usually computed as *effective radiated power* (*ERP*). ERP is calculated with respect to a reference antenna system — usually a dipole but occasionally an isotropic antenna — and answers the question, "How much power does my station radiate as compared to that if my antenna was a simple dipole?" Effective isotropic radiated power (EIRP) results when an isotropic antenna is used as the reference. If no antenna reference is specified, assume a dipole reference antenna.

ERP is especially useful in designing and coordinating repeater systems. The effective power radiated from the antenna helps establish the coverage area of the repeater. In addition, the height of the repeater antenna as compared to buildings and mountains in the surrounding area (*height above average terrain*, or *HAAT*) has a large effect on the repeater coverage. In general, for a given coverage area, with a greater antenna HAAT, less effective radiated power (ERP) is needed. A frequency coordinator may even specify a maximum ERP for a repeater, to help reduce interference between stations using the same frequencies.

ERP calculations begin with the *transmitter power output* (*TPO*). (This is assumed to be the output of the final power amplification stage if an external power amplifier is used.) Then the *system gain* of the entire antenna system including the transmission line and all transmission line components is applied to TPO to compute the entire station's output power. [E9H04]

There is always some power lost in the feed line and often there are other devices inserted in the line, such as a filter or an impedance-matching network. In the case of a repeater system, there is usually a duplexer so the transmitter and receiver can use the same antenna and perhaps a circulator to reduce the possibility of intermodulation interference. These devices also introduce some loss to the system. The antenna system then usually returns some gain to the system.

ERP = TPO × System Gain   (Equation 9-4A)

Since the system gains and losses are usually expressed in decibels, they can simply be added together, with losses written as negative values. System gain must then be converted back to a linear value from dB to calculate ERP.

$$ERP = TPO \times \log^{-1}\left(\frac{\text{system gain}}{10}\right)$$ (Equation 9-4B)

It is also common to work entirely in dBm and dB until the final result for ERP is obtained and then converted back to watts.

ERP (in dBm) = TPO (in dBm) + System Gain (in dB)   (Equation 9-4C)

Suppose we have a repeater station that uses a 50-W transmitter and a feed line with 4 dB of loss. There is a duplexer in the line that exhibits 2 dB of loss and a circulator that adds another 1 dB of loss. This repeater uses an antenna that has a gain of 6 dBd. Our total system gain looks like:

System gain = –4 dB + –2 dB + –1 dB + 6 dBd = –1 dB

Note that this is a loss of 1 dB total for the system from TPO to radiated power. The effect on the 50 W of TPO results in:

$$ERP = 50 \text{ W} \times \log^{-1}\left(\frac{\text{system gain}}{10}\right) = 50 \times \log^{-1}(-0.1) = 50 \times 0.79 = 39.7 \text{ W}$$

This is consistent with the expectation that with a 1 dB system loss we would have somewhat less ERP than transmitter output power.

As another example, suppose we have a transmitter that feeds a 100 W output signal into a feed line that has 1 dB of loss. The feed line connects to an antenna that has a gain of 6 dBd. What is the effective radiated power from the antenna? To calculate the total system gain (or loss) we add the decibel values given:

System gain = – 1 dB + 6 dBd = 5 dB

and

$$ERP = 100 \text{ W} \times \log^{-1}\left(\frac{\text{system gain}}{10}\right) = 100 \times \log^{-1}(0.5) = 100 \times 3.16 = 316 \text{ W}$$

The total system has positive gain, so we should have expected a larger value for ERP than TPO. Keep in mind that the gain antenna concentrates more of the signal in a desired direction, with less signal in undesired directions. So the antenna doesn't really increase the total available power. If directional antennas are used, ERP will change with direction.

Example 9-4

What is the effective radiated power of a repeater station with 150 watts transmitter power output, 2 dB feed line loss, 2.2 dB duplexer loss and 7 dBd antenna gain? [E9H01]

System gain = –2 dB – 2.2 dB + 7 dBd = 2.8 dB

$$ERP = 150 \text{ W} \times \log^{-1}\left(\frac{\text{system gain}}{10}\right) = 150 \times \log^{-1}(0.28) = 150 \times 1.9 = 285 \text{ W}$$

Example 9-5

What is the effective radiated power of a repeater station with 200 watts transmitter power output, 4 dB feed line loss, 3.2 dB duplexer loss, 0.8 dB circulator loss and 10 dBd antenna gain? [E9H02]

System gain = –4 – 3.2 – 0.8 + 10 = 2 dB

$$ERP = 200 \text{ W} \times \log^{-1}\left(\frac{\text{system gain}}{10}\right) = 200 \times \log^{-1}(0.2) = 200 \times 1.58 = 317 \text{ W}$$

Example 9-6

What is the effective isotropic radiated power of a repeater station with 200 watts transmitter power output, 2 dB feed line loss, 2.8 dB duplexer loss, 1.2 dB circulator loss and 7 dBi antenna gain? [E9H03]

System gain = –2 – 2.8 – 1.2 + 7 = 1 dB

$$EIRP = 200 \text{ W} \times \log^{-1}\left(\frac{\text{system gain}}{10}\right) = 100 \times \log^{-1}(0.1) = 200 \times 1.26 = 252 \text{ W}$$

Before you go on, study test questions E9H01 through E9H04. Review this section if you have difficulty.

SATELLITE ANTENNA SYSTEMS

Previous chapters discuss Amateur Radio satellites and their operation. In this section, we consider the special requirements for antennas used in space radio communications. These include gain, polarization and aiming abilities.

Gain and Antenna Size

Before discussing gain, you should be aware that communicating through amateur satellites doesn't take tremendous amounts of gain and huge, expensive antennas. Some of the satellites can be worked with a handheld VHF/UHF transceiver! Nevertheless, for consistent results and to access the widest variety of satellites a steerable antenna system with some gain is required.

You might think more gain is better. That may be true when you are transmitting in a fixed direction. If you are trying to communicate through a rapidly moving satellite, however, it is not true. The sharper pattern of a higher-gain antenna can cause aiming problems, so an antenna with less gain — and the wider beamwidth that goes with it — may be more to your advantage. You'll want to find the right compromise between gain and beamwidth.

"The larger the antenna (in wavelengths) the greater the gain" is a good guideline. At VHF and UHF and even the lower microwave frequencies, Yagi-style antennas are the most common for satellite contacts. For a properly designed Yagi, that means the longer the boom, the greater the gain.

Sufficient gain for most satellites can be obtained from a single antenna (subject to polarization effects discussed below) on each satellite band. At microwave frequencies, though, a dish antenna may be required because of the higher path loss at these frequencies and the difficulty of constructing Yagis with enough gain.

Parabolic dish antennas are so named because the cross section of the dish is a parabola with the feed point at its focus. While these antennas can be quite large at frequencies of more than a few GHz, at 10 GHz and up, they are quite reasonably sized. For example, at 10 GHz a typical dish antenna need only be a bit more than a foot wide to develop substantial amounts of gain. The gain of a parabolic antenna is directly proportional to the square of the dish diameter and directly proportional to the square of the frequency. That means the gain will increase by 6 dB if either the dish diameter or the operating frequency is doubled. [E9D01]

What About Polarization?

Best results in space radio communication are obtained not by using horizontal or vertical polarization, but by using a combination of the two called *circular polarization* as discussed in the AC Waveforms and Measurements chapter. When two equal signals, one horizontally polarized and one vertically polarized, are combined with a phase difference of 90°, the result is a circularly polarized wave.

A circularly polarized antenna can be constructed from two dipoles or Yagis mounted at 90° with respect to each other and fed 90° out of phase. [E9D02] **Figure 9-17** shows an example of a circularly polarized antenna made from two Yagi antennas. The two driven elements must be at the same position along the boom for this antenna. The driven elements are in the same plane, which is perpendicular to the boom and to the direction of maximum signal.

There is one other antenna factor that you should be aware of if you plan to operate through a satellite. For terrestrial communications, beam antennas are mounted with the boom parallel to the horizon, and a rotator turns the antenna to any desired azimuth. As a satellite passes your location, however, it may be high above the horizon for part of the time. The antenna's gain at that vertical angle may be quite low. To point your antenna at the satellite, then, you will also have to be able to elevate it above horizontal. A second rotator is used to change the antenna elevation angle. The combination of rotation in azimuth and elevation, called *az-el*, tracks the satellite as it orbits the Earth which is desirable because the antenna's full gain is always directed at the satellite. [E9D04]

Before you go on, study test questions E9D01, E9D02 and E9D04. Review this section if you have difficulty.

RECEIVING LOOP ANTENNAS

A simple receiving antenna at MF and HF is a small loop antenna consisting of one or more turns of wire wound in the shape of a large open inductor or coil. [E9H09] The loop is usually tuned to resonance with a capacitor. The loop must be small compared to the wavelength — in a single-turn loop, the conductor should be less than 0.08 wavelength long.

Loops used for receiving purposes (including direction-finding as described below) are used for their nulls rather than their gain, which is often quite low. The nulls can be used to reject a local source of noise or interference or they can

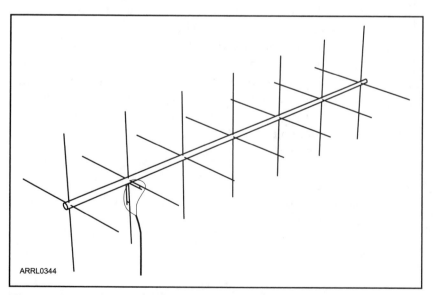

ARRL0344

Figure 9-17 — Two Yagi antennas built on the same boom, with elements placed perpendicular to each other create a circularly polarized antenna. The driven elements are located at the same position along the boom and are fed 90° out of phase.

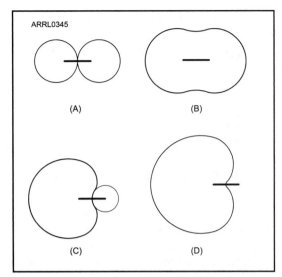

Figure 9-18 — The radiation patterns for a small loop antenna. The heavy line shows the plane of the loop. (A) shows the pattern for an ideal loop and (B) modifies the pattern showing the pattern with appreciable antenna effect present. (C) shows the pattern of a detuned loop with shifted phasing. In (D) the detuning is optimum to produce a single null.

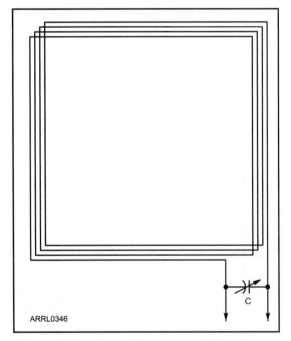

Figure 9-19 — A small loop consisting of several turns somewhat improved sensitivity. The total conductor length must still be less than 0.08 λ to have the same pattern as the single-turn loop.

be used to indicate the direction to a signal source.

An ideal loop antenna has maximum response in the plane of the loop, as **Figure 9-18** shows. The ideal loop has deep nulls at right angles to its plane. Because there are two nulls, the pattern is bidirectional.

If the loop is not balanced with respect to ground (meaning that voltages from vertically polarized electric fields will not cancel) it will exhibit two modes of operation. One is the mode of the true loop while the other is that of an omnidirectional, small vertical antenna. This second mode of operation is sometimes called *antenna effect*. The voltages picked up by the two modes are not in phase and may add or subtract, depending on the direction from which the wave is coming.

When the antenna effect is appreciable and the loop is tuned to resonance, the loop may exhibit little directivity as in Figure 9-18B. By detuning the loop to shift the phasing, a pattern similar to Figure 9-18C may be obtained. This pattern does exhibit a pair of nulls, although they are not symmetrical. The nulls may not be as sharp as those obtained with a well-balanced loop, and they may not be at right angles to the plane of the loop.

With suitable detuning, a unidirectional pattern may be approached as in Figure 9-18D. There is no complete null in the pattern but the loop is adjusted for the best null. An electrostatic balance can be obtained by shielding the loop. This eliminates the antenna effect and the response of a well-constructed shielded loop is quite close to the ideal pattern of Figure 9-18A.

For the lower-frequency amateur bands, single-turn loops are generally not sensitive enough to be effective. Therefore, multi-turn loops, such as shown in **Figure 9-19**, are generally used. This loop may also be shielded and if the total conductor length remains below 0.08 wavelengths, the pattern is that of the single-turn loop.

The voltage generated by the loop is proportional to the strength of the magnetic component of the radio wave passing through the coil, and to the number of turns in the coil. The action is much the same as in the secondary winding of a transformer. The output voltage of the loop can be increased by increasing the number of turns in the loop or the loop area. [E9H10]

Before you go on, study test questions E9H09 and E9H10. Review this section if you have difficulty.

DIRECTION-FINDING AND DF ANTENNAS

Radio direction finding (RDF) is as old as radio itself. RDF is just what the name implies — finding the direction or location of a transmitted signal. The practical aspects of RDF include radio navigation, location of downed aircraft and identification of sources of signals that are jamming communications or causing radio-frequency interference. In many countries, the hunting of hidden transmitters has become a sport, often called *fox hunting*. Participants in automobiles or on foot use receivers and direction-finding techniques to locate a hidden transmitter. RDF has become popular enough that a world championship is held every year, with numerous regional events!

The equipment required for an RDF system is a directive antenna and a device for detecting the radio signal. In Amateur Radio direction finding (ARDF), the signal detector is usually a receiver with a meter to indicate signal strength. Some form of RF attenuation is desirable to allow proper operation of the receiver under high signal conditions, such as when zeroing-in on the transmitter at close range. Otherwise the strong signals may overload the receiver. [E9H07]

The directive antenna can take many forms. In general, an antenna for direction-finding work should have good front-to-back and front-to-side ratios. Since the peak in an antenna pattern is often very broad, it can be difficult to identify the exact direction of a signal by its peak direction. Antenna-pattern nulls are usually very narrow, however. The normal technique for RDF is to use the pattern null to indicate the direction of the transmitted signal. A shielded loop has the additional advantage of being easier to balance with respect to ground, reducing antenna effect and giving deeper, sharper nulls. [E9H12]

The wire-loop antenna is a simple one to construct, but the bidirectional pattern is a major drawback. You can't tell which of the two directions points to the signal source! Thus, a single null reading with a small loop antenna will not indicate the exact direction toward the transmitter — only the line along which it lies. [E9H05]

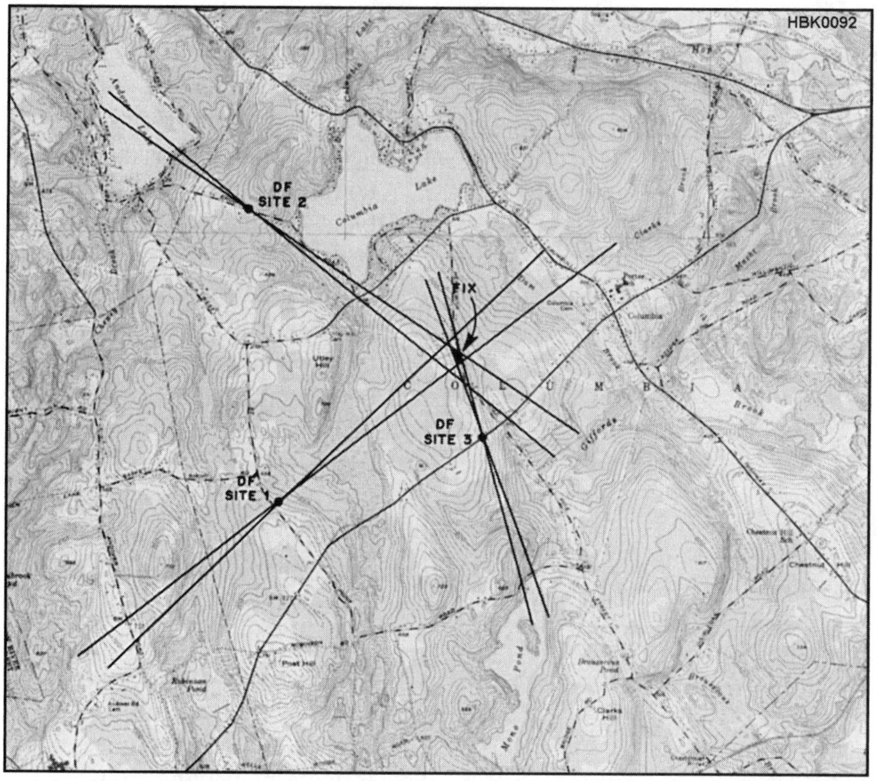

Triangulation

If two or more RDF bearing measurements are made at several locations separated by a significant distance, the bearing lines can be drawn from those positions as represented on a map as in **Figure 9-20**. This technique is called *triangulation*. [E9H06] It is important that the two DF sites not be on the same straight line with the signal you are trying to find. The point where the lines cross (assuming the bearings are not the same or 180° apart) will indicate a "fix" of the approximate transmitter location.

The word "approximate" is used because there is always some uncertainty in the bearings obtained. Propagation and terrain effects may add to the

Figure 9-20 — Bearings from three RDF positions are drawn on a map to perform triangulation of a signal source. This technique allows antennas with bidirectional patterns to be used since the lines on multiple bearings will only intersect at the signal source.

uncertainty. In order to most precisely identify the probable location of the transmitter, the bearings from each position should be drawn as narrow sectors instead of single lines. Figure 9-20 shows the effect of drawing bearings in sectors — the location of the transmitter is likely to be found in the area bounded by the intersection of the various sectors.

Sense Antennas

Because there are two nulls 180° apart in the directional patterns of loop antennas, an ambiguity exists as to which one indicates the true direction of the signal. If there is more than one receiving station, or if the single receiving station takes bearings from more than one position, the ambiguity may be resolved through triangulation.

It is better for the receiving antenna pattern to have just one null, so there is no question about where the transmitter's true direction lies. A loop may be made to have a single null if a second antenna element, called a *sense antenna*, is added. [E9H08] The sense antenna must be omnidirectional, such as a short vertical. If the signals from the loop and the sense antenna are combined with a 90° phase shift between the two, a cardioid radiation pattern results, similar to that of Figure 9-18D. This pattern has a single large lobe in one direction, with a deep, narrow null in the opposite direction. The deep null can help pinpoint the direction of the desired signal.

The loop and sensing-element patterns combine to form the cardioid pattern which has a very sharp single null. [E9H11] For the best null in the composite pattern, the signals from the loop and the sensing antenna must be of equal amplitude so that they can cancel completely in the direction of the null. The null of the cardioid will be 90° away from the nulls of the loop, so it is customary to first use the loop alone to obtain a precise bearing line, then switch in the sensing antenna to resolve the ambiguity.

Terrain Effects

Most ARDF activity is done on the VHF and UHF bands with the signal source nearby. The best accuracy in determining a bearing to a signal source is when the propagation path is over homogeneous terrain, and when only the vertically polarized component of the ground wave is present. (Homogeneous terrain means there are no hills, trees or buildings to block or reflect the signals.) If a boundary exists, such as between land and water, the different conductivities under the ground wave can cause bending (refraction) of the wave front, leading to false bearings. In addition, reflection of RF energy from vertical objects, such as mountains or buildings, can add to the direct wave and cause bearing errors.

The effects of refraction and reflection are shown in **Figure 9-21**. At A, the signal is actually arriving from a direction different from the true direction of the transmitter. This happens because the

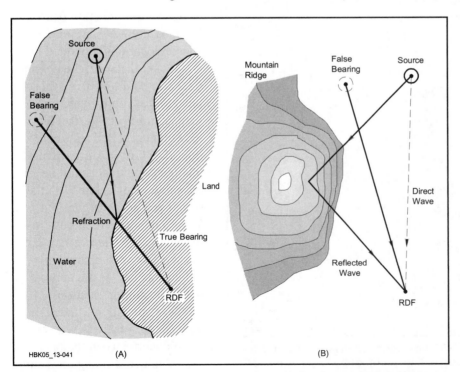

Figure 9-21 — This drawing shows RDF errors caused by refraction (A) and reflection (B).

wave is refracted at the shoreline. Even the most sophisticated equipment will not indicate the true bearing in this instance, as the equipment can only show the direction from which the signal is arriving.

In Figure 9-21B, there are two apparent sources for the incoming signal: a direct wave from the source itself, and another wave that is reflected from the mountain ridge. In this case, the two signals add at the antenna of the RDF equipment. The uninitiated observer would probably obtain a false bearing in a direction somewhere between the directions to the two sources. The experienced operator might notice that the null reading in this situation is not as sharp, or deep, as it usually is, but these indications would be subtle and easy to overlook.

Water towers, tall radio towers and similar objects can also lead to false bearings. The effects of these objects become significant when they are large in terms of a wavelength. Local objects, such as buildings of concrete and steel construction, power lines, and the like, also tend to distort the field. It is important that direction-finding antennas be in the clear, well away from surrounding objects.

> Before you go on, study test questions E9H05 through E9H08, E9H11 and E9H12. Review this section if you have difficulty.

9.3 Antenna Systems

An antenna system is much more than the pieces of metal dangling from a tower or tied off to the local leaf-bearers. It consists of the antenna, the supports, the connection to the feed line, the feed line itself, plus any metering and impedance-matching devices. Some of those are the subjects of this section.

IMPEDANCE MATCHING

It seems that antennas do everything possible to avoid presenting a 50-Ω impedance at their feed point! As many antennas as you can imagine, that's how many different impedances you'll encounter. Through the hard work of antenna designers over the years, there are many techniques for matching transmission lines to antennas. This section covers the basic principles of the delta, gamma, hairpin and stub-matching systems. The *ARRL Antenna Book* contains additional information about these and other impedance-matching techniques.

You may be wondering why impedance mismatches can't simply be corrected by an impedance matching unit at the transmitter. Why is impedance matching at the antenna important? You would be correct that matching the impedance to 50 Ω can be done anywhere along the feed line. Yet there are practical reasons to match the impedances at the antenna.

One reason is to prevent losses in the feed line as the power is reflected back and forth between the transmitter and the antenna. Eventually, the power will be transferred to the antenna, but each trip along the feed line results in some loss. The higher the feed line loss, the greater the additional loss due to impedance mismatches at the antenna. Remember that the impedance matching unit at the transmitter does not reduce the SWR in the feed line to the antenna! The additional loss is usually insignificant at HF, but can be substantial at VHF and higher frequencies.

Another reason is to eliminate the need for an impedance matching unit at the transmitter. Impedance matching hardware at the antenna is often less expensive than a piece of equipment. In a station with multiple antennas, matching their impedances to that of the feed line eliminates the need to retune an impedance matching unit when changing antennas.

These are reasons why it is important to know the antenna's feed point impedance — so that a matching system can be designed. When the antenna is matched to the feed line, maximum power transfer to the antenna is achieved and standing wave ratio is minimized. [E9A04] Station design and operation are also simplified.

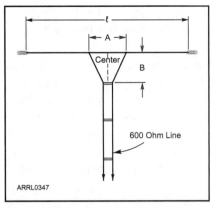

Figure 9-22 — The delta matching system is used to match a high-impedance transmission line to a lower-impedance antenna. The feed line attaches to the driven element in two places, spaced a fraction of a wavelength to each side of the element's center point.

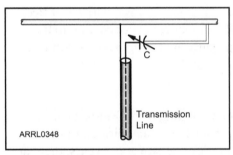

Figure 9-23 — The gamma matching system is used to match an unbalanced feed line to an antenna. The feed line attaches to the center of the driven element and to a point that is a fraction of a wavelength to one side of center.

The Delta Match

If you try to feed a half-wave dipole with an open-wire feed line, you will face a problem. The center impedance of the dipole is too low to be matched directly by any practical type of air-insulated parallel-conductor line. It is possible to find a value of impedance between two points on the antenna that can be matched to an open-wire line when a "fanned" section or *delta match* is used to couple the line and antenna. The antenna is not broken in the center, so there is no center insulator. Also, the delta-match connection is symmetrical about the center of the antenna. This principle is illustrated in **Figure 9-22**. The fanned-out section of feed line is triangular, similar to the Greek letter Δ (delta) that gives the technique its name.

The delta match gives us a way to match a high-impedance transmission line to a lower impedance antenna. The line connects to the driven element in two places, spaced a fraction of a wavelength on each side of the element center. [E9E01] When the proper dimensions are unknown, the delta match is awkward to adjust because both the length and width of the delta must be varied. An additional disadvantage is that there is always some radiation from the delta. This is because the conductors are not close enough together to meet the requirement (for negligible radiation) that the spacing should be very small in comparison with the wavelength.

The Gamma Match

A commonly used method for matching a coaxial feed line to the driven element of a parasitic array is the *gamma match*. Shown in **Figure 9-23**, and named for the Greek letter Γ, the gamma match has considerable flexibility in impedance matching ratio. Because this match is inherently unbalanced, no balun is needed. The gamma match gives us a way to match an unbalanced feed line to an antenna. The feed line attaches at the center of the driven element and at a fraction of a wavelength to one side of center. [E9E02]

Electrically speaking, the gamma conductor and the associated antenna conductor can be considered as a section of transmission line shorted at the end. Since it is shorter than ¼ wavelength the gamma matching section has inductive reactance. This means that if the antenna itself is exactly resonant at the operating frequency, the input impedance of the gamma section will show inductive reactance as well as resistance. The reactance must be tuned out to present a good match to the transmission line. This can be done in two ways. The antenna can be shortened so that its impedance contains capacitive reactance to cancel the inductive reactance of the gamma section, or a capacitance of the proper value can be inserted in series at the input terminals as shown in Figure 9-23. [E9E04]

Gamma matches have been widely used for matching coaxial cable to parasitic beams for a number of years. Because this technique is well suited to all-metal construction, the gamma match has become quite popular for amateur antennas. Gamma matches can also be used to match the impedance at the base of a grounded tower to be used as a vertical antenna. In this application, the driven element is turned on its side with the missing half supplied by the electrical image created by the ground system. [E9E09]

Because of the system's many variables — driven-element length, gamma rod length, rod diameter, spacing between rod and driven element, and value of series capacitance — a number of combinations will provide the desired match. A more detailed discussion of the

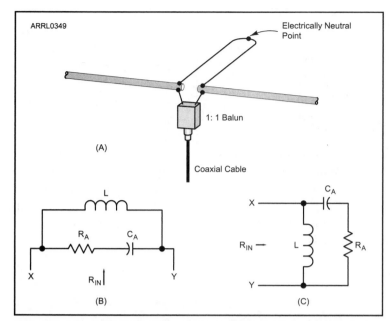

Figure 9-24 — The driven element of a Yagi antenna can be fed with a hairpin matching system, as shown in Part A. B shows the lumped-constant equivalent circuit, where R_A and C_A represent the antenna feed point impedance, and L represents the parallel inductance of the hairpin. Points X and Y represent the feed line connection. When the equivalent circuit is redrawn as shown in (C), L and C_A are seen to form an L-network to match the feed line characteristic impedance to the antenna resistance, R_A.

gamma match can be found in *The ARRL Antenna Book*.

The Hairpin Match

The *hairpin matching system* is a popular method of matching a feed line to a Yagi antenna. **Figure 9-24** illustrates this technique. The hairpin match is also referred to as a *beta match*. To use a hairpin match, the driven element must be split in the middle and insulated from its supporting structure. The driven element is tuned so it has a capacitive reactance at the desired operating frequency. [E9E05] (This means the element is a little too short for resonance.) The hairpin adds some inductive reactance, transforming the feed point impedance to that of the feed line.

Figure 9-24B shows the equivalent lumped-constant network for a typical hairpin matching system for a 3-element Yagi. R_A and C_A represent the antenna feed point impedance. L is the parallel inductance of the hairpin. When the network is redrawn, as shown in Figure 9-24C, you can see that the circuit is the equivalent of an L network. [E9E06] The center point of the hairpin is electrically neutral and is often attached to an antenna's metal boom for mechanical stability.

The Stub Match

In some cases, it is possible to match a transmission line and antenna by connecting an appropriate reactance in parallel with them at the antenna feed point. Reactances formed from sections of transmission line are called *matching stubs*. Those stubs are designated either as open or closed, depending on whether the free end is an open or short circuit. Using a stub in this way is called a *stub match*. [E9E03]

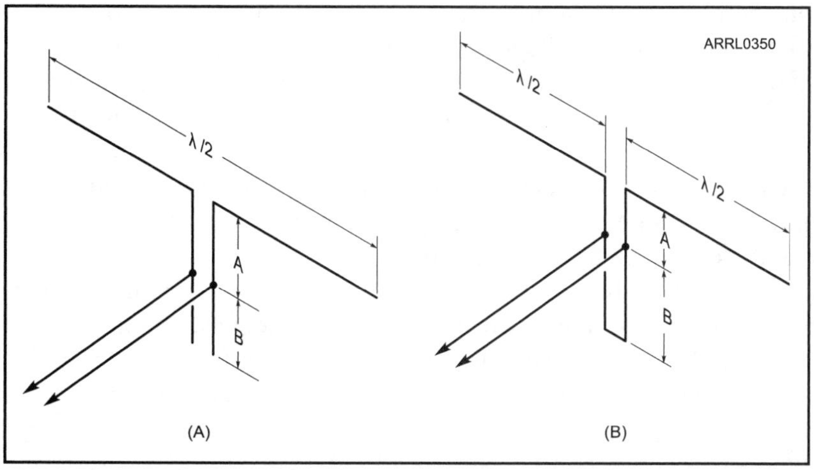

An impedance match can also be obtained by connecting the feed line at an appropriate point along the matching stub, as shown in **Figure 9-25**. The *universal stub* system illustrated here is used mostly at VHF and higher frequencies where the lengths of transmission lines are more manageable. It allows a feed line and antenna impedance to be matched, even if both impedances are unknown. [E9E11]

Matching stubs have the advantage that they can be used even when

Figure 9-25 — The stub matching system uses a short perpendicular section of transmission line connected to the feed line near the antenna. Dimensions A and B are adjusted to provide a match to the transmitter and feed line.

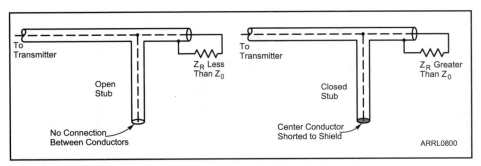

Figure 9-26 — Open and closed stubs can be used for matching to coaxial lines, as well.

the load is considerably reactive. That is a particularly useful characteristic when the antenna is not a multiple of a quarter-wavelength long, as in the case of a ⅝-wavelength radiator.

A stub match can also be made with coaxial cable, as illustrated in **Figure 9-26**. In a practical installation, the junction of the transmission line and stub would be a T-connector, to keep the stub and feed line perpendicular to each other.

Before you go on, study test questions E9A04, E9E01 through E9E06, E9E09 and E9E11. Review this section if you have difficulty.

TRANSMISSION LINE MECHANICS

As you've already noticed from operating radio equipment, transmission lines or feed lines seem to make up a significant portion of a radio station! And while they are the "silent partners" in the antenna system, not turning like antennas or pumping out power like a transmitter, every signal you send and receive goes through them. It pays to understand the basics of how they work!

Wavelength in a Feed Line

You've become familiar with wavelength and frequency. In free-space where radio waves move at 3×10^8 m/sec the wavelength of a radio wave is $\lambda = 300 / f$ (in MHz). In the AC Waveforms and Measurements chapter you also learned that a radio wave is a collection of wavefronts moving through space with the passing field strengths creating the wave observed by the receiver.

The same type of phenomenon also occurs along a feed line. An ac voltage applied to a feed line would give rise to the sort of current shown in **Figure 9-27**. If the frequency of the ac voltage is 10 MHz, each cycle will take 0.1 microsecond. Therefore, a complete current cycle will be present along each 30 meters of line (assuming free-space velocity). This distance is one wavelength.

Current observed at B occurs just one cycle later in time than the current at A. To put it another way, the current initiated at A does not appear at B, one wavelength away, until the applied voltage has had time to go through a complete cycle.

In Figure 9-27, the series of drawings shows how the instantaneous current might appear if we could take snapshots of it at quarter-cycle intervals in time. The current travels out from the input end of the line in waves. At any selected point on the line, the current goes through its complete range of ac values in the time of one cycle just as it does at the input end.

Velocity of Propagation

In the previous example, we assumed that energy traveled along the line at the velocity of light. The actual velocity is very close to that of light if the insulation between the conduc-

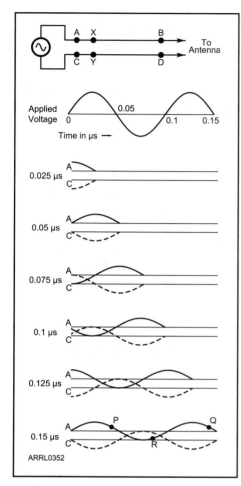

Figure 9-27 — Instantaneous current along a transmission line at successive time intervals. The period of the wave (the time for one cycle) is 0.1 microsecond.

tors of the line is solely air. The presence of dielectric materials other than air reduces the velocity since electromagnetic waves travel more slowly in materials other than in a vacuum. [E9F02] Because of this, the length of line that makes one wavelength will depend on the velocity of the wave as it moves along the line. The ratio of the actual velocity at which a signal travels along a line to the speed of light in a vacuum is called the *velocity factor*. [E9F01, E9F08]

$$VF = \frac{\text{speed of wave in line}}{\text{speed of light in vacuum}} \qquad \text{(Equation 9-5)}$$

where VF is the velocity factor.

The velocity factor is also related to the dielectric constant, ε, by:

$$VF = \frac{1}{\sqrt{\varepsilon}} \qquad \text{(Equation 9-6)}$$

For example, several popular types of coaxial cable have a polyethylene dielectric, which has a dielectric constant of 2.3. For those types of coaxial cable, we can use Equation 9-6 to calculate the velocity factor of the line. [E9F04]

$$VF = \frac{1}{\sqrt{\varepsilon}} = \frac{1}{\sqrt{2.3}} = \frac{1}{1.5} = 0.66$$

Electrical Length

The *electrical length* of a transmission line (or antenna) is not the same as its physical length. The electrical length is measured in wavelengths at a given frequency. Waves move slower in the line than in air, so the physical length of line will always be shorter than its electrical length. [E9F03] To calculate the physical length of a transmission line that is electrically one wavelength, use the formulas:

$$\text{Length (meters)} = VF \times \frac{300}{f \text{ (in MHz)}} \qquad \text{(Equation 9-7)}$$

or

$$\text{Length (feet)} = VF \times \frac{984}{f \text{ (in MHz)}} \qquad \text{(Equation 9-8)}$$

where:
   f = operating frequency (in MHz).
   VF = velocity factor.

Suppose you want a section of RG-8 coaxial cable that is ¼ wavelength long at 14.1 MHz. What is its physical length? The answer depends on the dielectric used in the coaxial cable. RG-8 is manufactured with polyethylene or foamed polyethylene dielectric; velocity factors for the two versions are 0.66 and 0.80, respectively. We'll use the polyethylene line with a velocity factor of 0.66 for our example. The physical length in meters of 1 wavelength of feed line is given by Equation 9-7:

$$\text{Length (meters)} = 0.66 \times \frac{300}{14.1} = 0.66 \times 21.3 = 14.1 \text{ m}$$

To find the physical length for a ¼-wavelength section of line, we must divide this value by 4. A ¼-wavelength section of this coax is 3.52 meters. [E9F05]

Example 9-7

What would be the physical length of a typical coaxial transmission line that is electrically one-quarter wavelength long at 7.2 MHz? (Assume a velocity factor of 0.66.)

$$\text{Length (meters)} = 0.66 \times \frac{300}{7.2} = 0.66 \times 41.7 = 27.5 \text{ m}$$

To find the length of the ¼-wavelength line, divide by 4 = 6.9 meters. [E9F09]

Table 9-1 lists velocity factors and other characteristics for some other common feed lines. You can calculate the physical length of a section of any type of feed line, including twin lead and ladder line, at some specific frequency as long as you know the velocity factor.

You may have noted in the table that some of the parallel conductor lines have velocity factors closer to 1 than those of coaxial cables. The air insulation used in the open-wire and ladder lines has a value for ε much closer to vacuum. This means that the electrical and physical lengths are more nearly equal in these feed lines.

Example 9-8

What is the physical length of a parallel conductor feed line that is electrically one-half wavelength long at 14.1 MHz? (Assume a velocity factor of 0.95.)

$$\text{Length (meters)} = \text{VF} \times \frac{300}{\text{f (in MHz)}} = 0.95 \times \frac{300}{14.1} = 0.95 \times 21.3 = 20.2 \text{ meters}$$

To find the length of the ½-wavelength line, divide by 2, so the length is 10 meters. [E9F06]

Feed Line Loss

When selecting a feed line, you must consider some conflicting factors and make a few trade-offs. For example, most amateurs want to use a relatively inexpensive feed line. We also want a feed line that does not lose an appreciable amount of signal energy. For many applications, coaxial cable seems to be a good choice but parallel-conductor feed lines generally have lower loss and may provide some other advantages such as being able to operate with high SWR on the line.

Line loss increases as the operating frequency increases, so on the lower-frequency HF bands you may decide to use a less-expensive coaxial cable with a higher loss than you would on 10 meters. Open-wire or ladder-line feed lines generally have lower loss than coaxial cables at any frequency. [E9F07] Table 9-1 includes approximate loss values for 100 feet of the various feed lines at 100 MHz. Again, these values vary significantly as the frequency changes. Note that there will be minor variations in specifications for similar cable types from different manufacturers.

While you're looking at Table 9-1 note the difference in *maximum rated voltage* between the coaxial cables with PE (solid polyethylene) and FPE (foamed polyethylene) dielectrics. The addition of air to the dielectric reduces loss and increases velocity factor but the tradeoff is a considerably lower ability to handle high voltages. [E9F16]

Reflection Coefficient and SWR

The *voltage reflection coefficient* is the ratio of the reflected voltage at some point on a feed line to the incident voltage at the same point. It is also equal to the ratio of reflected current to incident current at the same point on the line. The reflection coefficient is determined by the relationship between the feed line characteristic impedance, Z_0, and the actual load impedance, Z_L. The reflection coefficient is a good parameter to describe the interactions at

Table 9-1

Characteristics of Commonly Used Transmission Lines

Type of Line	Z_0 Ω	VF %	Cap. pF/ft	Diel. Type*	OD inches	Max V (RMS)	Loss (dB/100 ft) at 100 MHz
RG-8	50	82	24.8	FPE	0.405	600	1.5
RG-8	52	66	29.5	PE	0.405	3700	1.9
RG-8X	50	82	24.8	FPE	0.242	600	3.2
RG-58	52	66	28.5	PE	0.195	1400	4.3
RG-58A	53	73	26.5	FPE	0.195	300	4.5
RG-174	50	66	30.8	PE	0.110	1100	8.6
CATV Hardline (Aluminum Jacket)							
½ inch	75	81	16.7	FPE	0.500	2500	0.8
⅞ inch	75	81	16.7	FPE	0.875	4000	0.6
Parallel Lines							
Twin lead	300	80	4.4	PE	0.400	8000	1.1
Ladder line	450	91	2.5	PE	1.000	10000	0.3
Open-wire line	600	95-99	1.7	none	varies	12000	0.2

*Dielectric type: PE = solid polyethylene; FPE = foamed polyethylene
Excerpted from *The ARRL Antenna Book* (22nd edition), Chapter 23, Table 1

the load end of a mismatched transmission line. [E9E07]

The reflection coefficient is a complex quantity, having both amplitude and phase. It is generally designated by the lower case Greek letter ρ (rho), although some professional literature uses the capital Greek letter Γ (gamma). The formula for reflection coefficient is:

$$\rho = \frac{Z_L - Z_0}{Z_L + Z_0}$$ (Equation 9-9)

where Z_0 is the line's characteristic impedance and Z_L is the impedance of the load.

Evaluate this equation when $Z_L = Z_0$ or 0 Ω (shorted) or ∞ Ω (open). The only situation where ρ = 0, meaning no power is reflected and is all delivered to the load, occurs for $Z_L = Z_0$.

When ρ = 0, the voltage distribution along the line is constant or "flat." A line operating under these conditions is called either a *matched* or a *flat line*. If reflections do exist, a voltage standing-wave pattern will result from the interaction of the forward and reflected waves along the line. For a lossless transmission line, the ratio of the maximum peak voltage anywhere on the line to the minimum value anywhere on the line (which must be at least ¼ wavelength long) is defined as the voltage standing-wave ratio, or VSWR. Reflections from the load also produce a standing-wave pattern of currents flowing in the line. The ratio of maximum to minimum current, or ISWR, is identical to the VSWR in a given line.

In amateur literature, the abbreviation SWR is commonly used for standing-wave ratio, as the results are identical when taken from proper measurements of either current or voltage. Since SWR is a ratio of maximum to minimum, it can never be less than one-to-one. In other words, a perfectly flat line has an SWR of 1:1. For any impedance mismatch — high to low or low to high — the SWR will be greater than 1:1. [E9E08]

The SWR is related to the magnitude of the reflection coefficient by:

$$SWR = \frac{1+\rho}{1-\rho}$$ (Equation 9-10)

and conversely the reflection coefficient magnitude may be defined from a measurement of

SWR as

$$\rho = \frac{SWR - 1}{SWR + 1}$$ (Equation 9-11)

When both the line and load impedances are purely resistive, SWR can be computed directly from the impedances of the line and load:

For $Z_L > Z_0$, SWR $= \dfrac{Z_L}{Z_0}$ and for $Z_L < Z_0$, SWR $= \dfrac{Z_0}{Z_L}$ (Equation 9-12)

Power Measurement

You can use a variety of instruments to tell your relative power output. For example, as you tune a transmitter or antenna tuning unit, the increasing brightness of a neon bulb connected across the feed line or an increased current reading of an RF ammeter tells you that more power is going into the antenna. [E4B09]

It is much more convenient, however, to read the forward and reflected power directly, by using a *directional RF wattmeter* or *power meter*. This type of meter can determine total power flowing in either direction in the transmission line. This allows you to adjust an impedance-matching network by observing reflected power.

The reflection coefficient can also be computed from the forward and reflected power:

$$\rho = \sqrt{\frac{P_R}{P_F}}$$ (Equation 9-13)

where:
P_R = power in the reflected wave.
P_F = power in the forward wave.

Whatever the reflected and forward power may be, the difference between them is the net amount of power being transferred to the load; $P_{LOAD} = P_F - P_R$. Both forward and reflected power can be measured with a directional power meter or directional wattmeter in the transmission line. Remember that the net forward power ($P_F - P_R$) is the power delivered to the load.

Example 9-9

How much power is being absorbed by the load when a directional wattmeter connected between a transmitter and a terminating load reads 100 watts forward power and 25 watts reflected power? [E4B06]

$P_{LOAD} = P_F - P_R = 100 - 25 = 75$ watts

Before you go on, study test questions E4B06, E4B09, E9E07, E9E08, E9F01 through E9F09 and E9F16. Review this section if you have difficulty.

SMITH CHART

Before discussing the Smith Chart, let's back up a step. All impedances consist of two components: resistance and reactance. Graphically, these components are represented as a pair of axes — the rectangular coordinate system for graphing impedance. The horizontal axis represents resistance — positive to the right of the origin and negative to the left. The vertical axis represents reactance — positive (inductive) above the origin and negative (capacitive) below.

All possible impedances can be plotted as one point (Z) on that graph, corresponding to the values of resistance and reactance. Those two values are the rectangular coordinates of the impedance. When an impedance is connected to a transmission line and a signal of some frequency is applied to the other end of the line, the interaction between the energy in the line and that terminating impedance results in energy being reflected back and forth in the transmission line.

The ratios of voltage and current (that's the definition of impedance) also turn out to change with position along the line. So if an impedance-measuring meter is inserted at different points along the line, it would observe different values of impedance because of the different values of voltage and current.

Starting at the terminating impedance itself, as it moved farther and farther away along the line, the impedance measured by the meter would change until at ½ wavelength away from the terminating impedance, it would again report the terminating impedance's actual value and the cycle would begin again.

Smith Chart Construction

Plotted on rectangular coordinates, the path of that impedance measurement as the meter's position changed would be pretty messy, described by a fairly involved mathematical equation. What Phillip Smith discovered was that if you distort the rectangular graph in a certain way (called a *mapping*), the path of the impedance point along the transmission line becomes a circle! That concept is shown graphically on a Smith Chart.

Imagine the Smith Chart as a fun house mirror in reverse. Instead of taking your handsome image and distorting it to look bizarre, it takes the bizarre path of the impedance on the rectangular graph and makes a lovely circle out of it! This is a lot easier to work with. For this reason, the Smith Chart is used, among other things, to calculate impedances and SWR anywhere along a transmission line. [E9G01, E9G03]

What is this magic mapping? Imagine yourself standing at the origin of the rectangular graph with the positive resistance axis in front of you and the negative behind. The positive reactance axis starts at your feet and goes straight up and the negative straight down. All of the axes extend to infinity.

Now imagine reaching up over your head and bending the positive reactance axis down in front of you in a semicircle whose far end meets up with the far end of the positive resistance axis. Do the same for the negative reactance axis, bending it up instead. The negative resistance axis still extends behind you, as straight as ever. This process is sketched in **Figure 9-28**.

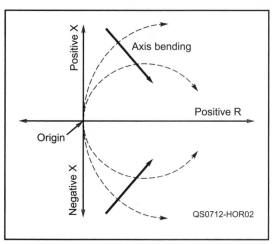

Figure 9-28 — Distorting or mapping the rectangular graph captures all of its right-hand side impedances inside the circle formed by the bent reactance axes. This is the basis of the Smith Chart.

You have created a circle from the two reactance axes, bisected by the only straight line on the chart, the horizontal *resistance axis* through the center. [E9G07] The infinity points join together with the infinite point of the positive resistance axis at the right of the chart. All of the points that were once in the right-hand side of the rectangular graph are now somewhere inside or on the boundary of that circle. Points on the left-hand side of the rectangular graph are now spread outside the circle. Nothing has been lost, just squashed or stretched.

The Smith Chart shown in **Figure 9-29** only contains the circle and what's inside. [E9G05] It ignores everything outside the circle because of the negative resistance value of those points originally. They were on the left side of the graph, remember? Those impedances cannot be present in a transmission line.

The circles and arcs on the Smith Chart show what happens to straight lines on the rectan-

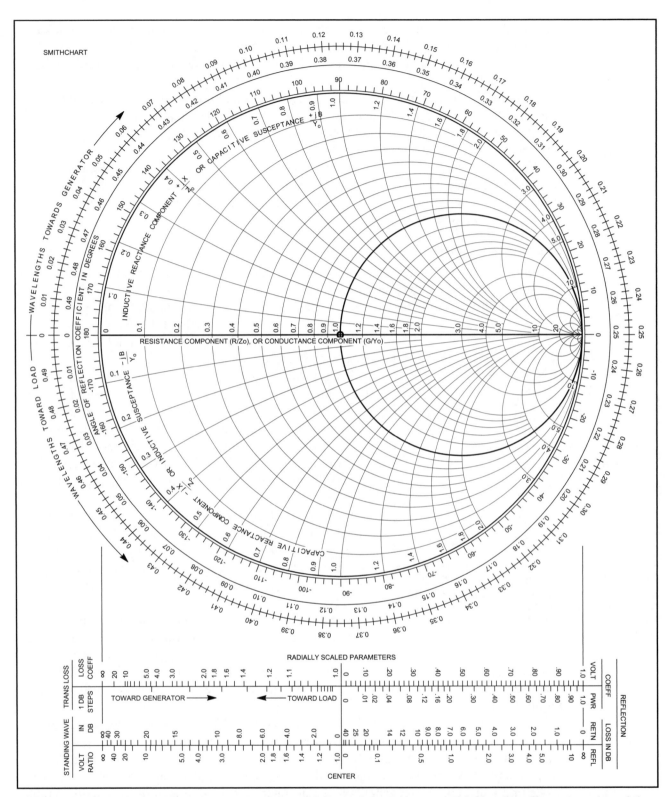

Figure 9-29 — The Smith Chart represents all possible impedances in a transmission line. The outer edge of the chart represents pure inductive and capacitive reactance. The horizontal axis through the center of the chart represents pure resistance. Impedances of any value are plotted at the intersection of constant-resistance circles and constant-reactance arcs.

gular graph after remapping. Lines of constant resistance, originally vertical and on which all points had the same value of resistance, are now nested *constant-resistance circles* that come together at the far right of the Smith Chart. That should make sense because all of those straight lines originally went where? To infinity — which is now the point at the right side of the Smith Chart. Horizontal constant-reactance lines that represented all the points having the same reactance are now bent into *constant-reactance arcs* with one end on the outer circle (the reactance axis) and the other end at infinity! [E9G02, E9G04, E9G06, E9G10] This distortion results in the path of the impedance point becoming a circle on the Smith Chart as we look at each point along the line.

Normalization

Take a close look at the Smith Chart in Figure 9-29. If you look for the impedance point of $50 + j0$ Ω, you will find it squashed way over in the nest of circles at the right-hand side of the chart — not very easy to use. Smith avoided the problem of big numbers by *normalizing* all of the coordinates to the characteristic impedance of the line, Z_0. That impedance point is the *prime center* of the Smith Chart.

Normalization reassigns the values of all points according to their ratio to Z_0 at the prime center, in this case dividing them by 50 Ω. [E9G08] So instead of 50 Ω being over in the hard-to-read section at the right, it's right in the middle of the chart at 1.0. Much better! From here on, all of the values you plot on the Smith Chart will be the value you read on the meter divided by 50 Ω.

Constant-SWR Circles

If you take all of the normalized impedance points on the Smith Chart that create a certain value of SWR in a 50-Ω transmission line, you will find that the points make a circle centered on the point $Z = 1.0 + j0$ that is at the center of the chart. This is called a *constant-SWR circle*. [E9G09] Lower SWR makes smaller and smaller circles until at SWR = 1.0, the circle is merely the point at the prime center of the chart, meaning that the terminating impedance is equal to $Z_0 = 50$ Ω.

As SWR increases, the circles increase in size until at SWR = ∞, the circle is the outside edge of the chart. The SWR caused by any impedance can be found by measuring the distance from the center of the chart to the impedance point, then translating that distance onto the linear SWR scale at the bottom of the chart. These scales are called *radially-scaled* because they represent measurements made radially from the center of the chart.

Wavelength Scales

Look carefully at the left side of the Smith Chart along the rim and you will see two arrows pointing in opposite directions, labeled "Wavelengths Toward Generator" and "Wavelengths Toward Load." The chart's outer scale, the reactance axis, is marked to show movement in wavelengths along the transmission line.

There are two scales, one starting at 0 and increasing clockwise and the other starting at 0.5 and decreasing clockwise. Both are calibrated in fractions of electrical wavelength inside the transmission line. [E9G11] These are used to work out problems that involve the changing impedance along a transmission line as described in the next section.

Before you go on, study test questions E9G01 through E9G11. Review this section if you have difficulty.

TRANSMISSION LINE STUBS AND TRANSFORMERS

In a transmission line, when a wave of RF voltage and current encounters an impedance different from the characteristic impedance of the transmission line, Z_0, some of the energy in the wave is reflected back toward the wave's source. The phase of the voltage and currents making up the reflected wave will differ from those in the incoming or incident wave depending on the value of the impedance causing the reflection.

The incident and reflected voltage and current waves combine at every point along the line. At each point, the combination results in voltage and current with a phase relationship different from either the incident or reflected waves. It is as if the same energy in the line had been applied to an impedance with values of resistance and reactance that create the same phase relationship. If you cut the line at that point and replace the section beyond the cut with actual components creating an equivalent impedance, there would be no change to the waves in the remaining section of the line!

The voltages and currents of both waves also vary with distance along the line because of the ac nature of the waves. This results in different combinations of incident and reflected voltages and currents and their equivalent impedance, as well. For example, if the equivalent impedance looks like 5 Ω of resistance and +20 Ω of reactance at one point, a bit farther along the line, the equivalent impedance might be 20 Ω of resistance and −5 Ω of reactance. For the Extra class exam we'll examine what impedance our impedance-meter would "see" if it looks into a transmission line with the other end shorted or open.

The first and easiest rule to remember is that if any transmission line is any integer multiple of ½ wavelength long, the impedance at one end will be the same as at the other. It doesn't matter what the terminating impedance is, how many ½-wavelengths of line are involved (neglecting line loss), or even what the characteristic impedance of the line is! Every ½ wavelength along the line, impedance repeats. If the terminat-

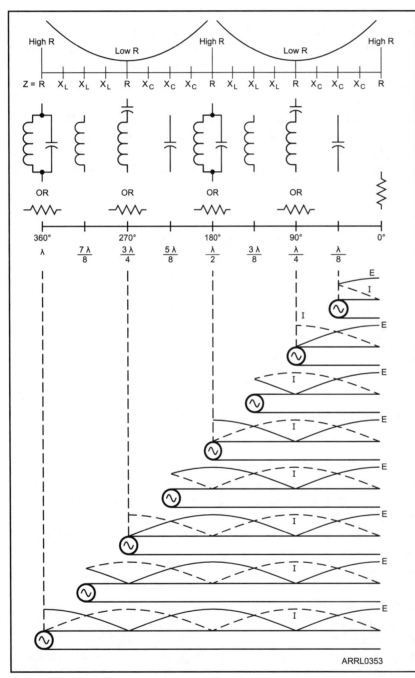

Figure 9-30 — This diagram summarizes the characteristics of open-circuited transmission lines. Voltage standing waves are shown as solid lines above each length of cable, and current standing waves are shown as dashed lines.

ing impedance is a short circuit, the impedance meter will see a short every ½ wavelength away and the same situation applies to an open circuit termination. [E9F14, E9F15] You can see this on the Smith Chart because the impedance point travels in one complete circle around the chart every ½ wavelength.

The second rule, almost as easy to remember, is that if the transmission line is an odd multiple of ¼-wavelength long, the impedance at one end is inverted from that at the other end. If the terminating impedance is an open, the impedance ¼ wavelength away will be a short, and vice versa. [E9F12, E9F13] This behavior repeats at ¾ wavelength, 1¼ wavelengths, 1¾ wavelengths away, and so forth.

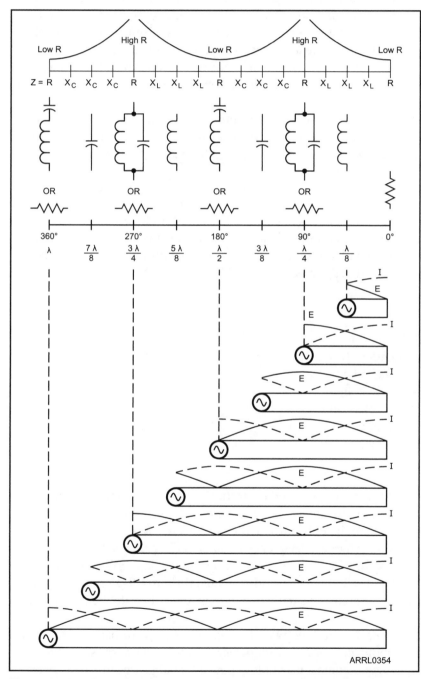

Figure 9-31 — This diagram summarizes the characteristics of short-circuited transmission lines. Voltage standing waves are shown as solid lines above each length of cable, and current standing waves are shown as dashed lines.

The remaining cases for ⅛-wavelength lines are not so easy to remember but with a little study, you'll be able to figure them out. Let's start with an open-circuited transmission line that is very, very short. An impedance meter will view this short piece of transmission line as an open circuit. As the line is lengthened it will exhibit a small capacitive reactance because that's what it is — a small capacitor formed by the inner and outer conductors. When the line reaches ⅛ wavelength, the capacitive reactance will reach the value of $-jZ_0$. In other words, a ⅛-wavelength piece of open 50-Ω transmission line will have an impedance of $-j50$ Ω at the other end. [E9F11]

Working with the opposite case, a shorted line, the impedance of the very, very short short-circuited line looks like — this shouldn't surprise you — a short-circuit. As the line is lengthened, it exhibits inductive reactance because the loop of inner and outer conductor form an inductor. When the line reaches ⅛ wavelength long, the impedance meter will read $+jZ_0$ Ω of inductive reactance. [E9F10]

Figures 9-30 and **9-31** illustrate the behavior of open and shorted transmission lines up to 2 λ long. You can clearly see the cyclic behavior of the impedances in the line. The impedance "seen" looking into various lengths of feed line is indicated directly above the chart. Curves above the axis marked with R, X_L and X_C indicate

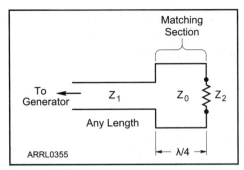

Figure 9-32 — The synchronous transformer creates two sets of reflections to power traveling in both directions in the transmission lines or between the transmission line and a load. These reflections cancel so that the net result is no increase in SWR from the impedance mismatch.

the relative value of the impedance presented at the input of the line. Circuit symbols indicate the equivalent circuits for the lines at that particular length. Standing waves of voltage (E) and current (I) are shown above each line. Remember that Z = E / I, by Ohm's Law, so you can use the curves above each piece of line to estimate the input impedance of a given line length.

Synchronous Transformers

There is one special technique of impedance matching using transmission lines that every Extra class ham should learn — the *synchronous transformer* shown in **Figure 9-32**. This method of matching involves setting up a series of reflections of just the right magnitude and phase so that two transmission lines or a transmission line and a load of two different impedances can be connected together without creating any SWR in the transmission lines!

To match two different impedances such as the transmission line and load shown in the figure, Z_0 and Z_{LOAD}, a ¼-wavelength section of transmission line with a characteristic impedance, Z_1, equal to the *geometric mean* of Z_0 and Z_{LOAD}. In mathematical form:

$$Z_1 = \sqrt{Z_0\,Z_{LOAD}}$$

(Equation 9-14)

The transformer is called *synchronous* because it must be a certain fraction of a wavelength long (1/4 λ) to function. For example, to match a 50-Ω transmission line to a quad antenna with a feed point impedance of 100 Ω, the transmission line used for the synchronous transformer should have a characteristic impedance of:

$$Z_1 = \sqrt{50 \times 100} = 70.7\ \Omega$$

A section of 75-Ω RG-59/U cable will work quite well in this application. [E9E10] The load impedance to be matched can also be the characteristic impedance of another transmission line. Other forms of synchronous transformers with different lengths can match different levels of impedances, too.

Before you go on, study test questions E9E10 and E9F10 through E9F15. Review this section if you have difficulty.

ANTENNA ANALYZERS

The impedance meter referred to in the preceding section is not just an imaginary instrument nor is it an expensive lab instrument. The *antenna analyzer* has made measuring impedance and SWR a simple task since its introduction to Amateur Radio in the 1990s. The analyzer consists of a tunable RF source, a frequency counter, an impedance bridge, displays and a microprocessor to run them. It is capable of measuring impedance, SWR, reactance and frequency from 160 meters through UHF for some models! **Figure 9-33** shows a typical model. Analyzers are battery powered and small enough to be taken into the field or up on an antenna tower. [E4A08]

The analyzer is used by connecting it directly to the impedance to be measured. [E4B11] The impedance can be a component, a circuit (with power removed), a transmission line or an antenna. Since the analyzer contains its own low-power signal source, it is not necessary to use a transmitter for testing antennas (as you would when an SWR meter is used). [E4A07] By using your knowledge of transmission lines (and the device's user manual), an antenna analyzer can be used to measure transmission line length and characteristic imped-

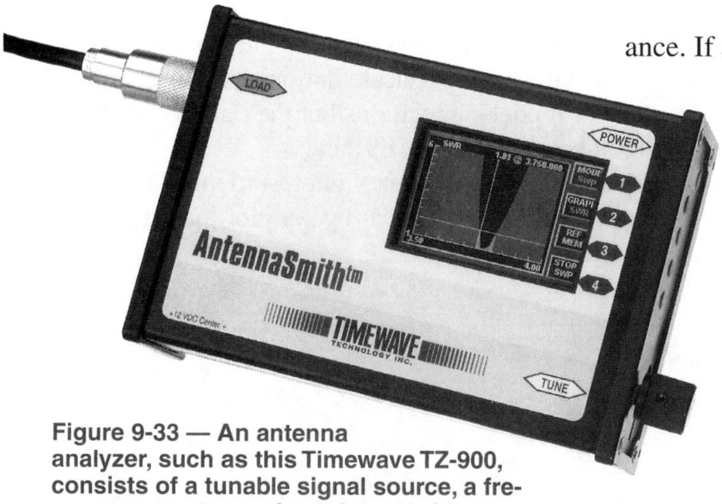

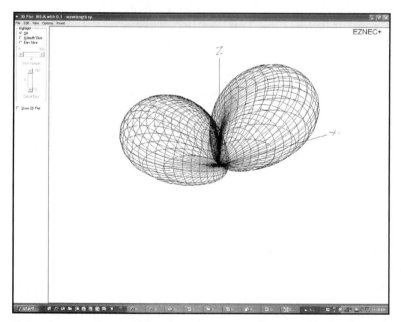

Figure 9-33 — An antenna analyzer, such as this Timewave TZ-900, consists of a tunable signal source, a frequency counter, an impedance bridge, and displays — all microprocessor controlled. The analyzer displays impedance (including reactance), SWR, and frequency.

ance. If a defective line is suspected, the analyzer can be used to find the location of a short or open circuit, as well. These handy accessories have become a fixture of ham radio test instrumentation.

Before you go on, study test questions E4A07, E4A08 and E4B11. Review this section if you have difficulty.

9.4 Antenna Design

The power of personal computers makes it possible to design and analyze antennas by mathematically modeling the antenna. Computer analysis allows us to study how performance changes as the height of the antenna changes or what effects different ground conditions will have. The antenna's characteristics can be adjusted until the design meets expectations. The accuracy of the programs is excellent in the hands of a reasonably skilled modeler. This saves a lot of time over the cut-and-try method!

Figure 9-34 — The *EZNEC* program is one of the best known antenna modeling programs used by hams. Written by Roy Lewallen, W7EL, the program is based on the mainframe computer *NEC* program and models antennas of any size and frequency.

ANTENNA MODELING AND DESIGN

There are a number of programs in common use for antenna analysis. Most of them are derived from a program developed at US government laboratories, called *NEC*, short for *Numerical Electromagnetics Code*. [E9B13] This complex program, originally written for mainframe computers, uses a modeling technique called the *method of moments*. [E9B09]

In the method of moments, the antenna wires (or tubing elements) are modeled as a series of *segments* and a uniform value of current in each segment is computed. [E9B10] The field resulting from the RF current in each segment is evaluated, along with the effects from other mutually coupled segments. The higher the number of segments, generally the more accurate the modeling results will be. However, most programs have a limit on the number of segments in a particular antenna because of the amount of memory and processing time required to perform the necessary calculations. A lower number of segments will reduce the time required for model calculations but the outputs, such as pattern shape or feed point impedance, will not be as accurate. [E9B11]

The result of an antenna model can take several forms. Of primary interest to most amateurs is the radiation pattern of the antenna. These are given in the standard polar plot format for far-field elevation and azimuthal patterns. Many programs provide a three-dimensional view of the pattern, as well. The programs compute, at a minimum, antenna gain, beamwidth, all the pattern ratios, feed point impedance, and SWR versus frequency "sweep" graphs. [E9B14] Example outputs from the *EZNEC* program by W7EL are shown in **Figure 9-34**.

Before you go on, study test questions E9B09, E9B10, E9B11, E9B13 and E9B14. Review this section if you have difficulty.

DESIGN TRADEOFFS AND OPTIMIZATION

Any antenna design represents some compromises. You may be able to modify the design of a particular antenna to improve some desired characteristic, if you are aware of the tradeoffs. As mentioned earlier, the method of moments computer modeling techniques that have become popular can be a great help in deciding which design modifications will produce the "best" antenna for your situation.

When you evaluate the gain of an antenna, you (or the computer modeling program) will have to take into account a number of parameters. You will have to include the antenna feed point impedance, any loss resistance in the elements and impedance-matching components, as well as the E-field and H-field radiation patterns.

You should also evaluate the antenna across the entire frequency band for which it is designed. You may discover that gain may change rapidly as you move away from its design frequency. [E9B04] (You may be willing to make that trade-off if all your operating on that band is within a narrow frequency range.) You may also discover that the feed point impedance varies widely as you change frequency across the band, making it difficult to design a single impedance-matching system for the antenna. You may also discover that the front-to-back ratio varies excessively across the band, resulting in too much variation in the rearward pattern lobes.

The forward gain of a Yagi antenna can be increased by using a longer boom, spreading the elements farther apart or adding more elements. Of course there are practical limitations on how long you can make the boom for any antenna! The element lengths will have to be adjusted to retune them as the boom length changes.[E9B06]

You may decide to optimize a Yagi antenna for maximum forward gain, but in that case the front-to-back ratio usually decreases, feed point impedance becomes very low, and the SWR bandwidth will decrease. [E9B05] Optimizing performance for one parameter often leads to a reduction in performance in other parameters. In general, the interdependency of gain, SWR bandwidth, and pattern ratios requires compromises by the antenna modeler to achieve realistic goals.

Before you go on, study test questions E9B04, E9B05 and E9B06. Review this section if you have difficulty.

Table 9-2

Questions Covered in This Chapter

9.1 Basics of Antennas	*9.2 Practical Antennas*	*9.3 Antenna Systems*
E9A01	E3C07	E4A07
E9A02	E3C10	E4A08
E9A03	E9A07	E4B06
E9A05	E9A12	E4B09
E9A06	E9C01	E4B11
E9A08	E9C02	E9A04
E9A09	E9C03	E9E01
E9A10	E9C04	E9E02
E9A11	E9C05	E9E03
E9A13	E9C06	E9E04
E9A14	E9C11	E9E05
E9A15	E9C12	E9E06
E9B01	E9C13	E9E07
E9B02	E9D01	E9E08
E9B03	E9D02	E9E09
E9B07	E9D04	E9E10
E9B08	E9D05	E9E11
E9B12	E9D06	E9F01
E9C07	E9D07	E9F02
E9C08	E9D08	E9F03
E9C09	E9D09	E9F04
E9C10	E9D10	E9F05
E9D03	E9D11	E9F06
	E9D12	E9F07
	E9D13	E9F08
	E9D14	E9F09
	E9D15	E9F10
	E9E12	E9F11
	E9E13	E9F12
	E9H01	E9F13
	E9H02	E9F14
	E9H03	E9F15
	E9H04	E9F16
	E9H05	E9G01
	E9H06	E9G02
	E9H07	E9G03
	E9H08	E9G04
	E9H09	E9G05
	E9H10	E9G06
	E9H11	E9G07
	E9H12	E9G08
		E9G09
		E9G10
		E9G11

9.4 Antenna Design

E9B04
E9B05
E9B06
E9B09
E9B10
E9B11
E9B13
E9B14

Chapter 10

Topics in Radio Propagation

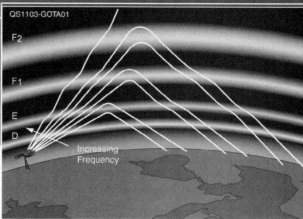

In this chapter, you'll learn about:
- **Ground-wave and sky-wave**
- **Long-path and gray-line**
- **The radio horizon**
- **Tropospheric and transequatorial propagation**
- **Auroral and meteor scatter communications**
- **Earth-Moon-Earth (moonbounce) techniques**

What happens after the signal leaves the antenna? Everything between the transmit and receive antennas involves propagation — the source of a great deal of radio's magic! This chapter is a collection of short discussions on various aspects of radio propagation at different frequencies. The topics are divided into those that matter most at HF and those of interest primarily at VHF and higher frequencies. There are a few questions about each in the Extra class exam. You can check your understanding of propagation by being able to answer all of the questions listed at the end of each topic.

10.1 HF Propagation

In nearly all cases, HF signals make the journey between stations by either traveling along the surface of the Earth (*ground-wave*) or by being returned to Earth after encountering the upper layers of the ionosphere (*sky-wave* or *skip*). The differences in frequency between the lowest current amateur band (1.8 MHz) and the highest HF band (28 MHz) cause the behavior of these modes of propagation to be quite different across the HF spectrum.

GROUND-WAVE PROPAGATION

The direction of waves of all types can be changed by both *diffraction* and *refraction*. Diffraction is created by the construction and reinforcement of wavefronts after the radio wave encounters a reflecting surface's corners or edges. Refraction is a gradual bending of the wave because of changes in the velocity of propagation in the medium through which the wave travels.

There is a special form of diffraction that primarily affects vertically polarized radio waves at HF and lower frequencies. This type of diffraction results from the lower part of the wave losing energy because of currents induced in the ground. This slows the lower portion of the wave, causing the entire wave to tilt forward slightly, following the curvature of the Earth.

This tilting results in *ground-wave propagation*, allowing low-frequency signals to be heard over distances well beyond line of sight. Although the term is often applied to any short-distance communication, the actual mechanism is unique to signals with longer wavelengths. Ground-wave propagation is most noticeable on the AM broadcast band and the 160 meter and 80 meter amateur bands. Practical ground-wave communication distances on

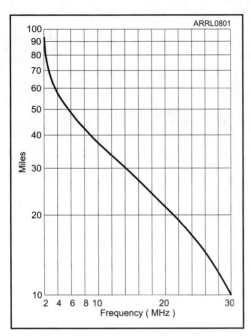

Figure 10-1 — Typical HF ground-wave range as a function of frequency.

these bands are often more than 100 miles.

Ground-wave propagation is lossy because the vertically polarized portion of the wave's electric field that extends into the ground is mostly absorbed. Over distance, the signal traveling along the ground is increasingly absorbed until the signal is too weak to be received. This loss increases significantly with frequency until at 28 MHz (10 meters), the maximum range of ground-wave is only a few miles. [E3C12] **Figure 10-1** shows typical ground-wave range at different frequencies.

Ground-wave propagation is most useful during the day at 1.8 and 3.5 MHz, when losses in the lower ionosphere make sky-wave propagation impossible. Vertically polarized antennas provide the best results. [E3C13] Ground-wave losses are reduced considerably over saltwater and are highest over dry and rocky land.

One simple way to observe the effects of ground-wave propagation is to listen to stations on the AM broadcast band. During the day you will regularly hear the high-power stations from 100 to 150 miles away. You won't hear stations much farther than 200 miles, however. At night, when sky-wave propagation becomes possible, you will begin to hear stations several hundred miles away. Of course AM broadcast stations usually have vertical antennas with excellent ground systems to radiate a strong signal!

Before you go on, study test questions E3C12 and E3C13. Review this section if you have difficulty.

SKY-WAVE PROPAGATION

Signals that travel into the ionosphere can be refracted (bent) by ionized gas in the ionosphere's E and F regions, returning to Earth some distance away. This refraction occurs because the region of ionized gases causes the radio wave to slow down, and this bends the wave. Refraction is primarily an HF propagation mode. Signals that follow a path away from the surface of the Earth are called *sky waves*. The path of a wave that returns to Earth after being bent by the ionosphere is called a *hop*.

The maximum one-hop skip distance for high-frequency radio signals via the F layer is usually considered to be about 2500 miles. (Skip via the E layer can extend to around 1500 miles.) Most HF communication beyond that distance takes place by means of several ionospheric hops in which the surface of the Earth reflects the signals back into the ionosphere for another hop. It is also possible that signals may reflect between the E and F regions, or even be reflected several times within the F region.

Propagation studies suggest radio waves may at times propagate for some distance through the uppermost F2 region of the ionosphere. In this type of propagation, a signal radiated at a medium elevation angle sometimes is returned to Earth at a greater distance than a wave radiated at a lower angle. The higher-angle wave, called the *Pedersen ray*, is believed to penetrate to the F2 region, farther than lower-angle rays. In the less-densely ionized F2 region, the amount of refraction is less, nearly equaling the curvature of the region itself as it encircles the Earth. **Figure 10-2** shows how the Pedersen ray could provide propagation beyond the normal single-hop distance. [E3C08]

This Pedersen-ray theory is further supported by studies of propagation times and signal strengths for signals that travel completely around the Earth. The time required is significantly less than would be necessary to hop between the Earth and the ionosphere 10 or more

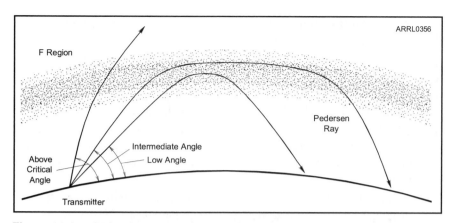

Figure 10-2 — This diagram shows a radio wave entering the F region at an intermediate angle, which penetrates higher than normal into the F region and then follows that region for some distance before being bent enough to return to Earth. A signal that travels for some distance through the F region is called a Pedersen ray.

times while circling the Earth. Return signal strengths are also significantly higher than should be expected otherwise. There is less attenuation for a signal that stays in the F region than for one that makes several additional trips through the E and D regions, in addition to the loss produced by reflections off the surface of the Earth.

Absorption

The lowest of the ionosphere's layers is the D layer, occupying from around 35 to 60 miles above the Earth's surface. The D layer exists in a relatively dense region, compared to the rest of the ionosphere. This means the ionized atoms and molecules are closer together and recombine quickly. As a result, the D layer is present only when illuminated by the Sun. Created at sunrise and reaching its strongest around local noon, the D layer disappears quickly after sunset.

When a passing wave causes D layer electrons to move, they collide with other electrons and ions so frequently that a great deal of the wave's energy is dissipated as heat. This is called *ionospheric absorption*. The longer the wavelength of the radio wave, the farther the electron travels under influence of the wave and the greater the portion of the wave's energy lost as heat. This means that absorption eliminates long-distance sky-wave propagation on the 1.8 and 3.5 MHz bands during the day, especially during periods of high solar activity. NVIS (near vertical incidence sky-wave) and ground-wave propagation can be used during daylight hours on these bands, however.

Before you go on, study test question E3C08. Review this section if you have difficulty.

LONG-PATH AND GRAY-LINE PROPAGATION

Most of the time, HF signals propagate over a *great circle* path between the transmitter and receiver. The great circle path is illustrated in **Figure 10-3**. A careful inspection shows that there are really two great circle paths, one shorter than the other. The longer of the two paths may also be useful for communications when conditions are favorable. This is called *long-path propagation*. Both stations must have directional antennas, such as beams, that can be pointed in the long path direction to make the best use of this propagation.

The long- and short-path directions usually differ by 180°. Since the circumference of the Earth is 25,000 miles, short-path propagation is always over a path length of less than 12,500 miles. The long-path distance is 25,000 miles minus the short-path distance. For example, the distance from Pittsburgh, Pennsylvania, to Tokyo, Japan is 6619 miles at a bearing of 328°. The long-path circuit would be a distance of 18,381 miles at a bearing of 148°.

Suppose you are in North America, receiving signals from a

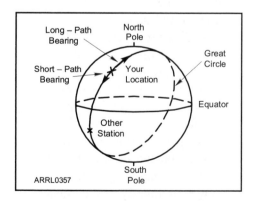

Figure 10-3 — A great circle drawn on the globe between two stations. The short-path and long-path bearings are shown from the Northern Hemisphere station.

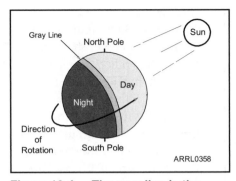

Figure 10-4 — The gray line is the transition region between daylight and darkness. On one side of the Earth the gray line is coming into sunrise, and on the other side just past sunset.

Persian Gulf station with your beam antenna pointed directly at that station. If you notice an echo on the signals (not polar flutter) delayed by a fraction of a second, you should turn your beam antenna 180°. If the signals from that station are significantly stronger when you turn your antenna 180°, you are receiving the station via the long path. [E3B04] The reason for the echo is that you are receiving two signals. One signal is coming in over short-path and the second one is coming in over long-path, which introduces a slight delay because the signal must travel a significantly longer distance. [E3B07]

Long-path propagation can occur on any band with sky-wave propagation, so you may hear long-path signals on the 160 to 10 meter bands. [E3B05] (Six meter long-path has been noted to occur on rare occasions.) Long-path enhancement occurs most often on the 20 meter band. [E3B06] All it takes is a modest beam antenna with a relatively high gain compared to a dipole, such as a three-element beam. The antenna should be at a height above ground that allows low takeoff angles.

For paths less than about 6000 miles, the short-path signal almost always will be stronger because of the increased losses caused by multiple-hop ground-reflection losses and ionospheric absorption over the long path. When the short-path is more than 6000 miles, however, long-path propagation usually will be observed either along the *gray-line* (the *terminator* between darkness and light that runs completely around the Earth) or over the nighttime side of the Earth.

The gray line is a band along the terminator that extends to either side for a number of miles. **Figure 10-4** illustrates the gray line. Notice that on one side of the Earth, the gray line is coming into daylight (sunrise) and on the other side it is coming into darkness (sunset).

Gray-line propagation exists when signals can propagate by sky-wave through this region. [E3B08] Gray-line propagation can be quite effective because the D layer, which absorbs HF signals, disappears rapidly on the sunset side of the gray line before it has had time to build up on the sunrise side. Meanwhile, the E and F layers, being at higher altitudes, are still illuminated and providing propagation. [E3B10] This mode of propagation lasts until D layer absorption or the loss of F layer refraction prevent the signal from reaching the receiving station.

Look for gray-line propagation around sunrise and sunset. [E3B09] Both long-path and short-path contacts can be made along the gray line. The four lowest-frequency amateur bands (160, 80, 40 and 30 meters) are the most likely to experience gray-line enhancement because they are the most affected by D layer absorption. [E3B11]

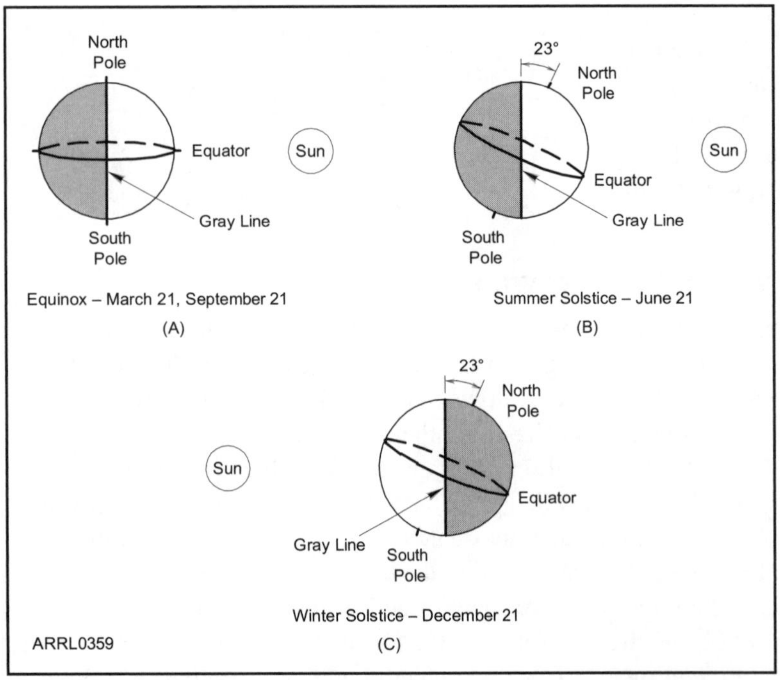

Figure 10-5 — The angle at which the gray line crosses the equator depends on the time of year, and on whether it is at sunset or sunrise. Part A shows that the gray line is perpendicular to the equator twice a year, at the vernal equinox and the autumnal equinox. B shows the North Pole tilted toward the Sun at a 23° angle at the summer solstice, and at C the North Pole tilted away from the Sun at a 23° angle at the winter solstice. This tilt causes the direction of the gray line to change so it is not North-South.

The gray line varies as much as 23° either side of a north-south line. This variation is caused by the tilt of the Earth's axis relative to its orbital plane around the Sun. The gray line will be aligned exactly north and south at the equinoxes (March 21 and September 21). On the first day of summer in the Northern Hemisphere, June 21, it is tilted a maximum of 23° at the equator; NW to SE at sunrise and NE to SW at sunset. The opposite occurs six months later at the beginning of winter.

Figure 10-5 illustrates the changing the tilt of the gray line. The tilt angle will be between these extremes during the rest of the year. Knowledge of the tilt angle will be helpful in determining what directions are likely to provide gray line propagation on a particular day.

Before you go on, study test questions E3B04 through E3B11. Review this section if you have difficulty.

FADING

Fading is a general term used to describe variations in the strength of a received signal. It may be caused by natural phenomena such as constantly changing ionospheric layer heights, variations in the amount of absorption, or random polarization shifts when the signal is refracted. Fading may also be caused by man-made phenomena such as reflections from passing aircraft and even from ionospheric disturbances caused by exhaust from large rocket engines.

Selective fading occurs when the wave path from a transmitting station to a receiving station varies with very small changes in frequency, even across a receiver's pass band. [E3C05] It is possible for components of the same signal only a few kilohertz apart to be acted upon differently by the ionosphere. This can even cause modulation sidebands (such as the carrier and the sidebands in an AM signal) to arrive at the receiver out of phase with each other, causing mild to severe distortion and a loss of signal strength.

Wideband signals, such as FM and double-sideband AM, suffer the most from selective fading. The sidebands may have different fading rates with respect to each other or the carrier. Distortion from selective fading is especially bad when the carrier of an FM or AM signal fades while the sidebands do not. SSB and CW signals, which have a narrower bandwidth, are affected less by selective fading.

Before you go on, study test question E3C05. Review this section if you have difficulty.

10.2 VHF/UHF/Microwave Propagation

Without regular and dependable sky-wave propagation, VHF and UHF operators utilize alternative modes of propagation to make contacts. There are plenty of options — more than on HF, in fact — and many support communications over very long distances. This section touches on a number of interesting propagation modes that you're likely to encounter, should you decide to try using SSB, CW or one of the digital modes above 30 MHz. We'll begin close to ground level and work our way up!

For space wave signals that travel essentially in a straight line between the transmitter and the receiver, antennas that are low-angle radiators (concentrate signals toward the horizon) are best. Signals radiated at angles above the horizon may pass over the receiving antenna. In general, the radiation takeoff angle from a horizontally polarized Yagi antenna decreases as the antenna height increases above flat ground. So, if you can raise the height of your antenna, the takeoff angle will decrease.

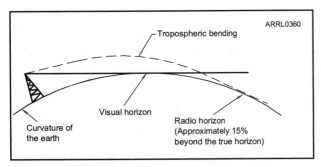

Figure 10-6 — Under normal conditions, tropospheric bending causes VHF and UHF radio waves to be returned to Earth beyond the visible horizon.

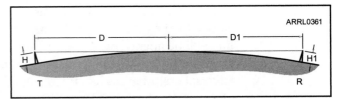

Figure 10-8 — The distance, D, to the radio horizon is greater from a higher antenna. The maximum distance over which two stations may communicate by space wave is equal to the sum of the distances to their respective radio horizons.

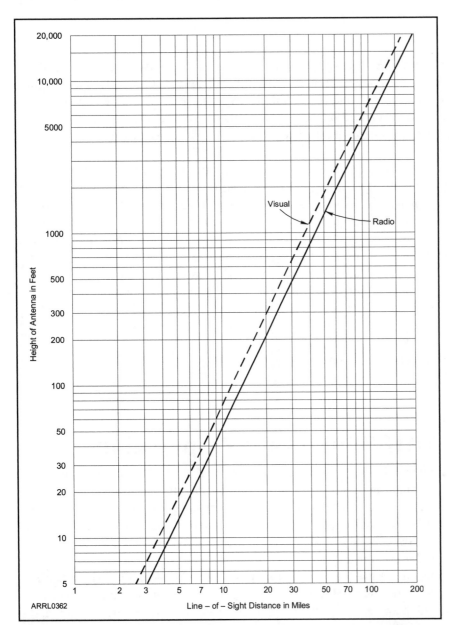

Figure 10-7 — Distance to the radio horizon from an antenna of given height above average terrain is indicated by the solid line. The broken line indicates the distance to the visual, or geometric, horizon. The radio horizon is approximately 15% farther than the visual horizon.

The polarization of both the receiving and transmitting antennas should be the same for VHF and UHF operation because the polarization of a space wave remains constant as it travels. There may be as much as 20 dB of signal loss between two stations that are using *cross-polarized* antennas.

RADIO HORIZON

In the early days of VHF amateur communications, it was generally believed that communications required direct line-of-sight paths between the antennas of the communicating stations. After some experiments with good equipment and antennas, however, it became clear that VHF radio waves are bent or scattered in several ways, making communications possible with stations beyond the *visual* or *geometric horizon*. The farthest point to which radio waves will travel directly, via *space-wave propagation*, is called the *radio horizon*.

Under normal conditions, density variations in the atmosphere near the Earth causes radio waves to bend into a curved path that keeps them nearer to the Earth than true straight-line travel would. **Figure 10-6** shows how this bending of the radio waves causes the distance to the radio horizon to exceed the distance to the visual horizon. [E3C14] The radio horizon is approximately 15% farther than the geometric horizon as shown graphically in **Figure 10-7**. [E3C06]

The point at the radio horizon is assumed to be on the ground. An antenna that is on a high hill or tall building well above any surrounding obstructions has a much farther radio horizon than an antenna located in a valley or shadowed by other obstructions. If the receiving antenna is also elevated, the maximum space-wave distance between the two antennas is equal to the sum of the distance to the radio horizon from the transmitting antenna plus the distance to the radio horizon from the receiving antenna. **Figure 10-8** illustrates this principle. Unless the two stations are identical, each will have a different radio horizon.

Multipath

A common cause of fading is an effect known as *multipath*. Several components of the same transmitted signal may arrive at the receiving antenna from different directions. The phase relationships between the multiple signals may cause them to cancel or reinforce each other. Multipath fading is responsible for the effect known as "picket fencing" in VHF communications, when signals from a station in motion have a rapid fluttering quality. This fluttering is caused by the change in the paths taken by the transmitted signal to reach the receiving station as the station moves. This effect is illustrated in **Figure 10-9**. Multipath effects can occur whenever the transmitted signal follows more than one path to the receiving station. Multipath and the distortion it causes to the received signal is a major challenge to providing high-speed digital service via wireless systems.

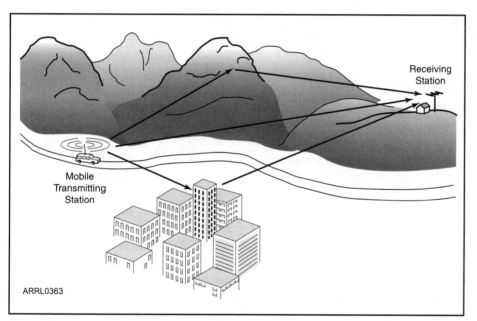

Figure 10-9 — If a signal travels from a transmitter to a receiver over several different paths, the signals may arrive at the receiver slightly out of phase. The out-of-phase signals alternately cancel and reinforce each other, and the result is a fading signal. This effect is known as multipath fading.

Before you go on, study test questions E3C06 and E3C14. Review this section if you have difficulty.

TROPOSPHERIC PROPAGATION

VHF propagation is usually limited to distances of approximately 500 miles. This is the normal limit for stations using high-gain antennas, high power and sensitive receivers. As a radio wave travels, it collides with air molecules and other particles, giving up some of its energy. This is why there is a limit to distances that may be covered by space-wave communications.

At times, weather conditions can create sharp transitions between air layers. These transitions can reflect radio waves, forming *ducts* in the troposphere, similar to propagation in a waveguide. Such ducts cause VHF radio waves to follow the curvature of the Earth for hundreds, or thousands, of miles. This form of propagation is called *tropospheric ducting*. [E3C09]

The possibility of propagating radio waves by tropospheric ducting increases with frequency. Ducting is rare on 50 MHz, fairly common on 144 MHz and more common on higher frequencies. Gulf Coast states experience it often, and the Atlantic Seaboard, Great Lakes and Mississippi Valley areas less frequently, usually in September and October.

Before you go on, study test question E3C09. Review this section if you have difficulty.

TRANSEQUATORIAL PROPAGATION

Transequatorial propagation (TE) is a form of F layer ionospheric propagation discovered by amateurs in the late 1940s. Amateurs on all continents reported the phenomenon almost simultaneously on various north-south paths on 50 MHz during the evening hours. At that time, the maximum predicted MUF was around 40 MHz for daylight hours. Since that time, research carried out by amateurs has shown that the TE mode works on 144 MHz and even to some degree at 432 MHz. TE occurs between mid-latitude stations approximately the same distance north and south of the Earth's magnetic equator. [E3B01] **Figure 10-10** shows the paths of a number of contacts made on 144 MHz using TE propagation.

You might expect the ionization of the ionosphere's upper layers to be at a maximum over the equator around the vernal (spring) and autumnal (fall) equinoxes. In fact, at the equinoxes there is not a single area of maximum ionization, but two. These maxima form in the morning, are well established by noon and last until after midnight. The high-density-ionization regions form approximately between 10° and 15° on either side of the Earth's magnetic equator — not the geographic equator — forming a pair of regions able to reflect VHF and even UHF signals.

As the relative position of the Sun moves away from the equator, the ionization levels in the Northern and Southern Hemispheres become unbalanced, lowering the MUF for TE propagation. So the best time of year to look for transequatorial propagation is around March 21 and September 21. The MUF for TE will also be higher during solar activity peaks. The best conditions for TE exist when the Earth's magnetic field is quiet, as well.

TE also enables very strong signals on the HF bands during the afternoon and early evening, so these are the best times

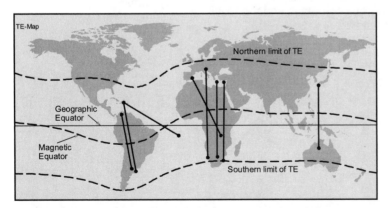

Figure 10-10 — This world map shows contacts made on 144 MHz over paths using transequatorial propagation (TE). Notice the symmetrical distribution of stations with respect to the magnetic equator.

to look for this propagation mode. [E3B03] Later at night, and sometimes in the early morning as well, you will hear weak and watery signals arriving by TE.

As the signal frequency increases, the communication zones become more restricted to those equidistant from, and perpendicular to, the magnetic equator. Further, the duration of the opening tends to be shorter and closer to 8 PM local time. The rate of flutter fading and the degree of frequency spreading increase with signal frequency. TE range extends to approximately 5000 miles — 2500 miles on each side of the magnetic equator. [E3B02]

Before you go on, study test questions E3B01, E3B02 and E3B03. Review this section if you have difficulty.

AURORAL PROPAGATION

Auroral propagation occurs when VHF radio waves are reflected from the ionization created by an auroral curtain. It is a VHF and UHF propagation mode that allows contacts up to about 1400 miles. Auroral propagation occurs for stations near the northern and southern polar regions but the discussion here is limited to auroral propagation in the Northern Hemisphere.

Aurora results from a large-scale interaction between the ionosphere and magnetic field of the Earth and electrically charged particles of the *solar wind*, ejected from the surface of the Sun. [E3C02] Visible aurora, often called the northern lights or *aurora borealis*, is caused by the collision of these solar-wind particles with oxygen and nitrogen molecules in the upper atmosphere. These collisions partially ionize the molecules by knocking loose some of their outer electrons.

When the electrons that were knocked loose from the oxygen and nitrogen recombine with the molecules, light is produced. The extent of the ionization determines how bright the aurora will appear. At times, the ionization is so strong that it is able to reflect radio signals with frequencies above about 20 MHz. This ionization occurs at an altitude of about 70 miles in the E layer of the ionosphere. [E3C03] Not all auroral activity is intense enough to reflect radio signals, so a distinction is made between a visible aurora and a radio aurora.

The number of auroras (both visible and radio) varies with geomagnetic latitude as described in **Figure 10-11**. Generally, auroral propagation is available only to stations in the northern states, but on occasion extremely intense auroras reflect signals from stations as far south as the Carolinas. Auroral propagation is most common for stations in the northeastern states and adjacent areas of Canada, closest to the Geomagnetic North Pole. Auroral propagation is rare below about 32° N latitude in the southeast and about 38° to 40° N in the southwest.

The number and distribution of auroras are related to the solar cycle. Auroras

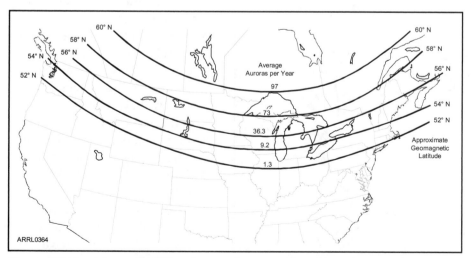

Figure 10-11 — The possibility of auroral propagation decreases as distance from the Geomagnetic North Pole increases. The Geomagnetic North Pole is currently near 78° N, 104° W. This map shows the approximate geomagnetic latitude for the northern US.

occur most often during sunspot peaks, but the peak of the auroral cycle appears to lag the peak of the solar cycle by about two years. Intense auroras can occur, however, at any point in the solar cycle.

Auroras also follow seasonal patterns, most common around the equinoxes in March and September. Auroral propagation is observed most often in the late afternoon and early evening hours, and it usually lasts from a few minutes to many hours. Often, it will disappear for a few hours and reappear around midnight. Major auroras often start in the early afternoon and last until early morning the next day.

Using Auroral Propagation

Most common on 10, 6 and 2 meters, auroral contacts have been made on frequencies as high as 222 and 432 MHz. The number and duration of openings decreases rapidly as the operating frequency rises.

The reflecting properties of an aurora vary rapidly, so signals received via this mode are badly distorted, making CW the most effective mode for auroral work. [E3C04] CW signals most often have a buzzing or raspy sound rather than a pure tone as reflection from the aurora makes them appear modulated by white noise. [E3C01] For this reason, auroral propagation is often called "buzz mode." SSB is usable for 6 meter auroral contacts if signals are strong; voices are often intelligible if the operator speaks slowly and distinctly. SSB is rarely usable at 2 meters or higher frequencies.

In addition to scattering radio signals, auroras have other effects on worldwide radio propagation. Communication below 20 MHz is disrupted in high latitudes, primarily by absorption, and the effect is especially noticeable over polar and near-polar paths. Signals on the AM broadcast band through the 40 meter band late in the afternoon may become weak and "watery" sounding. The 20 meter and higher HF bands may close down altogether to east-west and polar paths. At the same time, the MUF in equatorial regions may temporarily rise dramatically, providing transequatorial paths as described above.

Stations should point their antennas north (or toward the Geomagnetic North Pole) during the aurora and, in effect, "bounce" their signals off the auroral zone as illustrated by **Figure 10-12**. [E3C11] The optimum antenna heading varies with the position of the aurora and may change rapidly, just as the visible aurora does. Constant probing with the antenna is recommended to peak signals, especially for high-gain antennas with narrow beamwidths. Using the southernmost bearing that gives good signal strength is the best strategy, remembering that the optimum bearing will likely change.

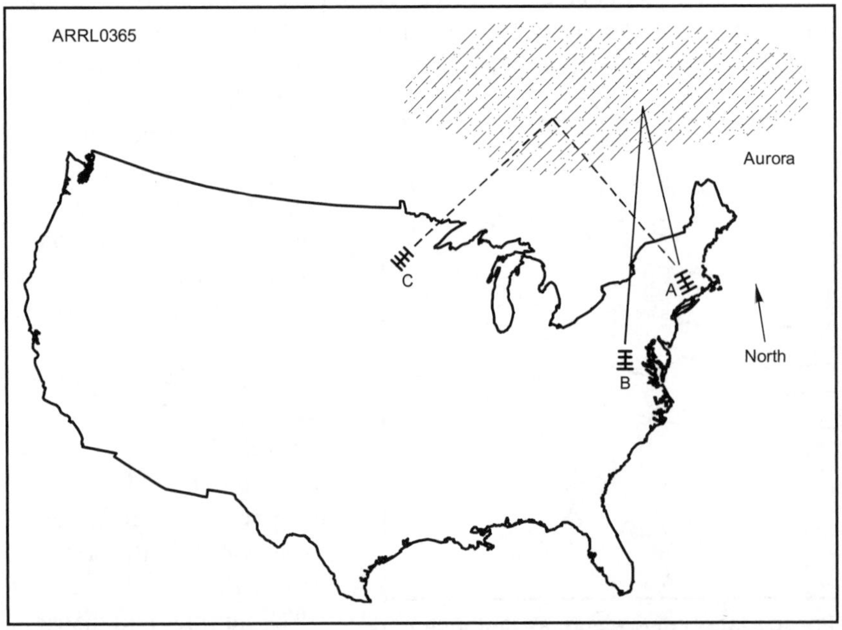

You can observe developing auroral conditions by monitoring signals in the region between the AM broadcast band and 5 MHz. If, for example, signals in the 75 meter band begin to waver suddenly (flutter or sound "watery") in the afternoon or early evening hours, a radio aurora may be beginning. Since auroras are associated with solar disturbances, you often can predict one

Figure 10-12 — To use auroral propagation, stations point their antennas toward the Geomagnetic North Pole. Station A may have to beam west of north to work station C.

by listening to the WWV Geo Alert broadcasts at 18 minutes after each hour. In particular, the *K index* may be used to indicate auroral activity. K-index values of 3 and rising indicate that conditions associated with auroral propagation are present in the Boulder, Colorado, area. Timing and severity may be different elsewhere, however. Maximum occurrence of radio aurora is for K-index values of 7 to 9. Web-sites like **www.spaceweather.com** provide information about auroras and conditions that might support auroral propagation.

Before you go on, study test questions E3C01 through E3C04 and E3C11. Review this section if you have difficulty.

METEOR SCATTER COMMUNICATIONS

Meteoroids are particles of mineral or metallic matter that travel in highly elliptical orbits about the Sun. Most of these are microscopic in size. Every day hundreds of millions of these meteoroids enter the Earth's atmosphere. Attracted by the Earth's gravitational field, they attain speeds from 6 to 60 mi/s (22,000 to 220,000 mi/h).

As a meteoroid speeds through the upper atmosphere, it heats up and begins to vaporize as it collides with air molecules. This action creates heat and light and leaves a trail of free electrons and positively charged ions called a *meteor*. Trail size is directly dependent on particle size and speed. A typical meteoroid the size of a grain of sand creates a trail about 3 feet in diameter and 12 miles or longer, depending on speed.

The duration of this meteor-produced ionization is directly related to electron density. Ionized air molecules contact and recombine with free electrons over time, gradually lowering the electron density until it returns to its previous state.

Radio waves can be reflected by the ionized trail of a meteor. The ability of a meteor trail to reflect radio signals depends on electron density. Greater density causes greater reflecting ability and reflection at higher frequencies. The electron density in a typical meteor trail will strongly affect radio waves on the upper HF and lower VHF bands. Signal frequencies as low as 20 MHz and as high as 432 MHz will be usable for meteor-scatter communication at times. The best range of frequencies for amateur meteor scatter communications is from 28 to 148 MHz. [E3A09]

Meteor trails are formed at approximately the altitude of the ionospheric E layer, 50 to 75 miles above the Earth. [E3A08] That means the range for meteor-scatter propagation is about the same as for single-hop E (or sporadic E) skip — a maximum of approximately 1200 miles, as **Figure 10-13** shows.

The propagation potential of a meteor trail is highly frequency dependent. For example, consider the result of a relatively large (about the size of a peanut) meteoroid entering the Earth's atmosphere. Using that ionized meteor trail, no 432-MHz propagation is possible. At 222 MHz, moderately strong signals are heard over a 1200-mile path for perhaps 12 seconds. The signals gradually fade into the noise over the next four or five seconds. At 144 MHz, the same conditions might result in very strong signals for 20 seconds that slowly fade

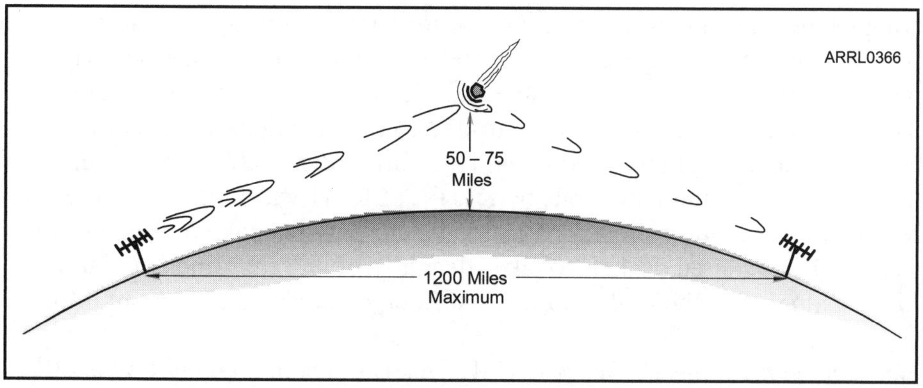

Figure 10-13 — Meteor-scatter communication makes extended-range VHF and UHF communications possible for short periods.

Table 10-1

Major Meteor Showers

Date	Name
January 3-5	Quadrantids
April 19-23	Lyrids
*June 8	Arietids
July 26-31	Aquarids
July 27-August 14	Perseids
October 18-23	Orionids
October 26 – November 16	Taurids
November 14-16	Leonids
December 10-14	Geminids
December 22	Ursids

All showers occur in the evening except those marked (*), which are daytime showers. Evening showers begin at approximately 2300 local standard time; daylight showers at approximately 0500 local standard time.

into the noise during the next 20 seconds. At 28 MHz, propagation might well last for a couple of minutes. Meteors of this size are relatively rare; most are much smaller. As a result, the typical burst of meteor-reflected signal is much shorter than that cited in the example.

Meteor-scatter communication (sometimes called *meteor-burst communication*) is best between midnight and noon. The part of the Earth between those hours, local time, is always on the leading edge as the Earth travels along its orbit. It is at that leading edge where most meteoroids enter the Earth's atmosphere.

Meteor Showers

At certain times of the year, the Earth encounters clouds of meteoroids. Most are the remnants of a comet as it orbits the Sun. The resulting increased meteor activity is known as a *meteor shower*. These showers greatly enhance meteor-scatter communications at VHF. The degree of enhancement depends on the time of day, shower intensity and the frequency in use. The largest meteor showers of the year are the Perseids in August and the Geminids in December. **Table 10-1** is a partial list of meteor showers throughout the year.

Meteor Scatter Techniques

The secret to successful meteor-scatter communication is short transmissions. To call CQ, for example, say (or send) CQ followed by your call sign, repeated two or three times. To answer a CQ, give the other station's call once followed by your call phonetically. Example: "CQ from Whiskey Five Lima Uniform Alfa." "W5LUA this is Whiskey One Juliet Romeo." W5LUA responds: "W1JR 59 Texas." W1JR immediately responds: "Roger 59 New Hampshire." W5LUA says: "Roger QRZ from W5LUA." The entire QSO with information exchanged and confirmed in both directions lasted only 12 seconds! (Many operators exchange grid squares instead of states.)

A single meteor may produce a strong enough path to sustain communication long enough to complete a short QSO. At other times multiple bursts are needed to complete the QSO, especially at higher frequencies. Remember, the key is short, concise transmissions. Do not repeat information unnecessarily, wasting time and propagation.

The accepted convention for transmission timing breaks each minute into four 15-second periods. The station at the western end of the path transmits during the first and third period of each minute. During the second and fourth 15-second periods, the eastern station transmits.

Several new digital modes provide quite effective meteor-scatter communications. These modes use the short meteor bursts that are available on a day-to-day basis rather than just during meteor showers. High-speed, computer-controlled communications systems exchange messages. A computer program called FSK441 is part of the *WSJT* software suite, which stands for "weak-signal communications by K1JT". FSK441 was written specifically for amateur meteor-scatter communications. In this mode, the stations make repeated short transmissions of specially formatted packets to take advantage of any short meteor "burn" within range of both stations. See the Radio Modes and Equipment chapter for more information on WSJT.

Another system is called *high-speed CW*, or *HSCW*. This isn't about copying CW at 50 or 60 WPM, though. Computer software uses a sound card to transmit and receive signals at effective speeds of 800 to 2000 WPM or more! [E3A10]

EARTH-MOON-EARTH COMMUNICATIONS

The concept of *Earth-Moon-Earth* (*EME*) communications, popularly known as *moon-bounce*, is straightforward: Stations that can simultaneously see the Moon communicate by reflecting VHF or UHF signals off the lunar surface. Those stations may be separated by nearly half the circumference of the Earth — a distance of nearly 12,000 miles — as long as they can both "see" the Moon. [E3A01] This is called their *mutual lunar window* in which the Moon is above the "radio horizon" for both stations at the same time.

Since the Moon's average distance from Earth is 239,000 miles, *path losses* are huge when compared to "local" VHF paths. (Path loss refers to the total signal loss between the transmitting and receiving stations relative to the total radiated signal energy.) Nevertheless, for any type of amateur communication over a distance of 500 miles or more at 432 MHz, for example, moonbounce comes out the winner over terrestrial propagation paths when all the factors limiting propagation are taken into account.

EME presents amateurs with the ultimate challenge in radio system performance. Today, most of the components for an EME station on 144, 222, 432 or 1296 MHz are commercially available. Advanced software helps recover signals that are inaudible by ear. Nevertheless, success at EME places a premium on station construction skill, equipment quality and operator ability.

A low-noise receiving setup is essential for successful EME work. Since many of the signals to be copied on EME are close to or in the noise, a low-noise-figure receiver is a must. [E3A04] The mark to shoot for at 144 MHz is less than 0.5 dB, as *cosmic noise* from outside the atmosphere will then be the limiting factor in the system. Noise figures of this level are relatively easy to achieve with inexpensive modern devices. As low a noise figure as can be attained is required at and above 432 and 1296 MHz. Noise figures on the order of 0.5 dB are possible with preamplifiers that use GaAsFET devices.

EME Scheduling

The best days to schedule an EME contact are usually when the Moon is at *perigee* (closest to the Earth) since the path loss is typically 2 dB less than when the Moon is at *apogee* (farthest from the Earth). [E3A03] The Moon's perigee and apogee dates may be determined from publications such as *The Nautical Almanac* or websites on EME communications. The Moon's orbit is slightly elliptical. Hence, the day-to-day path-loss changes at apogee and perigee are minor. The greatest changes take place at the time when the Moon is

Libration Fading

One of the most troublesome aspects of receiving a moonbounce signal, besides the enormous path loss and Faraday-rotation fading in the ionosphere, is *libration* (pronounced lie-BRAY-shun) *fading*. Libration fading of an EME signal is experienced as fluttery, rapid, irregular fading not unlike that observed in tropospheric-scatter propagation. [E3A02] Fading can be very deep, 20 dB or more, and the maximum fading is directly proportional to operating frequency. At 1296 MHz the maximum fading rate is about 10 Hz.

On a weak CW EME signal, libration fading gives the impression of a randomly keyed signal. In fact on very slow CW telegraphy the effect is as though the keying is being done at a much faster speed. On very weak signals only the peaks of libration fading are heard in the form of occasional short bursts or "pings."

What causes libration fading? Very simply, multipath scattering of the radio waves from the very large (2000-mile diameter) and rough Moon surface combined with the relative short-term motion between Earth and Moon. These short-term oscillations in the apparent aspect of the Moon relative to Earth are called librations.

traveling between apogee and perigee.

One or two days will be unusable near the time of a new Moon, since the Sun is behind the Moon, causing increased Sun noise to be received, masking weak signals. Avoid schedules when the Moon is within 10° of the Sun (and farther if your antenna has a wide beamwidth or strong side lobes). The Moon's orbit follows a cycle of 18 to 19 years, so the relationships between perigee and new Moon will not be the same from one year to the next.

Low Moon declinations (position in the sky) and aiming elevations generally produce poor results and should be avoided if possible. Generally, low elevation angles increase antenna-noise pickup and increase tropospheric absorption, especially above 420 MHz, where the cosmic noise is very low. This situation cannot be avoided when one station is unable to elevate the antenna above the horizon or when there is great terrestrial distance between stations. High Moon declinations and high antenna elevation angles should yield best results.

EME Frequencies

EME contacts are generally made by prearranged schedule, although some contacts are made at random. Many stations, especially those with marginal capability, prefer to set up a specific time and frequency in advance so that they will have a better chance of finding each other. The more capable stations, especially on 144 and 432 MHz where there is a good amount of activity, often call CQ during evenings and weekends when the Moon is at perigee, and listen for random replies. Most of the contacts on 50, 222, 1296 and 2304 MHz, where activity is lighter, are made by schedule.

Most amateur EME operation on 144 MHz takes place near the low edge of the band. You may find 2 meter EME activity anywhere between 144.000 and 144.100 MHz, although most of the activity tends to concentrate in the lower 50 kHz except during peak hours. [E3A06] Generally, random activity and CQ calling take place in the lower 10 kHz or so, and schedules are run higher in the band.

On the 70 cm band you may find EME activity anywhere from 432.000 to 432.100 MHz. [E3A07] Formal schedules (that is, schedules arranged well in advance) are run on 432.000, 432.025 and 432.030 MHz. Other schedules are normally run on 432.035 and up to 432.070. For this band, the EME random-calling frequency is 432.010, with random activity spread out between 432.005 and 432.020. Random SSB CQ calling is at 432.015. Terrestrial activity is centered on 432.100 and is, by agreement, limited to 432.075 and above in North America.

Moving up in frequency, formal schedules are run on 1296.000 and 1296.025 MHz. The EME random calling frequency is 1296.010 with random activity spread out between 1296.005 and 1296.020. Terrestrial activity is centered at 1296.100. There is some EME activity on 2304 MHz. Specific frequencies are dictated by equipment availability and are arranged by the stations involved. For EME SSB contacts on 144 and 432 MHz, contact is usually established on CW, and then the stations move up 100 kHz from the CW frequency. (This method was adopted because of the US requirement for CW only below 144.1 MHz.)

Of course, it is obvious that as the number of stations on EME increases, the frequency spread must become greater. Since the Moon is in convenient locations only a few days out of the month, and only a certain number of stations can be scheduled for EME during a given evening, expect for contacts to occur a few kilohertz apart.

EME Operating Techniques

EME operators using SSB and CW have agreed to a standard calling procedure that is time-synchronized with the stations alternating transmissions. [E3A05] For 144 MHz contacts, a two-minute calling sequence is used. The eastern-most station transmits first for two full minutes, and then that station receives for two full minutes while the western-most station transmits. For 432 MHz EME contacts, operators use two-and-a-half minute calling sequences. Again, the eastern-most station transmits first, for two and a half minutes.

Then the eastern station listens for the next two and a half minutes, while the western-most station transmits. Operation using digital protocols such as JT65 (see WSJT above) must synchronize their system clocks since the computer controls the transmit-receive sequence.

Before you go on, study test questions E3A01 through E3A07. Review this section if you have difficulty.

Table 10-2
Questions Covered in This Chapter

10.1 HF Propagation	10.2 VHF/UHF/Microwave Propagation
E3B04	E3A01
E3B05	E3A02
E3B06	E3A03
E3B07	E3A04
E3B08	E3A05
E3B09	E3A06
E3B10	E3A07
E3B11	E3A08
E3C05	E3A09
E3C08	E3A10
E3C12	E3B01
E3C13	E3B02
	E3B03
	E3C01
	E3C02
	E3C03
	E3C04
	E3C06
	E3C09
	E3C11
	E3C14

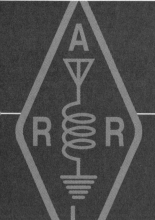

Chapter 11

Safety

In this chapter, you'll learn about:

- **Hazardous substances that may be found in the ham shack**
- **Ionizing and non-ionizing radiation**
- **Power density, duty cycle and absorption limits**
- **Controlled and uncontrolled environments**
- **Evaluating RF exposure**
- **Steps you can take to minimize RF exposure**

Amateur Radio is not a hazardous undertaking! Like driving a car, building and maintaining a station presents opportunities to act in an unsafe manner, but that doesn't mean the hazards are unavoidable. Quite the contrary! Learning about the hazards is the first step toward ham shack safety, and that's why the question pool touches on several topics involving hazardous materials and RF exposure. *The ARRL Handbook* contains a chapter on safety, as does *The ARRL Antenna Book*. RF exposure is covered thoroughly in the ARRL's *RF Exposure and You*. We recommend having a copy of each in your ham radio library!

11.1 Hazardous Materials

There aren't many materials considered hazardous that are required to communicate using Amateur Radio. To be honest, you probably have more hazardous materials in your garage and in your cleaning supplies than any ham shack ever will. Nevertheless, here are three that you ought to know about.

PCBs

Not printed-circuit boards, of course, but *polychlorinated biphenyls*, are the PCBs referred to by this section. You may have heard about PCBs contaminating industrial sites or locations where waste oils were dumped. PCBs are an additive to insulating oils used in electrical equipment with transformers and capacitors being the most common. [E0A10] PCBs helped the oil retain its insulating properties without breaking down and so became widely used until the hazard they presented became known. PCBs are known carcinogens — exposure to them, even in small amounts, elevates the risk of certain types of cancer.

In the ham shack, PCBs may be found in older oil-filled high-voltage capacitors. These would be found in dc power supplies for tube-type equipment, particularly amplifiers. Since any capacitor that contains PCBs would have to be fairly old, it is a good idea to replace the capacitor with a new one. If you are unsure whether a capacitor does or doesn't contain PCBs, you can contact the manufacturer with the model number of the capacitor (if the manufacturer is known and still in business).

If you remove such a capacitor from your equipment or find one in a junk box, wear protective gloves and wipe down the outside of the case with a paper towel. Place the paper towel and the capacitor in a plastic bag and take it to your local electric utility. They have procedures for safely and properly disposing of PCBs and may be able to handle the capacitor for you at no charge. Some local governments also have regular toxic-disposal opportunities and you can get rid of the capacitor there, as well.

BERYLLIUM AND BERYLLIUM OXIDE

Beryllium (Be) is the lightest metal next to lithium and a member of the alkali family of elements. On the periodic table (**en.wikipedia.org/wiki/Periodic_table**), you'll find beryllium at the upper left corner near hydrogen, lithium, sodium, magnesium and other similar elements. In electronics, beryllium is used in copper alloys to stiffen them and improve their conductivity. Beryllium copper is found in spring contacts and other flexible metal items that need to be both conductive and mechanically strong. In this form (or even as pure metal) beryllium is not dangerous. It is the dust and small particles that present a hazard and then only from chronic, extended exposure. To be safe, do not grind, weld or file metal containing beryllium in an unventilated area.

The oxide of beryllium (BeO) is a tough, durable ceramic that has the rare combination of being an excellent electrical insulator and an excellent thermal conductor (**en.wikipedia.org/wiki/Beryllium_oxide**). It is used inside some power semiconductors to insulate the transistor structure from the case and conduct heat away from the transistor. It is also used in larger vacuum tubes as a thermally conductive insulator. In general, BeO is only found inside the envelopes of tubes and inside transistor packages. Handling BeO in solid form is not dangerous but if a tube or package is broken, BeO pieces could become cracked or produce dust. The dust is toxic if inhaled. [E0A09]

It is difficult to tell whether a white ceramic is BeO or a more benign form of ceramic, so treat all such materials with caution. Vacuum any possible dust or small particles with a vacuum cleaner using HEPA-rated bags, place the vacuum cleaner bag in a sealed plastic bag, and contact your local recycling or solid waste disposal service for instructions.

LEAD AND SOLDERING

Soldering and electricity have been partners since before electronics! Solder, a mixture of lead and tin, has been used since antiquity. Lead is only dangerous when ingested or absorbed by the body. There are two ways of lead getting into your body during soldering: breathing lead fumes and lead contamination of food.

Inhaling lead fumes is fairly unlikely because the temperatures involved in soldering are not high enough to create much lead vapor. (The smoke from soldering is caused by burning flux, an organic resin that cleans the metal surface.) Small bits of solder can become dislodged and inhaled or eaten but airborne lead is not much of a hazard to hams. The main hazard from lead is contamination of the skin, followed by using the hands to handle food. Wash your hands carefully after soldering or handling soldered components.

In July of 2006, the European "Reduction of Hazardous Substances" (RoHS) regulations went into effect, requiring the elimination of lead in electronic solder. Alternative forms of solder containing higher levels of tin and silver, copper or bismuth are to be used. Appliances and equipment purchased after that date are made using the newer lead-free soldering techniques. If you are working on lead-free equipment, be sure to review lead-free soldering guidelines before doing any repairs or modifications.

CARBON MONOXIDE

The use of fossil fuel-powered generators and heaters during emergency and portable operation is becoming increasingly common. This presents several hazards of which the amateur should be aware, including electrical, fire and fuel storage hazards. A particularly worrisome hazard is caused by the carbon monoxide (CO) emitted by these devices.

Carbon monoxide is an odorless and colorless gas so there is no warning detectable by humans that concentrations of CO have risen to dangerous levels. For that reason, it is important for generators and heaters (including wood-burning stoves) to only be used in well-ventilated areas away from people. The only reliable method of sensing the presence of excessive levels of CO is by using a carbon monoxide detector — a smoke alarm will not respond to CO alone. [E0A07] An inexpensive CO detector should be placed in any area occupied by people in which CO from generator exhaust or heater vent can build up.

Before you go on, study test questions E0A07, E0A09 and E0A10. Review this section if you have difficulty.

11.2 RF Exposure

You've been exposed (so to speak) to various safety topics regarding exposure to RF in every license exam. The Extra class exam reviews familiar topics and introduces a couple of new ones. Although some of this material may be familiar from earlier exam studies (we hope!), it is worth covering again.

Let's start with a reminder that RF exposure at low levels — even continuously — is not hazardous in any way. It is only when the level of exposure is high enough to affect the temperature of the body that hazards occur. Although the power emitted by an antenna is called "radiation" and a hazard from RF exposure may be referred to as a "radiation hazard" it is not the same as radiation from a radioactive source. At RF, radiation does not have sufficient energy to break apart atoms and molecules, it can only cause heating. Radiation from radioactive sources does and that is why it is referred to as *ionizing radiation*. RF radiation is *nonionizing radiation* and is many orders of magnitude weaker than ionizing radiation in this regard. [E0A01]

Exposure to RF at low levels is not hazardous. At high power levels, for some frequencies the amount of energy that the body absorbs can be a problem. For example, exposure to high-power UHF or microwave RF can cause localized heating of the body. [E0A11] There are a number of factors to consider along with the power level, including frequency, average exposure, duty cycle of the transmission, and so forth. The two primary factors that determine how much RF the body will absorb are power density and frequency. This section discusses how to take into account the various factors and arrive at a reasonable estimate of what RF exposure results from your transmissions and whether any safety precautions are required.

POWER DENSITY

Heating from exposure to RF signals is caused by the body absorbing RF energy. The intensity of the RF energy is called *power density* and it is measured in mW/cm^2 (milliwatts per square centimeter) which is power per unit of area. For example, if the power density in an RF field is 10 mW/cm^2 and your hand's surface area is 75 cm^2, then your hand is exposed to a total of $10 \times 75 = 750$ mW of RF power in that RF field. Power density is highest near antennas and in the directions in which antennas have the most gain. Increasing transmitter power increases power density around the antenna. Increasing distance from an antenna lowers power density.

While RF exposure is measured in mW/cm^2 for most amateur requirements, the body's response to both E and H fields suggest that the RF field strengths can also be measured in V/m (for the E field) and A/m (for the H field). Depending on the source of the RF and the environment, either of these measurements may be more appropriate than power density. For example, around reflecting surfaces or conducting materials, the intensity of the E and H

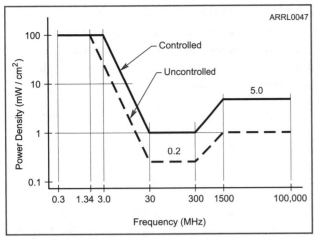

Figure 11-1 — Maximum Permissible Exposure (MPE) limits vary with frequency because the body responds differently to energy at different frequencies. The controlled and uncontrolled limits refer to the environment in which people are exposed to the RF energy.

Table 11-1
Maximum Permissible Exposure (MPE) Limits

Controlled Exposure (6-Minute Average)

Frequency Range (MHz)	Power Density (mW/cm²)
0.3-3.0	(100)*
3.0-30	(900/f²)*
30-300	1.0
300-1500	f/300
1500-100,000	5

Uncontrolled Exposure (30-Minute Average)

Frequency Range (MHz)	Magnetic Field Power Density (mW/cm²)
0.3-1.34	(100)*
1.34-30	(180/f²)*
30-300	0.2
300-1500	f/1500
1500-100,000	1.0

f = frequency in MHz
* = Plane-wave equivalent power density

fields can peak in different locations. Under and near antennas, ground reflections and scattering can make the field impedance (the ratio of E field to H field strength) vary with location, as well. [E0A06]

ABSORPTION AND LIMITS

The rate at which energy is absorbed from the power to which the body is exposed is called the *Specific Absorption Rate* (SAR). [E0A08] SAR is the best measure of RF exposure for amateur operators. The SAR varies with frequency, power density, average amount of exposure, and the duty cycle of transmission. Injury is only caused when the combination of frequency and power cause too much energy to be absorbed in too short a time.

SAR depends on the frequency and the size of the body or body part affected and is highest where the body and body parts are naturally resonant. The limbs (arms and legs) and torso have the highest SAR for RF fields in the VHF spectrum from 30 to 300 MHz. The head is most sensitive at UHF frequencies from 300 MHz to 3 GHz and the eyes are most affected by microwave signals above 1 GHz. The frequencies with highest SAR are between 30 and 1500 MHz. At frequencies above and below the ranges of highest absorption, the body responds less and less to the RF energy, just like an antenna responds poorly to signals away from its natural resonant frequency.

Safe levels of SAR based on demonstrated hazards have been established for amateurs by the FCC in the form of *Maximum Permissible Exposure* (MPE) limits that vary with frequency as shown in **Figure 11-1** and **Table 11-1**. These take into account the different sensitivity of the body to RF energy at different frequencies.

As you can see from the graph, safe exposure levels are much lower above 30 MHz. This means that extra caution is required around high-power RF sources in the amateur VHF, UHF and microwave bands. Of particular concern are amplifiers operating at these frequencies. Legal-limit power levels on these bands can create a significant hazard. When testing such equipment, take extra precautions to prevent accidental exposure, either near the transmitter or from an antenna. Radiation leaks above the MPE limits from klystron and magnetron transmitters, in particular, can present a significant hazard because of the power levels they can develop.

Table 11-2

Operating Duty Cycle of Modes Commonly Used by Amateurs

Mode	Duty Cycle	Notes
Conversational SSB	20%	1
Conversational SSB	40%	2
SSB AFSK	100%	
SSB SSTV	100%	
Voice AM, 50% modulation	50%	3
Voice AM, 100% modulation	25%	
Voice AM, no modulation	100%	
Voice FM	100%	
Digital FM	100%	
ATV, video portion, image	60%	
ATV, video portion, black screen	80%	
Conversational CW	40%	
Carrier	100%	4

Notes
1) Includes voice characteristics and syllabic duty factor. No speech processing.
2) Includes voice characteristics and syllabic duty factor. Heavy speech processing.
3) Full-carrier, double-sideband modulation, referenced to PEP. Typical for voice speech. Can range from 25% to 100% depending on modulation.
4) A full carrier is commonly used for tune-up purposes.

AVERAGING AND DUTY CYCLE

Exposure to RF energy is averaged over fixed time intervals because the response of the body to heating is different for short duration and long duration exposures. Time-averaging evaluates the total RF exposure over a fixed time interval. In addition, there are two types of environments with different averaging periods, controlled and uncontrolled.

Controlled and Uncontrolled Environments

People in *controlled environments* are considered to be aware of their exposure and are expected to take reasonable steps to minimize exposure. Examples of controlled environments are transmitting facilities (including Amateur Radio stations) and areas near antennas. In a controlled environment, access is restricted to authorized and informed individuals. The people expected to be in controlled environments would be station employees, licensed amateurs, and the families of licensed amateurs.

Uncontrolled environments are areas in which the general public has access, such as public roads and walkways, homes and schools, and even unfenced personal property. The homes of your neighbors are uncontrolled environments. [E0A02] People in uncontrolled environments are not aware of their exposure but are much less likely to receive continuous exposure. As a result, RF power density limits are higher for controlled environments and the averaging period is longer for uncontrolled environments. The averaging period is 6 minutes for controlled environments and 30 minutes for uncontrolled environments.

Duty Cycle

Duty cycle is the ratio of transmitter on time to total time during the exposure. Duty cycle has a maximum of 100%. (*Duty factor* is the same as duty cycle expressed as a fraction instead of percent. For example, a duty cycle of 25% is equivalent to a duty factor of 0.25.) The lower the transmission duty cycle (the less the transmitter is on), the lower the average exposure. A lower transmission duty cycle permits greater short-term exposure levels for a given average exposure. This is the *operational duty cycle*. For most amateur operations listening and transmitting time are about the same, so operational duty cycle is rarely higher than 50%.

Along with operational duty cycle, the different modes themselves have different *emission duty cycles* as shown in **Table 11-2**. For example, a normal SSB signal without speech processing to raise average power is considered to have an emission duty cycle of 20%. In contrast, FM is a constant-power mode so its emission duty cycle is 100%. Transmitter PEP multiplied by the emission duty cycle multiplied by the operating duty cycle gives the average power output.

For example, if a station is using SSB without speech processing, transmitting and listening for equal amounts of time and with a PEP of 150 W, then the average power output is 150 W × 20% × 50%, or 15 W.

For the AM entries, note that the table assumes the same PEP for all signals. If PEP is the same, an AM signal with 50% modulation has a higher duty factor (more carrier, less sidebands) than for a signal with 100% modulation. Because the average power of the signal depends on modulation, the duty cycle can range from 25% to 100%.

Before you go on, study test questions E0A01, E0A02, E0A06, E0A08 and E0A11. Review this section if you have difficulty.

ANTENNA SYSTEM

You must also take into account the amount of gain provided by your antenna and any significant losses from the feed line. High gain antennas increase a signal's average power considerably. For example, let's modify the example above by using an antenna with 6 dB of gain. If the transmitter PEP is increased from 150 W to 600 W by the 6 dB antenna gain, the average power is 600 W × 20% × 50%, or 60 W.

Including antenna gain in the field strength calculation is only required when the evaluation is being performed in the antenna's *far-field*. The far field begins approximately 10 wavelengths or so from the antenna and is generally considered to be the region in which the antenna's radiation pattern has assumed its final shape and does not change with increasing distance from the antenna. If the evaluation is to be performed in the *near-field* (anything closer than the far-field distance), then a different measure of antenna gain must be used.

ESTIMATING EXPOSURE AND STATION EVALUATION

All fixed amateur stations must evaluate their capability to cause RF exposure, no matter whether they use high or low power. (Mobile and handheld transceivers are exempt from having to calculate exposure because they do not stay in one location.) A routine evaluation must then be performed if the transmitter PEP and frequency are within the FCC rule limits. The limits vary with frequency and PEP as shown in **Table 11-3**. You are required to perform the RF exposure evaluation only if your transmitter output power exceeds the levels shown for any band. For example, if your HF transmitter cannot output more than 25 W, you are exempt from having to evaluate exposure caused by it.

You can perform the evaluation by actually measuring the RF field strength with calibrated field strength meters and calibrated antennas. You can also use computer modeling to determine the

Table 11-3

Power Thresholds for RF Exposure Evaluation

Band	Power (W)
160 meters	500
80	500
40	500
30	425
20	225
17	125
15	100
12	75
10	50
6	50
2	50
1.25	50
70 cm	70
33	150
23	200
13	250
SHF (all bands)	250
EHF (all bands)	250

exposure levels. However, it's easiest for most hams to use the tables provided by the ARRL (**www.arrl.org/rf-exposure**) or an online calculator, such as the one listed on the ARRL website. [E0A03]

If you choose to use the ARRL tables or calculators, you will need to know:

- Power at the antenna, including adjustments for duty cycle and feed line loss
- Antenna type (or gain) and height above ground
- Operating frequency

The ARRL tables are organized by frequency, antenna type and antenna height. They show the distance required from the antenna to comply with MPE limits for certain levels of transmitter output power.

Exposure can be evaluated in one of two ways. The first way is to determine the power density at a known distance to see if exposure at that distance meets the MPE limit. The second way is to determine the minimum distance from your antenna at which the MPE limit is satisfied. Either way, the goal is to determine if your station meets MPE limits for all controlled and uncontrolled environments present at your station.

If you make changes to your station, such as changing to a higher power transmitter, increasing antenna gain, or changing antenna height, then you must reevaluate the RF exposure from your station. If you reduce output power without making any other changes to a station already in compliance, you need not make any further changes.

In a multi-transmitter environment, such as at a commercial repeater site, each transmitter operator may be jointly responsible (with all other site operators) for ensuring that the total RF exposure from the site does not exceed the MPE limits. Any transmitter (including the antenna) that produces more than 5% of the total permissible exposure limit for transmissions at that frequency must be included in the site evaluation. (This is 5% of the permitted power density or 5% of the square of the E or H-field MPE limit. It is *not* 5% of the total exposure, which sometimes can be unknown.) [E0A04] The situation described by this question is common for amateur repeater installations, which often share a transmitting site.

EXPOSURE SAFETY MEASURES

The measures you can take if your evaluation results exceed MPE limits are summarized in **Figure 11-2**. These are all "good practice" suggestions that can save

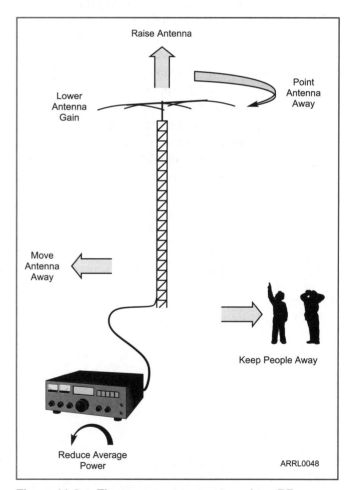

Figure 11-2 — **There are many ways to reduce RF exposure to nearby people. Whatever lowers the power density in areas where people are present will work. Raising the antenna will even benefit your signal strength to other stations as it lowers power density on the ground.**

time and expense if they are followed before doing your evaluation.

• Locate or move antennas away from where people can get close to them and be exposed to excessive RF fields. Raise the antenna or place it away from where people will be. Keep the ends (high voltage) and center (high current) of antennas away from people where people could come in contact with them. Locate the antenna away from property lines and place a fence around the base of ground-mounted antennas. This prevents people from encountering RF in excess of the MPE limits.

• Don't point gain antennas where people are likely to be. Use beam antennas to direct the RF energy away from people. Remember that high-gain antennas have a narrower beam, but exposure in the beam will be more intense. Take special care with high-gain VHF/UHF/ microwave antennas (such as Yagis and dishes used for EME) and transmitters — don't transmit when you or other people are close to the antenna or the antenna is pointed close to the horizon. [E0A05]

• If you have to use stealth or attic antennas, carefully evaluate whether MPE limits are exceeded in your home's living quarters.

• On VHF and UHF, place mobile antennas on the roof or trunk of the car to maximize shielding of the passengers. Use a remote microphone to hold a handheld transceiver away from your head while transmitting.

• When using microwave signals, take extra care around the high-gain antennas used on these frequencies. Even modest transmitter powers can result in significant RF levels when focused by a dish with more than 20 dB of gain! This is a particular concern when operating in a portable or rover configuration where antennas may not be far off the ground.

• From the transmitter's perspective, use a dummy load or dummy antenna when testing a transmitter. You can also reduce the power and duty cycle of your transmissions. This is often quite effective and has a minimal effect on your signal.

Table 11-4
Questions Covered in This Chapter

11.1 Hazardous Materials
E0A07
E0A09
E0A10

11.2 RF Exposure
E0A01
E0A02
E0A03
E0A04
E0A05
E0A06
E0A08
E0A11

Before you go on, study test questions E0A03, E0A04 and E0A05. Review this section if you have difficulty.

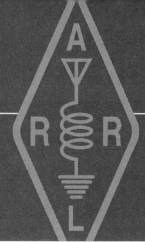

Chapter 12

Glossary

A

Absorption — The loss of energy from an electromagnetic wave as it travels through any material. The energy may be converted to heat or other forms. Absorption usually refers to energy lost as the wave travels through the ionosphere.

Accreditation — The process by which a Volunteer Examiner Coordinator (VEC) certifies that their Volunteer Examiners (VEs) are qualified to administer Amateur Radio license exams.

Adaptive filter — A filter that changes its parameters based on the input signals to the filter.

Admittance (Y) — The reciprocal of impedance.

Air link — The portion of a communication system that uses radio signals to transfer information.

Alias — A false signal created by an insufficient sampling rate.

Alien — Citizen of a country other than the United States.

Alpha (α) — The ratio of transistor collector current to emitter current. It is between 0.92 and 0.98 for a bipolar junction transistor.

Alpha cutoff frequency — The useful upper frequency limit of a transistor. The point at which the gain of a common-base amplifier is 0.707 times the gain at 1 kHz.

Alternator whine — A common form of conducted interference received as an audio tone in the received or transmitted signal. The pitch of the tone varies with alternator speed.

Amateur-satellite service — A radiocommunication service in which ground stations use satellites for the same purpose as stations in the amateur service.

Amplifier transfer function — A graph or equation that relates the input and output of an amplifier under various conditions.

Amplitude modulation (AM) — A method of superimposing an information signal on an RF carrier wave in which the amplitude of the RF envelope (carrier and sidebands) is varied in relation to the information signal strength.

Analog-to-digital converter (ADC) — A circuit that converts analog signals to digital values.

Anode — The terminal connected to the positive supply for current to flow through a device.

Antenna effect — A mode of a simple loop antenna in which the antenna exhibits the characteristics of a small, nondirectional vertical antenna.

Antenna efficiency — The ratio of the radiation resistance to the total resistance of an antenna system, including losses.

Apogee — That point in a satellite's orbit when it is farthest from the Earth.

Apparent power — The product in volt-amps (VA) of the RMS current and voltage values in a circuit without consideration of the phase angle between them.

ARES® — Amateur Radio Emergency Service

ASCII — American National Standard Code for Information Interchange

Ascending Pass — With respect to a particular ground station, a satellite pass during which the spacecraft is headed in a northerly direction while it is in range.

Astable (free-running) multivibrator — A circuit that alternates between two unstable states..

ATV (amateur television) — A fast-scan TV system that can use commercial transmission standards on the 70-cm band and higher frequencies.

Audio FSK (AFSK) — Generating a frequency shift keying (FSK) with tones input to the transmitter through the audio or voice input.

Aurora — A disturbance of the atmosphere at high latitudes resulting from an interaction between electrically charged particles from the Sun and the magnetic field of the Earth. *Auroral propagation* occurs when HF through UHF signals are reflected from the aurora to reach another station.

Automatic control — The operation of an amateur station without a control operator present at the control point. In §97.3(a)(6), the FCC defines automatic control as "The use of devices and procedures for control of a station when it is transmitting so that compliance with the FCC Rules is achieved without the control operator being present at a control point."

Automatic Packet Reporting System (APRS) — Also called Automatic Position Reporting System. A system of sending location and other data over packet radio to a common Web site for tracking and recording purposes.

Auxiliary station — A station that provides point-to-point communication with other stations in a communications system.

Avalanche point — That point on a diode characteristic curve where the amount of reverse current increases greatly for small increases in reverse bias voltage.

Average power — The product in watts (W) of the RMS current and voltage values in a purely resistive circuit.

Azimuthal pattern — An antenna's radiation pattern at different compass bearings.

B

Back EMF — An opposing electromotive force (voltage) produced by a changing current in a coil. It can be equal to (or greater than) the applied EMF under some conditions.

Balanced modulator — A circuit used in a single-sideband suppressed-carrier transmitter to combine a voice signal and an RF signal. The balanced modulator isolates the input signals from each other and the output, so that only the difference of the two input signals reaches the output.

Band-pass filter — A circuit that passes signals within a specified frequency range (the passband) and attenuates signals otherwise.

Bandwidth — (1) The frequency range over which a signal or the output of a circuit is within 3 dB of its peak strength within that range. (2) The frequency range over which a circuit or antenna meets a specified performance requirement.

Barkhausen criterion — To oscillate, a circuit's gain multiplied by the amount of feedback must be equal to or greater than 1 with a phase shift of 0° or 360°.

Base loading — The technique of adding series inductance at the base of an electrically short vertical antenna to cancel the capacitive reactance of the antenna.

Baseband video — A video signal with its lowest frequency component at or near dc. In an amateur television system, the video signal used to modulate the RF carrier or that is recovered from the modulated signal.

Baud — A unit of signaling speed equal to the number of discrete conditions or events per second. (For example, if an FSK signal changes frequency every 3.33 milliseconds, the signaling or baud rate is 300 bauds or the reciprocal of 0.00333 seconds.)

Baudot code — A digital code used for radioteletype operation, and also known as the International Telegraph Alphabet Number 2 (ITA2). Each character is represented by five data bits, plus additional start and stop bits.

Beacon station — A station that transmits continuously, allowing other stations to assess propagation to and from the location of the beacon station.

Beamwidth — The angular width of the major lobe of an antenna's radiation pattern within which the relative power is at least one-half (–3 dB) of the value at the peak of the lobe.

Beta cutoff frequency — The frequency at which a transistor's common-emitter current gain decreases to 0.707 times that at 1 kHz.

Bias stabilization — A technique of circuit design by which bias remains steady over temperature.

BiCMOS — A digital logic family that combines bipolar and CMOS technology in a single integrated circuit.

Bipolar junction transistor — A transistor made of three layers of alternating type material (N or P) creating two PN semiconductor junctions between them.

Bistable multivibrator — A type of *astable* circuit that has two stable output states.

Bit rate — The rate at which individual data bits are transferred by a communications system.

Blanking — Portion of a video signal that is "blacker than black," used to allow an electron beam to move across the image without being detected by the viewer.

Blocking — A receiver condition in which reception of a desired weak signal is prevented because of a nearby, unwanted strong signal that causes a reduction in gain.

Butterworth filter — A filter whose frequency response within its passband is as flat as possible.

C

Cabrillo format — A standardized digital file format for submitting information in a contest log.

Capacitive coupling — A method of transferring energy by means of an electric field.

Capture effect — An effect of FM and PM receivers whereby the strongest signal to reach the demodulator is the only one demodulated.

Cardioid radiation pattern — A heart-shaped antenna pattern characterized by a single main lobe and a deep, narrow null in the opposite direction.

Cathode — The terminal connected to the negative supply for current to flow through a device.

Center loading — A technique of adding series inductance at or near the center of an electrically short antenna in order to cancel the capacitive reactance of the antenna.

CEPT (European Conference of Postal and Telecommunications Administrations) agreement — A multilateral operating arrangement that allows US amateurs to operate in many European countries, and amateurs from many European countries to operate in the US.

Certificate of Successful Completion of Examination (CSCE) — A document issued by a Volunteer Examiner Team to certify that a candidate has passed specific exam elements at their test session. If the candidate qualified for a license upgrade at the exam session, the CSCE provides the authority to operate using the newly earned license privileges, with special identification procedures.

Certification — Equipment authorization granted by the FCC used to ensure that the equipment will function properly in the service for which it has been accepted.

Characteristic curves — A set of graphs that describe the operation of an electronic component in terms of input and output signals.

Charge-coupled device (CCD) — An integrated circuit that uses a combination of analog and digital circuitry to sample and store analog signal voltage levels.

Chebyshev filter — A filter whose passband and stop band frequency responses have an equal-amplitude ripple and a sharper transition to the stop band than does a Butterworth filter.

Chroma (chrominance) — Information in a composite video signal that carries the color information. A *chroma burst* is a short period of signal used to synchronize color processing circuitry.

Circular polarization — Describes an electromagnetic wave in which the electric and magnetic fields are rotating. If the electric field vector is rotating in a clockwise sense, then it is called *right-hand circular polarization* and if the electric field is rotating in a counterclockwise sense, it is called *left-hand circular polarization*. Polarization sense is determined looking in the direction in which the wave is traveling.

Circulator — A passive device with three or more ports that allows radio waves to travel between ports in only one direction.

Closed-loop gain — The gain of a feedback circuit with the feedback loop connected.

CMOS — Complementary Metal Oxide Semiconductor (digital logic family)

Code — (digital data) The method by which information is converted to and from digital data.

Code division multiple access (CDMA) — A method of using spread spectrum techniques to share a common frequency range by assigning each signal a different spreading code.

Common — In a transistor circuit (common-emitter/collector/base/source/gate/drain), the transistor electrode that is shared or used as a reference for both input and output circuits.

Common-mode — A signal that is present and in-phase on all conductors in a multi-conductor cable. Also used to refer to signals that are present on the outside of a cable shield.

Compensation — Adjustment of a probe or other device's frequency response to have the maximum bandwidth.

Complementary metal-oxide semiconductor (CMOS) — A type of construction used to make digital integrated circuits with both N-channel and P-channel MOS devices on the same chip.

Complex number — A number that includes both a real and an imaginary part. Complex numbers provide a convenient way to represent a quantity (such as impedance) that is made up of two different quantities (like resistance and reactance) or that has both amplitude and phase.

Complex waveform — A waveform made up of signals at more than one frequency.

Component — (signal) One of the signals making up a complex waveform. Components may or may not be harmonically related.

Composite video — A video signal that includes all of the image and control information.

Conductance (G) — The reciprocal of resistance. The real part of complex admittance.

Conducted noise — Electrical noise that is imparted to a radio receiver or transmitter through the connections to the radio.

Conduction angle — The portion of one cycle during which an amplifier circuit's output device is conducting current.

Contest — An Amateur Radio operating activity in which operators try to contact as many other stations as possible.

Control characters — Special characters that cause specific actions when received by a communications system.

Control link — A device used by a control operator to manipulate the station adjustment controls from a location other than the station location. A control link provides the means of control between a control point and a remotely controlled station.

Conventional current — Current considered to be the flow of positive charge from a positive voltage to a negative voltage. This is the opposite of *electronic current*, the flow of actual electrons. Both are equivalent.

Counter — (divider, divide-by-n counter) A circuit that is able to change from one state to the next each time it receives an input signal. A counter produces an output signal every time a predetermined number of input signals have been received.

Cross-modulation — See intermodulation distortion.

Cross-polarized — Antennas that are aligned with their polarization at right angles.

Crystal-lattice filter — A filter that employs quartz crystals as the reactive elements. They are most often used in the IF stages of a receiver or transmitter.

Current gain (beta, β) — The ratio of collector to base current in a bipolar junction transistor.

Curve tracer — An instrument that generates *characteristic curves*.

Cutoff frequency — The frequency at which (1) the output power of a passive circuit is reduced to half of its input or (2) the power gain of an active circuit is one-half its peak gain.

D

Data rate — The rate at which data is transferred over any specific portion of a communications system.

Data stream — The sequence of characters or values transferred by a communications system.

Data throughput — The net quantity of data transferred per second over a communications system.

Decibel (dB) — One tenth of a bel, denoting a logarithm of the ratio of two power levels: dB = 10 log (P2/P1). Power gains and losses are expressed in decibels.

Degenerative emitter feedback — The technique of causing increasing emitter current to reduce bias current, preventing thermal runaway. Also known as *self-bias*.

Delta match — A method for impedance matching between an open-wire transmission line and a half-wave radiator that is not split at the center. The feed line wires are fanned out to attach to the antenna wire symmetrically around the center point. The resulting connection looks somewhat like a capital Greek delta.

Depletion mode — Type of FET in which drain-source current is reduced by reverse bias on the gate.

Descending pass — With respect to a particular ground station, a satellite pass during which the spacecraft is headed in a southerly direction while it is in range.

Desensitization — A reduction in receiver sensitivity caused by the receiver front end being overloaded by noise or RF from a local transmitter. See also *blocking*.

Detector — A circuit used in a receiver to recover the modulation (voice or other information) signal from the RF signal.

Deviation — The peak difference between an instantaneous frequency of the modulated wave and the unmodulated-carrier frequency in an FM system.

Deviation ratio — The ratio of the maximum frequency deviation to the maximum modulating frequency in an FM system.

Dielectric — An insulating material in which energy can be stored by an electric field.

Dielectric constant (k) — Relative figure of merit for an insulating material used as a dielectric. This is the property that determines how much electric energy can be stored in a unit volume of the material per volt of applied potential.

Diffraction — Bending of waves by an edge or corner.

Digipeater — A station that relays digital data transmissions.

Digital-to-analog converter (DAC) — A circuit that converts digital values to analog signals.

Digital IC — An integrated circuit whose operation is characterized by one of two states, on or off.

Digital Radio Mondiale (DRM) — A digital modulation method used to transfer audio and data on HF bands.

Dip meter — A tunable RF oscillator that supplies energy to another circuit resonant at the frequency that the oscillator is tuned to. A meter indicates when the most energy is being coupled out of the circuit by showing a dip in indicated current.

Dipole antenna — An antenna with two electrical halves; literally, two poles. Most dipoles used by amateurs are ½ wavelength long at the operating frequency.

Direct digital synthesizer (DDS) — The technique of generating a signal from a sequence of digital values stored in a table.

Direct FSK — Generating a frequency shift keying (FSK) signal by shifting the transmitter frequency directly under the control of a digital signal.

Direct sequence — A spread-spectrum communications system where a very fast binary bit stream is used to shift the phase of an RF carrier.

Direction-finding — A method of using directional antennas to determine the location of a hidden or unknown transmitter. Also called DFing or fox hunting.

Directive antenna — An antenna that concentrates the radiated energy to form one or more major lobes in specific directions. The receiving pattern is the same as the transmitting pattern.

Discriminator — A circuit that translates changes in a signal's frequency to voltage variations.

Doping — The addition of impurities to a semiconductor material, with the intent to provide either excess electrons or positive charge carriers (holes) in the material.

Doppler shift — A change in the observed frequency of a signal, as compared with the transmitted frequency, caused by satellite movement toward or away from you. This is also called the *Doppler effect*.

Downlink — The frequency on which a satellite transmits information to Earth.

Drain — The point at which the charge carriers exit an FET.

Drift — (op amp) Long-term variations in the output signal amplitude or frequency of a circuit, particularly with temperature.

Driver — An amplifier that delivers power to the output amplifier in a transmitter.

DSB-SC — Double-sideband, suppressed carrier

DX — Distance. On HF, often used to describe stations in countries outside your own.

Dynamic range — The range input signal amplitudes over which a receiver responds linearly, without generating spurious signals. *Blocking dynamic range* and *intermodulation distortion (IMD) dynamic range* are the two most common dynamic range measurements used to predict receiver performance.

E

E plane — The plane of the electric field of an antenna's radiation.

Earth station — An amateur station located on, or within 50 km of the Earth's surface intended for communications with space stations or with other Earth stations by means of one or more other objects in space.

Edge-triggered — A digital logic device that changes state when the input signal changes state.

Effective isotropic radiated power (EIRP) — Same as ERP except the reference antenna is an isotropic radiator.

Effective radiated power (ERP) — A measure of the power radiated from an antenna system. ERP takes into account transmitter output power, feed line losses and other system losses, and antenna gain as compared to a dipole.

Electric field — The region through which an electric force will act on an electrically charged object.

Electric force — The push or pull exerted through an electric field by one electrically charged object on another.

Electromagnetic waves — Energy moving through space or materials in the form of changing electric and magnetic fields.

Electrostatic field — An electric field that does not change with time.

Element — (data transmission) The individual symbols that make up a digital data code.

Elevation pattern — An antenna's radiation pattern at all vertical angles.

Elliptical filter — A filter with equal-amplitude passband ripple and points of infinite attenuation in the stop band.

EME — Earth-Moon-Earth (see also *moonbounce*)

Emission designators — A method of identifying the characteristics of a signal from a radio transmitter using a series of three characters following the ITU system.

Emission types — A method of identifying the signals from a radio transmitter using a "plain English" format that simplifies the ITU emission designators.

Emissions — Any signals radiated by a transmitter.

Emitter follower — Another name for a *common-collector* circuit.

Enhancement mode — An FET in which drain-source current is increased by forward bias on the gate.

Equinoxes — One of two points in the orbit of the Earth around the Sun at which the Earth crosses a horizontal plane extending through the equator of the Sun. The vernal equinox marks the beginning of spring and the autumnal equinox marks the beginning of autumn.

Error correction — see *Forward error correction*

Error detection — The use of special codes to allow a receiving system to detect certain types of transmission errors.

Examination elements — Any of the exam sections required for an Amateur Radio license.

Exchange — Information that is exchanged during a contact in a contest.

F

Fall time — The time it takes for a waveform to reach a specified minimum value, often 10% of the steady-state maximum value.

Far field — The region in which an antenna's radiation pattern no longer changes with distance from the antenna.

Faraday rotation — A rotation of the polarization of radio waves when the waves travel through the ionized magnetic field of the ionosphere.

Fast-scan TV (FSTV) — Another name for amateur television (ATV), used because a new frame is transmitted
every $\frac{1}{30}$ of a second, as compared to every 8 seconds for slow-scan TV.

Ferrite — A ceramic material with magnetic properties.

Field — (video) In an interlaced video system, the image is made up of a sequence of fields that contain some of the image information. See *electric field* and *magnetic field*.

Field-effect transistor (FET) — A semiconductor device that uses voltage to control output current.

Fixed-point calculations — A type of computer arithmetic in which all values are represented by a fixed number of digits.

Flip-flop — (*bistable multivibrator*) A circuit that has two stable output states, and which can change from one state to the other when the proper input signals occur.

Floating-point calculations — A type of computer arithmetic in which values are represented by a fixed-point value (the mantissa) and an exponent.

FM — Frequency modulation

Forward bias — A voltage applied across a semiconductor junction to cause it to conduct current.

Forward error correction (FEC) — The method of adding special codes to a data stream so that a receiving system can detect and correct certain types of transmission errors.

Fox hunting — see *direction finding*

Frame — (video) A complete image.

Framing bits — Data bits added to a character or value's code to allow the receiving system to synchronize to the data transmission. Also called *start* and *stop* bits.

Free space — (antenna) A region far from any reflecting or absorbing surfaces.

Frequency counter — A digital electronic device that counts the cycles of a signal and displays the frequency of the signal.

Frequency division multiplexing (FDM) — Combining more than one stream of information in a single transmitted signal by using different modulating frequencies.

Frequency domain — Representing signals in terms of frequency and amplitude.

Frequency hopping — A spread-spectrum communications system where the center frequency of a conventional carrier is altered many times a second in accordance with a pseudorandom list of channels.

Frequency modulation — A method of superimposing an information signal on an RF carrier wave in which the instantaneous frequency of an RF carrier wave is varied in relation to the information signal strength.

Frequency shift keying (FSK) — A method of digital modulation in which individual bit values are represented by specific frequencies. If two frequencies are used, one is called *mark* and one *space*.

Frequency standard — A circuit or device used to produce a highly accurate reference frequency. The frequency standard may be a crystal oscillator in a marker generator or a radio broadcast, such as from WWV, with a carefully controlled transmit frequency.

Front-to-side/back/rear ratio — The ratio of field strength at the peak of the major lobe to that in the specified direction. Rear implies an average value over a specified angle centered on the back direction.

G

Gain — (amplifier) The increase in signal voltage or current provided by an amplifier circuit. (antenna) An increase in the effective power radiated by an antenna in a certain desired direction at the expense of power radiated in other directions.

Gain compression — A reduction in receiver gain caused by a strong input signal. See also see *blocking*

Gamma match — A method for matching the impedance of a feed line to a half-wave radiator that is split in the center (such as a dipole). It consists of an adjustable arm that is mounted close to the driven element and in parallel with it near the feed point. The connection looks somewhat like a capital Greek gamma.

Gate — (FET) Control terminal of an FET. (logic) A combinational logic element with two or more inputs and one output. The output state depends upon the state of the inputs.

Gateway — A digital station that provides connections between digital communications systems.

Geosynchronous — An orbit approximately 23,000 miles above the Earth in which a satellite takes the same time to make one orbit as one revolution of the Earth. If the satellite's orbit is aligned such that it remains above the same point on the Earth, it is a *geostationary orbit.*

Graticule (or graticle) — The calibrated markings on a display by which measurements are made.

Gray-line propagation — Propagation along the region on either side of the terminator, the dividing line on the Earth's surface between daylight and darkness.

Great circle — An imaginary circle around the surface of the Earth formed by the intersection of the surface with a plane passing through the center of the Earth.

Great-circle path — The most direct distance between two points on the surface of the Earth, which follows the arc of a great circle passing through both points.

Grid square locator — A 2° longitude by 1° latitude rectangle — part of the world wide Maidenhead locator system. Grid square locators are exchanged in some contests, and are used as the basis for some VHF/UHF awards.

Ground-wave propagation — Radio signals propagating along the ground rather than through the ionosphere or by some other means.

Gunn diode — A special, highly-doped diode that exhibits negative resistance over a portion of its characteristic curve.

H

H plane — The plane of the magnetic field of an antenna's radiation.

Half-power points — Those points on the response curve of an amplifier or filter or antenna radiation pattern where the power is one-half its maximum value. See *bandwidth*.

Height above average terrain (HAAT) — The height of an antenna above an average elevation of the surrounding terrain determined by measurements along several radial lines from the antenna.

High-pass filter — A filter that allows signals above the cutoff frequency to pass and attenuates signals below the cutoff frequency.

Horizontal polarization — An electromagnetic wave in which the electric field is horizontal, or parallel to the Earth's surface.

Hot-carrier diode — A type of diode in which a Schottky barrier is created by a small metal dot placed on a single semiconductor layer. See *Schottky barrier*.

I

IARP (International Amateur Radio Permit) — A multilateral operating arrangement that allows US amateurs to operate in many Central and South American countries, and amateurs from many Central and South American countries to operate in the US.

IF — Intermediate frequency.

Image signal — An unwanted signal that mixes with a receiver local oscillator to produce a signal at the desired intermediate frequency.

Impedance (Z) — The general term for opposition to current flow, either ac or dc. Impedance is made up of resistance and reactance.

Impulse noise — Noise that consists of short pulses of energy.

Inclination — The angle of a satellite's orbit with respect to the equator.

Inductive coupling — A method of transferring energy by means of a magnetic field between two coils.

Input impedance — The equivalent impedance at the input of a circuit or device.

Integrated circuit — A device composed of many transistors manufactured on the same chip, or wafer, of semiconducting material.

Intercept point — The level of a receiver input signal at which distortion products would be as strong as the desired output.

Interelectrode capacitance — The capacitance between the electrodes of a tube or transistor, particularly from the plate to the grid and from the grid to the cathode of a vacuum tube.

Intermodulation distortion (IMD) — A type of interference that results from the unwanted mixing of two strong signals, producing a signal on an unintended frequency. Often abbreviated as "intermod".

Inverter — A logic circuit with an output that is inverted from the input.

Ionizing radiation — Radiation, such as ultraviolet, X-rays, or gamma rays, from radioactive or other nuclear sources with sufficient energy to ionize atoms.

Isolator — A passive attenuator in which the loss in one direction is much greater than the loss in the other.

Isotropic radiator — An imaginary antenna in free space that radiates equally in all directions (a spherical radiation pattern). It is used as a reference to compare the gain of various real antennas.

J

Joule (J) — The unit of energy in the metric system of measure.

Junction field-effect transistor (JFET) — A field-effect transistor in which the gate electrode and channel are in direct contact and made of opposite types of semiconductor materials (N or P).

K

K index — A geomagnetic-field measurement used to indicate HF propagation conditions. Rising values generally indicate disturbed conditions while falling values indicate improving conditions.

Keplerian elements — Data that describes a satellite's orbit such that it can be located in the sky at any time.

Keyboard-to-keyboard — Digital communication in which characters typed by a user are transmitted to and displayed by the receiving station on a character-by-character or line-by-line basis.

Klystron — A type of vacuum tube that uses velocity modulation to amplify signals at UHF and microwave frequencies.

L

L network — A combination of two reactive components (L, C) used to transform or match impedances. One component is connected in series between the source and load and the other shunted across either the source or the load.

Latch — (noun) Another name for a flip-flop circuit that serves as a memory unit, storing a bit of information; (verb) the action of a circuit entering a state (often destructive) in which control signals no longer function.

Libration fading — A fluttery, rapid fading of EME signals, caused by short-term motion of the Moon's surface relative to an observer on Earth.

Light-emitting diode — A semiconductor junction that produces light when current flows through it.

Line A — A line parallel to and approximately 50 miles from the Canadian border, north of which US amateurs may not transmit on 420-430 MHz because of interference with Canadian stations.

Linear voltage regulator — A type of voltage-regulator circuit that varies the current through a series element so as to control the regulator's output voltage.

Linear IC — An integrated circuit with an output voltage that is a linear representation of its input voltage.

Linear polarization — A fixed orientation of the electric field of an electromagnetic wave, usually vertical or horizontal with respect to the Earth's surface. (Also called *plane polarization*.)

LO — Local oscillator

Load — (noun) The component, antenna, or circuit to which power is delivered; (verb) To apply a load to a circuit or a transmission line.

Loading coil — An inductor inserted in an antenna element or transmission line for the purpose of canceling capacitive reactance.

Load line — The graph of current and voltage that the output of a circuit can produce with a specific load.

Logic probe — A simple piece of test equipment used to indicate high or low logic states (voltage levels) in digital electronic circuits.

Long-path propagation — Propagation between two points on the Earth's surface that follows the longer of the two paths along the great circle between them.

Loop antenna — An antenna configured in the shape of an open shape such as a square or triangle.. If the current in the loop, or in multiple parallel turns, is essentially uniform, and if the loop circumference is small compared with a wavelength, the radiation pattern is symmetrical, with maximum response in the plane of the loop.

Low Earth Orbit (LEO) — Orbits from 200-500 miles above the Earth. The International Space Station is in LEO.

Low-pass filter — A filter that allows signals below the cutoff frequency to pass and attenuates signals above the cutoff frequency.

M

Magnetic field — The region through which a magnetic force will act on a magnetic object.

Magnetic force — The push or pull exerted through a magnetic field by one magnetic object on another.

Magnetron — A vacuum-tube diode that uses tuned cavities to produce oscillation.

Major lobe — (radiation pattern) A three-dimensional region that contains the maximum radiation peak in the space around an antenna surrounded by regions of minimum radiation. A *minor lobe* is a similar region whose peak field strength is weaker than that of the major lobe.

Marker generator — An oscillator circuit, usually crystal-controlled, with an output rich in harmonics that can be used to determine frequency on a receiver.

Matching stub — A section of transmission line used to tune an antenna element to resonance or to aid in obtaining an impedance match between the feed point and the feed line.

Maximum average forward current — The highest average current that can flow through the diode in the forward direction for a specified junction temperature.

Metal-oxide semiconductor FET (MOSFET) — A field-effect transistor with the gate insulated from the channel material. Also called an IGFET or insulated gate FET.

Meteor trail — The ionized trail that results from a particle of material entering the Earth's atmosphere from space. The particle is called a *meteoroid*, and if it reaches the ground, a *meteorite*.

Meteor-scatter communication — A method of radio communication that uses an ionized meteor trail to reflect radio signals back to Earth.

Mid-band gain — A circuit's peak gain between its upper and lower cutoff frequencies.

Minimum discernible signal (MDS) — The smallest input signal level that can just be detected separately from the receiver internal noise.

Mixer — A circuit that takes two or more input signals and produces an output that includes the sum and difference of those signal frequencies. *Single-balanced mixers* have one balanced port. *Double-balanced mixer* ports are both balanced.

Modulation index — (AM) The ratio of the maximum excursion of the modulated envelope above or below the unmodulated envelope to that of the unmodulated envelope. (FM) The ratio of the maximum frequency deviation of the modulated wave to the instantaneous frequency of the modulating signal.

Modulator — A circuit designed to superimpose an information signal on an RF carrier wave.

Monolithic microwave integrated circuit (MMIC) — An integrated circuit designed for operation at microwave frequencies. MMICs usually provide simple functions such as amplification.

Monostable multivibrator (one shot) — A circuit that has one stable state. It may be forced into an unstable state that persists for a time determined by external components but will revert to the stable state after that time.

Moonbounce — A common name for Earth-Moon-Earth (EME) communication in which signals are reflected from the Moon before being received.

Multipath — A fading effect caused by destructive interference that results from the transmitted signal traveling to the receiving station over more than one path.

Multimeter — A test instrument that can measure several different quantities, such as voltage, current and resistance. Also referred to as a *voltmeter*, *volt-ohm-meter*, or *VOM*.

Multiplier — In a contest, the special locations or other attributes by which the points from each contact are multiplied to obtain the final score. For example, states or sections or DXCC entities.

N

Near field — The region in which an antenna's radiation pattern changes with distance from the antenna. See *far field*.

Neutralization — Feeding part of the output signal from an amplifier back to the input out of phase with the input signal to prevent oscillation. This negative feedback neutralizes the effect of positive feedback caused by coupling between the input and output circuits in the amplifier.

Node — A point where a satellite crosses the plane passing through the Earth's equator. It is an *ascending node* if the satellite is moving from south to north and a *descending node* if the satellite is moving from north to south.

Noise blanker — A circuit that removes noise from the receiver output by muting the receiver during a noise pulse.

Noise figure — The ratio in dB of the noise output power to the noise input power with the input termination at a standard temperature of 290 K. It is a measure of the noise generated in the receiver circuitry.

Noise floor — The level of a receiver's internal noise.

Noise reduction — A type of *adaptive filtering* by DSP that removes unwanted components in a signal's passband.

Nonionizing radiation — Radiation, such as radio waves, with energy too low to cause the ionization of atoms.

Nonresonant rhombic antenna — A diamond-shaped antenna consisting of sides that are each at least one wavelength long. The feed line is connected to one end of the diamond and there is a terminating resistance at the opposite end. The antenna has a unidirectional radiation pattern.

Normalization (Smith chart) — Dividing all impedances by that assigned to the *prime center*.

NTSC — National Television Standard Committee. The US analog television standard.

N-type material — Semiconductor material that has been treated with impurities to give it an excess of electrons.

Nyquist sampling theorem — States that a signal can be completely characterized by a sequence of samples at or above twice the signal's maximum frequency.

O

Offset voltage — (op amp) The amplifier output voltage when the inputs are shorted.

Operating class — The category of amplifier operation based on the conduction angle of an amplifier is conducting current. Class A — 100%, Class B (push-pull) — 50%, Class AB — between 50 and 100%, Class C — less than 50%. Class D and higher — switch-mode amplifiers.

Operating point — The set of bias currents and voltages for a circuit without any signal applied.

Operational amplifier (op amp) — An amplifier with very high input impedance, very low output impedance and very high gain.

Optical shaft encoder — A device consisting of two pairs of photoemitters and photodetectors, used to sense the rotation speed and direction of a knob or dial. Optical shaft encoders are often used as controls for equipment controlled by a microprocessor.

Optocoupler (optoisolator) — A device consisting of a photoemitter and a photodetector used to transfer a signal between circuits.

Order — (intermodulation products) The sum of the integers by which the input signals are multiplied before being added or subtracted. For odd-order products, the sum is odd, and for even-order, the sum is even.

Oscillator — A amplifier with positive feedback that produces an alternating current output signal with no input signal except dc power.

Oscilloscope — A device that displays the waveform of a signal with respect to time or as compared with another signal.

Output impedance — The equivalent impedance of a signal source.

P

Packet radio — A form of digital communication using the AX.25 digital protocol.

Parabolic (dish) antenna — An antenna reflector that is a portion of a parabolic curve. Used mainly at UHF and higher frequencies to obtain high gain and narrow beamwidth when excited by one of a variety of driven elements placed at the dish focus to illuminate the reflector.

Parallel-resonant circuit — A circuit including a capacitor, an inductor and sometimes a resistor, connected in parallel, and in which the inductive and capacitive reactances are equal at the applied-signal frequency. The circuit impedance is a maximum and the current through the circuit is a minimum at the resonant frequency.

Parasitics — (amplifier) Undesired oscillations or other responses. (Component) Undesired inductance, capacitance, or resistance that occurs as a consequence of the device's construction.

Parity — A function of the number of data bits equal to 1 in a binary value. The value has even parity if the number of 1 bits is even, and odd if the number of 1 bits is odd. Parity is used for error detection.

Pass — The time during which a satellite is visible from a specific location on Earth.

Path loss — The total signal loss between transmitting and receiving stations relative to the total radiated signal energy.

PCB (hazardous materials) — Polychlorinated biphenyls, carcinogenic hydrocarbons once added to insulating oils

Peak envelope power (PEP) — The maximum average power level in a signal during one cycle during a modulation peak. (Used for modulated RF signals.)

Peak inverse voltage (PIV) — The maximum instantaneous anode-to-cathode reverse voltage that may be applied to a diode without damage.

Peak power — The product of peak voltage and peak current in a resistive circuit. (Used with sine wave signals.)

Peak voltage — A waveform's maximum positive or negative value of voltage.

Peak-to-peak (P-P) voltage — The voltage between the negative and positive peaks of a waveform.

Pedersen ray — A high-angle radio wave that penetrates deeper into the F region of the ionosphere so the wave is bent less than a lower-angle wave and thus travels for some distance through the F region, returning to Earth at a distance farther than normally expected for single-hop propagation.

Perigee — That point in the orbit of a satellite (such as the Moon) when it is closest to the Earth.

Period (satellite) — The time it takes for a complete orbit, usually measured from one equator crossing to the next. The higher the altitude of the orbit, the longer the period.

Period (T) — The time it takes to complete one cycle of an ac waveform.

Permeability (μ) — The ability of a material to store energy in a magnetic field.

Permittivity (ε) — The ability of a material to store energy in an electric field.

Phase angle — The phase relationship between two points on a waveform or between two waveforms expressed as an angle.

Phase-locked loop (PLL) — A servo loop consisting of a phase detector, low-pass filter, dc amplifier and voltage-controlled oscillator.

Phase modulation (PM) — A method of superimposing an information signal on an RF carrier wave in which the phase of an RF carrier wave is varied in relation to the information signal strength.

Phase modulator — A device capable of modulating the phase of an ac signal in response to the modulating signal.

Phase noise — Undesired variations in the phase of an oscillator signal..

Phase response — The difference between a filter's input and output signal phase with frequency.

Phase shift keying (PSK) — A method of modulation in which the phase of a carrier signal is varied to represent different digital values.

Phased array — An antenna in which the phase difference between currents in the elements determines the antenna's radiation pattern.

Photocell — A solid-state device in which the voltage and current-conducting characteristics change as the amount of light striking the cell changes.

Photoconductive effect — An increase in the electric conductivity of a material in response to illumination by light.

Photodetector — A device that produces a signal that changes with the amount of light striking a light-sensitive surface.

Photoelectric effect — An interaction between electromagnetic radiation and atoms resulting in an electron being freed from an atom after absorbing a photon of light.

Phototransistor — A bipolar transistor constructed so the base-emitter junction is exposed to incident light. When light strikes this surface, current is generated at the junction and amplified as in a regular transistor.

Photovoltaic effect — The conversion of light to electricity in a semiconductor material.

Pi network—A circuit in which reactances of one type (L or C) are in parallel with the input and output and a reactance of the opposite type is in series between the input and output.

Piezoelectric effect — The physical deformation of a material when a voltage is applied between its surfaces. Also, the generation of a voltage between a material's surfaces in response to physical deformation.

PIN diode — A diode consisting of a relatively thick layer of nearly pure semiconductor material (intrinsic semiconductor) with a layer of P-type material on one side and a layer of N-type material on the other.

Plane polarization — see *linear polarization*

PN junction — The contact area between two layers of opposite-type semiconductor material.

Point-contact diode — A diode that is made by a pressure contact between a semiconductor material and a metal point.

Polar-coordinate system — A method of representing the position of a point on a plane by specifying the radial distance from an origin and an angle measured counterclockwise from the $0°$ line.

Polarization — Describes whether the electric field of a wave is oriented vertically or horizontally with respect to the Earth.

Potential energy — Stored energy. This stored energy can do some work when it is "released." For example, electrical energy can be stored as an electric field in a capacitor or as a magnetic field in an inductor and produce energy when released.

Power — The time rate of transferring or transforming energy, or the rate at which work is done. In an electric circuit, power is calculated by multiplying the voltage applied to the circuit by the current through the circuit.

Power factor — The ratio of *real power* to *apparent power* in a circuit. Calculated as the cosine of the phase angle between current and voltage in a circuit.

Pre-emphasis — The technique of emphasizing high audio frequencies to improve a PM modulator's performance. *De-emphasis* restores the original signal's spectrum by de-emphasizing high frequencies at the receiver.

Prescaler — A divider circuit used to increase the useful range of a frequency counter.

Preselector — A tunable filter at the input to a receiver to reject out-of-band signals.

Prime center — The impedance assigned to the center of the *Smith chart.*

Product detector — A SSB detector that demodulates an SSB signal by multiplying it by the signal from a *beat-frequency oscillator (BFO)* tuned to the carrier frequency

Pseudonoise (PN) — A binary sequence designed to appear to be random (contain an approximately equal number of ones and zeros). Pseudonoise is generated by a digital circuit and mixed with digital information to produce a direct-sequence spread-spectrum signal.

P-type material — A semiconductor material that has been treated with impurities to give it an electron shortage. This creates excess positive charge carriers, or "holes."

Pulse modulation — Modulation of an RF carrier by a series of pulses that carry the information as amplitude, position, or width variations.

Pulse-amplitude modulation (PAM) — A pulse modulation system in which the amplitude of a standard pulse is varied in relation to the information signal amplitude at any instant.

Pulse-position modulation (PPM) — A pulse modulation system in which the position (timing) of the pulses is varied from a quiescent value in relation to the information-signal amplitude at any instant.

Pulse-width modulation (PWM) — A pulse modulation system in which the width of a pulse is varied from a quiescent value in relation to the information-signal amplitude at any instant.

Q

Q — (Component) A quality factor describing how much energy is lost in a component or circuit due to resistance compared to energy stored in reactance. (Frequency response) The ratio of center frequency of a filter to its bandwidth.

Q point — See *operating point*; also called quiescent point.

Quantization error — The uncertainty in value of a digital sample due to the range of analog values that all result in the same digital value.

Quiet hours — An FCC-imposed period during which no transmissions may be made.

Quiet zone — A region in which transmissions of all sorts are generally not allowed.

R

RACES — Radio Amateur Civil Emergency Service.

Radians — A unit of angular measurement. There are 2π radians in a circle and 1 radian = 57.3°

Radiation resistance — The equivalent resistance that would dissipate the same amount of power as is radiated from an antenna. It is calculated by dividing the radiated power by the square of the RMS antenna current.

Radio horizon — The position at which a direct wave radiated from an antenna becomes tangent to the surface of the Earth. Note that as the wave continues past the horizon, the wave gets higher and higher above the surface.

Raster — The set of video and control signals that make up a video image.

Reactance — The opposition to ac current due to capacitance or inductance.

Reactance modulator — A device capable of frequency or phase modulation by varying reactance in an oscillator or amplifier circuit in response to the modulating signal.

Reactive power — The apparent power in an inductor or capacitor. The product of RMS current through a reactive component and the RMS voltage across it. Also called *wattless power*.

Real power — The actual power dissipated in a circuit, calculated to be the product of the apparent power times the cosine of the phase angle between the voltage and current.

Reciprocal operating permission — Mutual agreement by the government of the United States and another country to recognize each other's amateur licenses as sufficient to grant operating privileges.

Rectangular-coordinate system — A method of representing the position of a point on a plane by specifying the distance from an origin in two perpendicular directions.

Reflection coefficient (ρ) — The ratio of the reflected voltage at a given point on a transmission line to the incident voltage at the same point. The reflection coefficient is also equal to the ratio of reflected and incident currents.

Refraction — The bending of waves due to changes in the velocity of propagation.

Remote control — The operation of an Amateur Radio station using a control link to manipulate the station operating controls from somewhere other than the station location.

Resonant frequency — (circuit) The frequency at which a circuit including capacitors and inductors presents a purely resistive impedance. The inductive reactance in the circuit is equal to the capacitive reactance. (antenna) The frequency at which an antenna's feed point impedance is purely resistive.

Resonant rhombic antenna — A diamond-shaped antenna consisting of sides that are each at least one wavelength long. The feed line is connected to one end of the diamond, and the opposite end is left open. The antenna has a bidirectional radiation pattern.

Reverse bias — A voltage applied to a device that prevents or opposes current flow..

Ripple — (filter) Amplitude response variations, (power supply) the ac component in the power supply's output voltage.

Rise time — The time it takes for a waveform to reach a specified maximum value. Usually 90% of a steady-state minimum value.

Roofing filter — A relatively wide receiving filter that prevents signals near the desired frequency from overloading gain stages.

Root-mean-square (RMS) voltage — A measure of the effective value of an ac voltage. The value of a dc voltage that would produce the same amount of heat in a resistance as the ac voltage.

Running — Operating in a contest by repeatedly calling CQ on a single frequency.

S

Sampling — Changing an analog signal into digital form by converting it to digital form at repeated intervals.

Sawtooth wave — A waveform consisting of a linear ramp followed by an abrupt return to the original value.

Schottky barrier diodes — A diode made by using a Schottky barrier, the junction between semiconductor material and metal. (See *hot-carrier diode* and *point-contact diode*.)

Search-and-pounce — Operating in a contest by tuning across a band to find new stations to contact.

Selective fading — A variation of radio-wave intensity that changes over small frequency changes. It may be caused by changes in the medium through which the wave is traveling or changes in transmission path, among other things.

Selectivity — A measure of the ability of a receiver to distinguish between a desired signal and an undesired one at a different frequency. Selectivity can be applied to the RF (*front-end selectivity*), IF and AF stages.

Semiconductor material — A material with conductivity between that of metals and insulators.

Sense antenna — An omnidirectional antenna used in conjunction with an antenna that exhibits a bidirectional pattern to produce a radio direction-finding system with a cardioid pattern.

Sensitivity — A measure of the minimum input signal level that will produce a specified output from a receiver.

Sequential logic — A type of circuit element that has at least one output and one or more input channels, and in which the output state depends on the previous input states. A flip-flop is a sequential-logic element.

Serial number — In a contest, the sequential number of a contact.

Series-resonant circuit — A circuit including a capacitor, an inductor and sometimes a resistor, connected in series, and in which the inductive and capacitive reactances are equal at the applied-signal frequency. The circuit impedance is at a minimum and the current through the circuit is a maximum at the resonant frequency.

Shannon's Information Theorems — The laws of information transmission relating information content, noise and bandwidth.

Shift — The frequency difference between mark and space frequencies in a frequency shift keying signal.

Signal-to-noise ratio (SNR) — The numeric ratio of signal power to noise power in a given bandwidth. *Signal-to-noise-plus-distortion (SINAD)* adds distortion product power to the noise power.

Sine wave — A single-frequency waveform expressed in terms of the mathematical sine function.

Single-sideband, suppressed-carrier signal (SSB) — A radio signal in which only one of the two sidebands generated by amplitude modulation is transmitted. The other sideband and the RF carrier wave are removed before the signal is transmitted.

Skin effect — A condition in which ac flows in the outer portions of a conductor. The higher the signal frequency, the less the electric and magnetic fields penetrate the conductor and the smaller the effective cross-sectional area of a given conductor for carrying the electrons.

Sky-wave — Radio signal propagation through the ionosphere to reach the receiving station.

Slew rate — The maximum rate at which the output voltage of an op amp can change.

Slow-scan television (SSTV) — A television system used by amateurs to transmit pictures within a voice signal's bandwidth allowed on the HF bands by the FCC. Each frame takes several seconds to transmit.

Smith chart — A coordinate system developed by Phillip Smith to represent complex impedances. This chart makes it easy to perform calculations involving antenna and transmission-line impedances and SWR.

Solar wind —Particles emitted by the Sun and traveling through space.

Solenoid — A coil wound in a single-layer around a central axis. Such a coil is *solenoidal*.

Source — The point at which the charge carriers enter the channel of a field-effect transistor (FET).

Space station — An amateur station located more than 50 km above the Earth's surface.

Specific absorption rate (SAR) — The rate at which the body absorbs electromagnetic energy.

Spectrum analyzer — A test instrument generally used to display the power (or amplitude) distribution of a signal with respect to frequency.

Spin modulation — Periodic amplitude fade-and-peak variations resulting from a satellite's spin.

Split — Transmitting on one frequency and listening on another, mostly used during DX operation

Spread-spectrum (SS) communication — A communications method in which the RF bandwidth of the transmitted signal is much larger than that needed for traditional modulation schemes, and in which the RF bandwidth is independent of the modulation content. Increasing the bandwidth of the signal by means of a randomizing sequence (*spreading code*) is called *spreading*.

Spurious emissions — Any emission that is not part of the desired signal. The FCC defines this term as "an emission, on frequencies outside the necessary bandwidth of a transmission, the level of which may be reduced without affecting the information being transmitted."

Square wave — A periodic waveform that alternates between two values and spends an equal time at each value.

Stage — Each separate circuit in a sequence of circuits.

Store-and-forward — A data system that receives and stores an entire message before relaying it to another station. This does not apply to systems that have internal digital connections between a receiving system and a transmitting system, such as a digital mode repeater.

Subcarrier — In a multiplexed signal, the subcarrier modulates the main carrier and carries a separate data or information stream.

Sun-synchronous orbit — An orbit that places the satellite above approximately the same location on Earth at the same time each day.

Surface-mount package — An electronic component without wire leads, designed to be soldered directly to copper-foil pads on a circuit board.

Susceptance (B) — The reciprocal of reactance. This is the imaginary part of complex admittance.

Switching regulator — A voltage-regulator circuit in which the output voltage is controlled by turning the pass element on and off at a high rate, usually 10 to 15 kHz or higher. The control-element duty cycle is proportional to the line or load conditions.

SWR bandwidth — The frequency range over which an antenna's SWR is less than a specified value.

Sync (video) — Pulses used to reset the horizontal and vertical position of the electron beam.

Synchronous logic — Digital circuits that operate based on a common clock. *Asynchronous logic* does not use a clock signal.

T

Takeoff angle — The elevation angle at the peak of an antenna's major lobe.

Tank circuit — A resonant circuit used to store energy and smooth or filter the output of an amplifier.

Telecommand operation — A one-way transmission to initiate, modify, or terminate functions of a device at a distance.

Telecommand station — An amateur station that transmits communications to initiate, modify, or terminate functions of a space station.

Telemetry — A one-way transmission of measurements to a receiving station at a distance from the measuring instrument.

Thermal runaway — An uncontrolled increase in current in a semiconductor due to heating

Thevenin's theorem — Any combination of voltage sources and impedances, no matter how complex, can be replaced by a single voltage source and a single impedance that will present the same voltage and current to a load circuit.

Time constant — The product of resistance and capacitance in a simple series or parallel RC circuit, or the inductance divided by the resistance in a simple series or parallel RL circuit. One time constant is the time required for a voltage across a capacitor or a current through an inductor to build up to 63.2% of its steady-state value, or to decay to 36.8% of the initial value.

Time division multiplexing (TDM) — Combining more than one stream of information in a single transmitted signal by using different time periods or "slots" for each stream.

Time domain — Representing signals in terms of time and amplitude.

Top loading — The addition of capacitive reactance (a capacitance hat) at the end opposite the feed point of an electrically short antenna in order to increase its electrical length.

Toroid — A coil wound on a ring-shaped form, usually made of ferrite or powdered-iron.

Transconductance (g_m) — The ratio of output current to input voltage, primarily used with FETs and vacuum tubes.

Transequatorial propagation — A form of F-layer ionospheric propagation, in which signals of higher frequency than the expected MUF are propagated across the Earth's magnetic equator.

Transponder — A repeater aboard a satellite that retransmits, on another frequency band, any type of signals it receives.

Traps — Parallel LC circuits inserted in an antenna element to provide multiband operation.

Triangulation — A radio direction-finding technique in which compass bearings from two or more locations are taken, and lines are drawn on a map to predict the location of a radio signal source.

Tropospheric ducting — A type of radio-wave propagation whereby weather conditions cause portions of the troposphere to act like a duct or waveguide for VHF and higher-frequency radio signals.

Truth table — A chart showing the outputs for all possible input combinations to a digital circuit.

U

Universal stub system — A matching network consisting of a pair of transmission line stubs that can transform any impedance to any other impedance.

Uplink — The frequency used to transmit information to a satellite.

V

Varactor diode — A diode with a PN junction optimized to act as a variable capacitance as reverse voltage across it is varied. Used in RF circuits as a means of tuning or modulation.

Vector — A quantity consisting of both a magnitude and a direction. For example, 20 MPH on a bearing of 244 degrees or 5.3 ohms with a phase angle of 21 degrees.

Velocity factor (velocity of propagation) — An expression of how fast a radio wave will travel through a material or transmission line. It is usually stated as a fraction of the speed the wave would have in free space (where the wave would have its maximum velocity). Velocity factor is also sometimes specified as a percentage of the speed of a radio wave in free space.

Velocity modulation — A type of amplification technique that works by changing the speed of electrons in a beam inside a vacuum tube.

Vertical Interval Signaling (VIS) — The method of identifying the type of SSTV signal by sending coded information during the vertical synchronization period.

Vertical polarization — Describes an electromagnetic wave in which the electric field is vertical, or perpendicular to the Earth's surface.

Vestigial sideband (VSB) — A signal-transmission method in which one sideband, the carrier and part of the second sideband are transmitted.

Virtual ground — A point in a circuit maintained at ground potential but without a direct connection to ground.

Viterbi encoding — A method of forward error correction in which the possible sequences of transmitted codes is restricted to allow the receiver to predict which sequence, called the *Viterbi path*, is the most likely received sequence.

Voltage divider — A circuit with two or more resistors in series whose function is to reduce an applied voltage by a specific fraction

Volunteer Examiner (VE) — A licensed amateur who is accredited by a Volunteer Examiner Coordinator (VEC) to administer amateur license exams.

Volunteer Examiner Coordinator (VEC) — An organization that has entered into an agreement with the FCC to coordinate amateur license examinations.

Voltage-controlled oscillator (VCO) — An oscillator whose frequency is varied by means of an applied control voltage.

W

Wavefront — A surface traveling through space on which an electromagnetic wave has a constant electric and magnetic field strength.

Weak signal — On the VHF/UHF bands, operation on modes that perform better at low signal-to-noise ratios, such as CW or SSB.

Z

Zener diode — A diode that is designed to be operated in the reverse-breakdown region of its characteristic curve. The *Zener voltage* is the reverse voltage at which the diode begins to conduct reverse current.

Chapter 13

Question Pool

Extra Class (Element 4) Syllabus

SUBELEMENT E1 — COMMISSION'S RULES

[6 Exam Questions — 6 Groups]

E1A Operating Standards: frequency privileges; emission standards; automatic message forwarding; frequency sharing; stations aboard ships or aircraft

E1B Station restrictions and special operations: restrictions on station location; general operating restrictions, spurious emissions, control operator reimbursement; antenna structure restrictions; RACES operations

E1C Station control: definitions and restrictions pertaining to local, automatic and remote control operation; control operator responsibilities for remote and automatically controlled stations

E1D Amateur Satellite service: definitions and purpose; license requirements for space stations; available frequencies and bands; telecommand and telemetry operations; restrictions, and special provisions; notification requirements

E1E Volunteer examiner program: definitions, qualifications, preparation and administration of exams; accreditation; question pools; documentation requirements

E1F Miscellaneous rules: external RF power amplifiers; national quiet zone; business communications; compensated communications; spread spectrum; auxiliary stations; reciprocal operating privileges; IARP and CEPT licenses; third party communications with foreign countries; special temporary authority

SUBELEMENT E2 — OPERATING PROCEDURES

[5 Exam Questions — 5 Groups]

E2A Amateur radio in space: amateur satellites; orbital mechanics; frequencies and modes; satellite hardware; satellite operations

E2B Television practices: fast scan television standards and techniques; slow scan television standards and techniques

E2C Operating methods: contest and DX operating; spread-spectrum transmissions; selecting an operating frequency

E2D Operating methods: VHF and UHF digital modes; APRS

E2E Operating methods: operating HF digital modes; error correction

SUBELEMENT E3 — RADIO WAVE PROPAGATION

[3 Exam Questions — 3 Groups]

E3A Propagation and technique, Earth-Moon-Earth communications; meteor scatter

E3B Propagation and technique, trans-equatorial; long path; gray-line; multi-path propagation

E3C Propagation and technique, Aurora propagation; selective fading; radio-path horizon; take-off angle over flat or sloping terrain; effects of ground on propagation; less common propagation modes

SUBELEMENT E4 — AMATEUR PRACTICES

[5 Exam Questions — 5 Groups]

E4A   Test equipment: analog and digital instruments; spectrum and network analyzers, antenna analyzers; oscilloscopes; testing transistors; RF measurements

E4B   Measurement technique and limitations: instrument accuracy and performance limitations; probes; techniques to minimize errors; measurement of "Q"; instrument calibration

E4C   Receiver performance characteristics, phase noise, capture effect, noise floor, image rejection, MDS, signal-to-noise-ratio; selectivity

E4D   Receiver performance characteristics, blocking dynamic range, intermodulation and cross-modulation interference; 3rd order intercept; desensitization; preselection

E4E   Noise suppression: system noise; electrical appliance noise; line noise; locating noise sources; DSP noise reduction; noise blankers

SUBELEMENT E5 — ELECTRICAL PRINCIPLES

[4 Exam Questions — 4 Groups]

E5A   Resonance and Q: characteristics of resonant circuits: series and parallel resonance; Q; half-power bandwidth; phase relationships in reactive circuits

E5B   Time constants and phase relationships: RLC time constants: definition; time constants in RL and RC circuits; phase angle between voltage and current; phase angles of series and parallel circuits

E5C   Impedance plots and coordinate systems: plotting impedances in polar coordinates; rectangular coordinates

E5D   AC and RF energy in real circuits: skin effect; electrostatic and electromagnetic fields; reactive power; power factor; coordinate systems

SUBELEMENT E6 — CIRCUIT COMPONENTS

[6 Exam Questions — 6 Groups]

E6A   Semiconductor materials and devices: semiconductor materials germanium, silicon, P-type, N-type; transistor types: NPN, PNP, junction, field-effect transistors: enhancement mode; depletion mode; MOS; CMOS; N-channel; P-channel

E6B   Semiconductor diodes

E6C   Integrated circuits: TTL digital integrated circuits; CMOS digital integrated circuits; gates

E6D   Optical devices and toroids: cathode-ray tube devices; charge-coupled devices (CCDs); liquid crystal displays (LCDs); toroids: permeability, core material, selecting, winding

E6E   Piezoelectric crystals and MMICs: quartz crystals; crystal oscillators and filters; monolithic amplifiers

E6F   Optical components and power systems: photoconductive principles and effects, photovoltaic systems, optical couplers, optical sensors, and optoisolators

SUBELEMENT E7 — PRACTICAL CIRCUITS

[8 Exam Questions — 8 Groups]

E7A   Digital circuits: digital circuit principles and logic circuits: classes of logic elements; positive and negative logic; frequency dividers; truth tables

E7B   Amplifiers: Class of operation; vacuum tube and solid-state circuits; distortion and intermodulation; spurious and parasitic suppression; microwave amplifiers

E7C   Filters and matching networks: filters and impedance matching networks: types of networks; types of filters; filter applications; filter characteristics; impedance matching; DSP filtering

E7D   Power supplies and voltage regulators

E7E   Modulation and demodulation: reactance, phase and balanced modulators; detectors; mixer stages; DSP modulation and demodulation; software defined radio systems

E7F   Frequency markers and counters: frequency divider circuits; frequency marker generators; frequency counters

E7G   Active filters and op-amps: active audio filters; characteristics; basic circuit design; operational amplifiers

E7H   Oscillators and signal sources: types of oscillators; synthesizers and phase-locked loops; direct digital synthesizers

SUBELEMENT E8 — SIGNALS AND EMISSIONS

[4 Exam Questions — 4 Groups]

E8A   AC waveforms: sine, square, sawtooth and irregular waveforms; AC measurements; average and PEP of RF signals; pulse and digital signal waveforms

E8B   Modulation and demodulation: modulation methods; modulation index and deviation ratio; pulse modulation; frequency and time division multiplexing

E8C   Digital signals: digital communications modes; CW; information rate vs. bandwidth; spread-spectrum communications; modulation methods

E8D   Waves, measurements, and RF grounding: peak-to-peak values, polarization; RF grounding

SUBELEMENT E9 — ANTENNAS AND TRANSMISSION LINES

[8 Exam Questions — 8 Groups]

E9A   Isotropic and gain antennas: definition; used as a standard for comparison; radiation pattern; basic antenna parameters: radiation resistance and reactance, gain, beamwidth, efficiency

E9B   Antenna patterns: E and H plane patterns; gain as a function of pattern; antenna design; Yagi antennas

E9C   Wire and phased vertical antennas: beverage antennas; terminated and resonant rhombic antennas; elevation above real ground; ground effects as related to polarization; take-off angles

E9D   Directional antennas: gain; satellite antennas; antenna beamwidth; losses; SWR bandwidth; antenna efficiency; shortened and mobile antennas; grounding

E9E   Matching: matching antennas to feed lines; power dividers

E9F   Transmission lines: characteristics of open and shorted feed lines: 1/8 wavelength; 1/4 wavelength; 1/2 wavelength; feed lines: coax versus open-wire; velocity factor; electrical length; transformation characteristics of line terminated in impedance not equal to characteristic impedance

E9G   The Smith chart

E9H   Effective radiated power; system gains and losses; radio direction finding antennas

SUBELEMENT E0 — SAFETY

[1 exam question — 1 group]

E0A   Safety: amateur radio safety practices; RF radiation hazards; hazardous materials

Element 4 — Extra Class Question Pool

Valid July 1, 2012 through June 30, 2016

SUBELEMENT E1
COMMISSION'S RULES
6 Exam Questions — 6 Groups

E1A — Operating Standards: frequency privileges; emission standards; automatic message forwarding; frequency sharing; stations aboard ships or aircraft

E1A01
(D)
[97.301, 97.305]
Page 3-4

E1A01
When using a transceiver that displays the carrier frequency of phone signals, which of the following displayed frequencies represents the highest frequency at which a properly adjusted USB emission will be totally within the band?
A. The exact upper band edge
B. 300 Hz below the upper band edge
C. 1 kHz below the upper band edge
D. 3 kHz below the upper band edge

E1A02
(D)
[97.301, 97.305]
Page 3-4

E1A02
When using a transceiver that displays the carrier frequency of phone signals, which of the following displayed frequencies represents the lowest frequency at which a properly adjusted LSB emission will be totally within the band?
A. The exact lower band edge
B. 300 Hz above the lower band edge
C. 1 kHz above the lower band edge
D. 3 kHz above the lower band edge

E1A03
(C)
[97.301, 97.305]
Page 3-4

E1A03
With your transceiver displaying the carrier frequency of phone signals, you hear a DX station's CQ on 14.349 MHz USB. Is it legal to return the call using upper sideband on the same frequency?
A. Yes, because the DX station initiated the contact
B. Yes, because the displayed frequency is within the 20 meter band
C. No, my sidebands will extend beyond the band edge
D. No, USA stations are not permitted to use phone emissions above 14.340 MHz

E1A04

With your transceiver displaying the carrier frequency of phone signals, you hear a DX station calling CQ on 3.601 MHz LSB. Is it legal to return the call using lower sideband on the same frequency?

A. Yes, because the DX station initiated the contact
B. Yes, because the displayed frequency is within the 75 meter phone band segment
C. No, my sidebands will extend beyond the edge of the phone band segment
D. No, USA stations are not permitted to use phone emissions below 3.610 MHz

E1A04
(C)
[97.301, 97.305]
Page 3-4

E1A05

What is the maximum power output permitted on the 60 meter band?

A. 50 watts PEP effective radiated power relative to an isotropic radiator
B. 50 watts PEP effective radiated power relative to a dipole
C. 100 watts PEP effective radiated power relative to the gain of a half-wave dipole
D. 100 watts PEP effective radiated power relative to an isotropic radiator

E1A05
(C)
[97.313]
Page 3-4

E1A06

Which of the following describes the rules for operation on the 60 meter band?

A. Working DX is not permitted
B. Operation is restricted to specific emission types and specific channels
C. Operation is restricted to LSB
D. All of these choices are correct

E1A06
(B)
[97.303]
Page 3-4

E1A07

What is the only amateur band where transmission on specific channels rather than a range of frequencies is permitted?

A. 12 meter band
B. 17 meter band
C. 30 meter band
D. 60 meter band

E1A07
(D)
[97.303]
Page 3-4

E1A08

If a station in a message forwarding system inadvertently forwards a message that is in violation of FCC rules, who is primarily accountable for the rules violation?

A. The control operator of the packet bulletin board station
B. The control operator of the originating station
C. The control operators of all the stations in the system
D. The control operators of all the stations in the system not authenticating the source from which they accept communications

E1A08
(B)
[97.219]
Page 3-6

E1A09

What is the first action you should take if your digital message forwarding station inadvertently forwards a communication that violates FCC rules?

A. Discontinue forwarding the communication as soon as you become aware of it
B. Notify the originating station that the communication does not comply with FCC rules
C. Notify the nearest FCC Field Engineer's office
D. Discontinue forwarding all messages

E1A09
(A)
[97.219]
Page 3-6

E1A10
(A)
[97.11]
Page 3-7

E1A10
If an amateur station is installed aboard a ship or aircraft, what condition must be met before the station is operated?
  A. Its operation must be approved by the master of the ship or the pilot in command of the aircraft
  B. The amateur station operator must agree to not transmit when the main ship or aircraft radios are in use
  C. It must have a power supply that is completely independent of the main ship or aircraft power supply
  D. Its operator must have an FCC Marine or Aircraft endorsement on his or her amateur license

E1A11
(B)
[97.5]
Page 3-7

E1A11
What authorization or licensing is required when operating an amateur station aboard a US-registered vessel in international waters?
  A. Any amateur license with an FCC Marine or Aircraft endorsement
  B. Any FCC-issued amateur license or a reciprocal permit for an alien amateur licensee
  C. Only General class or higher amateur licenses
  D. An unrestricted Radiotelephone Operator Permit

E1A12
(C)
[97.301,
97.305]
Page 3-4

E1A12
With your transceiver displaying the carrier frequency of CW signals, you hear a DX station's CQ on 3.500 MHz. Is it legal to return the call using CW on the same frequency?
  A. Yes, the DX station initiated the contact
  B. Yes, the displayed frequency is within the 80 meter CW band segment
  C. No, sidebands from the CW signal will be out of the band.
  D. No, USA stations are not permitted to use CW emissions below 3.525 MHz

E1A13
(B)
[97.5]
Page 3-7

E1A13
Who must be in physical control of the station apparatus of an amateur station aboard any vessel or craft that is documented or registered in the United States?
  A. Only a person with an FCC Marine Radio
  B. Any person holding an FCC-issued amateur license or who is authorized for alien reciprocal operation
  C. Only a person named in an amateur station license grant
  D. Any person named in an amateur station license grant or a person holding an unrestricted Radiotelephone Operator Permit

E1B — Station restrictions and special operations: restrictions on station location; general operating restrictions; spurious emissions, control operator reimbursement; antenna structure restrictions; RACES operations

E1B01
(D)
[97.3]
Page 3-8

E1B01
Which of the following constitutes a spurious emission?
  A. An amateur station transmission made at random without the proper call sign identification
  B. A signal transmitted to prevent its detection by any station other than the intended recipient
  C. Any transmitted bogus signal that interferes with another licensed radio station
  D. An emission outside its necessary bandwidth that can be reduced or eliminated without affecting the information transmitted

E1B02

Which of the following factors might cause the physical location of an amateur station apparatus or antenna structure to be restricted?

A. The location is near an area of political conflict

B. The location is of geographical or horticultural importance

C. The location is in an ITU zone designated for coordination with one or more foreign governments

D. The location is of environmental importance or significant in American history, architecture, or culture

E1B03

Within what distance must an amateur station protect an FCC monitoring facility from harmful interference?

A. 1 mile

B. 3 miles

C. 10 miles

D. 30 miles

E1B04

What must be done before placing an amateur station within an officially designated wilderness area or wildlife preserve, or an area listed in the National Register of Historical Places?

A. A proposal must be submitted to the National Park Service

B. A letter of intent must be filed with the National Audubon Society

C. An Environmental Assessment must be submitted to the FCC

D. A form FSD-15 must be submitted to the Department of the Interior

E1B05

What is the maximum bandwidth for a data emission on 60 meters?

A. 60 Hz

B. 170 Hz

C. 1.5 kHz

D. 2.8 kHz

E1B06

Which of the following additional rules apply if you are installing an amateur station antenna at a site at or near a public use airport?

A. You may have to notify the Federal Aviation Administration and register it with the FCC as required by Part 17 of FCC rules

B. No special rules apply if your antenna structure will be less than 300 feet in height

C. You must file an Environmental Impact Statement with the EPA before construction begins

D. You must obtain a construction permit from the airport zoning authority

E1B07

Where must the carrier frequency of a CW signal be set to comply with FCC rules for 60 meter operation?

A. At the lowest frequency of the channel

B. At the center frequency of the channel

C. At the highest frequency of the channel

D. On any frequency where the signal's sidebands are within the channel

E1B02
(D)
[97.13]
Page 3-8

E1B03
(A)
[97.13]
Page 3-8

E1B04
(C)
[97.13,
1.1305-
1.1319]
Page 3-8

E1B05
(D)
[97.303]
Page 3-4

E1B06
(A)
[97.15]
Page 3-9

E1B07
(B)
[97.15]
Page 3-4

E1B08
(D)
[97.121]
Page 3-8

E1B08

What limitations may the FCC place on an amateur station if its signal causes interference to domestic broadcast reception, assuming that the receiver(s) involved are of good engineering design?

A. The amateur station must cease operation
B. The amateur station must cease operation on all frequencies below 30 MHz
C. The amateur station must cease operation on all frequencies above 30 MHz
D. The amateur station must avoid transmitting during certain hours on frequencies that cause the interference

E1B09
(C)
[97.407]
Page 3-6

E1B09

Which amateur stations may be operated in RACES?

A. Only those club stations licensed to Amateur Extra class operators
B. Any FCC-licensed amateur station except a Technician class operator's station
C. Any FCC-licensed amateur station certified by the responsible civil defense organization for the area served
D. Any FCC-licensed amateur station participating in the Military Affiliate Radio System (MARS)

E1B10
(A)
[97.407]
Page 3-6

E1B10

What frequencies are authorized to an amateur station participating in RACES?

A. All amateur service frequencies authorized to the control operator
B. Specific segments in the amateur service MF, HF, VHF and UHF bands
C. Specific local government channels
D. Military Affiliate Radio System (MARS) channels

E1B11
(A)
[97.307]
Page 3-8

E1B11

What is the permitted mean power of any spurious emission relative to the mean power of the fundamental emission from a station transmitter or external RF amplifier installed after January 1, 2003, and transmitting on a frequency below 30 MHZ?

A. At least 43 dB below
B. At least 53 dB below
C. At least 63 dB below
D. At least 73 dB below

E1B12
(B)
[97.307]
Page 7-25

E1B12

What is the highest modulation index permitted at the highest modulation frequency for angle modulation?

A. .5
B. 1.0
C. 2.0
D. 3.0

E1C — Station Control: Definitions and restrictions pertaining to local, automatic and remote control operation; control operator responsibilities for remote and automatically controlled stations

E1C01
What is a remotely controlled station?
  A. A station operated away from its regular home location
  B. A station controlled by someone other than the licensee
  C. A station operating under automatic control
  D. A station controlled indirectly through a control link

E1C01
(D)
[97.3]
Page 3-10

E1C02
What is meant by automatic control of a station?
  A. The use of devices and procedures for control so that the control operator does not have to be present at a control point
  B. A station operating with its output power controlled automatically
  C. Remotely controlling a station's antenna pattern through a directional control link
  D. The use of a control link between a control point and a locally controlled station

E1C02
(A)
[97.3, 97.109]
Page 3-12

E1C03
How do the control operator responsibilities of a station under automatic control differ from one under local control?
  A. Under local control there is no control operator
  B. Under automatic control the control operator is not required to be present at the control point
  C. Under automatic control there is no control operator
  D. Under local control a control operator is not required to be present at a control point

E1C03
(B)
[97.3, 97.109]
Page 3-12

E1C04
When may an automatically controlled station retransmit third party communications?
  A. Never
  B. Only when transmitting RTTY or data emissions
  C. When specifically agreed upon by the sending and receiving stations
  D. When approved by the National Telecommunication and Information Administration

E1C04
(B)
[97.109]
Page 3-12

E1C05
When may an automatically controlled station originate third party communications?
  A. Never
  B. Only when transmitting an RTTY or data emissions
  C. When specifically agreed upon by the sending and receiving stations
  D. When approved by the National Telecommunication and Information Administration

E1C05
(A)
[97.109]
Page 3-12

E1C06
Which of the following statements concerning remotely controlled amateur stations is true?
  A. Only Extra Class operators may be the control operator of a remote station
  B. A control operator need not be present at the control point
  C. A control operator must be present at the control point
  D. Repeater and auxiliary stations may not be remotely controlled

E1C06
(C)
[97.109]
Page 3-10

E1C07
(C)
[97.3]
Page 3-10

E1C07
What is meant by local control?
A. Controlling a station through a local auxiliary link
B. Automatically manipulating local station controls
C. Direct manipulation of the transmitter by a control operator
D. Controlling a repeater using a portable handheld transceiver

E1C08
(B)
[97.213]
Page 3-10

E1C08
What is the maximum permissible duration of a remotely controlled station's transmissions if its control link malfunctions?
A. 30 seconds
B. 3 minutes
C. 5 minutes
D. 10 minutes

E1C09
(D)
[97.205]
Page 3-12

E1C09
Which of these frequencies are available for an automatically controlled repeater operating below 30 MHz?
A. 18.110 - 18.168 MHz
B. 24.940 - 24.990 MHz
C. 10.100 - 10.150 MHz
D. 29.500 - 29.700 MHz

E1C10
(B)
[97.113]
Page 3-12

E1C10
What types of amateur stations may automatically retransmit the radio signals of other amateur stations?
A. Only beacon, repeater or space stations
B. Only auxiliary, repeater or space stations
C. Only earth stations, repeater stations or model craft
D. Only auxiliary, beacon or space stations

E1D — Amateur Satellite service: definitions and purpose; license requirements for space stations; available frequencies and bands; telecommand and telemetry operations; restrictions and special provisions; notification requirements

E1D01
(A)
[97.3]
Page 3-13

E1D01
What is the definition of the term telemetry?
A. One-way transmission of measurements at a distance from the measuring instrument
B. Two-way radiotelephone transmissions in excess of 1000 feet
C. Two-way single channel transmissions of data
D. One-way transmission that initiates, modifies, or terminates the functions of a device at a distance

E1D02
(C)
[97.3]
Page 3-12

E1D02
What is the amateur satellite service?
A. A radio navigation service using satellites for the purpose of self training, intercommunication and technical studies carried out by amateurs
B. A spacecraft launching service for amateur-built satellites
C. A radio communications service using amateur radio stations on satellites
D. A radio communications service using stations on Earth satellites for public service broadcast

E1D03

What is a telecommand station in the amateur satellite service?

A. An amateur station located on the Earth's surface for communications with other Earth stations by means of Earth satellites

B. An amateur station that transmits communications to initiate, modify or terminate functions of a space station

C. An amateur station located more than 50 km above the Earth's surface

D. An amateur station that transmits telemetry consisting of measurements of upper atmosphere data from space

E1D04

What is an Earth station in the amateur satellite service?

A. An amateur station within 50 km of the Earth's surface intended for communications with amateur stations by means of objects in space

B. An amateur station that is not able to communicate using amateur satellites

C. An amateur station that transmits telemetry consisting of measurement of upper atmosphere data from space

D. Any amateur station on the surface of the Earth

E1D05

What class of licensee is authorized to be the control operator of a space station?

A. All except Technician Class

B. Only General, Advanced or Amateur Extra Class

C. All classes

D. Only Amateur Extra Class

E1D06

Which of the following special provisions must a space station incorporate in order to comply with space station requirements?

A. The space station must be capable of terminating transmissions by telecommand when directed by the FCC

B. The space station must cease all transmissions after 5 years

C. The space station must be capable of changing its orbit whenever such a change is ordered by NASA

D. All of these choices are correct

E1D07

Which amateur service HF bands have frequencies authorized to space stations?

A. Only 40m, 20m, 17m, 15m, 12m and 10m

B. Only 40m, 20m, 17m, 15m and 10m bands

C. 40m, 30m, 20m, 15m, 12m and 10m bands

D. All HF bands

E1D08

Which VHF amateur service bands have frequencies available for space stations?

A. 6 meters and 2 meters

B. 6 meters, 2 meters, and 1.25 meters

C. 2 meters and 1.25 meters

D. 2 meters

E1D03
(B)
[97.3]
Page 3-13

E1D04
(A)
[97.3]
Page 3-12

E1D05
(C)
[97.207]
Page 3-13

E1D06
(A)
[97.207]
Page 3-14

E1D07
(A)
[97.207]
Page 3-14

E1D08
(D)
[97.207]
Page 3-14

E1D09

E1D09 [Question withdrawn]

E1D10
(B)
[97.211]
Page 3-14

E1D10
Which amateur stations are eligible to be telecommand stations?
A. Any amateur station designated by NASA
B. Any amateur station so designated by the space station licensee, subject to the privileges of the class of operator license held by the control operator
C. Any amateur station so designated by the ITU
D. All of these choices are correct

E1D11
(D)
[97.209]
Page 3-14

E1D11
Which amateur stations are eligible to operate as Earth stations?
A. Any amateur station whose licensee has filed a pre-space notification with the FCC's International Bureau
B. Only those of General, Advanced or Amateur Extra Class operators
C. Only those of Amateur Extra Class operators
D. Any amateur station, subject to the privileges of the class of operator license held by the control operator

E1E — Volunteer examiner program: definitions; qualifications; preparation and administration of exams; accreditation; question pools; documentation requirements

E1E01
(D)
[97.509]
Page 3-17

E1E01
What is the minimum number of qualified VEs required to administer an Element 4 amateur operator license examination?
A. 5
B. 2
C. 4
D. 3

E1E02
(C)
[97.523]
Page 3-15

E1E02
Where are the questions for all written US amateur license examinations listed?
A. In FCC Part 97
B. In a question pool maintained by the FCC
C. In a question pool maintained by all the VECs
D. In the appropriate FCC Report and Order

E1E03
(C)
[97.521]
Page 3-14

E1E03
What is a Volunteer Examiner Coordinator?
A. A person who has volunteered to administer amateur operator license examinations
B. A person who has volunteered to prepare amateur operator license examinations
C. An organization that has entered into an agreement with the FCC to coordinate amateur operator license examinations
D. The person who has entered into an agreement with the FCC to be the VE session manager

E1E04

Which of the following best describes the Volunteer Examiner accreditation process?
  A. Each General, Advanced and Amateur Extra Class operator is automatically accredited as a VE when the license is granted
  B. The amateur operator applying must pass a VE examination administered by the FCC Enforcement Bureau
  C. The prospective VE obtains accreditation from the FCC
  D. The procedure by which a VEC confirms that the VE applicant meets FCC requirements to serve as an examiner

E1E04
(D)
[97.509,
97.525]
Page 3-15

E1E05

What is the minimum passing score on amateur operator license examinations?
  A. Minimum passing score of 70%
  B. Minimum passing score of 74%
  C. Minimum passing score of 80%
  D. Minimum passing score of 77%

E1E05
(B)
[97.503]
Page 3-17

E1E06

Who is responsible for the proper conduct and necessary supervision during an amateur operator license examination session?
  A. The VEC coordinating the session
  B. The FCC
  C. Each administering VE
  D. The VE session manager

E1E06
(C)
[97.509]
Page 3-17

E1E07

What should a VE do if a candidate fails to comply with the examiner's instructions during an amateur operator license examination?
  A. Warn the candidate that continued failure to comply will result in termination of the examination
  B. Immediately terminate the candidate's examination
  C. Allow the candidate to complete the examination, but invalidate the results
  D. Immediately terminate everyone's examination and close the session

E1E07
(B)
[97.509]
Page 3-17

E1E08

To which of the following examinees may a VE not administer an examination?
  A. Employees of the VE
  B. Friends of the VE
  C. Relatives of the VE as listed in the FCC rules
  D. All of these choices are correct

E1E08
(C)
[97.509]
Page 3-17

E1E09

What may be the penalty for a VE who fraudulently administers or certifies an examination?
  A. Revocation of the VE's amateur station license grant and the suspension of the VE's amateur operator license grant
  B. A fine of up to $1000 per occurrence
  C. A sentence of up to one year in prison
  D. All of these choices are correct

E1E09
(A)
[97.509]
Page 3-18

E1E10
(C)
[97.509]
Page 3-17

E1E10
What must the administering VEs do after the administration of a successful examination for an amateur operator license?
- A. They must collect and send the documents to the NCVEC for grading
- B. They must collect and submit the documents to the coordinating VEC for grading
- C. They must submit the application document to the coordinating VEC according to the coordinating VEC instructions
- D. They must collect and send the documents to the FCC according to instructions

E1E11
(B)
[97.509]
Page 3-17

E1E11
What must the VE team do if an examinee scores a passing grade on all examination elements needed for an upgrade or new license?
- A. Photocopy all examination documents and forward them to the FCC for processing
- B. Three VEs must certify that the examinee is qualified for the license grant and that they have complied with the administering VE requirements
- C. Issue the examinee the new or upgrade license
- D. All these choices are correct

E1E12
(A)
[97.509]
Page 3-17

E1E12
What must the VE team do with the application form if the examinee does not pass the exam?
- A. Return the application document to the examinee
- B. Maintain the application form with the VEC's records
- C. Send the application form to the FCC and inform the FCC of the grade
- D. Destroy the application form

E1E13
(A)
[97.519]
Page 3-18

E1E13
What are the consequences of failing to appear for re-administration of an examination when so directed by the FCC?
- A. The licensee's license will be cancelled
- B. The person may be fined or imprisoned
- C. The licensee is disqualified from any future examination for an amateur operator license grant
- D. All these choices are correct

E1E14
(A)
[97.527]
Page 3-17

E1E14
For which types of out-of-pocket expenses do the Part 97 rules state that VEs and VECs may be reimbursed?
- A. Preparing, processing, administering and coordinating an examination for an amateur radio license
- B. Teaching an amateur operator license examination preparation course
- C. No expenses are authorized for reimbursement
- D. Providing amateur operator license examination preparation training materials

E1F — Miscellaneous rules: external RF power amplifiers; national quiet zone; business communications; compensated communications; spread spectrum; auxiliary stations; reciprocal operating privileges; IARP and CEPT licenses; third party communications with foreign countries; special temporary authority

E1F01
On what frequencies are spread spectrum transmissions permitted?
  A. Only on amateur frequencies above 50 MHz
  B. Only on amateur frequencies above 222 MHz
  C. Only on amateur frequencies above 420 MHz
  D. Only on amateur frequencies above 144 MHz

E1F01
(B)
[97.305]
Page 3-21

E1F02
Which of the following operating arrangements allows an FCC-licensed US citizen to operate in many European countries, and alien amateurs from many European countries to operate in the US?
  A. CEPT agreement
  B. IARP agreement
  C. ITU reciprocal license
  D. All of these choices are correct

E1F02
(A)
[97.5]
Page 3-21

E1F03
Under what circumstances may a dealer sell an external RF power amplifier capable of operation below 144 MHz if it has not been granted FCC certification?
  A. It was purchased in used condition from an amateur operator and is sold to another amateur operator for use at that operator's station
  B. The equipment dealer assembled it from a kit
  C. It was imported from a manufacturer in a country that does not require certification of RF power amplifiers
  D. It was imported from a manufacturer in another country, and it was certificated by that country's government

E1F03
(A)
[97.315]
Page 3-19

E1F04
Which of the following geographic descriptions approximately describes "Line A"?
  A. A line roughly parallel to and south of the US-Canadian border
  B. A line roughly parallel to and west of the US Atlantic coastline
  C. A line roughly parallel to and north of the US-Mexican border and Gulf coastline
  D. A line roughly parallel to and east of the US Pacific coastline

E1F04
(A)
[97.3]
Page 3-20

E1F05
Amateur stations may not transmit in which of the following frequency segments if they are located in the contiguous 48 states and north of Line A?
  A. 440 - 450 MHz
  B. 53 - 54 MHz
  C. 222 - 223 MHz
  D. 420 - 430 MHz

E1F05
(D)
[97.303]
Page 3-20

E1F06
(C)
[97.3]
Page 3-20

E1F06
What is the National Radio Quiet Zone?
 A. An area in Puerto Rico surrounding the Aricebo Radio Telescope
 B. An area in New Mexico surrounding the White Sands Test Area
 C. An area surrounding the National Radio Astronomy Observatory
 D. An area in Florida surrounding Cape Canaveral

E1F07
(D)
[97.113]
Page 3-20

E1F07
When may an amateur station send a message to a business?
 A. When the total money involved does not exceed $25
 B. When the control operator is employed by the FCC or another government agency
 C. When transmitting international third-party communications
 D. When neither the amateur nor his or her employer has a pecuniary interest in the communications

E1F08
(A)
[97.113]
Page 3-21

E1F08
Which of the following types of amateur station communications are prohibited?
 A. Communications transmitted for hire or material compensation, except as otherwise provided in the rules
 B. Communications that have a political content, except as allowed by the Fairness Doctrine
 C. Communications that have a religious content
 D. Communications in a language other than English

E1F09
(D)
[97.311]
Page 3-21

E1F09
Which of the following conditions apply when transmitting spread spectrum emission?
 A. A station transmitting SS emission must not cause harmful interference to other stations employing other authorized emissions
 B. The transmitting station must be in an area regulated by the FCC or in a country that permits SS emissions
 C. The transmission must not be used to obscure the meaning of any communication
 D. All of these choices are correct

E1F10
(C)
[97.313]
Page 3-21

E1F10
What is the maximum transmitter power for an amateur station transmitting spread spectrum communications?
 A. 1 W
 B. 1.5 W
 C. 10 W
 D. 1.5 kW

E1F11
(D)
[97.317]
Page 3-19

E1F11
Which of the following best describes one of the standards that must be met by an external RF power amplifier if it is to qualify for a grant of FCC certification?
 A. It must produce full legal output when driven by not more than 5 watts of mean RF input power
 B. It must be capable of external RF switching between its input and output networks
 C. It must exhibit a gain of 0 dB or less over its full output range
 D. It must satisfy the FCC's spurious emission standards when operated at the lesser of 1500 watts, or its full output power

E1F12

Who may be the control operator of an auxiliary station?
  A. Any licensed amateur operator
  B. Only Technician, General, Advanced or Amateur Extra Class operators
  C. Only General, Advanced or Amateur Extra Class operators
  D. Only Amateur Extra Class operators

E1F12
(B)
[97.201]
Page 3-19

E1F13

What types of communications may be transmitted to amateur stations in foreign countries?
  A. Business-related messages for non-profit organizations
  B. Messages intended for connection to users of the maritime satellite service
  C. Communications incidental to the purpose of the amateur service and remarks of a personal nature
  D. All of these choices are correct

E1F13
(C)
[97.117]
Page 3-20

E1F14

Under what circumstances might the FCC issue a "Special Temporary Authority" (STA) to an amateur station?
  A. To provide for experimental amateur communications
  B. To allow regular operation on Land Mobile channels
  C. To provide additional spectrum for personal use
  D. To provide temporary operation while awaiting normal licensing

E1F14
(A)
[1.931]
Page 3-22

SUBELEMENT E2
OPERATING PROCEDURES
5 Exam Questions — 5 Groups

E2A — Amateur radio in space: amateur satellites; orbital mechanics; frequencies and modes; satellite hardware; satellite operations

E2A01
(C)
Page 2-14

E2A01
What is the direction of an ascending pass for an amateur satellite?
  A. From west to east
  B. From east to west
  C. From south to north
  D. From north to south

E2A02
(A)
Page 2-14

E2A02
What is the direction of a descending pass for an amateur satellite?
  A. From north to south
  B. From west to east
  C. From east to west
  D. From south to north

E2A03
(C)
Page 2-13

E2A03
What is the orbital period of an Earth satellite?
  A. The point of maximum height of a satellite's orbit
  B. The point of minimum height of a satellite's orbit
  C. The time it takes for a satellite to complete one revolution around the Earth
  D. The time it takes for a satellite to travel from perigee to apogee

E2A04
(B)
Page 2-16

E2A04
What is meant by the term mode as applied to an amateur radio satellite?
  A. The type of signals that can be relayed through the satellite
  B. The satellite's uplink and downlink frequency bands
  C. The satellite's orientation with respect to the Earth
  D. Whether the satellite is in a polar or equatorial orbit

E2A05
(D)
Page 2-17

E2A05
What do the letters in a satellite's mode designator specify?
  A. Power limits for uplink and downlink transmissions
  B. The location of the ground control station
  C. The polarization of uplink and downlink signals
  D. The uplink and downlink frequency ranges

E2A06
(A)
Page 2-17

E2A06
On what band would a satellite receive signals if it were operating in mode U/V?
  A. 435-438 MHz
  B. 144-146 MHz
  C. 50.0-50.2 MHz
  D. 29.5 to 29.7 MHz

E2A07
Which of the following types of signals can be relayed through a linear transponder?
  A. FM and CW
  B. SSB and SSTV
  C. PSK and Packet
  D. All of these choices are correct

E2A08
Why should effective radiated power to a satellite which uses a linear transponder be limited?
  A. To prevent creating errors in the satellite telemetry
  B. To avoid reducing the downlink power to all other users
  C. To prevent the satellite from emitting out of band signals
  D. To avoid interfering with terrestrial QSOs

E2A09
What do the terms L band and S band specify with regard to satellite communications?
  A. The 23 centimeter and 13 centimeter bands
  B. The 2 meter and 70 centimeter bands
  C. FM and Digital Store-and-Forward systems
  D. Which sideband to use

E2A10
Why may the received signal from an amateur satellite exhibit a rapidly repeating fading effect?
  A. Because the satellite is spinning
  B. Because of ionospheric absorption
  C. Because of the satellite's low orbital altitude
  D. Because of the Doppler Effect

E2A11
What type of antenna can be used to minimize the effects of spin modulation and Faraday rotation?
  A. A linearly polarized antenna
  B. A circularly polarized antenna
  C. An isotropic antenna
  D. A log-periodic dipole array

E2A12
What is one way to predict the location of a satellite at a given time?
  A. By means of the Doppler data for the specified satellite
  B. By subtracting the mean anomaly from the orbital inclination
  C. By adding the mean anomaly to the orbital inclination
  D. By calculations using the Keplerian elements for the specified satellite

E2A13
What type of satellite appears to stay in one position in the sky?
  A. HEO
  B. Geostationary
  C. Geomagnetic
  D. LEO

E2A07
(D)
Page 2-16

E2A08
(B)
Page 2-16

E2A09
(A)
Page 2-17

E2A10
(A)
Page 2-15

E2A11
(B)
Page 2-15

E2A12
(D)
Page 2-13

E2A13
(B)
Page 2-13

E2B — Television practices: fast scan television standards and techniques; slow scan television standards and techniques

E2B01
(A)
Page 8-13

E2B01
How many times per second is a new frame transmitted in a fast-scan (NTSC) television system?
A. 30
B. 60
C. 90
D. 120

E2B02
(C)
Page 8-13

E2B02
How many horizontal lines make up a fast-scan (NTSC) television frame?
A. 30
B. 60
C. 525
D. 1080

E2B03
(D)
Page 8-13

E2B03
How is an interlaced scanning pattern generated in a fast-scan (NTSC) television system?
A. By scanning two fields simultaneously
B. By scanning each field from bottom to top
C. By scanning lines from left to right in one field and right to left in the next
D. By scanning odd numbered lines in one field and even numbered ones in the next

E2B04
(B)
Page 8-13

E2B04
What is blanking in a video signal?
A. Synchronization of the horizontal and vertical sync pulses
B. Turning off the scanning beam while it is traveling from right to left or from bottom to top
C. Turning off the scanning beam at the conclusion of a transmission
D. Transmitting a black and white test pattern

E2B05
(C)
Page 8-16

E2B05
Which of the following is an advantage of using vestigial sideband for standard fast- scan TV transmissions?
A. The vestigial sideband carries the audio information
B. The vestigial sideband contains chroma information
C. Vestigial sideband reduces bandwidth while allowing for simple video detector circuitry
D. Vestigial sideband provides high frequency emphasis to sharpen the picture

E2B06
(A)
Page 8-16

E2B06
What is vestigial sideband modulation?
A. Amplitude modulation in which one complete sideband and a portion of the other are transmitted
B. A type of modulation in which one sideband is inverted
C. Narrow-band FM transmission achieved by filtering one sideband from the audio before frequency modulating the carrier
D. Spread spectrum modulation achieved by applying FM modulation following single sideband amplitude modulation

E2B07
What is the name of the signal component that carries color information in NTSC video?
  A. Luminance
  B. Chroma
  C. Hue
  D. Spectral Intensity

E2B08
Which of the following is a common method of transmitting accompanying audio with amateur fast-scan television?
  A. Frequency-modulated sub-carrier
  B. A separate VHF or UHF audio link
  C. Frequency modulation of the video carrier
  D. All of these choices are correct

E2B09
What hardware, other than a receiver with SSB capability and a suitable computer, is needed to decode SSTV using Digital Radio Mondiale (DRM)?
  A. A special IF converter
  B. A special front end limiter
  C. A special notch filter to remove synchronization pulses
  D. No other hardware is needed

E2B10
Which of the following is an acceptable bandwidth for Digital Radio Mondiale (DRM) based voice or SSTV digital transmissions made on the HF amateur bands?
  A. 3 KHz
  B. 10 KHz
  C. 15 KHz
  D. 20 KHz

E2B11
What is the function of the Vertical Interval Signaling (VIS) code transmitted as part of an SSTV transmission?
  A. To lock the color burst oscillator in color SSTV images
  B. To identify the SSTV mode being used
  C. To provide vertical synchronization
  D. To identify the call sign of the station transmitting

E2B12
How are analog SSTV images typically transmitted on the HF bands?
  A. Video is converted to equivalent Baudot representation
  B. Video is converted to equivalent ASCII representation
  C. Varying tone frequencies representing the video are transmitted using PSK
  D. Varying tone frequencies representing the video are transmitted using single sideband

E2B07
(B)
Page 8-15

E2B08
(D)
Page 8-16

E2B09
(D)
Page 8-19

E2B10
(A)
Page 8-19

E2B11
(B)
Page 8-18

E2B12
(D)
Page 8-17

E2B13
(C)
Page 8-18

E2B13
How many lines are commonly used in each frame on an amateur slow-scan color television picture?
A. 30 to 60
B. 60 or 100
C. 128 or 256
D. 180 or 360

E2B14
(A)
Page 8-18

E2B14
What aspect of an amateur slow-scan television signal encodes the brightness of the picture?
A. Tone frequency
B. Tone amplitude
C. Sync amplitude
D. Sync frequency

E2B15
(A)
Page 8-18

E2B15
What signals SSTV receiving equipment to begin a new picture line?
A. Specific tone frequencies
B. Elapsed time
C. Specific tone amplitudes
D. A two-tone signal

E2B16
(D)
Page 8-13

E2B16
Which of the following is the video standard used by North American Fast Scan ATV stations?
A. PAL
B. DRM
C. Scottie
D. NTSC

E2B17
(B)
Page 8-18

E2B17
What is the approximate bandwidth of a slow-scan TV signal?
A. 600 Hz
B. 3 kHz
C. 2 MHz
D. 6 MHz

E2B18
(D)
Page 8-16

E2B18
On which of the following frequencies is one likely to find FM ATV transmissions?
A. 14.230 MHz
B. 29.6 MHz
C. 52.525 MHz
D. 1255 MHz

E2B19
(C)
Page 8-19

E2B19
What special operating frequency restrictions are imposed on slow scan TV transmissions?
A. None; they are allowed on all amateur frequencies
B. They are restricted to 7.245 MHz, 14.245 MHz, 21.345, MHz, and 28.945 MHz
C. They are restricted to phone band segments and their bandwidth can be no greater than that of a voice signal of the same modulation type
D. They are not permitted above 54 MHz

E2C — Operating methods: contest and DX operating; spread-spectrum transmissions; selecting an operating frequency

E2C01
Which of the following is true about contest operating?
  A. Operators are permitted to make contacts even if they do not submit a log
  B. Interference to other amateurs is unavoidable and therefore acceptable
  C. It is mandatory to transmit the call sign of the station being worked as part of every transmission to that station
  D. Every contest requires a signal report in the exchange

E2C01
(A)
Page 2-7

E2C02
Which of the following best describes the term "self-spotting" in regards to contest operating?
  A. The generally prohibited practice of posting one's own call sign and frequency on a call sign spotting network
  B. The acceptable practice of manually posting the call signs of stations on a call sign spotting network
  C. A manual technique for rapidly zero beating or tuning to a station's frequency before calling that station
  D. An automatic method for rapidly zero beating or tuning to a station's frequency before calling that station

E2C02
(A)
Page 2-9

E2C03
From which of the following bands is amateur radio contesting generally excluded?
  A. 30 meters
  B. 6 meters
  C. 2 meters
  D. 33 cm

E2C03
(A)
Page 2-7

E2C04
On which of the following frequencies is an amateur radio contest contact generally discouraged?
  A. 3.525 MHz
  B. 14.020 MHz
  C. 28.330 MHz
  D. 146.52 MHz

E2C04
(D)
Page 2-7

E2C05
What is the function of a DX QSL Manager?
  A. To allocate frequencies for DXpeditions
  B. To handle the receiving and sending of confirmation cards for a DX station
  C. To run a net to allow many stations to contact a rare DX station
  D. To relay calls to and from a DX station

E2C05
(B)
Page 2-3

E2C06
During a VHF/UHF contest, in which band segment would you expect to find the highest level of activity?
  A. At the top of each band, usually in a segment reserved for contests
  B. In the middle of each band, usually on the national calling frequency
  C. In the weak signal segment of the band, with most of the activity near the calling frequency
  D. In the middle of the band, usually 25 kHz above the national calling frequency

E2C06
(C)
Page 2-3

E2C07
(A)
Page 2-7

E2C07
What is the Cabrillo format?
A. A standard for submission of electronic contest logs
B. A method of exchanging information during a contest QSO
C. The most common set of contest rules
D. The rules of order for meetings between contest sponsors

E2C08
(A)
Page 8-9

E2C08
Why are received spread-spectrum signals resistant to interference?
A. Signals not using the spectrum-spreading algorithm are suppressed in the receiver
B. The high power used by a spread-spectrum transmitter keeps its signal from being easily overpowered
C. The receiver is always equipped with a digital blanker circuit
D. If interference is detected by the receiver it will signal the transmitter to change frequencies

E2C09
(D)
Page 8-10

E2C09
How does the spread-spectrum technique of frequency hopping work?
A. If interference is detected by the receiver it will signal the transmitter to change frequencies
B. If interference is detected by the receiver it will signal the transmitter to wait until the frequency is clear
C. A pseudo-random binary bit stream is used to shift the phase of an RF carrier very rapidly in a particular sequence
D. The frequency of the transmitted signal is changed very rapidly according to a particular sequence also used by the receiving station

E2C10
(D)
Page 2-5

E2C10
Why might a DX station state that they are listening on another frequency?
A. Because the DX station may be transmitting on a frequency that is prohibited to some responding stations
B. To separate the calling stations from the DX station
C. To reduce interference, thereby improving operating efficiency
D. All of these choices are correct

E2C11
(A)
Page 2-5

E2C11
How should you generally identify your station when attempting to contact a DX station working a pileup or in a contest?
A. Send your full call sign once or twice
B. Send only the last two letters of your call sign until you make contact
C. Send your full call sign and grid square
D. Send the call sign of the DX station three times, the words this is, then your call sign three times

E2C12
(B)
Page 2-6

E2C12
What might help to restore contact when DX signals become too weak to copy across an entire HF band a few hours after sunset?
A. Switch to a higher frequency HF band
B. Switch to a lower frequency HF band
C. Wait 90 minutes or so for the signal degradation to pass
D. Wait 24 hours before attempting another communication on the band

E2D — Operating methods: VHF and UHF digital modes; APRS

E2D01
Which of the following digital modes is especially designed for use for meteor scatter signals?
A. WSPR
B. FSK441
C. Hellschreiber
D. APRS

E2D01
(B)
Page 8-8

E2D02
What is the definition of baud?
A. The number of data symbols transmitted per second
B. The number of characters transmitted per second
C. The number of characters transmitted per minute
D. The number of words transmitted per minute

E2D02
(A)
Page 8-2

E2D03
Which of the following digital modes is especially useful for EME communications?
A. FSK441
B. PACTOR III
C. Olivia
D. JT65

E2D03
(D)
Page 8-8

E2D04
What is the purpose of digital store-and-forward functions on an Amateur Radio satellite?
A. To upload operational software for the transponder
B. To delay download of telemetry between satellites
C. To store digital messages in the satellite for later download by other stations
D. To relay messages between satellites

E2D04
(C)
Page 2-11

E2D05
Which of the following techniques is normally used by low Earth orbiting digital satellites to relay messages around the world?
A. Digipeating
B. Store-and-forward
C. Multi-satellite relaying
D. Node hopping

E2D05
(B)
Page 2-11

E2D06
Which of the following is a commonly used 2-meter APRS frequency?
A. 144.39 MHz
B. 144.20 MHz
C. 145.02 MHz
D. 146.52 MHz

E2D06
(A)
Page 2-11

E2D07
Which of the following digital protocols is used by APRS?
A. PACTOR
B. 802.11
C. AX.25
D. AMTOR

E2D07
(C)
Page 2-12

E2D08
(A)
Page 2-12

E2D08
Which of the following types of packet frames is used to transmit APRS beacon data?
A. Unnumbered Information
B. Disconnect
C. Acknowledgement
D. Connect

E2D09
(D)
Page 8-7

E2D09
Under clear communications conditions, which of these digital communications modes has the fastest data throughput?
A. AMTOR
B. 170-Hz shift, 45 baud RTTY
C. PSK31
D. 300-baud packet

E2D10
(C)
Page 2-12

E2D10
How can an APRS station be used to help support a public service communications activity?
A. An APRS station with an emergency medical technician can automatically transmit medical data to the nearest hospital
B. APRS stations with General Personnel Scanners can automatically relay the participant numbers and time as they pass the check points
C. An APRS station with a GPS unit can automatically transmit information to show a mobile station's position during the event
D. All of these choices are correct

E2D11
(D)
Page 2-12

E2D11
Which of the following data are used by the APRS network to communicate your location?
A. Polar coordinates
B. Time and frequency
C. Radio direction finding LOPs
D. Latitude and longitude

E2D12
(A)
Page 8-8

E2D12
How does JT65 improve EME communications?
A. It can decode signals many dB below the noise floor using FEC
B. It controls the receiver to track Doppler shift
C. It supplies signals to guide the antenna to track the Moon
D. All of these choices are correct

E2E — Operating methods: operating HF digital modes; error correction

E2E01
(B)
Page 8-6

E2E01
Which type of modulation is common for data emissions below 30 MHz?
A. DTMF tones modulating an FM signal
B. FSK
C. Pulse modulation
D. Spread spectrum

E2E02

What do the letters FEC mean as they relate to digital operation?
A. Forward Error Correction
B. First Error Correction
C. Fatal Error Correction
D. Final Error Correction

E2E02
(A)
Page 8-11

E2E03

How is Forward Error Correction implemented?
A. By the receiving station repeating each block of three data characters
B. By transmitting a special algorithm to the receiving station along with the data characters
C. By transmitting extra data that may be used to detect and correct transmission errors
D. By varying the frequency shift of the transmitted signal according to a predefined algorithm

E2E03
(C)
Page 8-11

E2E04

What is indicated when one of the ellipses in an FSK crossed-ellipse display suddenly disappears?
A. Selective fading has occurred
B. One of the signal filters has saturated
C. The receiver has drifted 5 kHz from the desired receive frequency
D. The mark and space signal have been inverted

E2E04
(A)
Page 8-7

E2E05

How does ARQ accomplish error correction?
A. Special binary codes provide automatic correction
B. Special polynomial codes provide automatic correction
C. If errors are detected, redundant data is substituted
D. If errors are detected, a retransmission is requested

E2E05
(D)
Page 8-11

E2E06

What is the most common data rate used for HF packet communications?
A. 48 baud
B. 110 baud
C. 300 baud
D. 1200 baud

E2E06
(C)
Page 8-7

E2E07

What is the typical bandwidth of a properly modulated MFSK16 signal?
A. 31 Hz
B. 316 Hz
C. 550 Hz
D. 2.16 kHz

E2E07
(B)
Page 8-8

E2E08

Which of the following HF digital modes can be used to transfer binary files?
A. Hellschreiber
B. PACTOR
C. RTTY
D. AMTOR

E2E08
(B)
Page 8-7

E2E09
(D)
Page 8-4

E2E09
Which of the following HF digital modes uses variable-length coding for bandwidth efficiency?
A. RTTY
B. PACTOR
C. MT63
D. PSK31

E2E10
(C)
Page 8-7

E2E10
Which of these digital communications modes has the narrowest bandwidth?
A. MFSK16
B. 170-Hz shift, 45 baud RTTY
C. PSK31
D. 300-baud packet

E2E11
(A)
Page 8-6

E2E11
What is the difference between direct FSK and audio FSK?
A. Direct FSK applies the data signal to the transmitter VFO
B. Audio FSK has a superior frequency response
C. Direct FSK uses a DC-coupled data connection
D. Audio FSK can be performed anywhere in the transmit chain

E2E12
(A)
Page 8-8

E2E12
Which type of digital communication does not support keyboard-to-keyboard operation?
A. Winlink
B. RTTY
C. PSK31
D. MFSK

SUBELEMENT E3
RADIO WAVE PROPAGATION
3 Exam Questions — 3 Groups

E3A — Propagation and technique: Earth-Moon-Earth communications (EME), meteor scatter

E3A01
What is the approximate maximum separation measured along the surface of the Earth between two stations communicating by Moon bounce?
  A. 500 miles, if the Moon is at perigee
  B. 2000 miles, if the Moon is at apogee
  C. 5000 miles, if the Moon is at perigee
  D. 12,000 miles, as long as both can "see" the Moon

E3A02
What characterizes libration fading of an Earth-Moon-Earth signal?
  A. A slow change in the pitch of the CW signal
  B. A fluttery irregular fading
  C. A gradual loss of signal as the Sun rises
  D. The returning echo is several Hertz lower in frequency than the transmitted signal

E3A03
When scheduling EME contacts, which of these conditions will generally result in the least path loss?
  A. When the Moon is at perigee
  B. When the Moon is full
  C. When the Moon is at apogee
  D. When the MUF is above 30 MHz

E3A04
What type of receiving system is desirable for EME communications?
  A. Equipment with very wide bandwidth
  B. Equipment with very low dynamic range
  C. Equipment with very low gain
  D. Equipment with very low noise figures

E3A05
Which of the following describes a method of establishing EME contacts?
  A. Time synchronous transmissions with each station alternating
  B. Storing and forwarding digital messages
  C. Judging optimum transmission times by monitoring beacons from the Moon
  D. High speed CW identification to avoid fading

E3A06
What frequency range would you normally tune to find EME signals in the 2 meter band?
  A. 144.000 - 144.001 MHz
  B. 144.000 - 144.100 MHz
  C. 144.100 - 144.300 MHz
  D. 145.000 - 145.100 MHz

E3A01
(D)
Page 10-13

E3A02
(B)
Page 10-13

E3A03
(A)
Page 10-13

E3A04
(D)
Page 10-13

E3A05
(A)
Page 10-14

E3A06
(B)
Page 10-14

E3A07
(D)
Page 10-14

E3A07
What frequency range would you normally tune to find EME signals in the 70 cm band?
A. 430.000 - 430.150 MHz
B. 430.100 - 431.100 MHz
C. 431.100 - 431.200 MHz
D. 432.000 - 432.100 MHz

E3A08
(A)
Page 10-11

E3A08
When a meteor strikes the Earth's atmosphere, a cylindrical region of free electrons is formed at what layer of the ionosphere?
A. The E layer
B. The F1 layer
C. The F2 layer
D. The D layer

E3A09
(C)
Page 10-11

E3A09
Which of the following frequency ranges is well suited for meteor-scatter communications?
A. 1.8 - 1.9 MHz
B. 10 - 14 MHz
C. 28 - 148 MHz
D. 220 - 450 MHz

E3A10
(D)
Page 10-12

E3A10
Which of the following is a good technique for making meteor-scatter contacts?
A. 15 second timed transmission sequences with stations alternating based on location
B. Use of high speed CW or digital modes
C. Short transmission with rapidly repeated call signs and signal reports
D. All of these choices are correct

E3B — Propagation and technique: trans-equatorial, long path, gray-line; multi-path propagation

E3B01
(A)
Page 10-8

E3B01
What is transequatorial propagation?
A. Propagation between two mid-latitude points at approximately the same distance north and south of the magnetic equator
B. Propagation between any two points located on the magnetic equator
C. Propagation between two continents by way of ducts along the magnetic equator
D. Propagation between two stations at the same latitude

E3B02
(C)
Page 10-9

E3B02
What is the approximate maximum range for signals using transequatorial propagation?
A. 1000 miles
B. 2500 miles
C. 5000 miles
D. 7500 miles

E3B03
(C)
Page 10-9

E3B03
What is the best time of day for transequatorial propagation?
A. Morning
B. Noon
C. Afternoon or early evening
D. Late at night

E3B04
What type of propagation is probably occurring if an HF beam antenna must be pointed in a direction 180 degrees away from a station to receive the strongest signals?
  A. Long-path
  B. Sporadic-E
  C. Transequatorial
  D. Auroral

E3B05
Which amateur bands typically support long-path propagation?
  A. 160 to 40 meters
  B. 30 to 10 meters
  C. 160 to 10 meters
  D. 6 meters to 2 meters

E3B06
Which of the following amateur bands most frequently provides long-path propagation?
  A. 80 meters
  B. 20 meters
  C. 10 meters
  D. 6 meters

E3B07
Which of the following could account for hearing an echo on the received signal of a distant station?
  A. High D layer absorption
  B. Meteor scatter
  C. Transmit frequency is higher than the MUF
  D. Receipt of a signal by more than one path

E3B08
What type of HF propagation is probably occurring if radio signals travel along the terminator between daylight and darkness?
  A. Transequatorial
  B. Sporadic-E
  C. Long-path
  D. Gray-line

E3B09
At what time of day is gray-line propagation most likely to occur?
  A. At sunrise and sunset
  B. When the Sun is directly above the location of the transmitting station
  C. When the Sun is directly overhead at the middle of the communications path between the two stations
  D. When the Sun is directly above the location of the receiving station

E3B10
What is the cause of gray-line propagation?
  A. At midday, the Sun being directly overhead superheats the ionosphere causing increased refraction of radio waves
  B. At twilight, D-layer absorption drops while E-layer and F-layer propagation remain strong
  C. In darkness, solar absorption drops greatly while atmospheric ionization remains steady
  D. At mid afternoon, the Sun heats the ionosphere decreasing radio wave refraction and the MUF

E3B04
(A)
Page 10-4

E3B05
(C)
Page 10-4

E3B06
(B)
Page 10-4

E3B07
(D)
Page 10-4

E3B08
(D)
Page 10-4

E3B09
(A)
Page 10-4

E3B10
(B)
Page 10-4

E3B11
(C)
Page 10-4

E3B11
Which of the following describes gray-line propagation?
A. Backscatter contacts on the 10 meter band
B. Over the horizon propagation on the 6 and 2 meter bands
C. Long distance communications at twilight on frequencies less than 15 MHz
D. Tropospheric propagation on the 2 meter and 70 centimeter bands

E3C — Propagation and technique: Aurora propagation selective fading; radio-path horizon; take-off angle over flat or sloping terrain; effects of ground on propagation; less common propagation modes

E3C01
(D)
Page 10-10

E3C01
Which of the following effects does Aurora activity have on radio communications?
A. SSB signals are raspy
B. Signals propagating through the Aurora are fluttery
C. CW signals appear to be modulated by white noise
D. All of these choices are correct

E3C02
(C)
Page 10-9

E3C02
What is the cause of Aurora activity?
A. The interaction between the solar wind and the Van Allen belt
B. A low sunspot level combined with tropospheric ducting
C. The interaction of charged particles from the Sun with the Earth's magnetic field and the ionosphere
D. Meteor showers concentrated in the northern latitudes

E3C03
(D)
Page 10-9

E3C03
Where in the ionosphere does Aurora activity occur?
A. In the F1-region
B. In the F2-region
C. In the D-region
D. In the E-region

E3C04
(A)
Page 10-10

E3C04
Which emission mode is best for Aurora propagation?
A. CW
B. SSB
C. FM
D. RTTY

E3C05
(B)
Page 10-5

E3C05
Which of the following describes selective fading?
A. Variability of signal strength with beam heading
B. Partial cancellation of some frequencies within the received pass band
C. Sideband inversion within the ionosphere
D. Degradation of signal strength due to backscatter

E3C06
By how much does the VHF/UHF radio-path horizon distance exceed the geometric horizon?
  A. By approximately 15% of the distance
  B. By approximately twice the distance
  C. By approximately one-half the distance
  D. By approximately four times the distance

E3C06
(A)
Page 10-7

E3C07
How does the radiation pattern of a horizontally polarized 3-element beam antenna vary with its height above ground?
  A. The main lobe takeoff angle increases with increasing height
  B. The main lobe takeoff angle decreases with increasing height
  C. The horizontal beam width increases with height
  D. The horizontal beam width decreases with height

E3C07
(B)
Page 9-10

E3C08
What is the name of the high-angle wave in HF propagation that travels for some distance within the F2 region?
  A. Oblique-angle ray
  B. Pedersen ray
  C. Ordinary ray
  D. Heaviside ray

E3C08
(B)
Page 10-2

E3C09
Which of the following is usually responsible for causing VHF signals to propagate for hundreds of miles?
  A. D-region absorption
  B. Faraday rotation
  C. Tropospheric ducting
  D. Ground wave

E3C09
(C)
Page 10-8

E3C10
How does the performance of a horizontally polarized antenna mounted on the side of a hill compare with the same antenna mounted on flat ground?
  A. The main lobe takeoff angle increases in the downhill direction
  B. The main lobe takeoff angle decreases in the downhill direction
  C. The horizontal beam width decreases in the downhill direction
  D. The horizontal beam width increases in the uphill direction

E3C10
(B)
Page 9-10

E3C11
From the contiguous 48 states, in which approximate direction should an antenna be pointed to take maximum advantage of aurora propagation?
  A. South
  B. North
  C. East
  D. West

E3C11
(B)
Page 10-10

E3C12
(C)
Page 10-2

E3C12
How does the maximum distance of ground-wave propagation change when the signal frequency is increased?
  A. It stays the same
  B. It increases
  C. It decreases
  D. It peaks at roughly 14 MHz

E3C13
(A)
Page 10-2

E3C13
What type of polarization is best for ground-wave propagation?
  A. Vertical
  B. Horizontal
  C. Circular
  D. Elliptical

E3C14
(D)
Page 10-7

E3C14
Why does the radio-path horizon distance exceed the geometric horizon?
  A. E-region skip
  B. D-region skip
  C. Downward bending due to aurora refraction
  D. Downward bending due to density variations in the atmosphere

SUBELEMENT E4
AMATEUR PRACTICES
5 Exam Questions — 5 Groups

E4A — Test equipment: analog and digital instruments; spectrum and network analyzers, antenna analyzers; oscilloscopes; testing transistors; RF measurements

E4A01
How does a spectrum analyzer differ from an oscilloscope?
  A. A spectrum analyzer measures ionospheric reflection; an oscilloscope displays electrical signals
  B. A spectrum analyzer displays the peak amplitude of signals; an oscilloscope displays the average amplitude of signals
  C. A spectrum analyzer displays signals in the frequency domain; an oscilloscope displays signals in the time domain
  D. A spectrum analyzer displays radio frequencies; an oscilloscope displays audio frequencies

E4A02
Which of the following parameters would a spectrum analyzer display on the horizontal axis?
  A. SWR
  B. Q
  C. Time
  D. Frequency

E4A03
Which of the following parameters would a spectrum analyzer display on the vertical axis?
  A. Amplitude
  B. Duration
  C. SWR
  D. Q

E4A04
Which of the following test instruments is used to display spurious signals from a radio transmitter?
  A. A spectrum analyzer
  B. A wattmeter
  C. A logic analyzer
  D. A time-domain reflectometer

E4A05
Which of the following test instruments is used to display intermodulation distortion products in an SSB transmission?
  A. A wattmeter
  B. A spectrum analyzer
  C. A logic analyzer
  D. A time-domain reflectometer

E4A01
(C)
Page 7-17

E4A02
(D)
Page 7-18

E4A03
(A)
Page 7-18

E4A04
(A)
Page 7-18

E4A05
(B)
Page 7-19

E4A06
(D)
Page 7-19

E4A06
Which of the following could be determined with a spectrum analyzer?
  A. The degree of isolation between the input and output ports of a 2 meter duplexer
  B. Whether a crystal is operating on its fundamental or overtone frequency
  C. The spectral output of a transmitter
  D. All of these choices are correct

E4A07
(B)
Page 9-40

E4A07
Which of the following is an advantage of using an antenna analyzer compared to an SWR bridge to measure antenna SWR?
  A. Antenna analyzers automatically tune your antenna for resonance
  B. Antenna analyzers do not need an external RF source
  C. Antenna analyzers display a time-varying representation of the modulation envelope
  D. All of these choices are correct

E4A08
(D)
Page 9-40

E4A08
Which of the following instruments would be best for measuring the SWR of a beam antenna?
  A. A spectrum analyzer
  B. A Q meter
  C. An ohmmeter
  D. An antenna analyzer

E4A09
(A)
Page 8-8

E4A09
Which of the following describes a good method for measuring the intermodulation distortion of your own PSK signal?
  A. Transmit into a dummy load, receive the signal on a second receiver, and feed the audio into the sound card of a computer running an appropriate PSK program
  B. Multiply the ALC level on the transmitter during a normal transmission by the average power output
  C. Use an RF voltmeter coupled to the transmitter output using appropriate isolation to prevent damage to the meter
  D. All of these choices are correct

E4A10
(D)
Page 7-20

E4A10
Which of the following tests establishes that a silicon NPN junction transistor is biased on?
  A. Measure base-to-emitter resistance with an ohmmeter; it should be approximately 6 to 7 ohms
  B. Measure base-to-emitter resistance with an ohmmeter; it should be approximately 0.6 to 0.7 ohms
  C. Measure base-to-emitter voltage with a voltmeter; it should be approximately 6 to 7 volts
  D. Measure base-to-emitter voltage with a voltmeter; it should be approximately 0.6 to 0.7 volts

E4A11
(B)
Page 7-16

E4A11
Which of these instruments could be used for detailed analysis of digital signals?
  A. Dip meter
  B. Oscilloscope
  C. Ohmmeter
  D. Q meter

E4A12

Which of the following procedures is an important precaution to follow when connecting a spectrum analyzer to a transmitter output?

A. Use high quality double shielded coaxial cables to reduce signal losses
B. Attenuate the transmitter output going to the spectrum analyzer
C. Match the antenna to the load
D. All of these choices are correct

E4A12
(B)
Page 7-19

E4B — Measurement techniques: Instrument accuracy and performance limitations; probes; techniques to minimize errors; measurement of Q; instrument calibration

E4B01

Which of the following factors most affects the accuracy of a frequency counter?

A. Input attenuator accuracy
B. Time base accuracy
C. Decade divider accuracy
D. Temperature coefficient of the logic

E4B01
(B)
Page 7-14

E4B02

What is an advantage of using a bridge circuit to measure impedance?

A. It provides an excellent match under all conditions
B. It is relatively immune to drift in the signal generator source
C. The measurement is based on obtaining a signal null, which can be done very precisely
D. It can display results directly in Smith chart format

E4B02
(C)
Page 7-13

E4B03

If a frequency counter with a specified accuracy of ± 1.0 ppm reads 146,520,000 Hz, what is the most the actual frequency being measured could differ from the reading?

A. 165.2 Hz
B. 14.652 kHz
C. 146.52 Hz
D. 1.4652 MHz

E4B03
(C)
Page 7-15

E4B04

If a frequency counter with a specified accuracy of ± 0.1 ppm reads 146,520,000 Hz, what is the most the actual frequency being measured could differ from the reading?

A. 14.652 Hz
B. 0.1 MHz
C. 1.4652 Hz
D. 1.4652 kHz

E4B04
(A)
Page 7-15

E4B05

If a frequency counter with a specified accuracy of ± 10 ppm reads 146,520,000 Hz, what is the most the actual frequency being measured could differ from the reading?

A. 146.52 Hz
B. 10 Hz
C. 146.52 kHz
D. 1465.20 Hz

E4B05
(D)
Page 7-15

E4B06
(D)
Page 9-34

E4B06
How much power is being absorbed by the load when a directional power meter connected between a transmitter and a terminating load reads 100 watts forward power and 25 watts reflected power?
  A. 100 watts
  B. 125 watts
  C. 25 watts
  D. 75 watts

E4B07
(A)
Page 7-17

E4B07
Which of the following is good practice when using an oscilloscope probe?
  A. Keep the signal ground connection of the probe as short as possible
  B. Never use a high impedance probe to measure a low impedance circuit
  C. Never use a DC-coupled probe to measure an AC circuit
  D. All of these choices are correct

E4B08
(C)
Page 7-12

E4B08
Which of the following is a characteristic of a good DC voltmeter?
  A. High reluctance input
  B. Low reluctance input
  C. High impedance input
  D. Low impedance input

E4B09
(D)
Page 9-34

E4B09
What is indicated if the current reading on an RF ammeter placed in series with the antenna feed line of a transmitter increases as the transmitter is tuned to resonance?
  A. There is possibly a short to ground in the feed line
  B. The transmitter is not properly neutralized
  C. There is an impedance mismatch between the antenna and feed line
  D. There is more power going into the antenna

E4B10
(B)
Page 7-19

E4B10
Which of the following describes a method to measure intermodulation distortion in an SSB transmitter?
  A. Modulate the transmitter with two non-harmonically related radio frequencies and observe the RF output with a spectrum analyzer
  B. Modulate the transmitter with two non-harmonically related audio frequencies and observe the RF output with a spectrum analyzer
  C. Modulate the transmitter with two harmonically related audio frequencies and observe the RF output with a peak reading wattmeter
  D. Modulate the transmitter with two harmonically related audio frequencies and observe the RF output with a logic analyzer

E4B11
(D)
Page 9-40

E4B11
How should a portable antenna analyzer be connected when measuring antenna resonance and feed point impedance?
  A. Loosely couple the analyzer near the antenna base
  B. Connect the analyzer via a high-impedance transformer to the antenna
  C. Connect the antenna and a dummy load to the analyzer
  D. Connect the antenna feed line directly to the analyzer's connector

E4B12

What is the significance of voltmeter sensitivity expressed in ohms per volt?

A. The full scale reading of the voltmeter multiplied by its ohms per volt rating will provide the input impedance of the voltmeter

B. When used as a galvanometer, the reading in volts multiplied by the ohms/volt will determine the power drawn by the device under test

C. When used as an ohmmeter, the reading in ohms divided by the ohms/volt will determine the voltage applied to the circuit

D. When used as an ammeter, the full scale reading in amps divided by ohms/volt will determine the size of shunt needed

E4B12
(A)
Page 7-12

E4B13

How is the compensation of an oscilloscope probe typically adjusted?

A. A square wave is displayed and the probe is adjusted until the horizontal portions of the displayed wave are as nearly flat as possible

B. A high frequency sine wave is displayed and the probe is adjusted for maximum amplitude

C. A frequency standard is displayed and the probe is adjusted until the deflection time is accurate

D. A DC voltage standard is displayed and the probe is adjusted until the displayed voltage is accurate

E4B13
(A)
Page 7-17

E4B14

What happens if a dip meter is too tightly coupled to a tuned circuit being checked?

A. Harmonics are generated

B. A less accurate reading results

C. Cross modulation occurs

D. Intermodulation distortion occurs

E4B14
(B)
Page 7-13

E4B15

Which of the following can be used as a relative measurement of the Q for a series-tuned circuit?

A. The inductance to capacitance ratio

B. The frequency shift

C. The bandwidth of the circuit's frequency response

D. The resonant frequency of the circuit

E4B15
(C)
Page 4-37

E4C — Receiver performance characteristics: phase noise; capture effect; noise floor; image rejection; MDS; signal-to-noise-ratio; selectivity

E4C01

What is an effect of excessive phase noise in the local oscillator section of a receiver?

A. It limits the receiver's ability to receive strong signals

B. It reduces receiver sensitivity

C. It decreases receiver third-order intermodulation distortion dynamic range

D. It can cause strong signals on nearby frequencies to interfere with reception of weak signals

E4C01
(D)
Page 8-26

E4C02

Which of the following portions of a receiver can be effective in eliminating image signal interference?

A. A front-end filter or pre-selector

B. A narrow IF filter

C. A notch filter

D. A properly adjusted product detector

E4C02
(A)
Page 8-21

E4C03
(C)
Page 8-27

E4C03
What is the term for the blocking of one FM phone signal by another, stronger FM phone signal?
 A. Desensitization
 B. Cross-modulation interference
 C. Capture effect
 D. Frequency discrimination

E4C04
(D)
Page 8-20

E4C04
What is the definition of the noise figure of a receiver?
 A. The ratio of atmospheric noise to phase noise
 B. The noise bandwidth in Hertz compared to the theoretical bandwidth of a resistive network
 C. The ratio of thermal noise to atmospheric noise
 D. The ratio in dB of the noise generated by the receiver compared to the theoretical minimum noise

E4C05
(B)
Page 8-20

E4C05
What does a value of –174 dBm/Hz represent with regard to the noise floor of a receiver?
 A. The minimum detectable signal as a function of receive frequency
 B. The theoretical noise at the input of a perfect receiver at room temperature
 C. The noise figure of a 1 Hz bandwidth receiver
 D. The galactic noise contribution to minimum detectable signal

E4C06
(D)
Page 8-20

E4C06
A CW receiver with the AGC off has an equivalent input noise power density of –174 dBm/Hz. What would be the level of an unmodulated carrier input to this receiver that would yield an audio output SNR of 0 dB in a 400 Hz noise bandwidth?
 A. 174 dBm
 B. –164 dBm
 C. –155 dBm
 D. –148 dBm

E4C07
(B)
Page 8-19

E4C07
What does the MDS of a receiver represent?
 A. The meter display sensitivity
 B. The minimum discernible signal
 C. The multiplex distortion stability
 D. The maximum detectable spectrum

E4C08
(B)
Page 8-21

E4C08
How might lowering the noise figure affect receiver performance?
 A. It would reduce the signal to noise ratio
 B. It would improve weak signal sensitivity
 C. It would reduce bandwidth
 D. It would increase bandwidth

E4C09
(C)
Page 8-21

E4C09
Which of the following choices is a good reason for selecting a high frequency for the design of the IF in a conventional HF or VHF communications receiver?
 A. Fewer components in the receiver
 B. Reduced drift
 C. Easier for front-end circuitry to eliminate image responses
 D. Improved receiver noise figure

E4C10
Which of the following is a desirable amount of selectivity for an amateur RTTY HF receiver?
A. 100 Hz
B. 300 Hz
C. 6000 Hz
D. 2400 Hz

E4C10
(B)
Page 8-22

E4C11
Which of the following is a desirable amount of selectivity for an amateur SSB phone receiver?
A. 1 kHz
B. 2.4 kHz
C. 4.2 kHz
D. 4.8 kHz

E4C11
(B)
Page 8-22

E4C12
What is an undesirable effect of using too wide a filter bandwidth in the IF section of a receiver?
A. Output-offset overshoot
B. Filter ringing
C. Thermal-noise distortion
D. Undesired signals may be heard

E4C12
(D)
Page 8-22

E4C13
How does a narrow-band roofing filter affect receiver performance?
A. It improves sensitivity by reducing front end noise
B. It improves intelligibility by using low Q circuitry to reduce ringing
C. It improves dynamic range by attenuating strong signals near the receive frequency
D. All of these choices are correct

E4C13
(C)
Page 8-22

E4C14
On which of the following frequencies might a signal be transmitting which is generating a spurious image signal in a receiver tuned to 14.300 MHz and which uses a 455 kHz IF frequency?
A. 13.845 MHz
B. 14.755 MHz
C. 14.445 MHz
D. 15.210 MHz

E4C14
(D)
Page 8-21

E4C15
What is the primary source of noise that can be heard from an HF receiver with an antenna connected?
A. Detector noise
B. Induction motor noise
C. Receiver front-end noise
D. Atmospheric noise

E4C15
(D)
Page 8-20

E4D — Receiver performance characteristics: blocking dynamic range; intermodulation and cross-modulation interference; 3rd order intercept; desensitization; preselection

E4D01
(A)
Page 8-23

E4D01
What is meant by the blocking dynamic range of a receiver?
  A. The difference in dB between the noise floor and the level of an incoming signal which will cause 1 dB of gain compression
  B. The minimum difference in dB between the levels of two FM signals which will cause one signal to block the other
  C. The difference in dB between the noise floor and the third order intercept point
  D. The minimum difference in dB between two signals which produce third order intermodulation products greater than the noise floor

E4D02
(A)
Page 8-26

E4D02
Which of the following describes two problems caused by poor dynamic range in a communications receiver?
  A. Cross-modulation of the desired signal and desensitization from strong adjacent signals
  B. Oscillator instability requiring frequent retuning and loss of ability to recover the opposite sideband
  C. Cross-modulation of the desired signal and insufficient audio power to operate the speaker
  D. Oscillator instability and severe audio distortion of all but the strongest received signals

E4D03
(B)
Page 7-28

E4D03
How can intermodulation interference between two repeaters occur?
  A. When the repeaters are in close proximity and the signals cause feedback in the final amplifier of one or both transmitters
  B. When the repeaters are in close proximity and the signals mix in the final amplifier of one or both transmitters
  C. When the signals from the transmitters are reflected out of phase from airplanes passing overhead
  D. When the signals from the transmitters are reflected in phase from airplanes passing overhead

E4D04
(B)
Page 7-28

E4D04
Which of the following may reduce or eliminate intermodulation interference in a repeater caused by another transmitter operating in close proximity?
  A. A band-pass filter in the feed line between the transmitter and receiver
  B. A properly terminated circulator at the output of the transmitter
  C. A Class C final amplifier
  D. A Class D final amplifier

E4D05
(A)
Page 8-24

E4D05
What transmitter frequencies would cause an intermodulation-product signal in a receiver tuned to 146.70 MHz when a nearby station transmits on 146.52 MHz?
  A. 146.34 MHz and 146.61 MHz
  B. 146.88 MHz and 146.34 MHz
  C. 146.10 MHz and 147.30 MHz
  D. 173.35 MHz and 139.40 MHz

E4D06
What is the term for unwanted signals generated by the mixing of two or more signals?
A. Amplifier desensitization
B. Neutralization
C. Adjacent channel interference
D. Intermodulation interference

E4D06
(D)
Page 7-27

E4D07
Which of the following describes the most significant effect of an off-frequency signal when it is causing cross-modulation interference to a desired signal?
A. A large increase in background noise
B. A reduction in apparent signal strength
C. The desired signal can no longer be heard
D. The off-frequency unwanted signal is heard in addition to the desired signal

E4D07
(D)
Page 7-27

E4D08
What causes intermodulation in an electronic circuit?
A. Too little gain
B. Lack of neutralization
C. Nonlinear circuits or devices
D. Positive feedback

E4D08
(C)
Page 7-27

E4D09
What is the purpose of the preselector in a communications receiver?
A. To store often-used frequencies
B. To provide a range of AGC time constants
C. To increase rejection of unwanted signals
D. To allow selection of the optimum RF amplifier device

E4D09
(C)
Page 8-21

E4D10
What does a third-order intercept level of 40 dBm mean with respect to receiver performance?
A. Signals less than 40 dBm will not generate audible third-order intermodulation products
B. The receiver can tolerate signals up to 40 dB above the noise floor without producing third-order intermodulation products
C. A pair of 40 dBm signals will theoretically generate a third-order intermodulation product with the same level as the input signals
D. A pair of 1 mW input signals will produce a third-order intermodulation product which is 40 dB stronger than the input signal

E4D10
(C)
Page 8-25

E4D11
Why are third-order intermodulation products created within a receiver of particular interest compared to other products?
A. The third-order product of two signals which are in the band of interest is also likely to be within the band
B. The third-order intercept is much higher than other orders
C. Third-order products are an indication of poor image rejection
D. Third-order intermodulation produces three products for every input signal within the band of interest

E4D11
(A)
Page 8-24

E4D12
(A)
Page 8-23

E4D12
What is the term for the reduction in receiver sensitivity caused by a strong signal near the received frequency?
A. Desensitization
B. Quieting
C. Cross-modulation interference
D. Squelch gain rollback

E4D13
(B)
Page 8-23

E4D13
Which of the following can cause receiver desensitization?
A. Audio gain adjusted too low
B. Strong adjacent-channel signals
C. Audio bias adjusted too high
D. Squelch gain misadjusted

E4D14
(A)
Page 8-23

E4D14
Which of the following is a way to reduce the likelihood of receiver desensitization?
A. Decrease the RF bandwidth of the receiver
B. Raise the receiver IF frequency
C. Increase the receiver front end gain
D. Switch from fast AGC to slow AGC

E4E — Noise suppression: system noise; electrical appliance noise; line noise; locating noise sources; DSP noise reduction; noise blankers

E4E01
(A)
Page 7-33

E4E01
Which of the following types of receiver noise can often be reduced by use of a receiver noise blanker?
A. Ignition noise
B. Broadband white noise
C. Heterodyne interference
D. All of these choices are correct

E4E02
(D)
Page 7-34

E4E02
Which of the following types of receiver noise can often be reduced with a DSP noise filter?
A. Broadband white noise
B. Ignition noise
C. Power line noise
D. All of these choices are correct

E4E03
(B)
Page 7-33

E4E03
Which of the following signals might a receiver noise blanker be able to remove from desired signals?
A. Signals which are constant at all IF levels
B. Signals which appear across a wide bandwidth
C. Signals which appear at one IF but not another
D. Signals which have a sharply peaked frequency distribution

E4E04

How can conducted and radiated noise caused by an automobile alternator be suppressed?

A. By installing filter capacitors in series with the DC power lead and by installing a blocking capacitor in the field lead

B. By installing a noise suppression resistor and a blocking capacitor in both leads

C. By installing a high-pass filter in series with the radio's power lead and a low-pass filter in parallel with the field lead

D. By connecting the radio's power leads directly to the battery and by installing coaxial capacitors in line with the alternator leads

E4E04
(D)
Page 7-33

E4E05

How can noise from an electric motor be suppressed?

A. By installing a high pass filter in series with the motor's power leads

B. By installing a brute-force AC-line filter in series with the motor leads

C. By installing a bypass capacitor in series with the motor leads

D. By using a ground-fault current interrupter in the circuit used to power the motor

E4E05
(B)
Page 7-30

E4E06

What is a major cause of atmospheric static?

A. Solar radio frequency emissions

B. Thunderstorms

C. Geomagnetic storms

D. Meteor showers

E4E06
(B)
Page 7-29

E4E07

How can you determine if line noise interference is being generated within your home?

A. By checking the power line voltage with a time domain reflectometer

B. By observing the AC power line waveform with an oscilloscope

C. By turning off the AC power line main circuit breaker and listening on a battery operated radio

D. By observing the AC power line voltage with a spectrum analyzer

E4E07
(C)
Page 7-30

E4E08

What type of signal is picked up by electrical wiring near a radio antenna?

A. A common-mode signal at the frequency of the radio transmitter

B. An electrical-sparking signal

C. A differential-mode signal at the AC power line frequency

D. Harmonics of the AC power line frequency

E4E08
(A)
Page 7-31

E4E09

What undesirable effect can occur when using an IF noise blanker?

A. Received audio in the speech range might have an echo effect

B. The audio frequency bandwidth of the received signal might be compressed

C. Nearby signals may appear to be excessively wide even if they meet emission standards

D. FM signals can no longer be demodulated

E4E09
(C)
Page 7-34

E4E10

What is a common characteristic of interference caused by a touch controlled electrical device?

A. The interfering signal sounds like AC hum on an AM receiver or a carrier modulated by 60 Hz hum on a SSB or CW receiver

B. The interfering signal may drift slowly across the HF spectrum

C. The interfering signal can be several kHz in width and usually repeats at regular intervals across a HF band

D. All of these choices are correct

E4E10
(D)
Page 7-31

E4E11
(B)
Page 7-28

E4E11
Which of the following is the most likely cause if you are hearing combinations of local AM broadcast signals within one or more of the MF or HF ham bands?
    A. The broadcast station is transmitting an over-modulated signal
    B. Nearby corroded metal joints are mixing and re-radiating the broadcast signals
    C. You are receiving sky wave signals from a distant station
    D. Your station receiver IF amplifier stage is defective

E4E12
(A)
Page 7-34

E4E12
What is one disadvantage of using some types of automatic DSP notch-filters when attempting to copy CW signals?
    A. The DSP filter can remove the desired signal at the same time as it removes interfering signals
    B. Any nearby signal passing through the DSP system will overwhelm the desired signal
    C. Received CW signals will appear to be modulated at the DSP clock frequency
    D. Ringing in the DSP filter will completely remove the spaces between the CW characters

E4E13
(D)
Page 7-30

E4E13
What might be the cause of a loud roaring or buzzing AC line interference that comes and goes at intervals?
    A. Arcing contacts in a thermostatically controlled device
    B. A defective doorbell or doorbell transformer inside a nearby residence
    C. A malfunctioning illuminated advertising display
    D. All of these choices are correct

E4E14
(C)
Page 7-31

E4E14
What is one type of electrical interference that might be caused by the operation of a nearby personal computer?
    A. A loud AC hum in the audio output of your station receiver
    B. A clicking noise at intervals of a few seconds
    C. The appearance of unstable modulated or unmodulated signals at specific frequencies
    D. A whining type noise that continually pulses off and on

SUBELEMENT E5
ELECTRICAL PRINCIPLES
4 Exam Questions — 4 Groups

E5A — Resonance and Q: characteristics of resonant circuits; series and parallel resonance; Q; half-power bandwidth; phase relationships in reactive circuits

E5A01
What can cause the voltage across reactances in series to be larger than the voltage applied to them?
A. Resonance
B. Capacitance
C. Conductance
D. Resistance

E5A02
What is resonance in an electrical circuit?
A. The highest frequency that will pass current
B. The lowest frequency that will pass current
C. The frequency at which the capacitive reactance equals the inductive reactance
D. The frequency at which the reactive impedance equals the resistive impedance

E5A03
What is the magnitude of the impedance of a series RLC circuit at resonance?
A. High, as compared to the circuit resistance
B. Approximately equal to capacitive reactance
C. Approximately equal to inductive reactance
D. Approximately equal to circuit resistance

E5A04
What is the magnitude of the impedance of a circuit with a resistor, an inductor and a capacitor all in parallel, at resonance?
A. Approximately equal to circuit resistance
B. Approximately equal to inductive reactance
C. Low, as compared to the circuit resistance
D. Approximately equal to capacitive reactance

E5A05
What is the magnitude of the current at the input of a series RLC circuit as the frequency goes through resonance?
A. Minimum
B. Maximum
C. R/L
D. L/R

E5A01
(A)
Page 4-33

E5A02
(C)
Page 4-32

E5A03
(D)
Page 4-33

E5A04
(A)
Page 4-34

E5A05
(B)
Page 4-33

E5A06
(B)
Page 4-34

E5A06
What is the magnitude of the circulating current within the components of a parallel LC circuit at resonance?
    A. It is at a minimum
    B. It is at a maximum
    C. It equals 1 divided by the quantity 2 times Pi, multiplied by the square root of inductance L multiplied by capacitance C
    D. It equals 2 multiplied by Pi, multiplied by frequency "F", multiplied by inductance "L"

E5A07
(A)
Page 4-34

E5A07
What is the magnitude of the current at the input of a parallel RLC circuit at resonance?
    A. Minimum
    B. Maximum
    C. R/L
    D. L/R

E5A08
(C)
Page 4-35

E5A08
What is the phase relationship between the current through and the voltage across a series resonant circuit at resonance?
    A. The voltage leads the current by 90 degrees
    B. The current leads the voltage by 90 degrees
    C. The voltage and current are in phase
    D. The voltage and current are 180 degrees out of phase

E5A09
(C)
Page 4-35

E5A09
What is the phase relationship between the current through and the voltage across a parallel resonant circuit at resonance?
    A. The voltage leads the current by 90 degrees
    B. The current leads the voltage by 90 degrees
    C. The voltage and current are in phase
    D. The voltage and current are 180 degrees out of phase

E5A10
(A)
Page 4-37

E5A10
What is the half-power bandwidth of a parallel resonant circuit that has a resonant frequency of 1.8 MHz and a Q of 95?
    A. 18.9 kHz
    B. 1.89 kHz
    C. 94.5 kHz
    D. 9.45 kHz

E5A11
(C)
Page 4-37

E5A11
What is the half-power bandwidth of a parallel resonant circuit that has a resonant frequency of 7.1 MHz and a Q of 150?
    A. 157.8 Hz
    B. 315.6 Hz
    C. 47.3 kHz
    D. 23.67 kHz

E5A12
What is the half-power bandwidth of a parallel resonant circuit that has a resonant frequency of
3.7 MHz and a Q of 118?
  A. 436.6 kHz
  B. 218.3 kHz
  C. 31.4 kHz
  D. 15.7 kHz

E5A13
What is the half-power bandwidth of a parallel resonant circuit that has a resonant frequency of
14.25 MHz and a Q of 187?
  A. 38.1 kHz
  B. 76.2 kHz
  C. 1.332 kHz
  D. 2.665 kHz

E5A14
What is the resonant frequency of a series RLC circuit if R is 22 ohms, L is 50 microhenrys and
C is 40 picofarads?
  A. 44.72 MHz
  B. 22.36 MHz
  C. 3.56 MHz
  D. 1.78 MHz

E5A15
What is the resonant frequency of a series RLC circuit if R is 56 ohms, L is 40 microhenrys and
C is 200 picofarads?
  A. 3.76 MHz
  B. 1.78 MHz
  C. 11.18 MHz
  D. 22.36 MHz

E5A16
What is the resonant frequency of a parallel RLC circuit if R is 33 ohms, L is 50 microhenrys and
C is 10 picofarads?
  A. 23.5 MHz
  B. 23.5 kHz
  C. 7.12 kHz
  D. 7.12 MHz

E5A17
What is the resonant frequency of a parallel RLC circuit if R is 47 ohms, L is 25 microhenrys and
C is 10 picofarads?
  A. 10.1 MHz
  B. 63.2 MHz
  C. 10.1 kHz
  D. 63.2 kHz

E5A12
(C)
Page 4-37

E5A13
(B)
Page 4-37

E5A14
(C)
Page 4-33

E5A15
(B)
Page 4-34

E5A16
(D)
Page 4-35

E5A17
(A)
Page 4-35

E5B — Time constants and phase relationships: RLC time constants; definition; time constants in RL and RC circuits; phase angle between voltage and current; phase angles of series and parallel circuits

E5B01
(B)
Page 4-9

E5B01
What is the term for the time required for the capacitor in an RC circuit to be charged to 63.2% of the applied voltage?
   A. An exponential rate of one
   B. One time constant
   C. One exponential period
   D. A time factor of one

E5B02
(D)
Page 4-9

E5B02
What is the term for the time it takes for a charged capacitor in an RC circuit to discharge to 36.8% of its initial voltage?
   A. One discharge period
   B. An exponential discharge rate of one
   C. A discharge factor of one
   D. One time constant

E5B03
(D)
Page 4-9

E5B03
The capacitor in an RC circuit is discharged to what percentage of the starting voltage after two time constants?
   A. 86.5%
   B. 63.2%
   C. 36.8%
   D. 13.5%

E5B04
(D)
Page 4-11

E5B04
What is the time constant of a circuit having two 220-microfarad capacitors and two 1-megohm resistors, all in parallel?
   A. 55 seconds
   B. 110 seconds
   C. 440 seconds
   D. 220 seconds

E5B05
(A)
Page 4-10

E5B05
How long does it take for an initial charge of 20 V DC to decrease to 7.36 V DC in a 0.01-microfarad capacitor when a 2-megohm resistor is connected across it?
   A. 0.02 seconds
   B. 0.04 seconds
   C. 20 seconds
   D. 40 seconds

E5B06
(C)
Page 4-11

E5B06
How long does it take for an initial charge of 800 V DC to decrease to 294 V DC in a 450-microfarad capacitor when a 1-megohm resistor is connected across it?
   A. 4.50 seconds
   B. 9 seconds
   C. 450 seconds
   D. 900 seconds

E5B07
What is the phase angle between the voltage across and the current through a series RLC circuit if
XC is 500 ohms, R is 1 kilohm, and XL is 250 ohms?
  A. 68.2 degrees with the voltage leading the current
  B. 14.0 degrees with the voltage leading the current
  C. 14.0 degrees with the voltage lagging the current
  D. 68.2 degrees with the voltage lagging the current

E5B08
What is the phase angle between the voltage across and the current through a series RLC circuit if
XC is 100 ohms, R is 100 ohms, and XL is 75 ohms?
  A. 14 degrees with the voltage lagging the current
  B. 14 degrees with the voltage leading the current
  C. 76 degrees with the voltage leading the current
  D. 76 degrees with the voltage lagging the current

E5B09
What is the relationship between the current through a capacitor and the voltage across a capacitor?
  A. Voltage and current are in phase
  B. Voltage and current are 180 degrees out of phase
  C. Voltage leads current by 90 degrees
  D. Current leads voltage by 90 degrees

E5B10
What is the relationship between the current through an inductor and the voltage across an inductor?
  A. Voltage leads current by 90 degrees
  B. Current leads voltage by 90 degrees
  C. Voltage and current are 180 degrees out of phase
  D. Voltage and current are in phase

E5B11
What is the phase angle between the voltage across and the current through a series RLC circuit if
XC is 25 ohms, R is 100 ohms, and XL is 50 ohms?
  A. 14 degrees with the voltage lagging the current
  B. 14 degrees with the voltage leading the current
  C. 76 degrees with the voltage lagging the current
  D. 76 degrees with the voltage leading the current

E5B12
What is the phase angle between the voltage across and the current through a series RLC circuit if
XC is 75 ohms, R is 100 ohms, and XL is 50 ohms?
  A. 76 degrees with the voltage lagging the current
  B. 14 degrees with the voltage leading the current
  C. 14 degrees with the voltage lagging the current
  D. 76 degrees with the voltage leading the current

E5B13
What is the phase angle between the voltage across and the current through a series RLC circuit if
XC is 250 ohms, R is 1 kilohm, and XL is 500 ohms?
  A. 81.47 degrees with the voltage lagging the current
  B. 81.47 degrees with the voltage leading the current
  C. 14.04 degrees with the voltage lagging the current
  D. 14.04 degrees with the voltage leading the current

E5B07
(C)
Page 4-26

E5B08
(A)
Page 4-27

E5B09
(D)
Page 4-14

E5B10
(A)
Page 4-14

E5B11
(B)
Page 4-27

E5B12
(C)
Page 4-28

E5B13
(D)
Page 4-28

E5C — Impedance plots and coordinate systems: plotting impedances in polar coordinates; rectangular coordinates

E5C01
In polar coordinates, what is the impedance of a network consisting of a 100-ohm-reactance inductor in series with a 100-ohm resistor?
A. 121 ohms at an angle of 35 degrees
B. 141 ohms at an angle of 45 degrees
C. 161 ohms at an angle of 55 degrees
D. 181 ohms at an angle of 65 degrees

E5C02
In polar coordinates, what is the impedance of a network consisting of a 100-ohm-reactance inductor, a 100-ohm-reactance capacitor, and a 100-ohm resistor, all connected in series?
A. 100 ohms at an angle of 90 degrees
B. 10 ohms at an angle of 0 degrees
C. 10 ohms at an angle of 90 degrees
D. 100 ohms at an angle of 0 degrees

E5C03
In polar coordinates, what is the impedance of a network consisting of a 300-ohm-reactance capacitor, a 600-ohm-reactance inductor, and a 400-ohm resistor, all connected in series?
A. 500 ohms at an angle of 37 degrees
B. 900 ohms at an angle of 53 degrees
C. 400 ohms at an angle of 0 degrees
D. 1300 ohms at an angle of 180 degrees

E5C04
In polar coordinates, what is the impedance of a network consisting of a 400-ohm-reactance capacitor in series with a 300-ohm resistor?
A. 240 ohms at an angle of 36.9 degrees
B. 240 ohms at an angle of -36.9 degrees
C. 500 ohms at an angle of 53.1 degrees
D. 500 ohms at an angle of -53.1 degrees

E5C05
In polar coordinates, what is the impedance of a network consisting of a 400-ohm-reactance inductor in parallel with a 300-ohm resistor?
A. 240 ohms at an angle of 36.9 degrees
B. 240 ohms at an angle of -36.9 degrees
C. 500 ohms at an angle of 53.1 degrees
D. 500 ohms at an angle of -53.1 degrees

E5C06
In polar coordinates, what is the impedance of a network consisting of a 100-ohm-reactance capacitor in series with a 100-ohm resistor?
A. 121 ohms at an angle of -25 degrees
B. 191 ohms at an angle of -85 degrees
C. 161 ohms at an angle of -65 degrees
D. 141 ohms at an angle of -45 degrees

E5C07
In polar coordinates, what is the impedance of a network comprised of a 100-ohm-reactance capacitor in parallel with a 100-ohm resistor?
  A. 31 ohms at an angle of -15 degrees
  B. 51 ohms at an angle of -25 degrees
  C. 71 ohms at an angle of -45 degrees
  D. 91 ohms at an angle of -65 degrees

E5C08
In polar coordinates, what is the impedance of a network comprised of a 300-ohm-reactance inductor in series with a 400-ohm resistor?
  A. 400 ohms at an angle of 27 degrees
  B. 500 ohms at an angle of 37 degrees
  C. 500 ohms at an angle of 47 degrees
  D. 700 ohms at an angle of 57 degrees

E5C09
When using rectangular coordinates to graph the impedance of a circuit, what does the horizontal axis represent?
  A. Resistive component
  B. Reactive component
  C. The sum of the reactive and resistive components
  D. The difference between the resistive and reactive components

E5C10
When using rectangular coordinates to graph the impedance of a circuit, what does the vertical axis represent?
  A. Resistive component
  B. Reactive component
  C. The sum of the reactive and resistive components
  D. The difference between the resistive and reactive components

E5C11
What do the two numbers represent that are used to define a point on a graph using rectangular coordinates?
  A. The magnitude and phase of the point
  B. The sine and cosine values
  C. The coordinate values along the horizontal and vertical axes
  D. The tangent and cotangent values

E5C12
If you plot the impedance of a circuit using the rectangular coordinate system and find the impedance point falls on the right side of the graph on the horizontal axis, what do you know about the circuit?
  A. It has to be a direct current circuit
  B. It contains resistance and capacitive reactance
  C. It contains resistance and inductive reactance
  D. It is equivalent to a pure resistance

E5C07
(C)
Page 4-25

E5C08
(B)
Page 4-20

E5C09
(A)
Page 4-17

E5C10
(B)
Page 4-17

E5C11
(C)
Page 4-2

E5C12
(D)
Page 4-18

E5C13
(D)
Page 4-17

E5C13
What coordinate system is often used to display the resistive, inductive, and/or capacitive reactance components of an impedance?
A. Maidenhead grid
B. Faraday grid
C. Elliptical coordinates
D. Rectangular coordinates

E5C14
(D)
Page 4-17

E5C14
What coordinate system is often used to display the phase angle of a circuit containing resistance, inductive and/or capacitive reactance?
A. Maidenhead grid
B. Faraday grid
C. Elliptical coordinates
D. Polar coordinates

E5C15
(A)
Page 4-18

E5C15
In polar coordinates, what is the impedance of a circuit of 100 -j100 ohms impedance?
A. 141 ohms at an angle of -45 degrees
B. 100 ohms at an angle of 45 degrees
C. 100 ohms at an angle of -45 degrees
D. 141 ohms at an angle of 45 degrees

E5C16
(B)
Page 4-19

E5C16
In polar coordinates, what is the impedance of a circuit that has an admittance of 7.09 millisiemens at 45 degrees?
A. 5.03 E–06 ohms at an angle of 45 degrees
B. 141 ohms at an angle of -45 degrees
C. 19,900 ohms at an angle of -45 degrees
D. 141 ohms at an angle of 45 degrees

E5C17
(C)
Page 4-19

E5C17
In rectangular coordinates, what is the impedance of a circuit that has an admittance of 5 millisiemens at -30 degrees?
A. 173 - j100 ohms
B. 200 + j100 ohms
C. 173 + j100 ohms
D. 200 - j100 ohms

E5C18
(B)
Page 4-23

E5C18
In polar coordinates, what is the impedance of a series circuit consisting of a resistance of 4 ohms, an inductive reactance of 4 ohms, and a capacitive reactance of 1 ohm?
A. 6.4 ohms at an angle of 53 degrees
B. 5 ohms at an angle of 37 degrees
C. 5 ohms at an angle of 45 degrees
D. 10 ohms at an angle of -51 degrees

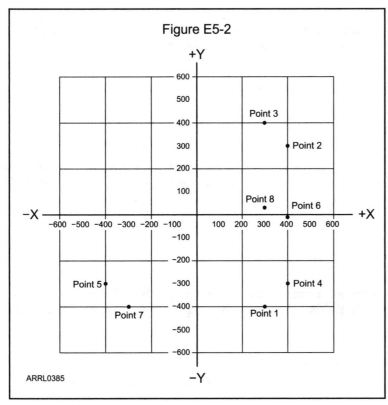

Figure E5-2

Figure E5-2 — Refer to this figure for questions E5C19 through E5C21 and E5C23.

ARRL0385

E5C19

Which point on Figure E5-2 best represents that impedance of a series circuit consisting of a 400 ohm resistor and a 38 picofarad capacitor at 14 MHz?

A. Point 2
B. Point 4
C. Point 5
D. Point 6

E5C19
(B)
Page 4-22

E5C20

Which point in Figure E5-2 best represents the impedance of a series circuit consisting of a 300 ohm resistor and an 18 microhenry inductor at 3.505 MHz?

A. Point 1
B. Point 3
C. Point 7
D. Point 8

E5C20
(B)
Page 4-21

E5C21

Which point on Figure E5-2 best represents the impedance of a series circuit consisting of a 300 ohm resistor and a 19 picofarad capacitor at 21.200 MHz?

A. Point 1
B. Point 3
C. Point 7
D. Point 8

E5C21
(A)
Page 4-22

E5C22
In rectangular coordinates, what is the impedance of a network consisting of a 10-microhenry inductor in series with a 40-ohm resistor at 500 MHz?
- A. 40 + j31,400
- B. 40 - j31,400
- C. 31,400 + j40
- D. 31,400 - j40

E5C23
Which point on Figure E5-2 best represents the impedance of a series circuit consisting of a 300-ohm resistor, a 0.64-microhenry inductor and an 85-picofarad capacitor at 24.900 MHz?
- A. Point 1
- B. Point 3
- C. Point 5
- D. Point 8

E5D — AC and RF energy in real circuits: skin effect; electrostatic and electromagnetic fields; reactive power; power factor; coordinate systems

E5D01
What is the result of skin effect?
- A. As frequency increases, RF current flows in a thinner layer of the conductor, closer to the surface
- B. As frequency decreases, RF current flows in a thinner layer of the conductor, closer to the surface
- C. Thermal effects on the surface of the conductor increase the impedance
- D. Thermal effects on the surface of the conductor decrease the impedance

E5D02
Why is the resistance of a conductor different for RF currents than for direct currents?
- A. Because the insulation conducts current at high frequencies
- B. Because of the Heisenburg Effect
- C. Because of skin effect
- D. Because conductors are non-linear devices

E5D03
What device is used to store electrical energy in an electrostatic field?
- A. A battery
- B. A transformer
- C. A capacitor
- D. An inductor

E5D04
What unit measures electrical energy stored in an electrostatic field?
- A. Coulomb
- B. Joule
- C. Watt
- D. Volt

E5D05
Which of the following creates a magnetic field?
  A. Potential differences between two points in space
  B. Electric current
  C. A charged capacitor
  D. A battery

E5D06
In what direction is the magnetic field oriented about a conductor in relation to the direction of electron flow?
  A. In the same direction as the current
  B. In a direction opposite to the current
  C. In all directions; omnidirectional
  D. In a direction determined by the left-hand rule

E5D07
What determines the strength of a magnetic field around a conductor?
  A. The resistance divided by the current
  B. The ratio of the current to the resistance
  C. The diameter of the conductor
  D. The amount of current

E5D08
What type of energy is stored in an electromagnetic or electrostatic field?
  A. Electromechanical energy
  B. Potential energy
  C. Thermodynamic energy
  D. Kinetic energy

E5D09
What happens to reactive power in an AC circuit that has both ideal inductors and ideal capacitors?
  A. It is dissipated as heat in the circuit
  B. It is repeatedly exchanged between the associated magnetic and electric fields, but is not dissipated
  C. It is dissipated as kinetic energy in the circuit
  D. It is dissipated in the formation of inductive and capacitive fields

E5D10
How can the true power be determined in an AC circuit where the voltage and current are out of phase?
  A. By multiplying the apparent power times the power factor
  B. By dividing the reactive power by the power factor
  C. By dividing the apparent power by the power factor
  D. By multiplying the reactive power times the power factor

E5D11
What is the power factor of an R-L circuit having a 60 degree phase angle between the voltage and the current?
  A. 1.414
  B. 0.866
  C. 0.5
  D. 1.73

E5D05
(B)
Page 4-5

E5D06
(D)
Page 4-6

E5D07
(D)
Page 4-7

E5D08
(B)
Page 4-5

E5D09
(B)
Page 4-29

E5D10
(A)
Page 4-30

E5D11
(C)
Page 4-31

E5D12
(B)
Page 4-31

E5D12
How many watts are consumed in a circuit having a power factor of 0.2 if the input is 100-V AC at 4 amperes?
   A. 400 watts
   B. 80 watts
   C. 2000 watts
   D. 50 watts

E5D13
(B)
Page 4-31

E5D13
How much power is consumed in a circuit consisting of a 100 ohm resistor in series with a 100 ohm inductive reactance drawing 1 ampere?
   A. 70.7 Watts
   B. 100 Watts
   C. 141.4 Watts
   D. 200 Watts

E5D14
(A)
Page 4-29

E5D14
What is reactive power?
   A. Wattless, nonproductive power
   B. Power consumed in wire resistance in an inductor
   C. Power lost because of capacitor leakage
   D. Power consumed in circuit Q

E5D15
(D)
Page 4-31

E5D15
What is the power factor of an RL circuit having a 45 degree phase angle between the voltage and the current?
   A. 0.866
   B. 1.0
   C. 0.5
   D. 0.707

E5D16
(C)
Page 4-31

E5D16
What is the power factor of an RL circuit having a 30 degree phase angle between the voltage and the current?
   A. 1.73
   B. 0.5
   C. 0.866
   D. 0.577

E5D17
(D)
Page 4-31

E5D17
How many watts are consumed in a circuit having a power factor of 0.6 if the input is 200V AC at 5 amperes?
   A. 200 watts
   B. 1000 watts
   C. 1600 watts
   D. 600 watts

E5D18
(B)
Page 4-32

E5D18
How many watts are consumed in a circuit having a power factor of 0.71 if the apparent power is 500 VA?
   A. 704 W
   B. 355 W
   C. 252 W
   D. 1.42 mW

SUBELEMENT E6
CIRCUIT COMPONENTS
6 Exam Questions — 6 Groups

E6A — Semiconductor materials and devices: semiconductor materials; germanium, silicon, P-type, N-type; transistor types: NPN, PNP, junction, field-effect transistors: enhancement mode; depletion mode; MOS; CMOS; N-channel; P-channel

E6A01
In what application is gallium arsenide used as a semiconductor material in preference to germanium or silicon?
  A. In high-current rectifier circuits
  B. In high-power audio circuits
  C. At microwave frequencies
  D. At very low frequency RF circuits

E6A01
(C)
Page 5-3

E6A02
Which of the following semiconductor materials contains excess free electrons?
  A. N-type
  B. P-type
  C. Bipolar
  D. Insulated gate

E6A02
(A)
Page 5-2

E6A03
What are the majority charge carriers in P-type semiconductor material?
  A. Free neutrons
  B. Free protons
  C. Holes
  D. Free electrons

E6A03
(C)
Page 5-3

E6A04
What is the name given to an impurity atom that adds holes to a semiconductor crystal structure?
  A. Insulator impurity
  B. N-type impurity
  C. Acceptor impurity
  D. Donor impurity

E6A04
(C)
Page 5-2

E6A05
What is the alpha of a bipolar junction transistor?
  A. The change of collector current with respect to base current
  B. The change of base current with respect to collector current
  C. The change of collector current with respect to emitter current
  D. The change of collector current with respect to gate current

E6A05
(C)
Page 5-9

E6A06
What is the beta of a bipolar junction transistor?
  A. The frequency at which the current gain is reduced to 1
  B. The change in collector current with respect to base current
  C. The breakdown voltage of the base to collector junction
  D. The switching speed of the transistor

E6A06
(B)
Page 5-9

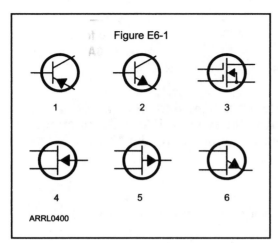

Figure E6-1

ARRL0400

Figure E6-1 — Refer to this figure for question E6A07.

E6A07
(A)
Page 5-8

E6A07
In Figure E6-1, what is the schematic symbol for a PNP transistor?
A. 1
B. 2
C. 4
D. 5

E6A08
(D)
Page 5-10

E6A08
What term indicates the frequency at which the common-base current gain of a transistor has decreased to 0.7 of the gain obtainable at 1 kHz?
A. Corner frequency
B. Alpha rejection frequency
C. Beta cutoff frequency
D. Alpha cutoff frequency

E6A09
(A)
Page 5-12

E6A09
What is a depletion-mode FET?
A. An FET that exhibits a current flow between source and drain when no gate voltage is applied
B. An FET that has no current flow between source and drain when no gate voltage is applied
C. Any FET without a channel
D. Any FET for which holes are the majority carriers

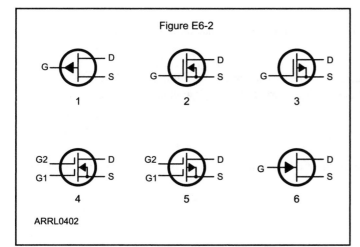

Figure E6-2

ARRL0402

Figure E6-2 — Refer to this figure for questions E6A10 and E6A11.

E6A10
In Figure E6-2, what is the schematic symbol for an N-channel dual-gate MOSFET?
  A. 2
  B. 4
  C. 5
  D. 6

E6A10
(B)
Page 5-11

E6A11
In Figure E6-2, what is the schematic symbol for a P-channel junction FET?
  A. 1
  B. 2
  C. 3
  D. 6

E6A11
(A)
Page 5-11

E6A12
Why do many MOSFET devices have internally connected Zener diodes on the gates?
  A. To provide a voltage reference for the correct amount of reverse-bias gate voltage
  B. To protect the substrate from excessive voltages
  C. To keep the gate voltage within specifications and prevent the device from overheating
  D. To reduce the chance of the gate insulation being punctured by static discharges or excessive voltages

E6A12
(D)
Page 5-11

E6A13
What do the initials CMOS stand for?
  A. Common Mode Oscillating System
  B. Complementary Mica-Oxide Silicon
  C. Complementary Metal-Oxide Semiconductor
  D. Common Mode Organic Silicon

E6A13
(C)
Page 5-28

E6A14
(C)
Page 5-11

E6A14
How does DC input impedance at the gate of a field-effect transistor compare with the DC input impedance of a bipolar transistor?
   A. They are both low impedance
   B. An FET has low input impedance; a bipolar transistor has high input impedance
   C. An FET has high input impedance; a bipolar transistor has low input impedance
   D. They are both high impedance

E6A15
(B)
Page 5-2

E6A15
Which of the following semiconductor materials contains an excess of holes in the outer shell of electrons?
   A. N-type
   B. P-type
   C. Superconductor-type
   D. Bipolar-type

E6A16
(B)
Page 5-3

E6A16
What are the majority charge carriers in N-type semiconductor material?
   A. Holes
   B. Free electrons
   C. Free protons
   D. Free neutrons

E6A17
(D)
Page 5-10

E6A17
What are the names of the three terminals of a field-effect transistor?
   A. Gate 1, gate 2, drain
   B. Emitter, base, collector
   C. Emitter, base 1, base 2
   D. Gate, drain, source

E6B — Semiconductor diodes

E6B01
(B)
Page 5-7

E6B01
What is the most useful characteristic of a Zener diode?
   A. A constant current drop under conditions of varying voltage
   B. A constant voltage drop under conditions of varying current
   C. A negative resistance region
   D. An internal capacitance that varies with the applied voltage

E6B02
(D)
Page 5-5

E6B02
What is an important characteristic of a Schottky diode as compared to an ordinary silicon diode when used as a power supply rectifier?
   A. Much higher reverse voltage breakdown
   B. Controlled reverse avalanche voltage
   C. Enhanced carrier retention time
   D. Less forward voltage drop

E6B03
What special type of diode is capable of both amplification and oscillation?
A. Point contact
B. Zener
C. Tunnel
D. Junction

E6B04
What type of semiconductor device is designed for use as a voltage-controlled capacitor?
A. Varactor diode
B. Tunnel diode
C. Silicon-controlled rectifier
D. Zener diode

E6B05
What characteristic of a PIN diode makes it useful as an RF switch or attenuator?
A. Extremely high reverse breakdown voltage
B. Ability to dissipate large amounts of power
C. Reverse bias controls its forward voltage drop
D. A large region of intrinsic material

E6B06
Which of the following is a common use of a hot-carrier diode?
A. As balanced mixers in FM generation
B. As a variable capacitance in an automatic frequency control circuit
C. As a constant voltage reference in a power supply
D. As a VHF / UHF mixer or detector

E6B07
What is the failure mechanism when a junction diode fails due to excessive current?
A. Excessive inverse voltage
B. Excessive junction temperature
C. Insufficient forward voltage
D. Charge carrier depletion

E6B08
Which of the following describes a type of semiconductor diode?
A. Metal-semiconductor junction
B. Electrolytic rectifier
C. CMOS-field effect
D. Thermionic emission diode

E6B09
What is a common use for point contact diodes?
A. As a constant current source
B. As a constant voltage source
C. As an RF detector
D. As a high voltage rectifier

E6B03
(C)
Page 5-7

E6B04
(A)
Page 5-7

E6B05
(D)
Page 5-8

E6B06
(D)
Page 5-6

E6B07
(B)
Page 5-5

E6B08
(A)
Page 5-5

E6B09
(C)
Page 5-6

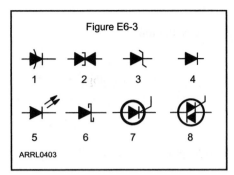

Figure E6-3 — Refer to this figure for question E6B10.

E6B10
In Figure E6-3, what is the schematic symbol for a light-emitting diode?
   A. 1
   B. 5
   C. 6
   D. 7

E6B11
What is used to control the attenuation of RF signals by a PIN diode?
   A. Forward DC bias current
   B. A sub-harmonic pump signal
   C. Reverse voltage larger than the RF signal
   D. Capacitance of an RF coupling capacitor

E6B12
What is one common use for PIN diodes?
   A. As a constant current source
   B. As a constant voltage source
   C. As an RF switch
   D. As a high voltage rectifier

E6B13
What type of bias is required for an LED to emit light?
   A. Reverse bias
   B. Forward bias
   C. Zero bias
   D. Inductive bias

E6C — Integrated circuits: TTL digital integrated circuits; CMOS digital integrated circuits; gates

E6C01
What is the recommended power supply voltage for TTL series integrated circuits?
   A. 12 volts
   B. 1.5 volts
   C. 5 volts
   D. 13.6 volts

E6C02

What logic state do the inputs of a TTL device assume if they are left open?

A. A logic-high state

B. A logic-low state

C. The device becomes randomized and will not provide consistent high or low-logic states

D. Open inputs on a TTL device are ignored

E6C02
(A)
Page 5-28

E6C03

Which of the following describes tri-state logic?

A. Logic devices with 0, 1, and high impedance output states

B. Logic devices that utilize ternary math

C. Low power logic devices designed to operate at 3 volts

D. Proprietary logic devices manufactured by Tri-State Devices

E6C03
(A)
Page 5-20

E6C04

Which of the following is the primary advantage of tri-state logic?

A. Low power consumption

B. Ability to connect many device outputs to a common bus

C. High speed operation

D. More efficient arithmetic operations

E6C04
(B)
Page 5-20

E6C05

Which of the following is an advantage of CMOS logic devices over TTL devices?

A. Differential output capability

B. Lower distortion

C. Immune to damage from static discharge

D. Lower power consumption

E6C05
(D)
Page 5-28

E6C06

Why do CMOS digital integrated circuits have high immunity to noise on the input signal or power supply?

A. Larger bypass capacitors are used in CMOS circuit design

B. The input switching threshold is about two times the power supply voltage

C. The input switching threshold is about one-half the power supply voltage

D. Input signals are stronger

E6C06
(C)
Page 5-29

E6C07
(A)
Page 5-19

E6C07
In Figure E6-5, what is the schematic symbol for an AND gate?
  A. 1
  B. 2
  C. 3
  D. 4

E6C08
(B)
Page 5-19

E6C08
In Figure E6-5, what is the schematic symbol for a NAND gate?
  A. 1
  B. 2
  C. 3
  D. 4

E6C09
(B)
Page 5-19

E6C09
In Figure E6-5, what is the schematic symbol for an OR gate?
  A. 2
  B. 3
  C. 4
  D. 6

E6C10
(D)
Page 5-19

E6C10
In Figure E6-5, what is the schematic symbol for a NOR gate?
  A. 1
  B. 2
  C. 3
  D. 4

E6C11
(C)
Page 5-18

E6C11
In Figure E6-5, what is the schematic symbol for the NOT operation (inverter)?
  A. 2
  B. 4
  C. 5
  D. 6

E6C12
(D)
Page 5-29

E6C12
What is BiCMOS logic?
  A. A logic device with two CMOS circuits per package
  B. An FET logic family based on bimetallic semiconductors
  C. A logic family based on bismuth CMOS devices
  D. An integrated circuit logic family using both bipolar and CMOS transistors

E6C13
(C)
Page 5-29

E6C13
Which of the following is an advantage of BiCMOS logic?
  A. Its simplicity results in much less expensive devices than standard CMOS
  B. It is totally immune to electrostatic damage
  C. It has the high input impedance of CMOS and the low output impedance of bipolar transistors
  D. All of these choices are correct

Figure E6-5

1   2   3

4   5   6

ARRL0401

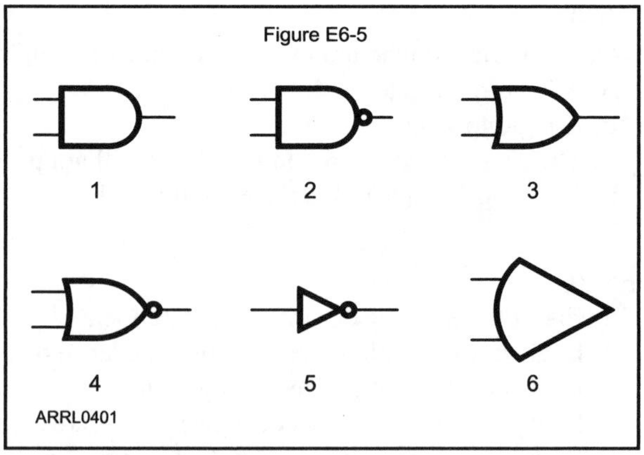

Figure E6-5 — Refer to this figure for questions E6C07 through E6C11.

E6D — Optical devices and toroids: cathode-ray tube devices; charge-coupled devices (CCDs); liquid crystal displays (LCDs) Toroids: permeability; core material; selecting; winding

E6D01
What is cathode ray tube (CRT) persistence?
  A. The time it takes for an image to appear after the electron beam is turned on
  B. The relative brightness of the display under varying conditions of ambient light
  C. The ability of the display to remain in focus under varying conditions
  D. The length of time the image remains on the screen after the beam is turned off

E6D01
(D)
Page 5-16

E6D02
Exceeding what design rating can cause a cathode ray tube (CRT) to generate X-rays?
  A. The heater voltage
  B. The anode voltage
  C. The operating temperature
  D. The operating frequency

E6D02
(B)
Page 5-16

E6D03
Which of the following is true of a charge-coupled device (CCD)?
  A. Its phase shift changes rapidly with frequency
  B. It is a CMOS analog-to-digital converter
  C. It samples an analog signal and passes it in stages from the input to the output
  D. It is used in a battery charger circuit

E6D03
(C)
Page 5-16

E6D04
What function does a charge-coupled device (CCD) serve in a modern video camera?
  A. It stores photogenerated charges as signals corresponding to pixels
  B. It generates the horizontal pulses needed for electron beam scanning
  C. It focuses the light used to produce a pattern of electrical charges corresponding to the image
  D. It combines audio and video information to produce a composite RF signal

E6D04
(A)
Page 5-17

E6D05
What is a liquid-crystal display (LCD)?
  A. A modern replacement for a quartz crystal oscillator which displays its fundamental frequency
  B. A display using a crystalline liquid which, in conjunction with polarizing filters, becomes opaque when voltage is applied
  C. A frequency-determining unit for a transmitter or receiver
  D. A display that uses a glowing liquid to remain brightly lit in dim light

E6D05
(B)
Page 5-14

E6D06
What core material property determines the inductance of a toroidal inductor?
  A. Thermal impedance
  B. Resistance
  C. Reactivity
  D. Permeability

E6D06
(D)
Page 4-38

E6D07
(B)
Page 4-39

E6D07
What is the usable frequency range of inductors that use toroidal cores, assuming a correct selection of core material for the frequency being used?
  A. From a few kHz to no more than 30 MHz
  B. From less than 20 Hz to approximately 300 MHz
  C. From approximately 10 Hz to no more than 3000 kHz
  D. From about 100 kHz to at least 1000 GHz

E6D08
(B)
Page 4-38

E6D08
What is one important reason for using powdered-iron toroids rather than ferrite toroids in an inductor?
  A. Powdered-iron toroids generally have greater initial permeability
  B. Powdered-iron toroids generally maintain their characteristics at higher currents
  C. Powdered-iron toroids generally require fewer turns to produce a given inductance value
  D. Powdered-iron toroids have higher power handling capacity

E6D09
(C)
Page 4-41

E6D09
What devices are commonly used as VHF and UHF parasitic suppressors at the input and output terminals of transistorized HF amplifiers?
  A. Electrolytic capacitors
  B. Butterworth filters
  C. Ferrite beads
  D. Steel-core toroids

E6D10
(A)
Page 4-39

E6D10
What is a primary advantage of using a toroidal core instead of a solenoidal core in an inductor?
  A. Toroidal cores confine most of the magnetic field within the core material
  B. Toroidal cores make it easier to couple the magnetic energy into other components
  C. Toroidal cores exhibit greater hysteresis
  D. Toroidal cores have lower Q characteristics

E6D11
(C)
Page 4-40

E6D11
How many turns will be required to produce a 1-mH inductor using a ferrite toroidal core that has an inductance index (A_L) value of 523 millihenrys/1000 turns?
  A. 2 turns
  B. 4 turns
  C. 43 turns
  D. 229 turns

E6D12
(A)
Page 4-40

E6D12
How many turns will be required to produce a 5-microhenry inductor using a powdered-iron toroidal core that has an inductance index (A_L) value of 40 microhenrys/100 turns?
  A. 35 turns
  B. 13 turns
  C. 79 turns
  D. 141 turns

E6D13
(D)
Page 5-15

E6D13
What type of CRT deflection is better when high-frequency waveforms are to be displayed on the screen?
  A. Electromagnetic
  B. Tubular
  C. Radar
  D. Electrostatic

E6D14
Which is NOT true of a charge-coupled device (CCD)?
  A. It uses a combination of analog and digital circuitry
  B. It can be used to make an audio delay line
  C. It is commonly used as an analog-to-digital converter
  D. It samples and stores analog signals

E6D14
(C)
Page 5-16

E6D15
What is the principle advantage of liquid-crystal display (LCD) devices over other types of display devices?
  A. They consume less power
  B. They can display changes instantly
  C. They are visible in all light conditions
  D. They can be easily interchanged with other display devices

E6D15
(A)
Page 5-14

E6D16
What is one reason for using ferrite toroids rather than powdered-iron toroids in an inductor?
  A. Ferrite toroids generally have lower initial permeabilities
  B. Ferrite toroids generally have better temperature stability
  C. Ferrite toroids generally require fewer turns to produce a given inductance value
  D. Ferrite toroids are easier to use with surface mount technology

E6D16
(C)
Page 4-38

E6E — Piezoelectric crystals and MMICs: quartz crystal oscillators and crystal filters; monolithic amplifiers

E6E01
What is a crystal lattice filter?
  A. A power supply filter made with interlaced quartz crystals
  B. An audio filter made with four quartz crystals that resonate at 1-kHz intervals
  C. A filter with wide bandwidth and shallow skirts made using quartz crystals
  D. A filter with narrow bandwidth and steep skirts made using quartz crystals

E6E01
(D)
Page 6-39

E6E02
Which of the following factors has the greatest effect in helping determine the bandwidth and response shape of a crystal ladder filter?
  A. The relative frequencies of the individual crystals
  B. The DC voltage applied to the quartz crystal
  C. The gain of the RF stage preceding the filter
  D. The amplitude of the signals passing through the filter

E6E02
(A)
Page 6-39

E6E03
What is one aspect of the piezoelectric effect?
  A. Physical deformation of a crystal by the application of a voltage
  B. Mechanical deformation of a crystal by the application of a magnetic field
  C. The generation of electrical energy by the application of light
  D. Reversed conduction states when a P-N junction is exposed to light

E6E03
(A)
Page 6-19

E6E04
(A)
Page 5-13

E6E04
What is the most common input and output impedance of circuits that use MMICs?
A. 50 ohms
B. 300 ohms
C. 450 ohms
D. 10 ohms

E6E05
(A)
Page 8-20

E6E05
Which of the following noise figure values is typical of a low-noise UHF preamplifier?
A. 2 dB
B. −10 dB
C. 44 dBm
D. −20 dBm

E6E06
(D)
Page 5-13

E6E06
What characteristics of the MMIC make it a popular choice for VHF through microwave circuits?
A. The ability to retrieve information from a single signal even in the presence of other strong signals.
B. Plate current that is controlled by a control grid
C. Nearly infinite gain, very high input impedance, and very low output impedance
D. Controlled gain, low noise figure, and constant input and output impedance over the specified frequency range

E6E07
(B)
Page 5-13

E6E07
Which of the following is typically used to construct a MMIC-based microwave amplifier?
A. Ground-plane construction
B. Microstrip construction
C. Point-to-point construction
D. Wave-soldering construction

E6E08
(A)
Page 5-12

E6E08
How is power-supply voltage normally furnished to the most common type of monolithic microwave integrated circuit (MMIC)?
A. Through a resistor and/or RF choke connected to the amplifier output lead
B. MMICs require no operating bias
C. Through a capacitor and RF choke connected to the amplifier input lead
D. Directly to the bias-voltage (VCC IN) lead

E6E09
(B)
Page 6-19

E6E09
Which of the following must be done to insure that a crystal oscillator provides the frequency specified by the crystal manufacturer?
A. Provide the crystal with a specified parallel inductance
B. Provide the crystal with a specified parallel capacitance
C. Bias the crystal at a specified voltage
D. Bias the crystal at a specified current

E6E10 [Question withdrawn]

E6E10

E6E11

Which of the following materials is likely to provide the highest frequency of operation when used in MMICs?
  A. Silicon
  B. Silicon nitride
  C. Silicon dioxide
  D. Gallium nitride

E6E11
(D)
Page 5-13

E6E12

What is a "Jones filter" as used as part of a HF receiver IF stage?
  A. An automatic notch filter
  B. A variable bandwidth crystal lattice filter
  C. A special filter that emphasizes image responses
  D. A filter that removes impulse noise

E6E12
(B)
Page 6-39

E6F — Optical components and power systems: photoconductive principles and effects, photovoltaic systems, optical couplers, optical sensors, and optoisolators

E6F01

What is photoconductivity?
  A. The conversion of photon energy to electromotive energy
  B. The increased conductivity of an illuminated semiconductor
  C. The conversion of electromotive energy to photon energy
  D. The decreased conductivity of an illuminated semiconductor

E6F01
(B)
Page 5-31

E6F02

What happens to the conductivity of a photoconductive material when light shines on it?
  A. It increases
  B. It decreases
  C. It stays the same
  D. It becomes unstable

E6F02
(A)
Page 5-31

E6F03

What is the most common configuration of an optoisolator or optocoupler?
  A. A lens and a photomultiplier
  B. A frequency modulated helium-neon laser
  C. An amplitude modulated helium-neon laser
  D. An LED and a phototransistor

E6F03
(D)
Page 5-32

E6F04
(B)
Page 5-33

E6F04
What is the photovoltaic effect?
    A. The conversion of voltage to current when exposed to light
    B. The conversion of light to electrical energy
    C. The conversion of electrical energy to mechanical energy
    D. The tendency of a battery to discharge when used outside

E6F05
(A)
Page 5-32

E6F05
Which of the following describes an optical shaft encoder?
    A. A device which detects rotation of a control by interrupting a light source with a patterned wheel
    B. A device which measures the strength a beam of light using analog to digital conversion
    C. A digital encryption device often used to encrypt spacecraft control signals
    D. A device for generating RTTY signals by means of a rotating light source.

E6F06
(A)
Page 5-31

E6F06
Which of these materials is affected the most by photoconductivity?
    A. A crystalline semiconductor
    B. An ordinary metal
    C. A heavy metal
    D. A liquid semiconductor

E6F07
(B)
Page 5-32

E6F07
What is a solid state relay?
    A. A relay using transistors to drive the relay coil
    B. A device that uses semiconductor devices to implement the functions of an electromechanical relay
    C. A mechanical relay that latches in the on or off state each time it is pulsed
    D. A passive delay line

E6F08
(C)
Page 5-32

E6F08
Why are optoisolators often used in conjunction with solid state circuits when switching 120 VAC?
    A. Optoisolators provide a low impedance link between a control circuit and a power circuit
    B. Optoisolators provide impedance matching between the control circuit and power circuit
    C. Optoisolators provide a very high degree of electrical isolation between a control circuit and the circuit being switched
    D. Optoisolators eliminate the effects of reflected light in the control circuit

E6F09
(D)
Page 5-33

E6F09
What is the efficiency of a photovoltaic cell?
    A. The output RF power divided by the input dc power
    B. The effective payback period
    C. The open-circuit voltage divided by the short-circuit current under full illumination
    D. The relative fraction of light that is converted to current

E6F10
(B)
Page 5-33

E6F10
What is the most common type of photovoltaic cell used for electrical power generation?
    A. Selenium
    B. Silicon
    C. Cadmium Sulfide
    D. Copper oxide

E6F11

Which of the following is the approximate open-circuit voltage produced by a fully-illuminated silicon photovoltaic cell?

A. 0.1 V
B. 0.5 V
C. 1.5 V
D. 12 V

E6F12

What absorbs the energy from light falling on a photovoltaic cell?

A. Protons
B. Photons
C. Electrons
D. Holes

E6F11
(B)
Page 5-33

E6F12
(C)
Page 5-33

SUBELEMENT E7
PRACTICAL CIRCUITS
8 Exam Questions — 8 Groups

E7A — Digital circuits: digital circuit principles and logic circuits: classes of logic elements; positive and negative logic; frequency dividers; truth tables

E7A01
(C)
Page 5-21

E7A01
Which of the following is a bistable circuit?
A. An "AND" gate
B. An "OR" gate
C. A flip-flop
D. A clock

E7A02
(C)
Page 5-23

E7A02
How many output level changes are obtained for every two trigger pulses applied to the input of a T flip-flop circuit?
A. None
B. One
C. Two
D. Four

E7A03
(B)
Page 5-24

E7A03
Which of the following can divide the frequency of a pulse train by 2?
A. An XOR gate
B. A flip-flop
C. An OR gate
D. A multiplexer

E7A04
(B)
Page 5-24

E7A04
How many flip-flops are required to divide a signal frequency by 4?
A. 1
B. 2
C. 4
D. 8

E7A05
(D)
Page 5-24

E7A05
Which of the following is a circuit that continuously alternates between two states without an external clock?
A. Monostable multivibrator
B. J-K flip-flop
C. T flip-flop
D. Astable multivibrator

E7A06
(A)
Page 5-24

E7A06
What is a characteristic of a monostable multivibrator?
A. It switches momentarily to the opposite binary state and then returns, after a set time, to its original state
B. It is a clock that produces a continuous square wave oscillating between 1 and 0
C. It stores one bit of data in either a 0 or 1 state
D. It maintains a constant output voltage, regardless of variations in the input voltage

E7A07

What logical operation does a NAND gate perform?

A. It produces a logic "0" at its output only when all inputs are logic "0"
B. It produces a logic "1" at its output only when all inputs are logic "1"
C. It produces a logic "0" at its output if some but not all of its inputs are logic "1"
D. It produces a logic "0" at its output only when all inputs are logic "1"

E7A07
(D)
Page 5-19

E7A08

What logical operation does an OR gate perform?

A. It produces a logic "1" at its output if any or all inputs are logic "1"
B. It produces a logic "0" at its output if all inputs are logic "1"
C. It only produces a logic "0" at its output when all inputs are logic "1"
D. It produces a logic "1" at its output if all inputs are logic "0"

E7A08
(A)
Page 5-19

E7A09

What logical operation is performed by a two-input exclusive NOR gate?

A. It produces a logic "0" at its output only if all inputs are logic "0"
B. It produces a logic "1" at its output only if all inputs are logic "1"
C. It produces a logic "0" at its output if any single input is a logic "1"
D. It produces a logic "1" at its output if any single input is a logic "1"

E7A09
(C)
Page 5-19

E7A10

What is a truth table?

A. A table of logic symbols that indicate the high logic states of an op-amp
B. A diagram showing logic states when the digital device's output is true
C. A list of inputs and corresponding outputs for a digital device
D. A table of logic symbols that indicates the low logic states of an op-amp

E7A10
(C)
Page 5-18

E7A11

What is the name for logic which represents a logic "1" as a high voltage?

A. Reverse Logic
B. Assertive Logic
C. Negative logic
D. Positive Logic

E7A11
(D)
Page 5-20

E7A12

What is the name for logic which represents a logic "0" as a high voltage?

A. Reverse Logic
B. Assertive Logic
C. Negative logic
D. Positive Logic

E7A12
(C)
Page 5-20

E7A13

What is an SR or RS flip-flop?

A. A speed-reduced logic device with high power capability
B. A set/reset flip-flop whose output is low when R is high and S is low, high when S is high and R is low, and unchanged when both inputs are low
C. A speed-reduced logic device with very low voltage operation capability
D. A set/reset flip-flop that toggles whenever the T input is pulsed, unless both inputs are high

E7A13
(B)
Page 5-23

E7A14
(A)
Page 5-24

E7A14
What is a JK flip-flop?
  A. A flip-flop similar to an RS except that it toggles when both J and K are high
  B. A flip-flop utilizing low power, low temperature Joule-Kelvin devices
  C. A flip-flop similar to a D flip-flop except that it triggers on the negative clock edge
  D. A flip-flop originally developed in Japan and Korea which has very low power consumption

E7A15
(A)
Page 5-23

E7A15
What is a D flip-flop?
  A. A flip-flop whose output takes on the state of the D input when the clock signal transitions from low to high
  B. A differential class D amplifier used as a flip-flop circuit
  C. A dynamic memory storage element
  D. A flip-flop whose output is capable of both positive and negative voltage excursions

E7B — Amplifiers: Class of operation; vacuum tube and solid-state circuits; distortion and intermodulation; spurious and parasitic suppression; microwave amplifiers

E7B01
(A)
Page 6-11

E7B01
For what portion of a signal cycle does a Class AB amplifier operate?
  A. More than 180 degrees but less than 360 degrees
  B. Exactly 180 degrees
  C. The entire cycle
  D. Less than 180 degrees

E7B02
(A)
Page 6-12

E7B02
What is a Class D amplifier?
  A. A type of amplifier that uses switching technology to achieve high efficiency
  B. A low power amplifier using a differential amplifier for improved linearity
  C. An amplifier using drift-mode FETs for high efficiency
  D. A frequency doubling amplifier

E7B03
(A)
Page 6-12

E7B03
Which of the following forms the output of a class D amplifier circuit?
  A. A low-pass filter to remove switching signal components
  B. A high-pass filter to compensate for low gain at low frequencies
  C. A matched load resistor to prevent damage by switching transients
  D. A temperature-compensated load resistor to improve linearity

E7B04
(A)
Page 6-11

E7B04
Where on the load line of a Class A common emitter amplifier would bias normally be set?
  A. Approximately half-way between saturation and cutoff
  B. Where the load line intersects the voltage axis
  C. At a point where the bias resistor equals the load resistor
  D. At a point where the load line intersects the zero bias current curve

E7B05

What can be done to prevent unwanted oscillations in an RF power amplifier?

A. Tune the stage for maximum SWR

B. Tune both the input and output for maximum power

C. Install parasitic suppressors and/or neutralize the stage

D. Use a phase inverter in the output filter

E7B05
(C)
Page 6-14

E7B06

Which of the following amplifier types reduces or eliminates even-order harmonics?

A. Push-push

B. Push-pull

C. Class C

D. Class AB

E7B06
(B)
Page 6-13

E7B07

Which of the following is a likely result when a Class C amplifier is used to amplify a single-sideband phone signal?

A. Reduced intermodulation products

B. Increased overall intelligibility

C. Signal inversion

D. Signal distortion and excessive bandwidth

E7B07
(D)
Page 6-13

E7B08

How can an RF power amplifier be neutralized?

A. By increasing the driving power

B. By reducing the driving power

C. By feeding a 180-degree out-of-phase portion of the output back to the input

D. By feeding an in-phase component of the output back to the input

E7B08
(C)
Page 6-14

E7B09

Which of the following describes how the loading and tuning capacitors are to be adjusted when tuning a vacuum tube RF power amplifier that employs a pi-network output circuit?

A. The loading capacitor is set to maximum capacitance and the tuning capacitor is adjusted for minimum allowable plate current

B. The tuning capacitor is set to maximum capacitance and the loading capacitor is adjusted for minimum plate permissible current

C. The loading capacitor is adjusted to minimum plate current while alternately adjusting the tuning capacitor for maximum allowable plate current

D. The tuning capacitor is adjusted for minimum plate current, while the loading capacitor is adjusted for maximum permissible plate current

E7B09
(D)
Page 6-43

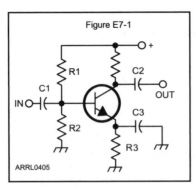

Figure E7-1 — Refer to this figure for questions E7B10 through E7B12.

ARRL0405

E7B10
In Figure E7-1, what is the purpose of R1 and R2?
  A. Load resistors
  B. Fixed bias
  C. Self bias
  D. Feedback

E7B11
In Figure E7-1, what is the purpose of R3?
  A. Fixed bias
  B. Emitter bypass
  C. Output load resistor
  D. Self bias

E7B12
What type of circuit is shown in Figure E7-1?
  A. Switching voltage regulator
  B. Linear voltage regulator
  C. Common emitter amplifier
  D. Emitter follower amplifier

E7B13
In Figure E7-2, what is the purpose of R?
  A. Emitter load
  B. Fixed bias
  C. Collector load
  D. Voltage regulation

E7B14
In Figure E7-2, what is the purpose of C2?
  A. Output coupling
  B. Emitter bypass
  C. Input coupling
  D. Hum filtering

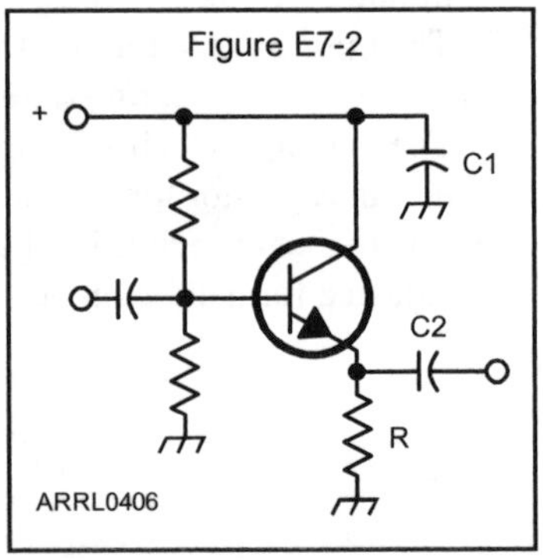

Figure E7-2

C1

C2

R

ARRL0406

Figure E7-2 — Refer to this figure for questions E7B13 and E7B14.

E7B15
What is one way to prevent thermal runaway in a bipolar transistor amplifier?
  A. Neutralization
  B. Select transistors with high beta
  C. Use a resistor in series with the emitter
  D. All of these choices are correct

E7B15
(C)
Page 6-3

E7B16
What is the effect of intermodulation products in a linear power amplifier?
  A. Transmission of spurious signals
  B. Creation of parasitic oscillations
  C. Low efficiency
  D. All of these choices are correct

E7B16
(A)
Page 6-12

E7B17
Why are third-order intermodulation distortion products of particular concern in linear power amplifiers?
  A. Because they are relatively close in frequency to the desired signal
  B. Because they are relatively far in frequency from the desired signal
  C. Because they invert the sidebands causing distortion
  D. Because they maintain the sidebands, thus causing multiple duplicate signals

E7B17
(A)
Page 6-12

E7B18
Which of the following is a characteristic of a grounded-grid amplifier?
  A. High power gain
  B. High filament voltage
  C. Low input impedance
  D. Low bandwidth

E7B18
(C)
Page 6-6

E7B19
What is a klystron?
  A. A high speed multivibrator
  B. An electron-coupled oscillator utilizing a pentode vacuum tube
  C. An oscillator utilizing ceramic elements to achieve stability
  D. A VHF, UHF, or microwave vacuum tube that uses velocity modulation

E7B19
(D)
Page 6-15

E7B20
What is a parametric amplifier?
  A. A type of bipolar operational amplifier with excellent linearity derived from use of very high voltage on the collector
  B. A low-noise VHF or UHF amplifier relying on varying reactance for amplification
  C. A high power amplifier for HF application utilizing the Miller effect to increase gain
  D. An audio push-pull amplifier using silicon carbide transistors for extremely low noise

E7B20
(B)
Page 6-16

E7B21
Which of the following devices is generally best suited for UHF or microwave power amplifier applications?
  A. Field effect transistor
  B. Nuvistor
  C. Silicon controlled rectifier
  D. Triac

E7B21
(A)
Page 6-16

E7C — Filters and matching networks: types of networks; types of filters; filter applications; filter characteristics; impedance matching; DSP filtering

E7C01
How are the capacitors and inductors of a low-pass filter Pi-network arranged between the network's input and output?
    A. Two inductors are in series between the input and output, and a capacitor is connected between the two inductors and ground
    B. Two capacitors are in series between the input and output and an inductor is connected between the two capacitors and ground
    C. An inductor is connected between the input and ground, another inductor is connected between the output and ground, and a capacitor is connected between the input and output
    D. A capacitor is connected between the input and ground, another capacitor is connected between the output and ground, and an inductor is connected between input and output

E7C02
A T-network with series capacitors and a parallel shunt inductor has which of the following properties?
    A. It is a low-pass filter
    B. It is a band-pass filter
    C. It is a high-pass filter
    D. It is a notch filter

E7C03
What advantage does a Pi-L-network have over a Pi-network for impedance matching between the final amplifier of a vacuum-tube transmitter and an antenna?
    A. Greater harmonic suppression
    B. Higher efficiency
    C. Lower losses
    D. Greater transformation range

E7C04
How does an impedance-matching circuit transform a complex impedance to a resistive impedance?
    A. It introduces negative resistance to cancel the resistive part of impedance
    B. It introduces transconductance to cancel the reactive part of impedance
    C. It cancels the reactive part of the impedance and changes the resistive part to a desired value
    D. Network resistances are substituted for load resistances and reactances are matched to the resistances

E7C05
Which filter type is described as having ripple in the passband and a sharp cutoff?
    A. A Butterworth filter
    B. An active LC filter
    C. A passive op-amp filter
    D. A Chebyshev filter

E7C06
What are the distinguishing features of an elliptical filter?
    A. Gradual passband rolloff with minimal stop band ripple
    B. Extremely flat response over its pass band with gradually rounded stop band corners
    C. Extremely sharp cutoff with one or more notches in the stop band
    D. Gradual passband rolloff with extreme stop band ripple

E7C07
What kind of filter would you use to attenuate an interfering carrier signal while receiving an SSB transmission?
  A. A band-pass filter
  B. A notch filter
  C. A Pi-network filter
  D. An all-pass filter

E7C07
(B)
Page 6-38

E7C08
What kind of digital signal processing audio filter might be used to remove unwanted noise from a received SSB signal?
  A. An adaptive filter
  B. A crystal-lattice filter
  C. A Hilbert-transform filter
  D. A phase-inverting filter

E7C08
(A)
Page 6-41

E7C09
What type of digital signal processing filter might be used to generate an SSB signal?
  A. An adaptive filter
  B. A notch filter
  C. A Hilbert-transform filter
  D. An elliptical filter

E7C09
(C)
Page 6-29

E7C10
Which of the following filters would be the best choice for use in a 2 meter repeater duplexer?
  A. A crystal filter
  B. A cavity filter
  C. A DSP filter
  D. An L-C filter

E7C10
(B)
Page 6-36

E7C11
Which of the following is the common name for a filter network which is equivalent to two L networks connected back-to-back with the inductors in series and the capacitors in shunt at the input and output?
  A. Pi-L
  B. Cascode
  C. Omega
  D. Pi

E7C11
(D)
Page 6-43

E7C12
Which of the following describes a Pi-L network used for matching a vacuum-tube final amplifier to a 50-ohm unbalanced output?
  A. A Phase Inverter Load network
  B. A Pi network with an additional series inductor on the output
  C. A network with only three discrete parts
  D. A matching network in which all components are isolated from ground

E7C12
(B)
Page 6-43

E7C13
What is one advantage of a Pi matching network over an L matching network consisting of a single inductor and a single capacitor?
A. The Q of Pi networks can be varied depending on the component values chosen
B. L networks can not perform impedance transformation
C. Pi networks have fewer components
D. Pi networks are designed for balanced input and output

E7C14
Which of these modes is most affected by non-linear phase response in a receiver IF filter?
A. Meteor Scatter
B. Single-Sideband voice
C. Digital
D. Video

E7D — Power supplies and voltage regulators

E7D01
What is one characteristic of a linear electronic voltage regulator?
A. It has a ramp voltage as its output
B. It eliminates the need for a pass transistor
C. The control element duty cycle is proportional to the line or load conditions
D. The conduction of a control element is varied to maintain a constant output voltage

E7D02
What is one characteristic of a switching electronic voltage regulator?
A. The resistance of a control element is varied in direct proportion to the line voltage or load current
B. It is generally less efficient than a linear regulator
C. The control device's duty cycle is controlled to produce a constant average output voltage
D. It gives a ramp voltage at its output

E7D03
What device is typically used as a stable reference voltage in a linear voltage regulator?
A. A Zener diode
B. A tunnel diode
C. An SCR
D. A varactor diode

E7D04
Which of the following types of linear voltage regulator usually make the most efficient use of the primary power source?
A. A series current source
B. A series regulator
C. A shunt regulator
D. A shunt current source

E7D05

Which of the following types of linear voltage regulator places a constant load on the unregulated voltage source?

A. A constant current source
B. A series regulator
C. A shunt current source
D. A shunt regulator

E7D05
(D)
Page 6-44

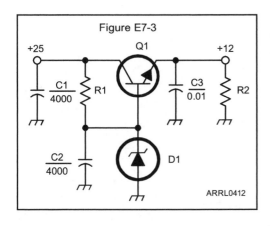

Figure E7-3

+25   Q1   +12

C1
4000   R1   C3
0.01   R2

C2
4000   D1

ARRL0412

Figure E7-3 — Refer to this figure for questions E7D06 through E7D13.

E7D06

What is the purpose of Q1 in the circuit shown in Figure E7-3?

A. It provides negative feedback to improve regulation
B. It provides a constant load for the voltage source
C. It increases the current-handling capability of the regulator
D. It provides D1 with current

E7D06
(C)
Page 6-45

E7D07

What is the purpose of C2 in the circuit shown in Figure E7-3?

A. It bypasses hum around D1
B. It is a brute force filter for the output
C. To self-resonate at the hum frequency
D. To provide fixed DC bias for Q1

E7D07
(A)
Page 6-45

E7D08

What type of circuit is shown in Figure E7-3?

A. Switching voltage regulator
B. Grounded emitter amplifier
C. Linear voltage regulator
D. Emitter follower

E7D08
(C)
Page 6-44

E7D09

What is the purpose of C1 in the circuit shown in Figure E7-3?

A. It resonates at the ripple frequency
B. It provides fixed bias for Q1
C. It decouples the output
D. It filters the supply voltage

E7D09
(D)
Page 6-45

E7D10
(A)
Page 6-45

E7D10
What is the purpose of C3 in the circuit shown in Figure E7-3?
    A. It prevents self-oscillation
    B. It provides brute force filtering of the output
    C. It provides fixed bias for Q1
    D. It clips the peaks of the ripple

E7D11
(C)
Page 6-45

E7D11
What is the purpose of R1 in the circuit shown in Figure E7-3?
    A. It provides a constant load to the voltage source
    B. It couples hum to D1
    C. It supplies current to D1
    D. It bypasses hum around D1

E7D12
(D)
Page 6-44

E7D12
What is the purpose of R2 in the circuit shown in Figure E7-3?
    A. It provides fixed bias for Q1
    B. It provides fixed bias for D1
    C. It decouples hum from D1
    D. It provides a constant minimum load for Q1

E7D13
(B)
Page 6-45

E7D13
What is the purpose of D1 in the circuit shown in Figure E7-3?
    A. To provide line voltage stabilization
    B. To provide a voltage reference
    C. Peak clipping
    D. Hum filtering

E7D14
(C)
Page 6-47

E7D14
What is one purpose of a "bleeder" resistor in a conventional (unregulated) power supply?
    A. To cut down on waste heat generated by the power supply
    B. To balance the low-voltage filament windings
    C. To improve output voltage regulation
    D. To boost the amount of output current

E7D15
(D)
Page 6-46

E7D15
What is the purpose of a "step-start" circuit in a high-voltage power supply?
    A. To provide a dual-voltage output for reduced power applications
    B. To compensate for variations of the incoming line voltage
    C. To allow for remote control of the power supply
    D. To allow the filter capacitors to charge gradually

E7D16
(D)
Page 6-46

E7D16
When several electrolytic filter capacitors are connected in series to increase the operating voltage of a power supply filter circuit, why should resistors be connected across each capacitor?
    A. To equalize, as much as possible, the voltage drop across each capacitor
    B. To provide a safety bleeder to discharge the capacitors when the supply is off
    C. To provide a minimum load current to reduce voltage excursions at light loads
    D. All of these choices are correct

E7D17

What is the primary reason that a high-frequency inverter type high-voltage power supply can be both less expensive and lighter in weight than a conventional power supply?

A. The inverter design does not require any output filtering

B. It uses a diode bridge rectifier for increased output

C. The high frequency inverter design uses much smaller transformers and filter components for an equivalent power output

D. It uses a large power-factor compensation capacitor to create "free power" from the unused portion of the AC cycle

E7E — Modulation and demodulation: reactance, phase and balanced modulators; detectors; mixer stages; DSP modulation and demodulation; software defined radio systems

E7E01

Which of the following can be used to generate FM phone emissions?

A. A balanced modulator on the audio amplifier

B. A reactance modulator on the oscillator

C. A reactance modulator on the final amplifier

D. A balanced modulator on the oscillator

E7E02

What is the function of a reactance modulator?

A. To produce PM signals by using an electrically variable resistance

B. To produce AM signals by using an electrically variable inductance or capacitance

C. To produce AM signals by using an electrically variable resistance

D. To produce PM signals by using an electrically variable inductance or capacitance

E7E03

How does an analog phase modulator function?

A. By varying the tuning of a microphone preamplifier to produce PM signals

B. By varying the tuning of an amplifier tank circuit to produce AM signals

C. By varying the tuning of an amplifier tank circuit to produce PM signals

D. By varying the tuning of a microphone preamplifier to produce AM signals

E7E04

What is one way a single-sideband phone signal can be generated?

A. By using a balanced modulator followed by a filter

B. By using a reactance modulator followed by a mixer

C. By using a loop modulator followed by a mixer

D. By driving a product detector with a DSB signal

E7E05

What circuit is added to an FM transmitter to boost the higher audio frequencies?

A. A de-emphasis network

B. A heterodyne suppressor

C. An audio prescaler

D. A pre-emphasis network

E7D17
(C)
Page 6-46

E7E01
(B)
Page 6-30

E7E02
(D)
Page 6-30

E7E03
(C)
Page 6-30

E7E04
(A)
Page 6-28

E7E05
(D)
Page 6-31

E7E06
(A)
Page 6-31

E7E06

Why is de-emphasis commonly used in FM communications receivers?
- A. For compatibility with transmitters using phase modulation
- B. To reduce impulse noise reception
- C. For higher efficiency
- D. To remove third-order distortion products

E7E07
(B)
Page 6-27

E7E07

What is meant by the term baseband in radio communications?
- A. The lowest frequency band that the transmitter or receiver covers
- B. The frequency components present in the modulating signal
- C. The unmodulated bandwidth of the transmitted signal
- D. The basic oscillator frequency in an FM transmitter that is multiplied to increase the deviation and carrier frequency

E7E08
(C)
Page 6-25

E7E08

What are the principal frequencies that appear at the output of a mixer circuit?
- A. Two and four times the original frequency
- B. The sum, difference and square root of the input frequencies
- C. The two input frequencies along with their sum and difference frequencies
- D. 1.414 and 0.707 times the input frequency

E7E09
(A)
Page 6-25

E7E09

What occurs when an excessive amount of signal energy reaches a mixer circuit?
- A. Spurious mixer products are generated
- B. Mixer blanking occurs
- C. Automatic limiting occurs
- D. A beat frequency is generated

E7E10
(A)
Page 6-31

E7E10

How does a diode detector function?
- A. By rectification and filtering of RF signals
- B. By breakdown of the Zener voltage
- C. By mixing signals with noise in the transition region of the diode
- D. By sensing the change of reactance in the diode with respect to frequency

E7E11
(C)
Page 6-32

E7E11

Which of the following types of detector is well suited for demodulating SSB signals?
- A. Discriminator
- B. Phase detector
- C. Product detector
- D. Phase comparator

E7E12
(D)
Page 6-33

E7E12

What is a frequency discriminator stage in a FM receiver?
- A. An FM generator circuit
- B. A circuit for filtering two closely adjacent signals
- C. An automatic band-switching circuit
- D. A circuit for detecting FM signals

E7E13

Which of the following describes a common means of generating an SSB signal when using digital signal processing?

A. Mixing products are converted to voltages and subtracted by adder circuits
B. A frequency synthesizer removes the unwanted sidebands
C. Emulation of quartz crystal filter characteristics
D. The quadrature method

E7E13
(D)
Page 6-29

E7E14

What is meant by direct conversion when referring to a software defined receiver?

A. Software is converted from source code to object code during operation of the receiver
B. Incoming RF is converted to the IF frequency by rectification to generate the control voltage for a voltage controlled oscillator
C. Incoming RF is mixed to "baseband" for analog-to-digital conversion and subsequent processing
D. Software is generated in machine language, avoiding the need for compilers

E7E14
(C)
Page 6-32

E7F — Frequency markers and counters: frequency divider circuits; frequency marker generators; frequency counters

E7F01

What is the purpose of a prescaler circuit?

A. It converts the output of a JK flip flop to that of an RS flip-flop
B. It multiplies a higher frequency signal so a low-frequency counter can display the operating frequency
C. It prevents oscillation in a low-frequency counter circuit
D. It divides a higher frequency signal so a low-frequency counter can display the input frequency

E7F01
(D)
Page 5-25

E7F02

Which of the following would be used to reduce a signal's frequency by a factor of ten?

A. A preamp
B. A prescaler
C. A marker generator
D. A flip-flop

E7F02
(B)
Page 5-25

E7F03

What is the function of a decade counter digital IC?

A. It produces one output pulse for every ten input pulses
B. It decodes a decimal number for display on a seven-segment LED display
C. It produces ten output pulses for every input pulse
D. It adds two decimal numbers together

E7F03
(A)
Page 5-25

E7F04

What additional circuitry must be added to a 100-kHz crystal-controlled marker generator so as to provide markers at 50 and 25 kHz?

A. An emitter-follower
B. Two frequency multipliers
C. Two flip-flops
D. A voltage divider

E7F04
(C)
Page 5-26

E7F05
(D)
Page 5-26

E7F05
Which of the following is a technique for providing high stability oscillators needed for microwave transmission and reception?
   A. Use a GPS signal reference
   B. Use a rubidium stabilized reference oscillator
   C. Use a temperature-controlled high Q dielectric resonator
   D. All of these choices are correct

E7F06
(C)
Page 5-26

E7F06
What is one purpose of a marker generator?
   A. To add audio markers to an oscilloscope
   B. To provide a frequency reference for a phase locked loop
   C. To provide a means of calibrating a receiver's frequency settings
   D. To add time signals to a transmitted signal

E7F07
(A)
Page 5-25

E7F07
What determines the accuracy of a frequency counter?
   A. The accuracy of the time base
   B. The speed of the logic devices used
   C. Accuracy of the AC input frequency to the power supply
   D. Proper balancing of the mixer diodes

E7F08
(C)
Page 5-25

E7F08
Which of the following is performed by a frequency counter?
   A. Determining the frequency deviation with an FM discriminator
   B. Mixing the incoming signal with a WWV reference
   C. Counting the number of input pulses occurring within a specific period of time
   D. Converting the phase of the measured signal to a voltage which is proportional to the frequency

E7F09
(A)
Page 5-25

E7F09
What is the purpose of a frequency counter?
   A. To provide a digital representation of the frequency of a signal
   B. To generate a series of reference signals at known frequency intervals
   C. To display all frequency components of a transmitted signal
   D. To provide a signal source at a very accurate frequency

E7F10
(B)
Page 5-26

E7F10
What alternate method of determining frequency, other than by directly counting input pulses, is used by some counters?
   A. GPS averaging
   B. Period measurement plus mathematical computation
   C. Prescaling
   D. D/A conversion

E7F11
(C)
Page 5-26

E7F11
What is an advantage of a period-measuring frequency counter over a direct-count type?
   A. It can run on battery power for remote measurements
   B. It does not require an expensive high-precision time base
   C. It provides improved resolution of low-frequency signals within a comparable time period
   D. It can directly measure the modulation index of an FM transmitter

E7G — Active filters and op-amps: active audio filters; characteristics; basic circuit design; operational amplifiers

E7G01
What primarily determines the gain and frequency characteristics of an op-amp RC active filter?
A. The values of capacitors and resistors built into the op-amp
B. The values of capacitors and resistors external to the op-amp
C. The input voltage and frequency of the op-amp's DC power supply
D. The output voltage and smoothness of the op-amp's DC power supply

E7G01
(B)
Page 6-39

E7G02
What is the effect of ringing in a filter?
A. An echo caused by a long time delay
B. A reduction in high frequency response
C. Partial cancellation of the signal over a range of frequencies
D. Undesired oscillations added to the desired signal

E7G02
(D)
Page 6-38

E7G03
Which of the following is an advantage of using an op-amp instead of LC elements in an audio filter?
A. Op-amps are more rugged
B. Op-amps are fixed at one frequency
C. Op-amps are available in more varieties than are LC elements
D. Op-amps exhibit gain rather than insertion loss

E7G03
(D)
Page 6-39

E7G04
Which of the following is a type of capacitor best suited for use in high-stability op-amp RC active filter circuits?
A. Electrolytic
B. Disc ceramic
C. Polystyrene
D. Paper

E7G04
(C)
Page 6-40

E7G05
How can unwanted ringing and audio instability be prevented in a multi-section op-amp RC audio filter circuit?
A. Restrict both gain and Q
B. Restrict gain, but increase Q
C. Restrict Q, but increase gain
D. Increase both gain and Q

E7G05
(A)
Page 6-40

E7G06
Which of the following is the most appropriate use of an op-amp active filter?
A. As a high-pass filter used to block RFI at the input to receivers
B. As a low-pass filter used between a transmitter and a transmission line
C. For smoothing power-supply output
D. As an audio filter in a receiver

E7G06
(D)
Page 6-40

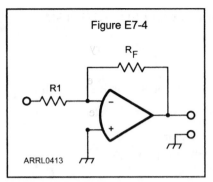

Figure E7-4

R_F

R1

ARRL0413

Figure E7-4 — Refer to this figure for questions E7G07 and E7G09 through E7G11.

E7G07
What magnitude of voltage gain can be expected from the circuit in Figure E7-4 when R1 is 10 ohms and RF is 470 ohms?
  A. 0.21
  B. 94
  C. 47
  D. 24

E7G08
How does the gain of an ideal operational amplifier vary with frequency?
  A. It increases linearly with increasing frequency
  B. It decreases linearly with increasing frequency
  C. It decreases logarithmically with increasing frequency
  D. It does not vary with frequency

E7G09
What will be the output voltage of the circuit shown in Figure E7-4 if R1 is 1000 ohms, RF is 10,000 ohms, and 0.23 volts dc is applied to the input?
  A. 0.23 volts
  B. 2.3 volts
  C. –0.23 volts
  D. –2.3 volts

E7G10
What absolute voltage gain can be expected from the circuit in Figure E7-4 when R1 is 1800 ohms and RF is 68 kilohms?
  A. 1
  B. 0.03
  C. 38
  D. 76

E7G11
What absolute voltage gain can be expected from the circuit in Figure E7-4 when R1 is 3300 ohms and RF is 47 kilohms?
  A. 28
  B. 14
  C. 7
  D. 0.07

E7G12

What is an integrated circuit operational amplifier?

A. A high-gain, direct-coupled differential amplifier with very high input and very low output impedance

B. A digital audio amplifier whose characteristics are determined by components external to the amplifier

C. An amplifier used to increase the average output of frequency modulated amateur signals to the legal limit

D. An RF amplifier used in the UHF and microwave regions

E7G12
(A)
Page 6-6

E7G13

What is meant by the term op-amp input-offset voltage?

A. The output voltage of the op-amp minus its input voltage

B. The difference between the output voltage of the op-amp and the input voltage required in the immediately following stage

C. The differential input voltage needed to bring the open-loop output voltage to zero

D. The potential between the amplifier input terminals of the op-amp in an open-loop condition

E7G13
(C)
Page 6-8

E7G14

What is the typical input impedance of an integrated circuit op-amp?

A. 100 ohms

B. 1000 ohms

C. Very low

D. Very high

E7G14
(D)
Page 6-7

E7G15

What is the typical output impedance of an integrated circuit op-amp?

A. Very low

B. Very high

C. 100 ohms

D. 1000 ohms

E7G15
(A)
Page 6-7

E7H — Oscillators and signal sources: types of oscillators; synthesizers and phase-locked loops; direct digital synthesizers

E7H01

What are three oscillator circuits used in Amateur Radio equipment?

A. Taft, Pierce and negative feedback

B. Pierce, Fenner and Beane

C. Taft, Hartley and Pierce

D. Colpitts, Hartley and Pierce

E7H01
(D)
Page 6-18

E7H02

What condition must exist for a circuit to oscillate?

A. It must have at least two stages

B. It must be neutralized

C. It must have positive feedback with a gain greater than 1

D. It must have negative feedback sufficient to cancel the input signal

E7H02
(C)
Page 6-17

E7H03
(A)
Page 6-18

E7H03
How is positive feedback supplied in a Hartley oscillator?
    A. Through a tapped coil
    B. Through a capacitive divider
    C. Through link coupling
    D. Through a neutralizing capacitor

E7H04
(C)
Page 6-18

E7H04
How is positive feedback supplied in a Colpitts oscillator?
    A. Through a tapped coil
    B. Through link coupling
    C. Through a capacitive divider
    D. Through a neutralizing capacitor

E7H05
(D)
Page 6-18

E7H05
How is positive feedback supplied in a Pierce oscillator?
    A. Through a tapped coil
    B. Through link coupling
    C. Through a neutralizing capacitor
    D. Through a quartz crystal

E7H06
(B)
Page 6-20

E7H06
Which of the following oscillator circuits are commonly used in VFOs?
    A. Pierce and Zener
    B. Colpitts and Hartley
    C. Armstrong and deForest
    D. Negative feedback and balanced feedback

E7H07
(C)
Page 6-20

E7H07
What is a magnetron oscillator?
    A. An oscillator in which the output is fed back to the input by the magnetic field of a transformer
    B. A crystal oscillator in which variable frequency is obtained by placing the crystal in a strong magnetic field
    C. A UHF or microwave oscillator consisting of a diode vacuum tube with a specially shaped anode, surrounded by an external magnet
    D. A reference standard oscillator in which the oscillations are synchronized by magnetic coupling to a rubidium gas tube

E7H08
(A)
Page 6-20

E7H08
What is a Gunn diode oscillator?
    A. An oscillator based on the negative resistance properties of properly-doped semiconductors
    B. An oscillator based on the argon gas diode
    C. A highly stable reference oscillator based on the tee-notch principle
    D. A highly stable reference oscillator based on the hot-carrier effect

E7H09
(A)
Page 6-35

E7H09
What type of frequency synthesizer circuit uses a phase accumulator, lookup table, digital to analog converter and a low-pass anti-alias filter?
    A. A direct digital synthesizer
    B. A hybrid synthesizer
    C. A phase locked loop synthesizer
    D. A diode-switching matrix synthesizer

E7H10

What information is contained in the lookup table of a direct digital frequency synthesizer?
  A. The phase relationship between a reference oscillator and the output waveform
  B. The amplitude values that represent a sine-wave output
  C. The phase relationship between a voltage-controlled oscillator and the output waveform
  D. The synthesizer frequency limits and frequency values stored in the radio memories

E7H10
(B)
Page 6-35

E7H11

What are the major spectral impurity components of direct digital synthesizers?
  A. Broadband noise
  B. Digital conversion noise
  C. Spurious signals at discrete frequencies
  D. Nyquist limit noise

E7H11
(C)
Page 6-36

E7H12

Which of the following is a principal component of a direct digital synthesizer (DDS)?
  A. Phase splitter
  B. Hex inverter
  C. Chroma demodulator
  D. Phase accumulator

E7H12
(D)
Page 6-35

E7H13

What is the capture range of a phase-locked loop circuit?
  A. The frequency range over which the circuit can lock
  B. The voltage range over which the circuit can lock
  C. The input impedance range over which the circuit can lock
  D. The range of time it takes the circuit to lock

E7H13
(A)
Page 6-34

E7H14

What is a phase-locked loop circuit?
  A. An electronic servo loop consisting of a ratio detector, reactance modulator, and voltage-controlled oscillator
  B. An electronic circuit also known as a monostable multivibrator
  C. An electronic servo loop consisting of a phase detector, a low-pass filter, a voltage-controlled oscillator, and a stable reference oscillator
  D. An electronic circuit consisting of a precision push-pull amplifier with a differential input

E7H14
(C)
Page 6-34

E7H15

Which of these functions can be performed by a phase-locked loop?
  A. Wide-band AF and RF power amplification
  B. Comparison of two digital input signals, digital pulse counter
  C. Photovoltaic conversion, optical coupling
  D. Frequency synthesis, FM demodulation

E7H15
(D)
Page 6-35

E7H16
(B)
Page 6-35

E7H16
Why is the short-term stability of the reference oscillator important in the design of a phase locked loop (PLL) frequency synthesizer?
A. Any amplitude variations in the reference oscillator signal will prevent the loop from locking to the desired signal
B. Any phase variations in the reference oscillator signal will produce phase noise in the synthesizer output
C. Any phase variations in the reference oscillator signal will produce harmonic distortion in the modulating signal
D. Any amplitude variations in the reference oscillator signal will prevent the loop from changing frequency

E7H17
(C)
Page 6-34

E7H17
Why is a phase-locked loop often used as part of a variable frequency synthesizer for receivers and transmitters?
A. It generates FM sidebands
B. It eliminates the need for a voltage controlled oscillator
C. It makes it possible for a VFO to have the same degree of frequency stability as a crystal oscillator
D. It can be used to generate or demodulate SSB signals by quadrature phase synchronization

E7H18
(A)
Page 6-35

E7H18
What are the major spectral impurity components of phase-locked loop synthesizers?
A. Phase noise
B. Digital conversion noise
C. Spurious signals at discrete frequencies
D. Nyquist limit noise

SUBELEMENT E8
SIGNALS AND EMISSIONS
4 Exam Questions — 4 Groups

E8A — AC waveforms: sine, square, sawtooth and irregular waveforms; AC measurements; average and PEP of RF signals; pulse and digital signal waveforms

E8A01
What type of wave is made up of a sine wave plus all of its odd harmonics?
  A. A square wave
  B. A sine wave
  C. A cosine wave
  D. A tangent wave

E8A01
(A)
Page 7-4

E8A02
What type of wave has a rise time significantly faster than its fall time (or vice versa)?
  A. A cosine wave
  B. A square wave
  C. A sawtooth wave
  D. A sine wave

E8A02
(C)
Page 7-4

E8A03
What type of wave is made up of sine waves of a given fundamental frequency plus all its harmonics?
  A. A sawtooth wave
  B. A square wave
  C. A sine wave
  D. A cosine wave

E8A03
(A)
Page 7-4

E8A04
What is equivalent to the root-mean-square value of an AC voltage?
  A. The AC voltage found by taking the square of the average value of the peak AC voltage
  B. The DC voltage causing the same amount of heating in a given resistor as the corresponding peak AC voltage
  C. The DC voltage causing the same amount of heating in a resistor as the corresponding RMS AC voltage
  D. The AC voltage found by taking the square root of the average AC value

E8A04
(C)
Page 7-5

E8A05
What would be the most accurate way of measuring the RMS voltage of a complex waveform?
  A. By using a grid dip meter
  B. By measuring the voltage with a D'Arsonval meter
  C. By using an absorption wavemeter
  D. By measuring the heating effect in a known resistor

E8A05
(D)
Page 7-5

E8A06
What is the approximate ratio of PEP-to-average power in a typical single-sideband phone signal?
  A. 2.5 to 1
  B. 25 to 1
  C. 1 to 1
  D. 100 to 1

E8A06
(A)
Page 7-9

E8A07
(B)
Page 7-8

E8A07
What determines the PEP-to-average power ratio of a single-sideband phone signal?
  A. The frequency of the modulating signal
  B. The characteristics of the modulating signal
  C. The degree of carrier suppression
  D. The amplifier gain

E8A08
(A)
Page 7-2

E8A08
What is the period of a wave?
  A. The time required to complete one cycle
  B. The number of degrees in one cycle
  C. The number of zero crossings in one cycle
  D. The amplitude of the wave

E8A09
(C)
Page 7-3

E8A09
What type of waveform is produced by human speech?
  A. Sinusoidal
  B. Logarithmic
  C. Irregular
  D. Trapezoidal

E8A10
(B)
Page 7-4

E8A10
Which of the following is a distinguishing characteristic of a pulse waveform?
  A. Regular sinusoidal oscillations
  B. Narrow bursts of energy separated by periods of no signal
  C. A series of tones that vary between two frequencies
  D. A signal that contains three or more discrete tones

E8A11
(D)
Page 7-25

E8A11
What is one use for a pulse modulated signal?
  A. Linear amplification
  B. PSK31 data transmission
  C. Multiphase power transmission
  D. Digital data transmission

E8A12
(D)
Page 8-5

E8A12
What type of information can be conveyed using digital waveforms?
  A. Human speech
  B. Video signals
  C. Data
  D. All of these choices are correct

E8A13
(C)
Page 8-5

E8A13
What is an advantage of using digital signals instead of analog signals to convey the same information?
  A. Less complex circuitry is required for digital signal generation and detection
  B. Digital signals always occupy a narrower bandwidth
  C. Digital signals can be regenerated multiple times without error
  D. All of these choices are correct

E8A14

Which of these methods is commonly used to convert analog signals to digital signals?

A. Sequential sampling
B. Harmonic regeneration
C. Level shifting
D. Phase reversal

E8A14
(A)
Page 6-21

E8A15

What would the waveform of a stream of digital data bits look like on a conventional oscilloscope?

A. A series of sine waves with evenly spaced gaps
B. A series of pulses with varying patterns
C. A running display of alpha-numeric characters
D. None of the above; this type of signal cannot be seen on a conventional oscilloscope

E8A15
(B)
Page 8-2

E8B — Modulation and demodulation: modulation methods; modulation index and deviation ratio; pulse modulation; frequency and time division multiplexing

E8B01

What is the term for the ratio between the frequency deviation of an RF carrier wave, and the modulating frequency of its corresponding FM-phone signal?

A. FM compressibility
B. Quieting index
C. Percentage of modulation
D. Modulation index

E8B01
(D)
Page 7-24

E8B02

How does the modulation index of a phase-modulated emission vary with RF carrier frequency (the modulated frequency)?

A. It increases as the RF carrier frequency increases
B. It decreases as the RF carrier frequency increases
C. It varies with the square root of the RF carrier frequency
D. It does not depend on the RF carrier frequency

E8B02
(D)
Page 7-24

E8B03

What is the modulation index of an FM-phone signal having a maximum frequency deviation of 3000 Hz either side of the carrier frequency, when the modulating frequency is 1000 Hz?

A. 3
B. 0.3
C. 3000
D. 1000

E8B03
(A)
Page 7-24

E8B04

What is the modulation index of an FM-phone signal having a maximum carrier deviation of plus or minus 6 kHz when modulated with a 2-kHz modulating frequency?

A. 6000
B. 3
C. 2000
D. 1/3

E8B04
(B)
Page 7-24

E8B05
What is the deviation ratio of an FM-phone signal having a maximum frequency swing of plus-or-minus 5 kHz when the maximum modulation frequency is 3 kHz?
A. 60
B. 0.167
C. 0.6
D. 1.67

E8B06
What is the deviation ratio of an FM-phone signal having a maximum frequency swing of plus or minus 7.5 kHz when the maximum modulation frequency is 3.5 kHz?
A. 2.14
B. 0.214
C. 0.47
D. 47

E8B07
When using a pulse-width modulation system, why is the transmitter's peak power greater than its average power?
A. The signal duty cycle is less than 100%
B. The signal reaches peak amplitude only when voice modulated
C. The signal reaches peak amplitude only when voltage spikes are generated within the modulator
D. The signal reaches peak amplitude only when the pulses are also amplitude modulated

E8B08
What parameter does the modulating signal vary in a pulse-position modulation system?
A. The number of pulses per second
B. The amplitude of the pulses
C. The duration of the pulses
D. The time at which each pulse occurs

E8B09
What is meant by deviation ratio?
A. The ratio of the audio modulating frequency to the center carrier frequency
B. The ratio of the maximum carrier frequency deviation to the highest audio modulating frequency
C. The ratio of the carrier center frequency to the audio modulating frequency
D. The ratio of the highest audio modulating frequency to the average audio modulating frequency

E8B10
Which of these methods can be used to combine several separate analog information streams into a single analog radio frequency signal?
A. Frequency shift keying
B. A diversity combiner
C. Frequency division multiplexing
D. Pulse compression

E8B11
Which of the following describes frequency division multiplexing?
A. The transmitted signal jumps from band to band at a predetermined rate
B. Two or more information streams are merged into a "baseband", which then modulates the transmitter
C. The transmitted signal is divided into packets of information
D. Two or more information streams are merged into a digital combiner, which then pulse position modulates the transmitter

E8B11
(B)
Page 7-26

E8B12
What is digital time division multiplexing?
A. Two or more data streams are assigned to discrete sub-carriers on an FM transmitter
B. Two or more signals are arranged to share discrete time slots of a data transmission
C. Two or more data streams share the same channel by transmitting time of transmission as the sub-carrier
D. Two or more signals are quadrature modulated to increase bandwidth efficiency

E8B12
(B)
Page 7-26

E8C — Digital signals: digital communications modes; CW; information rate vs. bandwidth; spread-spectrum communications; modulation methods

E8C01
Which one of the following digital codes consists of elements having unequal length?
A. ASCII
B. AX.25
C. Baudot
D. Morse code

E8C01
(D)
Page 8-4

E8C02
What are some of the differences between the Baudot digital code and ASCII?
A. Baudot uses four data bits per character, ASCII uses seven or eight; Baudot uses one character as a shift code, ASCII has no shift code
B. Baudot uses five data bits per character, ASCII uses seven or eight; Baudot uses two characters as shift codes, ASCII has no shift code
C. Baudot uses six data bits per character, ASCII uses seven or eight; Baudot has no shift code, ASCII uses two characters as shift codes
D. Baudot uses seven data bits per character, ASCII uses eight; Baudot has no shift code, ASCII uses two characters as shift codes

E8C02
(B)
Page 8-4

E8C03
What is one advantage of using the ASCII code for data communications?
A. It includes built-in error-correction features
B. It contains fewer information bits per character than any other code
C. It is possible to transmit both upper and lower case text
D. It uses one character as a shift code to send numeric and special characters

E8C03
(C)
Page 8-4

E8C04
What technique is used to minimize the bandwidth requirements of a PSK31 signal?
A. Zero-sum character encoding
B. Reed-Solomon character encoding
C. Use of sinusoidal data pulses
D. Use of trapezoidal data pulses

E8C04
(C)
Page 8-7

E8C05
What is the necessary bandwidth of a 13-WPM international Morse code transmission?
  A. Approximately 13 Hz
  B. Approximately 26 Hz
  C. Approximately 52 Hz
  D. Approximately 104 Hz

E8C06
What is the necessary bandwidth of a 170-hertz shift, 300-baud ASCII transmission?
  A. 0.1 Hz
  B. 0.3 kHz
  C. 0.5 kHz
  D. 1.0 kHz

E8C07
What is the necessary bandwidth of a 4800-Hz frequency shift, 9600-baud ASCII FM transmission?
  A. 15.36 kHz
  B. 9.6 kHz
  C. 4.8 kHz
  D. 5.76 kHz

E8C08
What term describes a wide-bandwidth communications system in which the transmitted carrier frequency varies according to some predetermined sequence?
  A. Amplitude compandored single sideband
  B. AMTOR
  C. Time-domain frequency modulation
  D. Spread-spectrum communication

E8C09
Which of these techniques causes a digital signal to appear as wide-band noise to a conventional receiver?
  A. Spread-spectrum
  B. Independent sideband
  C. Regenerative detection
  D. Exponential addition

E8C10
What spread-spectrum communications technique alters the center frequency of a conventional carrier many times per second in accordance with a pseudo-random list of channels?
  A. Frequency hopping
  B. Direct sequence
  C. Time-domain frequency modulation
  D. Frequency compandored spread-spectrum

E8C11
What spread-spectrum communications technique uses a high speed binary bit stream to shift the phase of an RF carrier?
  A. Frequency hopping
  B. Direct sequence
  C. Binary phase-shift keying
  D. Phase compandored spread-spectrum

E8C12

What is the advantage of including a parity bit with an ASCII character stream?
A. Faster transmission rate
B. The signal can overpower interfering signals
C. Foreign language characters can be sent
D. Some types of errors can be detected

E8C12
(D)
Page 8-4

E8C13

What is one advantage of using JT-65 coding?
A. Uses only a 65 Hz bandwidth
B. The ability to decode signals which have a very low signal to noise ratio
C. Easily copied by ear if necessary
D. Permits fast-scan TV transmissions over narrow bandwidth

E8C13
(B)
Page 8-8

E8D — Waveforms: measurement, peak-to-peak, RMS, average; Electromagnetic Waves: definition, characteristics, polarization

E8D01

Which of the following is the easiest voltage amplitude parameter to measure when viewing a pure sine wave signal on an analog oscilloscope?
A. Peak-to-peak voltage
B. RMS voltage
C. Average voltage
D. DC voltage

E8D01
(A)
Page 7-16

E8D02

What is the relationship between the peak-to-peak voltage and the peak voltage amplitude of a symmetrical waveform?
A. 0.707:1
B. 2:1
C. 1.414:1
D. 4:1

E8D02
(B)
Page 7-5

E8D03

What input-amplitude parameter is valuable in evaluating the signal-handling capability of a Class A amplifier?
A. Peak voltage
B. RMS voltage
C. Average power
D. Resting voltage

E8D03
(A)
Page 7-5

E8D04

What is the PEP output of a transmitter that develops a peak voltage of 30 volts into a 50-ohm load?
A. 4.5 watts
B. 9 watts
C. 16 watts
D. 18 watts

E8D04
(B)
Page 7-8

E8D05
(D)
Page 7-6

E8D05
If an RMS-reading AC voltmeter reads 65 volts on a sinusoidal waveform, what is the peak-to-peak voltage?
A. 46 volts
B. 92 volts
C. 130 volts
D. 184 volts

E8D06
(B)
Page 7-8

E8D06
What is the advantage of using a peak-reading wattmeter to monitor the output of a SSB phone transmitter?
A. It is easier to determine the correct tuning of the output circuit
B. It gives a more accurate display of the PEP output when modulation is present
C. It makes it easier to detect high SWR on the feed line
D. It can determine if any flat-topping is present during modulation peaks

E8D07
(C)
Page 7-9

E8D07
What is an electromagnetic wave?
A. Alternating currents in the core of an electromagnet
B. A wave consisting of two electric fields at right angles to each other
C. A wave consisting of an electric field and a magnetic field oscillating at right angles to each other
D. A wave consisting of two magnetic fields at right angles to each other

E8D08
(D)
Page 7-9

E8D08
Which of the following best describes electromagnetic waves traveling in free space?
A. Electric and magnetic fields become aligned as they travel
B. The energy propagates through a medium with a high refractive index
C. The waves are reflected by the ionosphere and return to their source
D. Changing electric and magnetic fields propagate the energy

E8D09
(B)
Page 7-11

E8D09
What is meant by circularly polarized electromagnetic waves?
A. Waves with an electric field bent into a circular shape
B. Waves with a rotating electric field
C. Waves that circle the Earth
D. Waves produced by a loop antenna

E8D10
(D)
Page 7-8

E8D10
What type of meter should be used to monitor the output signal of a voice-modulated single-sideband transmitter to ensure you do not exceed the maximum allowable power?
A. An SWR meter reading in the forward direction
B. A modulation meter
C. An average reading wattmeter
D. A peak-reading wattmeter

E8D11
(A)
Page 7-6

E8D11
What is the average power dissipated by a 50-ohm resistive load during one complete RF cycle having a peak voltage of 35 volts?
A. 12.2 watts
B. 9.9 watts
C. 24.5 watts
D. 16 watts

E8D12
What is the peak voltage of a sinusoidal waveform if an RMS-reading voltmeter reads 34 volts?
  A. 123 volts
  B. 96 volts
  C. 55 volts
  D. 48 volts

E8D13
Which of the following is a typical value for the peak voltage at a standard U.S. household electrical outlet?
  A. 240 volts
  B. 170 volts
  C. 120 volts
  D. 340 volts

E8D14
Which of the following is a typical value for the peak-to-peak voltage at a standard U.S. household electrical outlet?
  A. 240 volts
  B. 120 volts
  C. 340 volts
  D. 170 volts

E8D15
Which of the following is a typical value for the RMS voltage at a standard U.S. household electrical power outlet?
  A. 120V AC
  B. 340V AC
  C. 85V AC
  D. 170V AC

E8D16
What is the RMS value of a 340-volt peak-to-peak pure sine wave?
  A. 120V AC
  B. 170V AC
  C. 240V AC
  D. 300V AC

E8D12
(D)
Page 7-6

E8D13
(B)
Page 7-6

E8D14
(C)
Page 7-6

E8D15
(A)
Page 7-6

E8D16
(A)
Page 7-6

SUBELEMENT E9
ANTENNAS AND TRANSMISSION LINES
8 Exam Questions — 8 Groups

E9A — Isotropic and gain antennas: definitions; uses; radiation patterns; Basic antenna parameters: radiation resistance and reactance, gain, beamwidth, efficiency

E9A01
(C)
Page 9-2

E9A01
Which of the following describes an isotropic antenna?
   A. A grounded antenna used to measure earth conductivity
   B. A horizontally polarized antenna used to compare Yagi antennas
   C. A theoretical antenna used as a reference for antenna gain
   D. A spacecraft antenna used to direct signals toward the earth

E9A02
(B)
Page 9-3

E9A02
How much gain does a ½-wavelength dipole in free space have compared to an isotropic antenna?
   A. 1.55 dB
   B. 2.15 dB
   C. 3.05 dB
   D. 4.30 dB

E9A03
(D)
Page 9-3

E9A03
Which of the following antennas has no gain in any direction?
   A. Quarter-wave vertical
   B. Yagi
   C. Half-wave dipole
   D. Isotropic antenna

E9A04
(A)
Page 9-27

E9A04
Why would one need to know the feed point impedance of an antenna?
   A. To match impedances in order to minimize standing wave ratio on the transmission line
   B. To measure the near-field radiation density from a transmitting antenna
   C. To calculate the front-to-side ratio of the antenna
   D. To calculate the front-to-back ratio of the antenna

E9A05
(B)
Page 9-6

E9A05
Which of the following factors may affect the feed point impedance of an antenna?
   A. Transmission-line length
   B. Antenna height, conductor length/diameter ratio and location of nearby conductive objects
   C. Constant feed point impedance
   D. Sunspot activity and time of day

E9A06
(D)
Page 9-5

E9A06
What is included in the total resistance of an antenna system?
   A. Radiation resistance plus space impedance
   B. Radiation resistance plus transmission resistance
   C. Transmission-line resistance plus radiation resistance
   D. Radiation resistance plus ohmic resistance

E9A07

What is a folded dipole antenna?

A. A dipole one-quarter wavelength long
B. A type of ground-plane antenna
C. A dipole constructed from one wavelength of wire forming a very thin loop
D. A dipole configured to provide forward gain

E9A08

What is meant by antenna gain?

A. The ratio relating the radiated signal strength of an antenna in the direction of maximum radiation to that of a reference antenna
B. The ratio of the signal in the forward direction to that in the opposite direction
C. The ratio of the amount of power radiated by an antenna compared to the transmitter output power
D. The final amplifier gain minus the transmission-line losses, including any phasing lines present

E9A09

What is meant by antenna bandwidth?

A. Antenna length divided by the number of elements
B. The frequency range over which an antenna satisfies a performance requirement
C. The angle between the half-power radiation points
D. The angle formed between two imaginary lines drawn through the element ends

E9A10

How is antenna efficiency calculated?

A. (radiation resistance / transmission resistance) × 100%
B. (radiation resistance / total resistance) × 100%
C. (total resistance / radiation resistance) × 100%
D. (effective radiated power / transmitter output) × 100%

E9A11

Which of the following choices is a way to improve the efficiency of a ground-mounted quarter-wave vertical antenna?

A. Install a good radial system
B. Isolate the coax shield from ground
C. Shorten the radiating element
D. Reduce the diameter of the radiating element

E9A12

Which of the following factors determines ground losses for a ground-mounted vertical antenna operating in the 3-30 MHz range?

A. The standing-wave ratio
B. Distance from the transmitter
C. Soil conductivity
D. Take-off angle

E9A07
(C)
Page 9-14

E9A08
(A)
Page 9-3

E9A09
(B)
Page 9-8

E9A10
(B)
Page 9-6

E9A11
(A)
Page 9-7

E9A12
(C)
Page 9-9

E9A13
How much gain does an antenna have compared to a 1/2-wavelength dipole when it has 6 dB gain over an isotropic antenna?
A. 3.85 dB
B. 6.0 dB
C. 8.15 dB
D. 2.79 dB

E9A14
How much gain does an antenna have compared to a 1/2-wavelength dipole when it has 12 dB gain over an isotropic antenna?
A. 6.17 dB
B. 9.85 dB
C. 12.5 dB
D. 14.15 dB

E9A15
What is meant by the radiation resistance of an antenna?
A. The combined losses of the antenna elements and feed line
B. The specific impedance of the antenna
C. The value of a resistance that would dissipate the same amount of power as that radiated from an antenna
D. The resistance in the atmosphere that an antenna must overcome to be able to radiate a signal

E9B — Antenna patterns: E and H plane patterns; gain as a function of pattern; antenna design (comput Yagi antennas

E9B01
In the antenna radiation pattern shown in Figure E9-1, what is the 3-dB beamwidth?
A. 75 degrees
B. 50 degrees
C. 25 degrees
D. 30 degrees

E9B02
In the antenna radiation pattern shown in Figure E9-1, what is the front-to-back ratio?
A. 36 dB
B. 18 dB
C. 24 dB
D. 14 dB

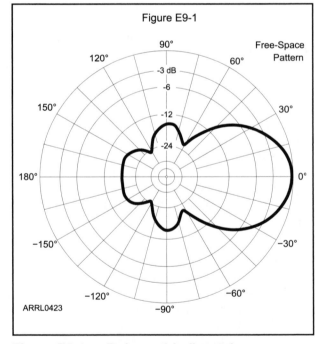

Figure E9-1 — Refer to this figure for questions E9B01 through E9B03.

E9B03

In the antenna radiation pattern shown in Figure E9-1, what is the front-to-side ratio?

A. 12 dB
B. 14 dB
C. 18 dB
D. 24 dB

E9B03
(B)
Page 9-5

E9B04

What may occur when a directional antenna is operated at different frequencies within the band for which it was designed?

A. Feed point impedance may become negative
B. The E-field and H-field patterns may reverse
C. Element spacing limits could be exceeded
D. The gain may change depending on frequency

E9B04
(D)
Page 9-42

E9B05

What usually occurs if a Yagi antenna is designed solely for maximum forward gain?

A. The front-to-back ratio increases
B. The front-to-back ratio decreases
C. The frequency response is widened over the whole frequency band
D. The SWR is reduced

E9B05
(B)
Page 9-42

E9B06

If the boom of a Yagi antenna is lengthened and the elements are properly retuned, what usually occurs?

A. The gain increases
B. The SWR decreases
C. The front-to-back ratio increases
D. The gain bandwidth decreases rapidly

E9B06
(A)
Page 9-42

E9B07

How does the total amount of radiation emitted by a directional gain antenna compare with the total amount of radiation emitted from an isotropic antenna, assuming each is driven by the same amount of power?

A. The total amount of radiation from the directional antenna is increased by the gain of the antenna
B. The total amount of radiation from the directional antenna is stronger by its front to back ratio
C. They are the same
D. The radiation from the isotropic antenna is 2.15 dB stronger than that from the directional antenna

E9B07
(C)
Page 9-3

E9B08

How can the approximate beamwidth in a given plane of a directional antenna be determined?

A. Note the two points where the signal strength of the antenna is 3 dB less than maximum and compute the angular difference
B. Measure the ratio of the signal strengths of the radiated power lobes from the front and rear of the antenna
C. Draw two imaginary lines through the ends of the elements and measure the angle between the lines
D. Measure the ratio of the signal strengths of the radiated power lobes from the front and side of the antenna

E9B08
(A)
Page 9-4

E9B09
(B)
Page 9-41

E9B09
What type of computer program technique is commonly used for modeling antennas?
    A. Graphical analysis
    B. Method of Moments
    C. Mutual impedance analysis
    D. Calculus differentiation with respect to physical properties

E9B10
(A)
Page 9-41

E9B10
What is the principle of a Method of Moments analysis?
    A. A wire is modeled as a series of segments, each having a uniform value of current
    B. A wire is modeled as a single sine-wave current generator
    C. A wire is modeled as a series of points, each having a distinct location in space
    D. A wire is modeled as a series of segments, each having a distinct value of voltage across it

E9B11
(C)
Page 9-42

E9B11
What is a disadvantage of decreasing the number of wire segments in an antenna model below the guideline of 10 segments per half-wavelength?
    A. Ground conductivity will not be accurately modeled
    B. The resulting design will favor radiation of harmonic energy
    C. The computed feed point impedance may be incorrect
    D. The antenna will become mechanically unstable

E9B12
(D)
Page 9-2

E9B12
What is the far-field of an antenna?
    A. The region of the ionosphere where radiated power is not refracted
    B. The region where radiated power dissipates over a specified time period
    C. The region where radiated field strengths are obstructed by objects of reflection
    D. The region where the shape of the antenna pattern is independent of distance

E9B13
(B)
Page 9-41

E9B13
What does the abbreviation NEC stand for when applied to antenna modeling programs?
    A. Next Element Comparison
    B. Numerical Electromagnetics Code
    C. National Electrical Code
    D. Numeric Electrical Computation

E9B14
(D)
Page 9-42

E9B14
What type of information can be obtained by submitting the details of a proposed new antenna to a modeling program?
    A. SWR vs. frequency charts
    B. Polar plots of the far-field elevation and azimuth patterns
    C. Antenna gain
    D. All of these choices are correct

E9C — Wire and phased vertical antennas: beverage antennas; rhombic antennas; elevation above real ground; ground effects as related to polarization; take-off angles

E9C01
What is the radiation pattern of two ¼-wavelength vertical antennas spaced ½-wavelength apart and fed 180 degrees out of phase?
  A. A cardioid
  B. Omnidirectional
  C. A figure-8 broadside to the axis of the array
  D. A figure-8 oriented along the axis of the array

E9C01
(D)
Page 9-20

E9C02
What is the radiation pattern of two ¼-wavelength vertical antennas spaced ¼-wavelength apart and fed 90 degrees out of phase?
  A. A cardioid
  B. A figure-8 end-fire along the axis of the array
  C. A figure-8 broadside to the axis of the array
  D. Omnidirectional

E9C02
(A)
Page 9-20

E9C03
What is the radiation pattern of two ¼-wavelength vertical antennas spaced ½-wavelength apart and fed in phase?
  A. Omnidirectional
  B. A cardioid
  C. A Figure-8 broadside to the axis of the array
  D. A Figure-8 end-fire along the axis of the array

E9C03
(C)
Page 9-18

E9C04
Which of the following describes a basic unterminated rhombic antenna?
  A. Unidirectional; four-sides, each side one quarter-wavelength long; terminated in a resistance equal to its characteristic impedance
  B. Bidirectional; four-sides, each side one or more wavelengths long; open at the end opposite the transmission line connection
  C. Four-sides; an LC network at each corner except for the transmission connection;
  D. Four-sides, each of a different physical length

E9C04
(B)
Page 9-16

E9C05
What are the disadvantages of a terminated rhombic antenna for the HF bands?
  A. The antenna has a very narrow operating bandwidth
  B. The antenna produces a circularly polarized signal
  C. The antenna requires a large physical area and 4 separate supports
  D. The antenna is more sensitive to man-made static than any other type

E9C05
(C)
Page 9-17

E9C06
What is the effect of a terminating resistor on a rhombic antenna?
  A. It reflects the standing waves on the antenna elements back to the transmitter
  B. It changes the radiation pattern from bidirectional to unidirectional
  C. It changes the radiation pattern from horizontal to vertical polarization
  D. It decreases the ground loss

E9C06
(B)
Page 9-16

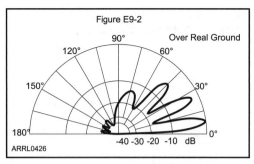

Figure E9-2 — Refer to this figure for questions E9C07 through E9C10.

E9C07
What type of antenna pattern over real ground is shown in Figure E9-2?
   A. Elevation
   B. Azimuth
   C. Radiation resistance
   D. Polarization

E9C08
What is the elevation angle of peak response in the antenna radiation pattern shown in Figure E9-2?
   A. 45 degrees
   B. 75 degrees
   C. 7.5 degrees
   D. 25 degrees

E9C09
What is the front-to-back ratio of the radiation pattern shown in Figure E9-2?
   A. 15 dB
   B. 28 dB
   C. 3 dB
   D. 24 dB

E9C10
How many elevation lobes appear in the forward direction of the antenna radiation pattern shown in Figure E9-2?
   A. 4
   B. 3
   C. 1
   D. 7

E9C11
How is the far-field elevation pattern of a vertically polarized antenna affected by being mounted over seawater versus rocky ground?
   A. The low-angle radiation decreases
   B. The high-angle radiation increases
   C. Both the high- and low-angle radiation decrease
   D. The low-angle radiation increases

E9C12

When constructing a Beverage antenna, which of the following factors should be included in the design to achieve good performance at the desired frequency?

A. Its overall length must not exceed 1/4 wavelength
B. It must be mounted more than 1 wavelength above ground
C. It should be configured as a four-sided loop
D. It should be one or more wavelengths long

E9C12
(D)
Page 9-17

E9C13

What is the main effect of placing a vertical antenna over an imperfect ground?

A. It causes increased SWR
B. It changes the impedance angle of the matching network
C. It reduces low-angle radiation
D. It reduces losses in the radiating portion of the antenna

E9C13
(C)
Page 9-9

E9D — Directional antennas: gain; satellite antennas; antenna beamwidth; stacking antennas; antenna efficiency; traps; folded dipoles; shortened and mobile antennas; grounding

E9D01

How does the gain of an ideal parabolic dish antenna change when the operating frequency is doubled?

A. Gain does not change
B. Gain is multiplied by 0.707
C. Gain increases by 6 dB
D. Gain increases by 3 dB

E9D01
(C)
Page 9-23

E9D02

How can linearly polarized Yagi antennas be used to produce circular polarization?

A. Stack two Yagis, fed 90 degrees out of phase, to form an array with the respective elements in parallel planes
B. Stack two Yagis, fed in phase, to form an array with the respective elements in parallel planes
C. Arrange two Yagis perpendicular to each other with the driven elements at the same point on the boom and fed 90 degrees out of phase
D. Arrange two Yagis collinear to each other, with the driven elements fed 180 degrees out of phase

E9D02
(C)
Page 9-23

E9D03

How does the beamwidth of an antenna vary as the gain is increased?

A. It increases geometrically
B. It increases arithmetically
C. It is essentially unaffected
D. It decreases

E9D03
(D)
Page 9-4

E9D04

Why is it desirable for a ground-mounted satellite communications antenna system to be able to move in both azimuth and elevation?

A. In order to track the satellite as it orbits the Earth
B. So the antenna can be pointed away from interfering signals
C. So the antenna can be positioned to cancel the effects of Faraday rotation
D. To rotate antenna polarization to match that of the satellite

E9D04
(A)
Page 9-23

E9D05
(A)
Page 9-12

E9D05
Where should a high-Q loading coil be placed to minimize losses in a shortened vertical antenna?
    A. Near the center of the vertical radiator
    B. As low as possible on the vertical radiator
    C. As close to the transmitter as possible
    D. At a voltage node

E9D06
(C)
Page 9-12

E9D06
Why should an HF mobile antenna loading coil have a high ratio of reactance to resistance?
    A. To swamp out harmonics
    B. To maximize losses
    C. To minimize losses
    D. To minimize the Q

E9D07
(A)
Page 9-14

E9D07
What is a disadvantage of using a multiband trapped antenna?
    A. It might radiate harmonics
    B. It radiates the harmonics and fundamental equally well
    C. It is too sharply directional at lower frequencies
    D. It must be neutralized

E9D08
(B)
Page 9-11

E9D08
What happens to the bandwidth of an antenna as it is shortened through the use of loading coils?
    A. It is increased
    B. It is decreased
    C. No change occurs
    D. It becomes flat

E9D09
(D)
Page 9-13

E9D09
What is an advantage of using top loading in a shortened HF vertical antenna?
    A. Lower Q
    B. Greater structural strength
    C. Higher losses
    D. Improved radiation efficiency

E9D10
(A)
Page 9-15

E9D10
What is the approximate feed point impedance at the center of a two-wire folded dipole antenna?
    A. 300 ohms
    B. 72 ohms
    C. 50 ohms
    D. 450 ohms

E9D11
(D)
Page 9-11

E9D11
What is the function of a loading coil as used with an HF mobile antenna?
    A. To increase the SWR bandwidth
    B. To lower the losses
    C. To lower the Q
    D. To cancel capacitive reactance

E9D12
What is one advantage of using a trapped antenna?
  A. It has high directivity in the higher-frequency bands
  B. It has high gain
  C. It minimizes harmonic radiation
  D. It may be used for multiband operation

E9D13
What happens to feed point impedance at the base of a fixed-length HF mobile antenna as the frequency of operation is lowered?
  A. The radiation resistance decreases and the capacitive reactance decreases
  B. The radiation resistance decreases and the capacitive reactance increases
  C. The radiation resistance increases and the capacitive reactance decreases
  D. The radiation resistance increases and the capacitive reactance increases

E9D14
Which of the following types of conductor would be best for minimizing losses in a station's RF ground system?
  A. A resistive wire, such as a spark plug wire
  B. A wide flat copper strap
  C. A cable with 6 or 7 18-gauge conductors in parallel
  D. A single 12 or 10-gauge stainless steel wire

E9D15
Which of the following would provide the best RF ground for your station?
  A. A 50-ohm resistor connected to ground
  B. An electrically-short connection to a metal water pipe
  C. An electrically-short connection to 3 or 4 interconnected ground rods driven into the Earth
  D. An electrically-short connection to 3 or 4 interconnected ground rods via a series RF choke

E9E — Matching: matching antennas to feed lines; power dividers

E9E01
What system matches a high-impedance transmission line to a lower impedance antenna by connecting the line to the driven element in two places spaced a fraction of a wavelength each side of element center?
  A. The gamma matching system
  B. The delta matching system
  C. The omega matching system
  D. The stub matching system

E9E02
What is the name of an antenna matching system that matches an unbalanced feed line to an antenna by feeding the driven element both at the center of the element and at a fraction of a wavelength to one side of center?
  A. The gamma match
  B. The delta match
  C. The epsilon match
  D. The stub match

E9D12
(D)
Page 9-13

E9D13
(B)
Page 9-11

E9D14
(B)
Page 9-10

E9D15
(C)
Page 9-11

E9E01
(B)
Page 9-28

E9E02
(A)
Page 9-28

E9E03
(D)
Page 9-29

E9E03
What is the name of the matching system that uses a section of transmission line connected in parallel with the feed line at or near the feed point?
A. The gamma match
B. The delta match
C. The omega match
D. The stub match

E9E04
(B)
Page 9-28

E9E04
What is the purpose of the series capacitor in a gamma-type antenna matching network?
A. To provide DC isolation between the feed line and the antenna
B. To cancel the inductive reactance of the matching network
C. To provide a rejection notch to prevent the radiation of harmonics
D. To transform the antenna impedance to a higher value

E9E05
(A)
Page 9-29

E9E05
How must the driven element in a 3-element Yagi be tuned to use a hairpin matching system?
A. The driven element reactance must be capacitive
B. The driven element reactance must be inductive
C. The driven element resonance must be lower than the operating frequency
D. The driven element radiation resistance must be higher than the characteristic impedance of the transmission line

E9E06
(C)
Page 9-29

E9E06
What is the equivalent lumped-constant network for a hairpin matching system on a 3-element Yagi?
A. Pi network
B. Pi-L network
C. L network
D. Parallel-resonant tank

E9E07
(B)
Page 9-33

E9E07
What term best describes the interactions at the load end of a mismatched transmission line?
A. Characteristic impedance
B. Reflection coefficient
C. Velocity factor
D. Dielectric constant

E9E08
(D)
Page 9-33

E9E08
Which of the following measurements is characteristic of a mismatched transmission line?
A. An SWR less than 1:1
B. A reflection coefficient greater than 1
C. A dielectric constant greater than 1
D. An SWR greater than 1:1

E9E09
(C)
Page 9-28

E9E09
Which of these matching systems is an effective method of connecting a 50-ohm coaxial cable feed line to a grounded tower so it can be used as a vertical antenna?
A. Double-bazooka match
B. Hairpin match
C. Gamma match
D. All of these choices are correct

E9E10

Which of these choices is an effective way to match an antenna with a 100-ohm feed point impedance to a 50-ohm coaxial cable feed line?

A. Connect a ¼-wavelength open stub of 300-ohm twin-lead in parallel with the coaxial feed line where it connects to the antenna
B. Insert a ½ wavelength piece of 300-ohm twin-lead in series between the antenna terminals and the 50-ohm feed cable
C. Insert a ¼-wavelength piece of 75-ohm coaxial cable transmission line in series between the antenna terminals and the 50-ohm feed cable
D. Connect ½ wavelength shorted stub of 75-ohm cable in parallel with the 50-ohm cable where it attaches to the antenna

E9E10
(C)
Page 9-40

E9E11

What is an effective way of matching a feed line to a VHF or UHF antenna when the impedances of both the antenna and feed line are unknown?

A. Use a 50-ohm 1:1 balun between the antenna and feed line
B. Use the universal stub matching technique
C. Connect a series-resonant LC network across the antenna feed terminals
D. Connect a parallel-resonant LC network across the antenna feed terminals

E9E11
(B)
Page 9-29

E9E12

What is the primary purpose of a phasing line when used with an antenna having multiple driven elements?

A. It ensures that each driven element operates in concert with the others to create the desired antenna pattern
B. It prevents reflected power from traveling back down the feed line and causing harmonic radiation from the transmitter
C. It allows single-band antennas to operate on other bands
D. It makes sure the antenna has a low-angle radiation pattern

E9E12
(A)
Page 9-20

E9E13

What is the purpose of a Wilkinson divider?

A. It divides the operating frequency of a transmitter signal so it can be used on a lower frequency band
B. It is used to feed high-impedance antennas from a low-impedance source
C. It divides power equally among multiple loads while preventing changes in one load from disturbing power flow to the others
D. It is used to feed low-impedance loads from a high-impedance source

E9E13
(C)
Page 9-20

E9F — Transmission lines: characteristics of open and shorted feed lines; ⅛ wavelength; ¼ wavelength; ½ wavelength; feed lines: coax versus open-wire; velocity factor; electrical length; coaxial cable dielectrics; velocity factor

E9F01

What is the velocity factor of a transmission line?

A. The ratio of the characteristic impedance of the line to the terminating impedance
B. The index of shielding for coaxial cable
C. The velocity of the wave in the transmission line multiplied by the velocity of light in a vacuum
D. The velocity of the wave in the transmission line divided by the velocity of light in a vacuum

E9F01
(D)
Page 9-31

E9F02
(C)
Page 9-31

E9F02
Which of the following determines the velocity factor of a transmission line?
  A. The termination impedance
  B. The line length
  C. Dielectric materials used in the line
  D. The center conductor resistivity

E9F03
(D)
Page 9-31

E9F03
Why is the physical length of a coaxial cable transmission line shorter than its electrical length?
  A. Skin effect is less pronounced in the coaxial cable
  B. The characteristic impedance is higher in a parallel feed line
  C. The surge impedance is higher in a parallel feed line
  D. Electrical signals move more slowly in a coaxial cable than in air

E9F04
(B)
Page 9-31

E9F04
What is the typical velocity factor for a coaxial cable with solid polyethylene dielectric?
  A. 2.70
  B. 0.66
  C. 0.30
  D. 0.10

E9F05
(C)
Page 9-32

E9F05
What is the approximate physical length of a solid polyethylene dielectric coaxial transmission line that is electrically one-quarter wavelength long at 14.1 MHz?
  A. 20 meters
  B. 2.3 meters
  C. 3.5 meters
  D. 0.2 meters

E9F06
(C)
Page 9-32

E9F06
What is the approximate physical length of an air-insulated, parallel conductor transmission line that is electrically one-half wavelength long at 14.10 MHz?
  A. 15 meters
  B. 20 meters
  C. 10 meters
  D. 71 meters

E9F07
(A)
Page 9-32

E9F07
How does ladder line compare to small-diameter coaxial cable such as RG-58 at 50 MHz?
  A. Lower loss
  B. Higher SWR
  C. Smaller reflection coefficient
  D. Lower velocity factor

E9F08
(A)
Page 9-31

E9F08
What is the term for the ratio of the actual speed at which a signal travels through a transmission line to the speed of light in a vacuum?
  A. Velocity factor
  B. Characteristic impedance
  C. Surge impedance
  D. Standing wave ratio

E9F09

What is the approximate physical length of a solid polyethylene dielectric coaxial transmission line that is electrically one-quarter wavelength long at 7.2 MHz?

A. 10 meters
B. 6.9 meters
C. 24 meters
D. 50 meters

E9F09
(B)
Page 9-32

E9F10

What impedance does a ⅛-wavelength transmission line present to a generator when the line is shorted at the far end?

A. A capacitive reactance
B. The same as the characteristic impedance of the line
C. An inductive reactance
D. The same as the input impedance to the final generator stage

E9F10
(C)
Page 9-39

E9F11

What impedance does a ⅛-wavelength transmission line present to a generator when the line is open at the far end?

A. The same as the characteristic impedance of the line
B. An inductive reactance
C. A capacitive reactance
D. The same as the input impedance of the final generator stage

E9F11
(C)
Page 9-39

E9F12

What impedance does a ¼-wavelength transmission line present to a generator when the line is open at the far end?

A. The same as the characteristic impedance of the line
B. The same as the input impedance to the generator
C. Very high impedance
D. Very low impedance

E9F12
(D)
Page 9-39

E9F13

What impedance does a ¼-wavelength transmission line present to a generator when the line is shorted at the far end?

A. Very high impedance
B. Very low impedance
C. The same as the characteristic impedance of the transmission line
D. The same as the generator output impedance

E9F13
(A)
Page 9-39

E9F14

What impedance does a ½-wavelength transmission line present to a generator when the line is shorted at the far end?

A. Very high impedance
B. Very low impedance
C. The same as the characteristic impedance of the line
D. The same as the output impedance of the generator

E9F14
(B)
Page 9-39

E9F15
(A)
Page 9-39

E9F15
What impedance does a 1/2-wavelength transmission line present to a generator when the line is open at the far end?
A. Very high impedance
B. Very low impedance
C. The same as the characteristic impedance of the line
D. The same as the output impedance of the generator

E9F16
(D)
Page 9-32

E9F16
Which of the following is a significant difference between foam-dielectric coaxial cable and solid-dielectric cable, assuming all other parameters are the same?
A. Reduced safe operating voltage limits
B. Reduced losses per unit of length
C. Higher velocity factor
D. All of these choices are correct

E9G — The Smith chart

E9G01
(A)
Page 9-35

E9G01
Which of the following can be calculated using a Smith chart?
A. Impedance along transmission lines
B. Radiation resistance
C. Antenna radiation pattern
D. Radio propagation

E9G02
(B)
Page 9-37

E9G02
What type of coordinate system is used in a Smith chart?
A. Voltage circles and current arcs
B. Resistance circles and reactance arcs
C. Voltage lines and current chords
D. Resistance lines and reactance chords

E9G03
(C)
Page 9-35

E9G03
Which of the following is often determined using a Smith chart?
A. Beam headings and radiation patterns
B. Satellite azimuth and elevation bearings
C. Impedance and SWR values in transmission lines
D. Trigonometric functions

E9G04
(C)
Page 9-37

E9G04
What are the two families of circles and arcs that make up a Smith chart?
A. Resistance and voltage
B. Reactance and voltage
C. Resistance and reactance
D. Voltage and impedance

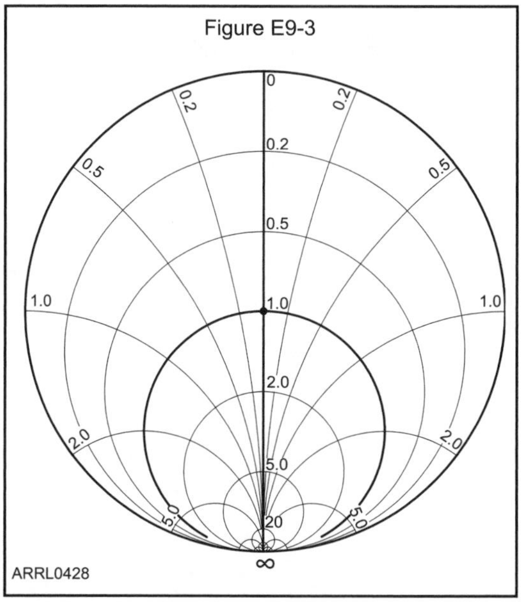

Figure E9-3

ARRL0428

Figure E9-3 — Refer to this figure for questions E9G05 through E9G07.

E9G05
What type of chart is shown in Figure E9-3?
 A. Smith chart
 B. Free-space radiation directivity chart
 C. Elevation angle radiation pattern chart
 D. Azimuth angle radiation pattern chart

E9G06
On the Smith chart shown in Figure E9-3, what is the name for the large outer circle on which the reactance arcs terminate?
 A. Prime axis
 B. Reactance axis
 C. Impedance axis
 D. Polar axis

E9G07
On the Smith chart shown in Figure E9-3, what is the only straight line shown?
 A. The reactance axis
 B. The current axis
 C. The voltage axis
 D. The resistance axis

E9G08
What is the process of normalization with regard to a Smith chart?
 A. Reassigning resistance values with regard to the reactance axis
 B. Reassigning reactance values with regard to the resistance axis
 C. Reassigning impedance values with regard to the prime center
 D. Reassigning prime center with regard to the reactance axis

E9G05
(A)
Page 9-35

E9G06
(B)
Page 9-37

E9G07
(D)
Page 9-35

E9G08
(C)
Page 9-37

E9G09
(A)
Page 9-37

E9G09
What third family of circles is often added to a Smith chart during the process of solving problems?
A. Standing-wave ratio circles
B. Antenna-length circles
C. Coaxial-length circles
D. Radiation-pattern circles

E9G10
(D)
Page 9-37

E9G10
What do the arcs on a Smith chart represent?
A. Frequency
B. SWR
C. Points with constant resistance
D. Points with constant reactance

E9G11
(B)
Page 9-37

E9G11
How are the wavelength scales on a Smith chart calibrated?
A. In fractions of transmission line electrical frequency
B. In fractions of transmission line electrical wavelength
C. In fractions of antenna electrical wavelength
D. In fractions of antenna electrical frequency

E9H — Effective radiated power; system gains and losses; radio direction finding antennas

E9H01
(D)
Page 9-21

E9H01
What is the effective radiated power relative to a dipole of a repeater station with 150 watts transmitter power output, 2-dB feed line loss, 2.2-dB duplexer loss and 7-dBd antenna gain?
A. 1977 watts
B. 78.7 watts
C. 420 watts
D. 286 watts

E9H02
(A)
Page 9-21

E9H02
What is the effective radiated power relative to a dipole of a repeater station with 200 watts transmitter power output, 4-dB feed line loss, 3.2-dB duplexer loss, 0.8-dB circulator loss and 10-dBd antenna gain?
A. 317 watts
B. 2000 watts
C. 126 watts
D. 300 watts

E9H03
(B)
Page 9-22

E9H03
What is the effective isotropic radiated power of a repeater station with 200 watts transmitter power output, 2-dB feed line loss, 2.8-dB duplexer loss, 1.2-dB circulator loss and 7-dBi antenna gain?
A. 159 watts
B. 252 watts
C. 632 watts
D. 63.2 watts

E9H04

What term describes station output, including the transmitter, antenna and everything in between, when considering transmitter power and system gains and losses?

A. Power factor
B. Half-power bandwidth
C. Effective radiated power
D. Apparent power

E9H04
(C)
Page 9-20

E9H05

What is the main drawback of a wire-loop antenna for direction finding?

A. It has a bidirectional pattern
B. It is non-rotatable
C. It receives equally well in all directions
D. It is practical for use only on VHF bands

E9H05
(A)
Page 9-25

E9H06

What is the triangulation method of direction finding?

A. The geometric angle of sky waves from the source are used to determine its position
B. A fixed receiving station plots three headings from the signal source on a map
C. Antenna headings from several different receiving locations are used to locate the signal source
D. A fixed receiving station uses three different antennas to plot the location of the signal source

E9H06
(C)
Page 9-25

E9H07

Why is it advisable to use an RF attenuator on a receiver being used for direction finding?

A. It narrows the bandwidth of the received signal to improve signal to noise ratio
B. It compensates for the effects of an isotropic antenna, thereby improving directivity
C. It reduces loss of received signals caused by antenna pattern nulls, thereby increasing sensitivity
D. It prevents receiver overload which could make it difficult to determine peaks or nulls

E9H07
(D)
Page 9-25

E9H08

What is the function of a sense antenna?

A. It modifies the pattern of a DF antenna array to provide a null in one direction
B. It increases the sensitivity of a DF antenna array
C. It allows DF antennas to receive signals at different vertical angles
D. It provides diversity reception that cancels multipath signals

E9H08
(A)
Page 9-26

E9H09

Which of the following describes the construction of a receiving loop antenna?

A. A large circularly-polarized antenna
B. A small coil of wire tightly wound around a toroidal ferrite core
C. One or more turns of wire wound in the shape of a large open coil
D. A vertical antenna coupled to a feed line through an inductive loop of wire

E9H09
(C)
Page 9-23

E9H10

How can the output voltage of a multi-turn receiving loop antenna be increased?

A. By reducing the permeability of the loop shield
B. By increasing the number of wire turns in the loop and reducing the area of the loop structure
C. By winding adjacent turns in opposing directions
D. By increasing either the number of wire turns in the loop or the area of the loop structure or both

E9H10
(D)
Page 9-24

E9H11
(B)
Page 9-26

E9H11
What characteristic of a cardioid-pattern antenna is useful for direction finding?
    A. A very sharp peak
    B. A very sharp single null
    C. Broad band response
    D. High-radiation angle

E9H12
(B)
Page 9-25

E9H12
What is an advantage of using a shielded loop antenna for direction finding?
    A. It automatically cancels ignition noise pickup in mobile installations
    B. It is electro-statically balanced against ground, giving better nulls
    C. It eliminates tracking errors caused by strong out-of-band signals
    D. It allows stations to communicate without giving away their position

SUBELEMENT E0
SAFETY
1 Exam Question — 1 Group

E0A — Safety: amateur radio safety practices; RF radiation hazards; hazardous materials

E0A01
What, if any, are the differences between the radiation produced by radioactive materials and the electromagnetic energy radiated by an antenna?
A. There is no significant difference between the two types of radiation
B. Only radiation produced by radioactivity can injure human beings
C. Radioactive materials emit ionizing radiation, while RF signals have less energy and can only cause heating
D. Radiation from an antenna will damage unexposed photographic film but ordinary radioactive materials do not cause this problem

E0A01
(C)
Page 11-3

E0A02
When evaluating RF exposure levels from your station at a neighbor's home, what must you do?
A. Make sure signals from your station are less than the controlled MPE limits
B. Make sure signals from your station are less than the uncontrolled MPE limits
C. You need only evaluate exposure levels on your own property
D. Advise your neighbors of the results of your tests

E0A02
(B)
Page 11-5

E0A03
Which of the following would be a practical way to estimate whether the RF fields produced by an amateur radio station are within permissible MPE limits?
A. Use a calibrated antenna analyzer
B. Use a hand calculator plus Smith-chart equations to calculate the fields
C. Use an antenna modeling program to calculate field strength at accessible locations
D. All of the choices are correct

E0A03
(C)
Page 11-7

E0A04
When evaluating a site with multiple transmitters operating at the same time, the operators and licensees of which transmitters are responsible for mitigating over-exposure situations?
A. Only the most powerful transmitter
B. Only commercial transmitters
C. Each transmitter that produces 5% or more of its MPE exposure limit at accessible locations
D. Each transmitter operating with a duty-cycle greater than 50%

E0A04
(C)
Page 11-7

E0A05
What is one of the potential hazards of using microwaves in the amateur radio bands?
A. Microwaves are ionizing radiation
B. The high gain antennas commonly used can result in high exposure levels
C. Microwaves often travel long distances by ionospheric reflection
D. The extremely high frequency energy can damage the joints of antenna structures

E0A05
(B)
Page 11-8

E0A06
(D)
Page 11-4

E0A06
Why are there separate electric (E) and magnetic (H) field MPE limits?
  A. The body reacts to electromagnetic radiation from both the E and H fields
  B. Ground reflections and scattering make the field impedance vary with location
  C. E field and H field radiation intensity peaks can occur at different locations
  D. All of these choices are correct

E0A07
(B)
Page 11-3

E0A07
How may dangerous levels of carbon monoxide from an emergency generator be detected?
  A. By the odor
  B. Only with a carbon monoxide detector
  C. Any ordinary smoke detector can be used
  D. By the yellowish appearance of the gas

E0A08
(C)
Page 11-4

E0A08
What does SAR measure?
  A. Synthetic Aperture Ratio of the human body
  B. Signal Amplification Rating
  C. The rate at which RF energy is absorbed by the body
  D. The rate of RF energy reflected from stationary terrain

E0A09
(C)
Page 11-2

E0A09
Which insulating material commonly used as a thermal conductor for some types of electronic devices is extremely toxic if broken or crushed and the particles are accidentally inhaled?
  A. Mica
  B. Zinc oxide
  C. Beryllium Oxide
  D. Uranium Hexaflouride

E0A10
(A)
Page 11-1

E0A10
What material found in some electronic components such as high-voltage capacitors and transformers is considered toxic?
  A. Polychlorinated biphenyls
  B. Polyethylene
  C. Polytetrafluroethylene
  D. Polymorphic silicon

E0A11
(C)
Page 11-3

E0A11
Which of the following injuries can result from using high-power UHF or microwave transmitters?
  A. Hearing loss caused by high voltage corona discharge
  B. Blood clotting from the intense magnetic field
  C. Localized heating of the body from RF exposure in excess of the MPE limits
  D. Ingestion of ozone gas from the cooling system

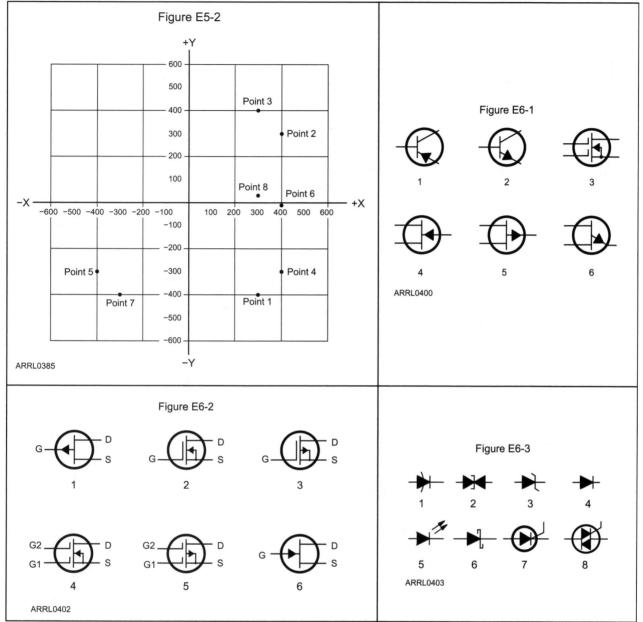

Figure E5-2

ARRL0385

Figure E6-1

ARRL0400

Figure E6-2

ARRL0402

Figure E6-3

ARRL0403

ARRL0431

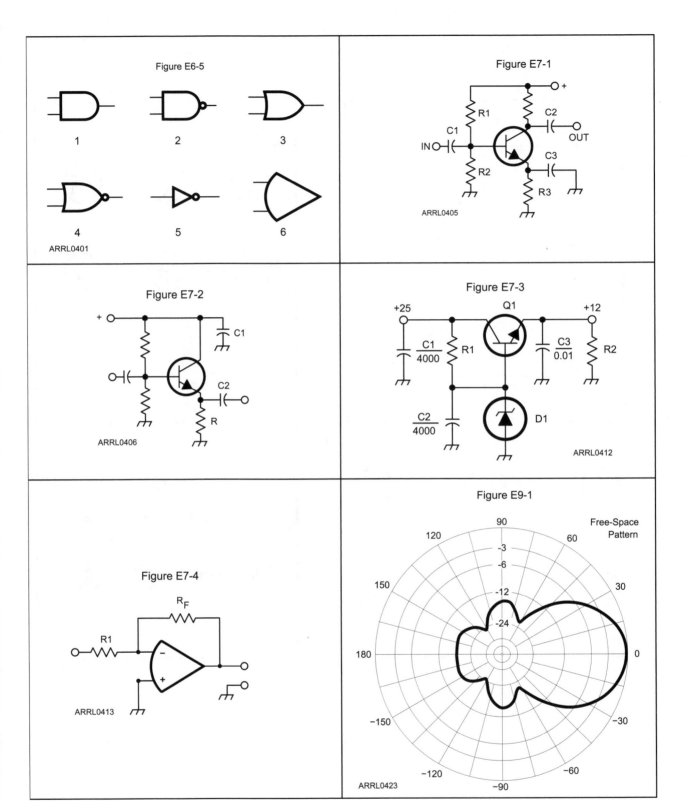

Figure E6-5

1   2   3

4   5   6

ARRL0401

Figure E7-1

ARRL0405

Figure E7-2

ARRL0406

Figure E7-3

ARRL0412

Figure E7-4

ARRL0413

Figure E9-1

Free-Space Pattern

ARRL0423

ARRL0432

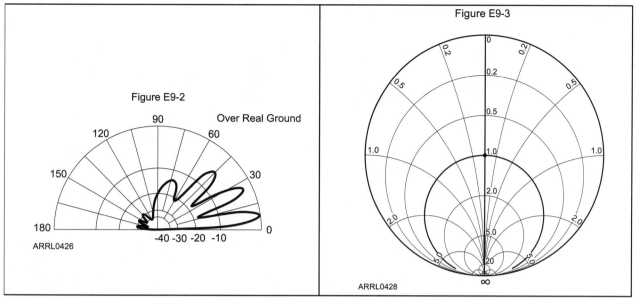

Figure E9-2

Figure E9-3

INDEX

Notes

Notes

Notes

Notes

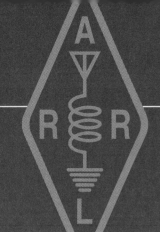

Notes

Notes

Notes

Notes

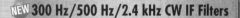

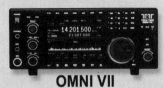

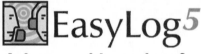
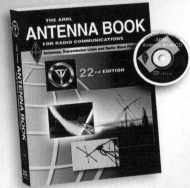

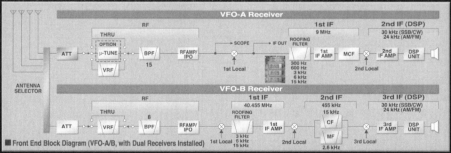

FEEDBACK

Please use this form to give us your comments on this book and what you'd like to see in future editions, or e-mail us at **pubsfdbk@arrl.org** (publications feedback). If you use e-mail, please include your name, call, e-mail address and the book title, edition and printing in the body of your message. Also indicate whether or not you are an ARRL member.

Where did you purchase this book? ☐ From ARRL directly   ☐ From an ARRL dealer

Is there a dealer who carries ARRL publications within:

☐ 5 miles   ☐ 15 miles   ☐ 30 miles   of your location?   ☐ Not sure.

License class:

☐ Technician   ☐ General   ☐ Amateur Extra

Name _____   ARRL member? ☐ Yes ☐ No

_____   Call Sign _____

Address _____

City, State/Province, ZIP/Postal Code _____

Daytime Phone   (   ) _____   Age _____

If licensed, how long? _____

Other hobbies _____   E-mail _____

_____

Occupation _____

From _____

_____

_____

EDITOR, EXTRA CLASS LICENSE MANUAL
ARRL—THE NATIONAL ASSOCIATION FOR AMATEUR RADIO
225 MAIN STREET
NEWINGTON CT 06111-1494

please fold and tape